MW01641548

ALGEBRA

The Addison-Wesley Mathematics Series

ALGEBRA

Richard E. Johnson
University of New Hampshire

Lona Lee Lendsey
Oak Park and River Forest High School
Oak Park, Illinois

William E. Slesnick
Dartmouth College

Addison-Wesley Publishing Company
Menlo Park, California • Reading, Massachusetts
London • Don Mills, Ontario • Sydney

Books in the
Addison-Wesley Mathematics Series

ALGEBRA
R. Johnson, L. Lendsey, W. Slesnick

GEOMETRY
E. Moise, F. Downs

ALGEBRA AND TRIGONOMETRY
R. Johnson, L. Lendsey, W. Slesnick, G. Bates

PRE-CALCULUS MATHEMATICS
M. Shanks, C. Fleenor, C. Brumfiel

ELEMENTS OF CALCULUS
AND ANALYTIC GEOMETRY
G. Thomas, R. Finney

 Printed in the United States of America. Published simultaneously in Canada.

ISBN 0-201-04652-0

ABCDEFGHIJKL-VH-8987654321

PREFACE

Algebra is one of the oldest branches of mathematics. There is historical evidence that the Babylonians were versed in its methods 4000 years ago. Until the end of the eighteenth century, algebra could be roughly described as the branch of mathematics which dealt with the solution of equations. In the nineteenth century, the long-overdue formalization of the structure of the real and complex number systems marked the beginning of what is known as modern algebra. Modern algebra, in addition to its concern with solving equations, supplies the language and patterns of reasoning used in other branches of mathematics.

The content of this book has been noticeably influenced by the report of the Commission on Mathematics of the College Entrance Examination Board and by the School Mathematics Study Group. Both of these groups have done an outstanding service in focusing the attention of the country on the need for changes in the high-school mathematics curriculum. We wish to acknowledge our indebtedness to their many publications.

The mathematical goals of the various chapters can be briefly stated as follows. Chapter 1 is primarily devoted to the language and symbolism of algebra. Properties of the real number system are discussed in Chapter 2 and then are used to solve equations in Chapter 3 and inequalities in Chapter 4. Functions, as relations and as mappings, are presented in Chapter 5. The study of polynomial systems is started for polynomials in one variable in Chapter 6, and then reinforced for polynomials in two variables in Chapter 9. Chapters 7 and 8 are concerned with solving linear equations and linear inequalities, and with the geometric meaning of the solutions. Rational algebraic expressions are studied in Chapter 10, and radical expressions in Chapter 11. Both the algebraic and the geometric significance of quadratic equations are studied in Chapter 12. Finally, Chapter 13 is devoted to right-triangle trigonometry.

The homework exercises are separated into three levels, with some discovery exercises included in the third level. At the end of many of the sets of exercises are Preparation Exercises for the following sections.

Enrichment material is strategically placed throughout the book. In Chapter 1, there are enrichment sections on games of chance, probability, and flow charts. In other chapters, there are sections on number theory, proof in algebra, and so on. There also are historical notes and challenging paragraphs marked EXTRA!

The structure of algebra is emphasized throughout the book. This is the first of many tiers of abstraction which typifies modern mathematics.

Among the many people, including colleagues, who have directly or indirectly helped shape this book, we would like especially to thank the Editorial Staff at Addison-Wesley Publishing Company.

R. E. J.
L. L. L.
W. E. S.

CONTENTS

1 The Language of Mathematics *1*

2 Axioms of Algebra *47*

3 Equations *111*

4 Inequalities 145

5 Polynomials in One Variable 181

6 Functions 239

7 Graphs 275

8 Systems of Linear Equations 331

9 Polynomials in Two Variables 377

10 Rational Algebraic Expressions 405

11 Roots 455

12 Quadratic Equations *505*

13 Trigonometry *553*

Appendix *589*

Glossary *599*

Index *607*

CHAPTER 1

The Language of Mathematics

Objectives . . .

- To use basic set operations to denote number relationships.
- To find products of factors expressed in exponential form.
- To use variables to write mathematical expressions for phrases involving numerical quantities.
- To calculate the probability of an outcome in a simple experiment.
- To write a flow chart to show steps in a familiar process.

1–1 NUMBERS AND NUMERALS

Symbols that are used to denote numbers such as "17" and "$\frac{3}{4}$" are called *numerals.* You never see the number seventeen or the number three-fourths, but you do see the numerals "17" and "$\frac{3}{4}$." Numbers are abstract quantities that you can only think about but not see. You use numerals as an aid in arithmetic, that is, in computing with numbers. Actually, it is possible to do arithmetic without ever writing or saying anything. For example, you can mentally compute 202×7. Then you write the numeral "1414" or say the number words "fourteen hundred fourteen" to describe your result to someone else.

The symbols used universally today for numbers are the Hindu-Arabic numerals. These numerals probably originated in India about 1400 years ago and were adopted by the Arabs about two centuries later. The Hindu-Arabic system is made up of the following ten basic numerals, called *digits:*

$$0, 1, 2, 3, 4, 5, 6, 7, 8, 9.$$

All other numerals are combinations of these digits. Since the relative value of a digit depends upon its position in the numeral, this system is called a positional system.

The positional system for denoting numbers based on ten digits is called the *decimal* system (from the Latin word *decimus* meaning tenth). In the decimal system, every number is described as a sum of so many ones, so many tens, so many hundreds, and so forth, as far as necessary. For example, the number two hundred thirty-seven is the sum of two hundreds, three tens, and seven ones:

$$237 = (2 \times 100) + (3 \times 10) + (7 \times 1).$$

On the other hand, the number three hundred seventy-two, consisting of the same three digits in a different order, is the sum of three hundreds, seven tens, and two ones:

$$372 = (3 \times 100) + (7 \times 10) + (2 \times 1).$$

In this text, the number named two hundred thirty-seven is written "the number 237" rather than "the number with the numeral '237,'" because the meaning is clear. Similarly, you will find numbers such as two-thirds referred to as "the rational number $\frac{2}{3}$" rather than as "the rational number with the numeral '$\frac{2}{3}$.'"

The numbers 10, 100, and 1000, are called *powers* of 10 and are indicated by the use of *exponents:*

$$\begin{aligned} 10^1 &= 10, \\ 10^2 &= 10 \times 10 = 100, \\ 10^3 &= 10 \times 10 \times 10 = 1000. \end{aligned}$$

You read "10^3" as "ten to the third power" or "ten cubed." The numeral "3" in "10^3" is an exponent.

You may use exponents in describing numbers in the decimal notation. For example,

$$\begin{aligned} 237 &= (2 \times 10^2) + (3 \times 10^1) + 7, \\ 304{,}681 &= (3 \times 10^5) + (4 \times 10^3) + (6 \times 10^2) + (8 \times 10^1) + 1. \end{aligned}$$

In scientific work, there is a need for simple methods of indicating large numbers. It is cumbersome to have to use the numeral

$$93{,}000{,}000$$

to indicate the approximate number of miles between the earth and the sun. However, if you realize that

$$93{,}000{,}000 = 93 \times 1{,}000{,}000,$$

you can say that the sun is approximately

$$93 \times 10^6$$

miles from the earth since $10^6 = 1{,}000{,}000$.

The closest star, Alpha Centauri, is approximately

$$25{,}000{,}000{,}000{,}000$$

miles from the earth. This number can also be written as

$$25 \times 1{,}000{,}000{,}000{,}000.$$

Therefore, you can say that Alpha Centauri is

$$25 \times 10^{12}$$

miles from the earth. You can also say that it is

$$2.5 \times 10^{13}$$

miles from the earth. Can you explain why this is so?

If a number is expressed as the product of a number between one and ten and a power of 10, the number is said to be given in *scientific notation.* After learning more about exponents, you will be able to express any number, large or small, in scientific notation.

Examples of scientific notation are given below.

$$3000 = 3 \times 10^3$$
$$34{,}500 = 3.45 \times 10^4$$
$$186{,}000 = 1.86 \times 10^5$$

Exponents may also be used to indicate powers of numbers other than 10. For example,

$$7^2 = 7 \times 7 = 49,$$
$$9^4 = 9 \times 9 \times 9 \times 9 = 6561.$$

Exercises

1. (a) Give four different numerals for the number eighteen.
(b) Give four different numerals for the number one-third.

Express each of the following numbers as the sum of its parts in ones, tens, hundreds, and so forth.

2. (a) 603 (b) 306

3. (a) 22,222 (b) 2,350,000

Write each of the following as a numeral without exponents.

4. (a) 8^2 (b) 13^2

5. (a) 30^5 (b) 4^4

6. (a) 1^5 (b) 0^2

7. (a) $(.2)^3$ (b) $(.03)^2$

8. (a) 7.5×10^3 (b) 2.7×10^2

For Exercises 9–13, write each as a numeral with an exponent.

9. (a) 125 (b) 64

10. (a) 625 (b) 729

11. (a) $\frac{25}{36}$ (b) $\frac{81}{121}$

12. (a) 10,000 (b) 1

13. (a) 1.44 (b) .0049

14. (a) Make a table of powers of 3 through 3^{10}.
(b) Make a table of powers of 5 through 5^{10}.

Complete the writing of each of the following numbers in scientific notation by filling in the missing exponent.

15. (a) $831 = 8.31 \times 10^?$ (b) $579 = 5.79 \times 10^?$

16. (a) $33.4 = 3.34 \times 10^?$ (b) $71.2 = 7.12 \times 10^?$

17. (a) $9735 = 9.735 \times 10^?$ (b) $1141 = 1.141 \times 10^?$

18. (a) $623.9 = 6.239 \times 10^?$ (b) $399.1 = 3.991 \times 10^?$

Express each of the following numbers in scientific notation.

19. (a) 186,000 (b) 309,000,000 **20.** (a) 8635 (b) 23,712

21. (a) 456.7 (b) 38.79 **22.** (a) 20^4 (b) 30^5

A product such as $3^2 \times 3^4$ may be given in exponential form as follows.

$$3^2 \times 3^4 = (3 \times 3) \times (3 \times 3 \times 3 \times 3) = 3^6.$$

Use this method to find each of the following products in exponential form.

23. (a) $2^3 \times 2^4$ (b) $11^5 \times 11^3$

24. (a) $10^4 \times 10^5$ (b) $10^{21} \times 10^{10}$

25. (a) $10^8 \times 10 \times 10^2$ (b) $10^{14} \times 10^4 \times 10$

---

26. There are approximately one hundred thousand million stars in our galaxy and countless more in other galaxies of our universe. According to one estimate, there are 100,000,000,000,000,000,000 (one hundred billion billion) stars that are visible, that is, whose light comes to the earth.

(a) Express the number of stars in our galaxy as a power of 10.

(b) Express the number of visible stars as a power of 10.

(c) Each star is conceivably a sun for a number of planets rotating about it. Suppose that one star in a thousand has a planet with some form of life on it. How many planets of the visible stars would have some form of life on them?

(d) Suppose that one-tenth of the planets having some form of life on them have a technical culture like ours. How many planets of the visible stars would have a technical culture?

Shown here are nebulae in the constellation Sagittarius. Quantities and distances in space are so vast that mathematicians and astronomers use scientific notation to express them numerically. (Official U.S. Navy photograph)

27. It is commonplace in astronomy to describe distances between stars in units of *light-years*. By definition, one light-year is the distance light travels in one year (365 days). If light travels at a speed of 186,000 miles per second, approximately how many miles are there in one light-year?

28. The 200-inch telescope at Mount Palomar Observatory in the United States photographed a celestial object at a distance of six billion light-years from the earth. This was three times the distance of any object previously identified. How many miles is this new object from the earth?

29. A radar system sends out and receives certain radio waves which travel with the speed of light. The distance of an object from the radar set can be determined by measuring how long it takes radio waves emitted by the set to travel to the object, reflect off it, and return to the set.

(a) If it takes $2\frac{1}{2}$ seconds for radio waves to go to the moon and back, how far is the moon from the earth?

(b) If the sun is 93 million miles from the earth, how long does it take a radio wave from earth to reach the sun?

30. Quasars are celestial objects far beyond our own star system which emit long-wave radiation. If the radio waves emitted travel with the speed of light, how many miles from earth was the first quasar to be discovered if it took its waves 1.5×10^9 years to reach the earth?

31. In 1967 British astonomers detected regular and rapidly pulsating signals from outer space and named them "pulsars." How many seconds did a signal, traveling at the speed of light, take to reach the earth if it came from a pulsar 200 light-years away?

32. Cornell University astronomers found that the pulse rate of a pulsar in the Crab nebula was slowing down by about one beat in 2000 every year. About how many beats in all will it have slowed down by the end of one year if it is now pulsing 30 times per second?

1–2 REAL NUMBERS

Numbers, such as

$$1,\ {}^{-}1,\ 2,\ {}^{-}54,\ 87,\ 1970,\ {}^{-}5320,$$

are called *integers.* Fractional numbers such as

$$7\tfrac{2}{3},\ {}^{-}\tfrac{3}{5},\ 2.37,\ \tfrac{22}{7},\ {}^{-}5.83,$$

are called *rational numbers.* Each such number can be represented as a quotient of two integers: $7\frac{2}{3}$ as $\frac{23}{3}$, and 2.37 as $\frac{237}{100}$. Every integer is also a rational number, since it can be represented as the quotient of itself and 1.

Rational numbers can also be written in decimal notation. For example,

$$5 = 5.0,\ \tfrac{1}{3} = .333\ldots,\ {}^{-}(\tfrac{7}{8}) = {}^{-}.875,\ \tfrac{13}{6} = 2.1666.\ldots$$

As shown above, some rational numbers such as 5 and ${}^{-}(\frac{7}{8})$ have terminating decimals; others such as $\frac{1}{3}$ and $\frac{13}{6}$ have nonterminating decimals. Their decimal representations contain an infinite number of digits.

Every rational number which has a nonterminating decimal has a *repeating* decimal. That is, its infinite decimal has a block of digits which is repeated indefinitely from some digit on. You can find the decimal representation of a rational number by division.

Example. The decimal representation of $\frac{7}{33}$ can be found as shown at the right. The division process never ends, because the remainders 4 and 7 are repeated endlessly. Therefore,

$$\tfrac{7}{33} = .21212121\ldots$$

The three dots indicate that the shaded block of digits, 21, is to be repeated indefinitely.

```
     .2121
 33)7.00000
    6 6
    ---
      40
      33
      --
       70
       66
       --
        40
        33
        --
         7
```

Certain numbers cannot be represented as quotients of integers. Two examples of such numbers are π and the square root of 2. The square root of 2 is written $\sqrt{2}$ ($\sqrt{\ }$ is called a radical sign) and it denotes the length of a diagonal of a unit square (see Fig. 1-1). The number π is the circumference of a circle of diameter 1 (see Fig. 1-2). These numbers are called irrational numbers. That is, an *irrational number* is a number which cannot be represented as a quotient of two integers. (Not all numerals written with a radical sign represent irrational numbers. The number $\sqrt{9}$ is the positive square root of 9 which is 3.)

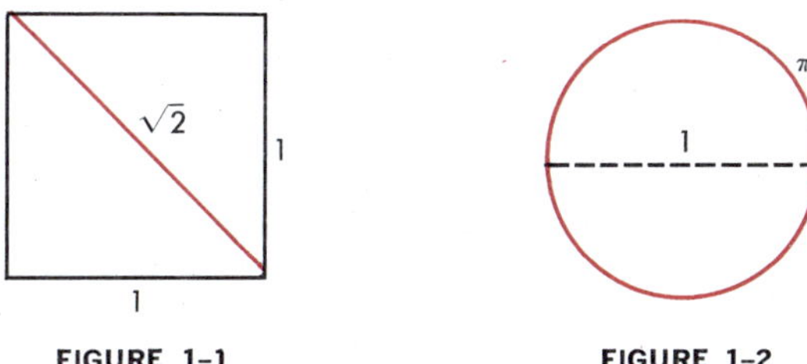

FIGURE 1-1 FIGURE 1-2

Every irrational number can be represented as an infinite, nonrepeating decimal. For example, the decimal representations of $\sqrt{2}$ and π begin as follows.

$$\sqrt{2} = 1.414214\ldots$$
$$\pi = 3.1415927\ldots$$

A difference between rational and irrational numbers is that rational numbers have terminating or repeating decimals whereas irrational numbers have nonterminating, nonrepeating decimals. Each irrational number may be *approximated* by a rational number. For example, 1.414 is an approximation of $\sqrt{2}$ correct to the nearest thousandth. The collection of all rational and irrational numbers is called the collection of *real numbers.*

There are precisely enough real numbers and enough points on a line so that each real number is assigned to a unique point on the line, and each point is assigned to a unique real number.

A line having a real-number scale on it is called a *number line.* The point assigned to a number is called the *graph* of the number, and the number assigned to a point is called the *coordinate* of the point.

The real number system can be represented by the set of all the points on a line. You may think of the line as an infinitely long ruler with an arbitrarily chosen point representing zero. Then, the positive real numbers are marked off on one side of zero and the negative real numbers on the other. The numbers are marked off uniformly. For example, you might put the integers an inch apart and then mark off the other real numbers proportionally, as on a yardstick. Part of such a line is shown in Fig. 1–3.

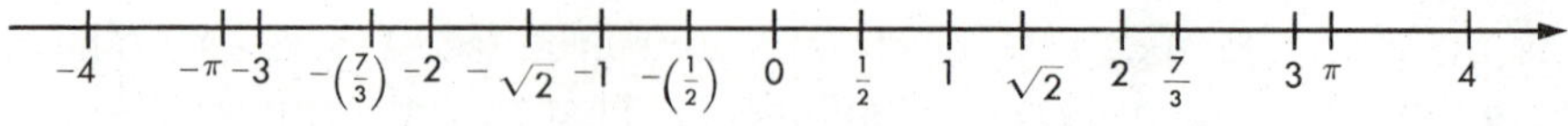

FIGURE 1–3

Exercises

1. Tell which of the following numerals denote rational numbers and which denote irrational numbers.
(a) 3, $\sqrt{3}$, 2.3, $2\frac{1}{3}$, 0, $^{-}(\frac{1}{5})$, 7.981, $\sqrt{4}$, 1.732
(b) 3.1416, $\frac{22}{7}$, π, $\sqrt{25}$, $\sqrt{1}$, $^{-}11$, 1.1, $^{-}6\frac{1}{6}$, $\sqrt{5}$

2. If possible, write each of the following numerals as a quotient of two integers. Tell which numerals denote rational numbers and which denote irrational numbers.
(a) 1001.0, 1, $6\frac{1}{2}$, 10.01, $^{-}7\frac{1}{7}$
(b) $\sqrt{100}$, 0, $6\frac{5}{7}$, $^{-}7.125$, 18.0

3. (a) Give five rational numbers that are not integers.
(b) Are there any integers that are not rational numbers? Why?

Find the decimal representation of each of the following rational numbers. For repeating decimals, shade the shortest repeating block of digits.

4. (a) $\frac{2}{9}$ (b) $\frac{5}{9}$

5. (a) $\frac{1}{11}$ (b) $\frac{2}{11}$

6. (a) $\frac{3}{16}$ (b) $\frac{7}{125}$

7. (a) $\frac{7}{5}$ (b) $\frac{5}{7}$

8. (a) $\frac{13}{9}$ (b) $\frac{23}{11}$

9. (a) $\frac{32}{33}$ (b) $\frac{1}{99}$

10. (a) $\frac{12}{37}$ (b) $\frac{2}{7}$

---

11. Find the decimal representation of $\frac{21}{23}$.

12. (a) In finding the decimal representation of $\frac{2}{13}$ by division, what are the possible remainders you may encounter?
(b) What is the maximum number of digits in the block that repeats?
(c) Find the decimal representation of $\frac{2}{13}$ to see what actually happens.

13. The decimal representation of a certain number begins like this:

$$.10100100010000100000\ldots.$$

The first 1 is followed by one zero, the second by two zeros, the third by three zeros, and this pattern continues indefinitely. Is there a block of digits that repeats? Is the number rational or irrational?

14. What is an approximation of $\sqrt{3}$ correct to the nearest thousandth?

15. Give an approximation of π correct to the nearest millionth.

Which of the following numerals denote rational numbers?

16. $\sqrt{4}$	**17.** $\sqrt{5}$	**18.** $\sqrt{6}$	**19.** $\sqrt{7}$
20. $\sqrt{15}$	**21.** $\sqrt{20}$	**22.** $\sqrt{25}$	**23.** $\sqrt{30}$
24. $\sqrt{36}$	**25.** $\sqrt{42}$	**26.** $\sqrt{59}$	

27. (a) Give six rational numbers between 0 and 1.
(b) Give six rational numbers between 0 and $\frac{1}{2}$.
(c) How many rational numbers do you think there are between 0 and 1?

28. (a) Name an integer between 4 and 10.
(b) Name an integer between 3 and 5.
(c) Is there always an integer between two integers? Explain your answer.

---

29. Do you think that the sum of two integers is always an integer? Is the product of two integers always an integer? Give examples.

30. Do you think that the sum of two rational numbers is always a rational number? Is the product of two rational numbers always a rational number? Give examples.

31. Give six irrational numbers between 0 and 1; between 0 and $\frac{1}{2}$.

1–3 SETS

A *set* is a collection of objects. Some examples of sets are the following:

The set of all animals in a zoo.
The set of all integers between 0 and 10.

The objects in a set are most often called *elements.* Thus, the elements of the first set are animals, and the elements of the second set are numbers.

A set is *well-defined* if its elements are *listed* specifically or if its elements are *described* in a way that makes it possible to decide whether an object in question is or is not an element of the set.

For example, you might define a set as being made up of the letters

a, e, i, o, u.

You might describe the same set as

The set of all vowels in the English alphabet.

In each case, you have a well-defined set. In fact, these two sets are one and the same set; only their descriptions differ.

Braces, { }, are used to denote a set. Thus,

$$\{a, e, i, o, u\}$$

denotes the set of all vowels, and

$$\{1, 2, 3, 4, 5, 6, 7, 8, 9\}$$

denotes the set of integers between 0 and 10. Note that a number is said to be *between* two numbers if it is greater than the smaller and less than the larger of the two numbers.

Sets are usually denoted by capital letters. For example, you might let P denote the set of all planets in the solar system or let E denote the set of all even positive integers,

$$E = \{2, 4, 6, 8, \ldots\}.$$

You can perform operations on sets just as you can perform operations, such as addition and multiplication, on numbers. The operations of union and intersection of sets are defined below.

Definition of union of sets

The union of set A and set B, denoted by $A \cup B$, consists of all elements in either set A or set B or both.

Definition of intersection of sets

The intersection of set A and set B, denoted by $A \cap B$, consists of all elements that are in both set A and set B.

For example,

$$\{2, 4, 6, 8, 10, 12\} \cup \{3, 6, 9, 12\} = \{2, 3, 4, 6, 8, 9, 10, 12\},$$
$$\{2, 4, 6, 8, 10, 12\} \cap \{3, 6, 9, 12\} = \{6, 12\}.$$

Every element of the set $\{4, 8, 12\}$ is also an element of the set $\{2, 4, 6, 8, 10, 12\}$. For this reason, $\{4, 8, 12\}$ is called a *subset* of $\{2, 4, 6, 8, 10, 12\}$. The symbol "$\subset$" is used to denote "is a subset of." Thus, you write

$$\{4, 8, 12\} \subset \{2, 4, 6, 8, 10, 12\}.$$

Definition of subset

Set A is a subset of set B, written $A \subset B$, if every element of set A is also an element of set B.

It is important to note that a definition consists of two parts. For example, the definition of subset consists of the following two parts.

1. If every element of set A is also an element of set B, then $A \subset B$.
2. If $A \subset B$, then every element of set A is also an element of set B.

Every definition in this text is a *two-part* statement.

It is convenient to be able to talk about the *empty set,* a set having no elements. The special symbol

$$\emptyset$$

is used to denote the empty set. Often { } is also used.

As an example of the empty set, consider the intersection of two sets which have no elements in common. That is,

$$\{4, 8, 12\} \cap \{5, 10, 15\} = \emptyset.$$

Exercises

Define, in words, the sets whose elements are listed.

1. (a) {2, 4, 6, 8} (b) {5, 10, 15, 20, 25}

2. (a) {3, 9, 27, 81} (b) {2, 4, 8, 16, 32, 64}

3. (a) {Pizarro, Cortez, Hudson, Champlain}
(b) {Armstrong, Aldrin, Collins}

List the elements of the following sets.

4. (a) The set of powers of 6 with not more than five digits
(b) The set of powers of 2 between 0 and 100

5. (a) The set of three-digit numbers whose digits are all the same
(b) The set of all three-digit numbers for which the sum of the digits is 3

Let sets A, B, and C be defined as follows:

$$A = \{2, 4, 8, 16, 32\}, \qquad B = \{4, 8, 12, 16, 20, 24, 28, 32\},$$
$$C = \{3, 6, 9, 12, 15, 18, 21, 24, 27, 30\}.$$

6. (a) List the elements in $A \cup B$. (b) List the elements in $A \cup C$.

7. (a) List the elements in $A \cap B$. (b) List the elements in $B \cap C$.

8. Describe $A \cap C$.

Name the set which is the intersection of each pair of sets below.

9. (a) A: all positive integers
B: all even positive numbers
(b) A: all positive integers
B: all odd positive integers

10. (a) A: all rational numbers
B: all real numbers which can be represented as the quotient of two integers
(b) A: all positive even integers
B: all positive multiples of 2

11. (a) A: all even positive integers
B: all odd positive integers
(b) A: all rational numbers
B: all irrational numbers

12. (a) Name the set which is the union of each pair of sets in Exercises 9(a), 10(a), and 11(a).
(b) Name the set which is the union of each pair of sets in Exercises 9(b), 10(b), and 11(b).

13. If $A = \{2, 4, 6, 8\}$ and $B = \{3, 6, 9, 12\}$, list the elements of the following sets.
(a) $A \cap B$ and $B \cap A$
(b) $A \cup B$ and $B \cup A$

Given that $A = \{0, 1, 3, 7\}$, $B = \{0\}$, and $C = \{0, 1, 2, 3, 4, 5, 6\}$, tell whether each of the following statements is true or false.

14. (a) $B \subset A$ (b) $A \subset B$

15. (a) $A \subset C$ (b) $B \subset C$

16. (a) B is the empty set (b) $A \cap B = \emptyset$

17. (a) $B \cup C = C$ (b) $A \cup C = C$

---

18. (a) Give examples of two sets P and Q such that P is a subset of Q and Q is not a subset of P.
(b) Give examples of two sets R and S such that R is not a subset of S and S is not a subset of R.
(c) Give examples of two sets T and U such that T is a subset of U and U is a subset of T. (Are T and U the same set?)

---

19. If $A = \{1, 2, 3\}$, is $A \subset A$? is $\emptyset \subset A$? How many different subsets of A are there?

20. Set P consists of the last five letters of the English alphabet, thus

$$P = \{v, w, x, y, z\}.$$

(a) List all of the subsets of P that contain exactly one element.
(b) List all of the subsets of P that contain exactly two elements.
(c) List all of the subsets of P that contain exactly three elements.
(d) List all of the subsets of P that contain exactly four elements.
(e) Remembering the definition of subset, do you think $P \subset P$? Explain your answer.
(f) Do you think $\emptyset \subset P$? Explain your answer.

21. If $B = \{0, 2, 4, 6\}$, how many subsets of B are there? List the elements of each subset of B.

22. Suppose you know that 10 is an element of a set A and also of a set B. Are you justified in concluding that $A \subset B$? $B \subset A$? $A \cap B \neq \emptyset$?

23. If $A \subset B$, what can you say about $A \cap B$? $A \cup B$? Illustrate with specific sets.

24. What is the largest subset of an arbitrary set A? What is the smallest subset of A?

1–4 VARIABLES AND OPEN SENTENCES

The statement

$$3 + 4 = 7$$

is true. The statement

$$4 + 5 = 13$$

is false. However, the statement

$$x + 4 = 10$$

is neither true nor false. The last statement becomes either true or false when you let x have a value. You could let $x = 3$ to get

$$3 + 4 = 10$$

which is false. If you let $x = 6$, the statement

$$6 + 4 = 10$$

is true.

Symbols, such as x in the example above, which can be assigned various values are called *variables.*

Definition of variable

A variable is a symbol used to denote any element of a given set. The set is called the domain of the variable. Each element of the domain is called a value of the variable.

In the statement "$x + 4 = 10$," x is the variable, the set of all real numbers is the domain of x, and any real number is a value for x.

Variables may be used to give brief expressions about numbers, as illustrated below.

Expression	*Expression using variables*
The sum of two numbers	$x + y$
A number divided by three	$n \div 3$
The product of two consecutive integers	$k(k + 1)$

The domain of each of the variables x, y, and n, shown above, is the set of all real numbers; however, the domain of k is the set of all integers.

When you assign the variables in the expressions above specific values, you obtain specific numbers. For example, if x is equal to 2 and y is equal to 7, then $x + y$ is equal to 9. You can express this fact in mathematical language as follows:

$$\text{If } x = 2 \text{ and } y = 7, \text{ then } x + y = 9.$$

Similarly,

$$\text{if } n = 11, \text{ then } n \div 3 = \tfrac{11}{3};$$
$$\text{if } k = 9, \text{ then } k(k + 1) = 9 \times 10, \text{ or } 90.$$

Statements, such as

$$x + 4 = 10,$$

that are neither true nor false but become either true or false when you assign the variable a value from its domain are called *open sentences.*

Definition of open sentence

A statement containing one or more variables is called an open sentence if it becomes either a true statement or a false statement when the variables are given specific values from their domains.

As another example,

$$x \text{ is an even integer}$$

is an open sentence. It is a true statement if x is assigned an even number and a false statement for any other value of x.

The statement

$$x + 3 \neq y + 9$$

is another example of an open sentence (the symbol $\neq$ means *is not equal to*). If $x = 2$ and $y = 5$, then the resulting statement

$$2 + 3 \neq 5 + 9$$

is true. If $x = 12$ and $y = 6$, then the resulting statement

$$12 + 3 \neq 6 + 9$$

is false.

Open sentences are frequently used in defining sets. For example, you might define set C as follows:

$$C = \{x, \text{ such that } x \text{ is an even integer}\}.$$

This statement is read, "C is the set of all x's such that x is an even integer." Here x is a variable and C consists of the set of values of x for which the open sentence

$$x \text{ is an even integer}$$

is true. You can further simplify the notation above by replacing the words "such that" with a vertical bar. Thus,

$$C = \{x \mid x \text{ is an even integer}\}.$$

The set

$$\{x \mid \underline{\qquad\qquad}\}$$

consists of all values of x which make the open sentence after the vertical bar true.

For example, the set S of all odd integers between 0 and 100 can be defined as follows:

$$S = \{x \mid x \text{ is an odd integer between 0 and 100}\}.$$

As another example,

$$T = \{n \mid n \text{ is an integral multiple of 9 between 0 and } 10^3\}.$$

The elements of T are 9, 18, 27, 36, 45, and so on, up to 999. That is, T consists of 9×1, 9×2, 9×3, and so on, up to 9×111. Thus, you could define T also as

$$T = \{9n \mid n \text{ is an integer between 0 and } 112\}.$$

(Remember that a number is said to be *between* two numbers if it is greater than the smaller and less than the larger of the two numbers.)

Exercises

Find the values of each expression in Exercises 1–3 when x is assigned the values 2, 4, 0, 1 in turn.

1. (a) $5x + 1$ (b) $5(x + 1)$

2. (a) $2x^3$ (b) $(2x)^3$

3. (a) $(x + 3)^2$ (b) $x^2 + 9$

4. (a) Find the values of the expression $2 + 3x$ when x has the values 0, 4, and $5\frac{1}{3}$.
(b) Find the values of the expression $3x^2$ when x has the values 0, $\frac{1}{3}$, and 4.

Use variables to give algebraic expressions for each of the phrases below.

5. (a) Nine more than a number (b) Nine less than a number

6. (a) One-third of a number (b) A number divided by five

7. (a) Twice the third power of a number
(b) Six more than twice a number

State in your own words the meaning of each of the algebraic expressions in Exercises 8–10.

8. (a) $4x$ (b) $\frac{2}{3}n$

9. (a) $2 + 3n$ (b) $2n - 9$

10. (a) $3x^2$ (b) $(3x)^2$

Tell which statements are true and which are false.

11. (a) $2 + 2 = 2 \cdot 2$ (b) $5 \div 1 = 5 \cdot 1$

12. (a) $\dfrac{12 + 3}{2 + 3} = 6 + 1$ (b) $\dfrac{12 + 3}{2 + 3} = 6$

13. (a) $6 \cdot 0 \neq 8 \cdot 0$ (b) $\frac{2}{9} \cdot \frac{9}{2} = 1$

In each of the open sentences, assign x the values, 5, 12, 0, 1, and 3, in turn, and tell which of the resulting statements are true and which are false.

14. (a) $2x + 1 = 7$ (b) $2x - 9 = 15$

15. (a) $x^2 - x = 0$ (b) $3x^2 = 12$

16. (a) $3x - 2 = x + 5$ (b) $4x - 3 = 2x - 1$

Each of the following is an open sentence. If possible, assign the variable a value so that

(i) the resulting statement is true,
(ii) the resulting statement is false.

17. (a) The product of N and every number is zero.
(b) The product of N and every number is the number itself.

18. (a) N is an even integer greater than 12 but less than 16.
(b) N is an odd integer divisible by 13.

19. (a) $x + 17 = 23$ (b) $x - 9 \neq 10$

20. (a) $x^2 \div 1 = x \cdot x$ (b) $x \cdot 0 = y \cdot 0$

21. (a) $3 + y \neq y + 3$ (b) $7(x - 6) \neq (7 \cdot x) - (7 \cdot 6)$

In Exercises 22–24, list the elements of each of the given sets. Which sets in parts (a) are equal, and which sets in parts (b) are equal?

22. (a) $A = \{x \mid x \text{ is an even integer between 30 and 50}\}$
(b) $B = \{x \mid x \text{ is a multiple of 5 between 20 and 40}\}$

23. (a) $C = \left\{\frac{x}{2} \,\middle|\, x \text{ is an even integer between 1 and 10}\right\}$

(b) $D = \left\{\frac{x}{3} \,\middle|\, x \text{ is a multiple of 3 between 10 and 20}\right\}$

24. (a) $E = \{x^2 \mid x \text{ is an integer between 0 and 5}\}$
(b) $F = \{5x \mid x \text{ is an integer between 4 and 8}\}$

---

25. For what domain of the variable n will the expression $\frac{3}{5}n$ be
(a) an integer? (b) a rational number?

26. (a) If notebooks sell for 30¢ each, then $30n$ represents the cost of n notebooks. What is the domain of the variable n?
(b) If high-octane gasoline sells for 35¢ a gallon, then $35x$ is the cost of x gallons of this gasoline. What is the domain of the variable x?

27. What is the domain of the variable x if
(a) $x - 3$ denotes the number of chairs in a room?
(b) x^2 denotes a number greater than 100?
(c) $x^2 - 10$ has a value between 100 and 1000?

28. If A denotes John's present age in years, what mathematical expression containing A will denote
(a) his age two years ago?
(b) his age three years hence?
(c) four times his present age?
(d) twice his age last year?

29. A collection of coins contains d dimes, n nickels, and q quarters. Give an expression for
(a) the number of coins.
(b) the value in cents of the dimes.
(c) the value in cents of the nickels and quarters.
(d) the value in dollars of the entire collection.

In Exercises 30–34, make a complete list of the elements in each of the sets, if possible. If not possible, list six elements of the set. Which of the sets are equal?

30. $A = \{x \mid x \text{ is an integral multiple of 3 between 0 and 30}\}$

31. $B = \{5x \mid x \text{ is an integer greater than 10}\}$

32. $C = \{3n \mid n \text{ is an integer between 0 and 10}\}$

33. $D = \{4n - 3 \mid n \text{ is an integer between 2 and 8}\}$

34. $H = \{x^2 \mid x \text{ is an integer greater than zero}\}$

---

35. Use braces, a variable, and an appropriate open sentence to describe each of the following sets.
(a) A is the set of two-digit integers divisible by 6.
(b) B is the set of powers of 4 that are less than 1000.
(c) C is the set of odd integers between 100 and 1000.
(d) D is the set of integers greater than 10 that leave a remainder of 2 when divided by 5.

Review for Sections 1–1 through 1–4

Express each of the following without exponents.

1. 2^3 **2.** 3^2

3. 30^2 **4.** 10^3

5. 0^4 **6.** 1^6

7. 3.6×10^2 **8.** 6.4×10^3

Express each of the following numbers in scientific notation.

9. 743,000 **10.** 269,000

11. 5,400,000

12. 75,200,000

13. 856

14. 972

Which of the following numerals denote rational numbers? Which denote irrational numbers? Write each of the rational numbers as a quotient of two integers.

15. $\sqrt{4}$

16. $3\frac{1}{2}$

17. $7\frac{2}{3}$

18. $\sqrt{5}$

19. $\sqrt{2}$

20. $^{-}5.6$

21. Find the decimal representation of $\frac{7}{9}$.

22. Find the decimal representation of $\frac{9}{11}$.

List the elements of the following sets.

23. The set of all even numbers between 5 and 15

24. The set of all odd numbers between 10 and 20

25. The set of all two-digit numbers for which the sum of the digits in each number is 5

26. The set of all two-digit numbers having 3 as its tens digit

Let sets A, F, and R be defined as follows.

$$A = \{2, 4, 6, 8, 10, 12\}$$
$$F = \{4, 8, 12, 16\}$$
$$R = \{1, 2, 3, 4, 5, 6, 7, 8, 9, 10, 11, 12\}$$

27. Is F a subset of A?

28. Is A a subset of R?

29. List the elements in $A \cup F$.

30. List the elements in $A \cup R$.

31. List the elements in $A \cap F$.

32. List the elements in $F \cap R$.

Let x have the values 3, 2, and 0, in turn, and find the corresponding values of the following expressions.

33. $3x + 4$

34. $2x + 1$

35. $3(x + 4)$

36. $2(x + 1)$

37. $(1 + x)^2$

38. $(3 + x)^2$

39. $1 + x^2$

40. $3 + x^2$

Use variables to give mathematical expressions for each of the phrases below.

41. A number multiplied by seven

42. A number divided by six

43. Eight more than a number

44. Five less than a number

45. Four more than three times a number

46. Nine less than twice a number

In each of the following open sentences, let x be assigned the values 2, 10, 7, and 0, in turn, and tell whether the resulting statements are true or false.

47. $x^2 + 4 = 8$

48. $3x + 4 = 25$

49. $2x + 5 = x + 5 + x$

50. $4(x + 3) = 4x + 12$

Answers to Review for Sections 1–1 through 1–4

1. 8
2. 9
3. 900
4. 1000
5. 0
6. 1
7. 360
8. 6400
9. 7.43×10^5
10. 2.69×10^5
11. 5.4×10^6
12. 7.52×10^7
13. 8.56×10^2
14. 9.72×10^2
15. Rational, for example $\frac{2}{1}$
16. Rational, for example $\frac{7}{2}$
17. Rational, for example $\frac{23}{3}$
18. Irrational
19. Irrational
20. Rational, for example $^{-}\frac{56}{10}$
21. .777 . . .
22. .8181 . . .
23. {6, 8, 10, 12, 14}
24. {11, 13, 15, 17, 19}
25. {14, 23, 32, 41, 50}
26. {30, 31, 32, 33, 34, 35, 36, 37, 38, 39}
27. No
28. Yes
29. {2, 4, 6, 8, 10, 12, 16}
30. {1, 2, 3, 4, 5, 6, 7, 8, 9, 10, 11, 12}
31. {4, 8, 12}
32. {4, 8, 12}
33. for $x = 3$, 13; for $x = 2$, 10; for $x = 0$, 4
34. for $x = 3$, 7; for $x = 2$, 5; for $x = 0$, 1
35. for $x = 3$, 21; for $x = 2$, 18; for $x = 0$, 12
36. for $x = 3$, 8; for $x = 2$, 6; for $x = 0$, 2
37. for $x = 3$, 16; for $x = 2$, 9; for $x = 0$, 1
38. for $x = 3$, 36; for $x = 2$, 25; for $x = 0$, 9
39. for $x = 3$, 10; for $x = 2$, 5; for $x = 0$, 1
40. for $x = 3$, 12; for $x = 2$, 7; for $x = 0$, 3
41. $7x$
42. $\frac{x}{6}$
43. $x + 8$
44. $x - 5$
45. $3x + 4$
46. $2x - 9$
47. True for $x = 2$; false for $x = 10$, 7, and 0
48. True for $x = 7$; false for $x = 2$, 10, and 0
49. True for all values of x
50. True for all values of x

*1–5 NUMBER THEORY I

For thousands of years, people have been fascinated by properties of the integers,

$$1, 2, 3, 4, 5, 6, 7, 8, 9, 10, 11, 12,$$

and so on. The part of mathematics which deals with their properties is called *number theory.* Some simple aspects of number theory will be given in this and following chapters.

The integer 14 can be *factored* in two different ways,

$$14 = 1 \times 14 \qquad \text{or} \qquad 14 = 2 \times 7.$$

The numbers

$$1, 2, 7, 14$$

are called the *divisors,* or *factors,* of 14. In turn, 14 is called a *multiple* of 1, 2, 7, and 14.

If you wish to find the divisors of 24, you express 24 in every possible way as a product of two integers:

$$24 = 1 \times 24, \quad 24 = 2 \times 12, \quad 24 = 3 \times 8, \quad 24 = 4 \times 6.$$

From these expressions, you can conclude that

$$\{1, 2, 3, 4, 6, 8, 12, 24\}$$

is the set of divisors of 24. Each number in this set has 24 as a multiple.

These examples illustrate the following definition.

Definition of divisor and multiple

If a, b, and c are integers different from zero such that $c = ab$, then a and b are called divisors, or factors, of c, and c is called a multiple of a and b.

Some integers have many divisors while others have few. For example, 24 has eight divisors, whereas 13 has only two, 1 and 13. Every integer greater than 1 has at least two divisors, 1 and the integer itself. An integer such as 13 is called a *prime number,* according to the following definition.

Definition of prime number

An integer greater than 1 is called a prime number if it has exactly two positive divisors, 1 and itself.

Since the only factors of 2 are 1 and 2, 2 is a prime. It is the smallest prime. The only factors of 3 are 1 and 3, so 3 is also a prime. The first nonprime greater than 1 is 4; it has three factors, 1, 2, and 4. The first eight primes are

$$2, 3, 5, 7, 11, 13, 17, 19.$$

The integers which are multiples of 2 are called *even integers;* the integers which are not even are called *odd integers.* It is interesting that 2 is the only even prime number. Can you explain why?

The numeral of an integer is composed of digits. Each of the digits is, in turn, the numeral of an integer less than 10. That is, the integers represented by the individual numerals are called the *digits* of the number. For example, the number 853 has as its digits the numbers 8, 5, and 3, since the numeral "853" has "8," "5," and "3" as its digits.

The units, or ones, digit of an even number is always one of the five digits

$$0, 2, 4, 6, 8$$

and of an odd number one of the five digits

$$1, 3, 5, 7, 9.$$

There is nothing unusual about the units digits of the multiples of 3. It can be any digit. However, there is something unusual about the sum of the digits of a multiple of 3.

RULE FOR DIVISIBILITY BY 3

If the sum of the digits of an integer is a number divisible by 3, the integer itself is divisible by 3. If the sum is not divisible by 3, the integer is not divisible by 3.

For example, the sum of the digits of the integer 348 is

$$3 + 4 + 8 = 15,$$

and 15 is divisible by 3. Therefore 348 is divisible by 3. Actually,

$$348 \div 3 = 116.$$

Is the integer

$$28{,}513$$

divisible by 3? Since the sum of its digits is given by

$$2 + 8 + 5 + 1 + 3 = 19$$

and 19 is not divisible by 3, the integer 28,513 is not divisible by 3.

The rule for divisibility by 3 contains two statements. If you wish, you can combine these into one statement.

An integer is divisible by 3 if, and only if, the sum of its digits is divisible by 3.

Do you see that both statements in the rule for divisibility by 3 are contained in this single statement?

Exercises

Express each of the following integers as a product of two integers in as many ways as possible. Then list the set of divisors of each integer.

1. (a) 12 (b) 16 **2.** (a) 26 (b) 22

3. (a) 18 (b) 45 **4.** (a) 42 (b) 50

5. (a) 84 (b) 100 **6.** (a) 76 (b) 72

7. (a) 101 (b) 103 **8.** (a) 132 (b) 96

In Exercises 9–13, each of the integers is a power of a prime number. List its set of divisors. State a rule for relating the number of divisors of a power of a prime and the value of the exponent of that prime.

9. (a) 32 (b) 27 **10.** (a) 25 (b) 49

11. (a) 81 (b) 64 **12.** (a) 625 (b) 1331

13. (a) 2048 (b) 729

14. Make a list of all prime numbers between 1 and 100.

---

15. In order to find a rule for divisibility of an integer by 9, list several different multiples of 9. Is there anything unusual about the sum of the digits of your listed numbers? Is there a rule for divisibility by 9 similar to that for 3? If so, test your rule on several more integers.

16. List the multiples of 4 up to 100. Without listing them, describe the multiples of 4 between 100 and 200 and between 200 and 300. Give a rule for divisibility of an integer by 4.

17. Give a rule for divisibility of an integer by 6. (Hint: An integer divisible by 6 is also divisible by 2 and 3. An integer divisible by both 2 and 3 is also divisible by 6.)

---

18. If an integer has two or more different prime divisors, show that it has at least four different divisors. Give some examples.

19. If $A = \{n \mid n$ an integer with exactly three divisors$\}$, then list all the elements of A between 1 and 30. Can an element of A have more than one prime divisor? Is each element of A a power of a prime? Explain.

20. If an integer has three or more different prime divisors, then show that it has at least 8 divisors. Give some examples.

21. If $B = \{n \mid n$ an integer with exactly four divisors$\}$, then list all the elements of B between 1 and 30. What powers of primes belong to B? Describe the elements of B which are not powers of a prime number.

22. The divisors of 6 which are smaller than 6 are 1, 2, and 3. The sum of these divisors, $1 + 2 + 3$, is 6, the number with which we started. When the sum of the divisors of an integer equals the integer, we call the integer a *perfect number.* An example of an integer that is not perfect is 12, since the sum of its divisors which are less than 12 is $1 + 2 + 3 + 4 + 6$, or 16.

(a) Are there any perfect numbers smaller than 6?

(b) What is the next perfect number larger than 6?

23. What positive integer less than 100 has the greatest number of divisors? Is there more than one answer?

1-6 MATHEMATICAL GAMES OF CHANCE

You are undoubtedly familiar with games whose outcomes depend entirely upon chance, not skill. A game of this type is called *fair* if each player has an equal chance of winning. Some mathematical games of chance will be presented below.

How do you decide if a particular game is fair? Sometimes a careful study of the rules will enable you to decide. Another possibility is to play the game many times and keep track of the results. This may give you a "feel" as to whether the game is fair. If the game is grossly unfair, you shouldn't have to play it too many times to convince yourself of this fact. On the other hand, if the game is almost fair, you might have to play it many times to convince yourself of its fairness or unfairness.

Consider the following games for two players:

Coin games

(1) Toss a penny. If it comes up heads, you win. If it comes up tails, your opponent wins.

(2) Toss two pennies. If they come up the same, you win. If they come up a head and a tail, your opponent wins.

(3) Toss two pennies. If they both come up heads, you win. If one comes up heads and the other tails, your opponent wins. (If both come up tails, no one wins.)

Analysis

(1) Since heads and tails are equally likely, this game appears to be fair.

(2) Does it seem to you equally likely that the two coins will come up the same as that they will come up different? You might toss two pennies 30 or 40 times to help you decide. We did. In 25 trials, we got "heads heads" 5 times, "tails tails" 6 times, and "heads tails" 14 times. Thus, they came up the same 11 times and different 14 times. It seems that this game is almost fair, if not fair.

(3) If game (2) is fair, then this game is not fair. In our 25 trials in game (2), we got "heads heads" only 5 times and "heads tails" 14 times. However, you shouldn't jump to conclusions with so few trials. How did *your* trials turn out?

Dice games. A die has six faces marked 1, 2, 3, 4, 5, and 6.

(1) A die is tossed. You win if an odd number is thrown. Your opponent wins if an even number is thrown.

(2) A die is tossed. You win if a number greater than 3 is thrown and he wins if a number less than 3 is thrown (neither wins if 3 is thrown).

(3) Two dice are tossed. You win if the sum of the numbers tossed is even and he wins if the sum is odd.

Analysis

(1) Each number on the die is equally likely to be thrown (assuming the die is "fair"). Since there are three odd-numbered faces and three even-numbered faces, the game appears to be fair.

(2) There are three faces of the die having numbers greater than 3 and only two faces having numbers less than 3. Therefore, this is not a fair game.

(3) Any number from 2 to 12 can be the sum of the numbers on the dice. Thus, there are six possible even sums and five possible odd sums. So, you might think that you will win more often than your opponent. Beware! Any good dice player knows a sum of 7 is more likely than 6 or 8 (or any other even number), and a sum of 5 is more likely than 4. So perhaps the game is fair! You might play the game several times to help you decide if it is fair.

Marble games. A bag contains 4 marbles of the same size: two red ones and two green ones.

(1) Mix the marbles and then, without looking, pick one out. If it is red, you win; if it is green, your opponent wins.

(2) Again, mix the four marbles in the bag. This time pick out two. If they are both red, you win; if one is red and one is green, your opponent wins. (If they are both green, neither wins.)

(3) Mix the four marbles in the bag again and pick out two. If they are of the same color, you win. If they are of opposite colors, your opponent wins.

Analysis

(1) Since red and green are equally likely, this appears to be a fair game.

(2) We played the game 50 times, winning 6 games ("red red"), losing 34 ("red green"), and tying 10 ("green green"). From these trials, the game certainly seems unfair. You play the game 50 times and record your results.

(3) Using our trials of game (2), we win 16 times ("red red" or "green green") and lose 34 times ("red green"). The game still seems unfair. Use your trials of game (2) to decide if this game is fair.

Exercises

1. (a) Toss a coin 100 times and record whether it comes up heads or tails each time. When you have recorded all 100 outcomes, consider the following questions.

(i) How many times did the coin come up heads and how many times tails?

(ii) Out of 100 tosses of a coin, approximately how many do you *expect* will come up heads? tails?

(b) Thow a die 100 times and record which face is up each time.
  (i) How many times did each of the numbers 1 through 6 appear?
  (ii) Out of 100 throws of a die, approximately how many do you *expect* will show 1? show 2?

Consider the following games. For each game a rule is given which tells whether you or your opponent wins. Decide whether each game is fair.

2. A bag contains 3 marbles of the same size: 1 green, 1 red, and 1 yellow. Mix the marbles and then, without looking, pick one out.
  (a) If it is red or green, you win. If it is yellow, your opponent wins.
  (b) If it is green, you win. Otherwise, your opponent wins.

A die has six faces marked 1, 2, 3, 4, 5, 6. The die is tossed.

3. (a) You win if a 2 is thrown. Your opponent wins if a 4 is thrown.
  (b) You win if a 1 or a 2 is thrown. You opponent wins if a 4 or a 6 is thrown.

4. (a) You win if a 2 or a 3 is thrown; your opponent wins if a number greater than 3 is thrown.
  (b) You win if a 3 or 4 is thrown; your opponent wins if a number greater than 4 is thrown.

5. A die has "1" marked on one face, "2" on two other faces, and "3" on the remaining three faces.
  (a) You win if 3 is thrown. Otherwise your opponent wins.
  (b) You win if 2 is thrown. Your opponent wins if 3 is thrown.

Two dice marked as in Exercise 3 are tossed.

6. (a) You win if 2 is thrown on each die. Your opponent wins if 6 is thrown on each die.
  (b) You win if one die shows 1 and the other shows 2. Your opponent wins if one die shows 3 and the other 5.

7. (a) You win if the sum of the numbers tossed is 7, and he wins if the sum of the numbers tossed is either 2 or 6.
  (b) You win if the sum of the numbers tossed is less than or equal to 9, and he wins if the sum of the numbers tossed is greater than 9.

A bag contains 5 marbles of the same size: 1 red, 1 green, 1 yellow, 1 blue, and 1 black. Mix them and, without looking, pick one out.

8. (a) If it is red or green, you win. Otherwise your opponent wins.
  (b) If it is red, green, or blue, you win. Otherwise your opponent wins.

9. (a) If you pick one marble and record its color, then return the marble to the bag and repeat the process until you have drawn 50 times in all, approximately how many times do you expect to draw each color?
  (b) If you repeat the process described in Exercise 9(a) 100 times, approximately how many times do you expect to draw each color?

Decide whether each of the following games is fair.

10. A bag contains 1 red marble and 3 green ones. Mix them and pick out two.
   (a) If both are green, you win. Otherwise your opponent wins.
   (b) If one is green and the other red, you win. Otherwise, your opponent wins.

A card is drawn from a well-shuffled bridge deck containing 52 cards.

11. (a) If it is either a spade or a club, you win. Otherwise your opponent wins.
   (b) If it is a diamond, you win. Otherwise your opponent wins.

12. (a) If it is either a diamond or the ace of spades, you win. If it is a heart, your opponent wins. Otherwise it is a tie.
   (b) If it is a jack, king, or queen, you win. If it is a ten or an ace, your opponent wins. Otherwise it is a tie.

---

13. Three coins are tossed.
   (a) If at least two of them are heads, you win. Otherwise your opponent wins.
   (b) If exactly one of them is heads, you win. Otherwise your opponent wins.

14. A bag contains 3 red marbles, 2 blue ones, and 1 white one, all of the same size.
   (a) Mix them and pick out two. If they are both the same color, you win. Otherwise your opponent wins.
   (b) Mix them and pick out two. If they are both red, you win. Otherwise your opponent wins.

15. Two cards are drawn from a well-shuffled bridge deck from which the red kings and red jacks have been removed.
   (a) If one is a king and the other a queen, you win. If one is a jack and the other an ace, your opponent wins. Otherwise it is a tie.
   (b) If one is a king and the other a queen, you win. If both are aces, your opponent wins. Otherwise it is a tie.

16. A bag contains 10 marbles: 2 red, 2 green, and 6 white. Two marbles are picked from the bag.
   (a) You win if both are green and your opponent wins if one is red and one is green. Otherwise it is a tie.
   (b) You win if both are the same color and your opponent wins if the two marbles are of different colors.

1–7 PROBABILITY

Each mathematical game of chance described in Section 1–6 can be thought of as an *experiment*. In the first coin game, the experiment consists of tossing a penny. The two possible *outcomes* are "heads" and "tails." In the second and third coin games, the experiments consist of tossing two pennies. There are three possible outcomes, "both heads," "one head and one tail," and "both tails."

In the dice games, the experiments consist of tossing dice. The outcomes in games (1) and (2) are the numbers 1, 2, 3, 4, 5, 6. In game (3), the outcomes are 2, 3, 4, 5, 6, 7, 8, 9, 10, 11, 12, the possible sums of the numbers showing on the two dice.

In trying to analyze such experiments mathematically, you might assign a number to each outcome, the *probability* of the outcome. The letter "P" is used for "probability," and

$$P(\text{outcome})$$

is used for the probability of an outcome. $P(\text{outcome})$ should measure the likelihood of the outcome occurring. Thus, *equally likely outcomes should have equal probabilities.*

For example, in the experiment of tossing one penny, the outcomes "heads" and "tails" are equally likely. Since each outcome is expected to occur 1 time in 2 trials, let

$$P(\text{heads}) = \tfrac{1}{2},\ P(\text{tails}) = \tfrac{1}{2}.$$

In the experiment of tossing a die, each of the outcomes is again equally likely. Since each outcome is expected to occur 1 time in 6 trials, let

$$P(1) = \tfrac{1}{6},\ P(2) = \tfrac{1}{6},\ P(3) = \tfrac{1}{6},\ P(4) = \tfrac{1}{6},\ P(5) = \tfrac{1}{6},\ P(6) = \tfrac{1}{6}.$$

Consider the following additional experiments.

Experiment 1. A bag contains 3 marbles of the same size: 2 red and 1 green. Mix the marbles and then, without looking, pick one out. Record its color and return it to the bag. Repeat the process 30 times. How many times did you expect to pick a red marble? a green marble?

Analysis. You have 2 chances in 3 of picking a red marble, and 1 chance in 3 of picking a green marble. Thus, it seems reasonable to let

$$P(\text{red}) = \tfrac{2}{3},\ P(\text{green}) = \tfrac{1}{3}.$$

In 30 trials, you would expect 20 red marbles and 10 green marbles on the average. This does not mean you will get exactly 20 red and 10 green, but over a large number of trials you will pick approximately twice as many red marbles as green.

Experiment 2. A bag contains 4 marbles of the same size: 2 red and 2 green. This time, pick out two marbles, record their colors, and return them to the bag. Repeat the process 30 times. How many times do you expect the outcome "red red"? the outcome "red green"? the outcome "green green"?

Analysis. Imagine picking out the marbles one at a time. After picking out one marble, 2 of the 3 remaining marbles are of a different color than the one you selected. Thus, the probability that the second marble picked will be of a different color than the first is $\frac{2}{3}$. That is,

$$P(\text{red green}) = \tfrac{2}{3}.$$

On the other hand, the probability that the two marbles picked will have the same color is $\frac{1}{3}$. Clearly $P(\text{red red}) = P(\text{green green})$. Therefore, $P(\text{red red})$ should be one-half the probability of picking two marbles the same color,

$$P(\text{red red}) = \tfrac{1}{2} \cdot \tfrac{1}{3} = \tfrac{1}{6}.$$

Similarly, $P(\text{green green}) = \frac{1}{6}$. In 30 trials, you expect "red red" 5 times, "green green" 5 times, and "red green" 20 times on the average.

Any set of outcomes of an experiment is called an *event*. For example, the set of outcomes where the two marbles picked are of the same color, {red red, green green}, is an event in Experiment 2. You can assign a probability to an event,

$$P(\text{event}),$$

which measures the likelihood of some one of the outcomes in the event occurring. In Experiment 2, {red red, green green} is the event that the two marbles picked have the same color, and

$$P(\{\text{red red, green green}\}) = \tfrac{1}{3}.$$

The probability of an event is the sum of the probabilities of the outcomes in the event. For example, in Experiment 2,

$$P(\{\text{red red, green green}\}) = P(\text{red red}) + P(\text{green green}).$$

The event of all possible outcomes is always taken to have probability 1. Thus, the sum of the probabilities of all outcomes of an experiment is always 1. For example, in Experiment 1 the possible outcomes are "red" and "green," and

$$P(\text{red}) + P(\text{green}) = \tfrac{2}{3} + \tfrac{1}{3} = 1.$$

In Experiment 2,

$$P(\text{red red}) + P(\text{green green}) + P(\text{red green}) = \tfrac{1}{6} + \tfrac{1}{6} + \tfrac{2}{3} = 1.$$

Experiment 3. Two dice are tossed. What is the probability of the event "the sum of the numbers on top is 7"?

Analysis. Imagine one die is red and one is green. Then the possible outcomes of tossing the dice are (using "R" for "red" and "G" for "green"):

R1G1, R1G2, R1G3, R1G4, R1G5, R1G6,
R2G1, R2G2, R2G3, R2G4, R2G5, R2G6,

and so on, to

R6G1, R6G2, R6G3, R6G4, R6G5, R6G6.

In all, there are 36 possible outcomes. Each outcome is equally likely. Since the sum of the probabilities of the 36 outcomes is 1, for each outcome

$$P(\text{outcome}) = \tfrac{1}{36}.$$

The event "the sum of the numbers on top is 7" is the set

{R1G6, R2G5, R3G4, R4G3, R5G2, R6G1}.

Hence,

$$\begin{aligned} P(\text{event}) &= P(\text{R1G6}) + P(\text{R2G5}) + P(\text{R3G4}) + P(\text{R4G3}) \\ &\quad + P(\text{R5G2}) + P(\text{R6G1}), \\ &= \tfrac{1}{36} + \tfrac{1}{36} + \tfrac{1}{36} + \tfrac{1}{36} + \tfrac{1}{36} + \tfrac{1}{36} = \tfrac{1}{6}. \end{aligned}$$

Therefore, 7 will be the sum of the numbers appearing on the dice in approximately one-sixth of the tosses.

Exercises

1. (a) A bag contains 5 marbles of the same size: 2 blue and 3 green. Mix the marbles and then, without looking, pick one out. What is the probability of picking a blue marble? a green marble?
   (b) A bag contains 7 marbles of the same size, 3 red, 2 green, and 2 yellow. Mix the marbles and then, without looking, pick one out. What is the probability of picking a red marble? a green one? a yellow one?
2. A die is tossed.
   (a) (i) What is $P(2)$, the probability that a 2 is thrown? What is $P(4)$?
   (ii) What is the probability that an even number is thrown?
   (b) (i) What is $P(3)$? $P(5)$? $P(1)$?
   (ii) What is the probability that a number less than 4 is thrown?
3. A card is drawn from a well-shuffled bridge deck containing 52 cards.
   (a) What is $P(\text{heart})$, the probability of drawing a heart? $P(\text{diamond})$? $P(\text{red card})$?
   (b) What is $P(\text{spade})$? $P(\text{club})$? What is the probability of drawing either a spade or a club?
4. (a) A bag contains 100 marbles of the same size: 50 black, 30 white, and 20 red. The marbles are mixed and one is drawn from the bag.
   (i) What is $P(\text{black})$, the probability that the marble drawn is black? $P(\text{white})$? $P(\text{red})$?
   (ii) What is the probability that the marble drawn is either white oı red?
   (iii) What is the probability that the marble drawn is either white, black, or red?
   (b) A bag contains 100 marbles of the same size: 10 black, 20 white, 30 red, and 40 green. The marbles are mixed and one is drawn from the bag.
   (i) What is $P(\text{black})$? $P(\text{white})$? $P(\text{red})$? $P(\text{green})$?
   (ii) What is the probability that the marble drawn is either black or white?
   (iii) What is the probability that the marble drawn is either black, white, red, or green?
5. There are five coats in a dark closet. Two belong to Mr. Jones and three belong to his guests. Mr. Jones reaches in the closet and takes out a coat.
   (a) What is the probability that he picked one of his own coats?
   (b) What is the probability that he picked a coat belonging to a guest?
6. (a) A coin is tossed 10 times.
   (i) What is the probability that the coin will show heads on the eleventh toss?
   (ii) Does the outcome of the first 10 tosses have any effect on the outcome of the eleventh toss?

(b) In a family of 14 children, what is the probability that the fifteenth child will be a girl?

7. A bag contains 10 marbles of the same size numbered from 1 to 10. One marble is picked and its number recorded.

(a) What is the probability that the number on the marble is even? odd? even or odd? greater than 10?

(b) What is the probability that the number on the marble is greater than 5? less than 5? less than or equal to 5?

8. From a group of 2 boys and 2 girls, a committee of two members is to be chosen. If each person has an equal chance of being selected, what is the probability of a committee consisting of

(a) a boy and a girl?

(b) two girls?

9. A die has "1" marked on one face, "2" on two other faces, and "3" on the remaining three faces. The die is tossed.

(a) What is the probability that a 3 is thrown? that a 1 or a 2 is thrown?

(b) What is the probability that a 2 is thrown? that a 1 or a 3 is thrown?

A green die and a red die are each tossed once and their top faces noted. Make a complete list of all the possible outcomes, writing them in a rectangular array as indicated in Experiment 3.

10. (a) What is the probability of a 6 showing on each die?

(b) What is the probability of the same number showing on each die?

11. (a) What is the probability of a 1 showing on the red die and a 2 on the green die?

(b) What is the probability of a 1 showing on one die and a 2 on the other?

12. (a) What is the probability of a 3 showing on the red die?

(b) What is the probability of a 3 showing on at least one die?

13. (a) What is the probability that the sum of the numbers showing is 10?

(b) What is the probability that the sum of the numbers showing is 3?

14. (a) What is the probability that the sum of the numbers showing is greater than 10?

(b) What is the probability that the sum of the numbers showing is less than 3?

15. A bag contains 4 marbles: 1 red and 3 green. The marbles are mixed and are picked out.

(a) What is the probability that both marbles will be green?

(b) What is the probability that one will be green and the other red?

16. A card is drawn from a well-shuffled bridge deck containing 52 cards.
    (a) What is P(queen)? P(red queen)? P(queen of hearts)?
    (b) What is the probability of drawing a jack or a king? a red jack or a red king? the jack or king of hearts?
17. Cards are drawn one at a time from a well-shuffled bridge deck.
    (a) (i) What is the probability that the first card drawn is the ace of spades?
        (ii) Assume that the queen of hearts is drawn on the first draw and put aside. What is the probability that the second card drawn is the ace of spades?
    (b) Suppose that the first two cards drawn are the queen of hearts and the king of diamonds and that they are put aside. What is the probability that the third card drawn is the ace of spades?
18. Two coins are tossed. List all possible outcomes.
    (a) What is the probability of heads on both coins?
    (b) What is the probability that one will show heads and the other tails?

---

19. Assume that the events "birth of a girl" and "birth of a boy" are equally likely. In a family of two children, what is the probability that
    (a) both are boys? (b) one is a boy and one is a girl?
20. Two marbles are picked from a bag containing 7 marbles: 3 red and 4 green. What is the probability of selecting
    (a) two red marbles? (b) two green marbles?

1-8 ALGORITHMS AND FLOW CHARTS

Whenever you have a job to do, be it taking tickets at a football game, tying your shoelaces, or solving an arithmetic problem, the first step is to decide on a procedure for doing it. For example, in taking tickets at the football game, you may decide to stand at the entrance gate and, as each person passes, ask for his ticket. If he gives you the proper ticket, you would allow him to enter. If not, you would send him away.

A procedure which specifically states, step-by-step, how to perform a given task and how to proceed in all circumstances is called an *algorithm.*

Algorithms are used extensively in mathematics. For example, in multiplication of fractions, you know that

$$\frac{a}{b} \times \frac{c}{d} = \frac{ac}{bd}.$$

In the following problem, the algorithm for multiplying fractions is used.

Problem. Find $\frac{a}{b} \times \frac{c}{d}$.

Solution. The algorithm is:

1. Multiply $a \times c$.
2. Multiply $b \times d$.
3. Put the product from step 1 in the numerator's position, and put the product from step 2 in the denominator's position. This is your answer.

Following the algorithm step-by-step, you can solve the following.

Problem 1. $\frac{2}{3} \times \frac{4}{5}$

Solution. 1. $2 \times 4 = 8$

2. $3 \times 5 = 15$

3. $\frac{8}{15}$

Problem 2. $\frac{2}{5} \times \frac{1}{4}$

Solution. 1. $2 \times 1 = 2$

2. $5 \times 4 = 20$

3. $\frac{2}{20}$, or $\frac{1}{10}$

You may have noticed that, while the answer is not always given in the lowest terms, you do get a correct answer.

Sometimes it is convenient to draw a picture of the algorithm you are using. A picture of this type is called a *flow chart.*

Whenever you have a definite method for doing something that can be broken down into definite steps, you can make a flow chart to picture these steps.

Some standard symbols used in flow-charting are shown in Fig. 1–4.

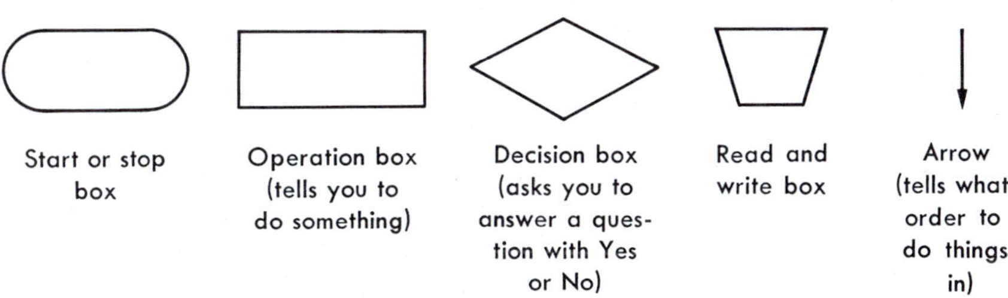

FIGURE 1–4

You can construct a flow chart by using these symbols, writing instructions in the boxes, and connecting the boxes with arrows, as shown in Fig. 1–5.

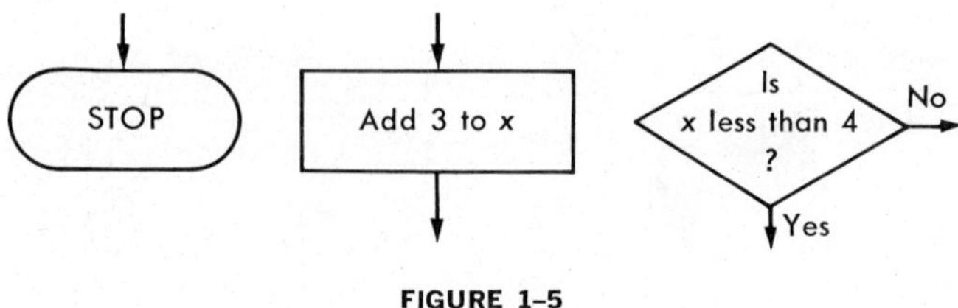

FIGURE 1–5

In Fig. 1–6, the flow chart for the algorithm, given on page 35, on multiplication of fractions is shown.

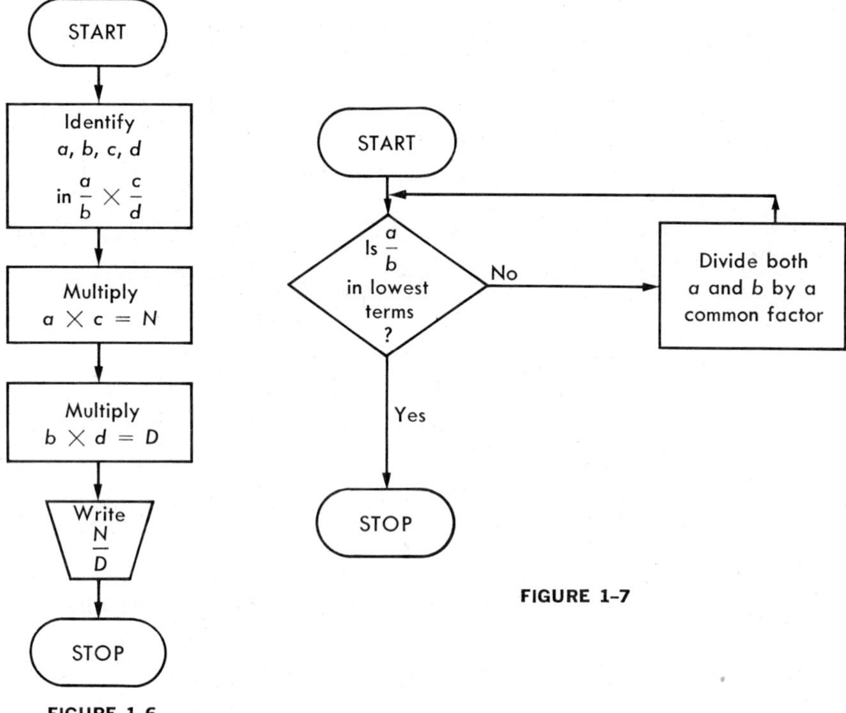

FIGURE 1–6

FIGURE 1–7

You could also make a flow chart for reducing fractions. First, outline the step-by-step procedure:

1. Is N/D in lowest terms?
   (a) If so, you are finished.
   (b) If not, go to step 2.
2. Divide the numerator, N, and the denominator, D, by a common factor. Go back to step 1.

The flow chart for this algorithm is illustrated in Fig. 1–7. You can combine this flow chart with the flow chart for multiplication of fractions to get a flow chart which will always give the answer in lowest terms:

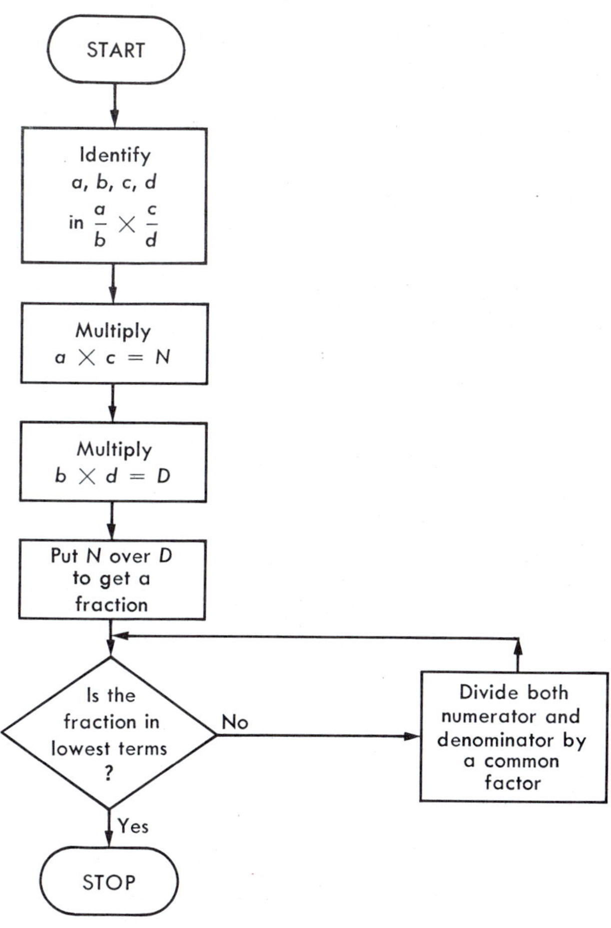

FIGURE 1-8

Exercises

Using the flow chart above, solve the following problems.

1. (a) $\frac{2}{3} \times \frac{3}{5}$ (b) $\frac{1}{4} \times \frac{2}{5}$
2. (a) $\frac{3}{10} \times \frac{1}{2}$ (b) $\frac{4}{5} \times \frac{1}{3}$
3. (a) $\frac{6}{7} \times \frac{2}{3}$ (b) $\frac{7}{10} \times \frac{4}{5}$
4. (a) $\frac{2}{3} \times \frac{9}{10}$ (b) $\frac{3}{5} \times \frac{5}{9}$
5. (a) $\frac{4}{7} \times \frac{3}{2}$ (b) $\frac{2}{5} \times \frac{7}{4}$

In the following problems, write an algorithm for each of the procedures named.

6. (a) Brushing your teeth
(b) Sharpening a pencil

7. (a) Adding two 2-digit numbers. (Hint: Let one number have the form $10a + b$ where a is the tens' digit and b is the units' digit. Let the other by $10c + d$ where c is the tens digit and d is the units digit. Then

$$\begin{array}{r} 10a + b \\ + \ 10c + d \\ \hline 10e + f \end{array}$$

where $f = b + d$ if $b + d$ is less than 10, and $e = a + c$ if $b + d$ is less than 10. Remember to allow for the case where $b + d$ is greater than 10.)

(b) Multiplying a one-digit number by a two-digit number. (Hint: Again, let $10a + b$ be the two-digit number where a is the tens digit and b is the units digit. Let c be the one-digit number.)

8. Draw the flow chart illustrating the algorithm you wrote for:
(a) Problem 6(a)
(b) Problem 6(b)

9. Draw the flow chart illustrating the algorithm you wrote for:
(a) Problem 7(a)
(b) Problem 7(b)

Use the flow charts you drew in 9(a) and 9(b) to compute the following.

10. (a) $32 + 24$
(b) 2×31

11. (a) $15 + 33$
(b) 3×23

12. (a) $27 + 46$
(b) 4×18

13. (a) $36 + 29$
(b) 3×26

14. (a) $72 + 43$
(b) 4×61

15. (a) $59 + 84$
(b) 5×63

EXTRA!

A group of 6 soldiers can march in rectangular formation in 4 different ways as shown in the figure.

Formation (1) consists of 1 row and 6 columns; (2) of 2 rows and 3 columns; (3) of 3 rows and 2 columns; (4) of 6 rows and 1 column. That is, (1) is a 1×6 formation; (2) is a 2×3 formation; (3) is a 3×2 formation; (4) is a 6×1 formation.

1. Show the different formations in which a group of 12 soldiers can march.
2. In how many different formations can a group of each of the following number of soldiers march?
   (a) 5 (b) 9 (c) 15 (d) 24
3. What is the smallest group of soldiers which can march in 12 different formations?
4. Each formation in which a group of n soldiers can march corresponds to a factorization of n. What is the relationship between the number of different formations and the number of factors of n? Illustrate the case for $n = 60$.
5. In how many different formations can a group of n soldiers march in each of the following cases?
   (a) n is a prime. (b) n is the square of a prime.
   (c) n is the product of two different primes.

HISTORICAL NOTE

In 2000 B.C., the Babylonians used algebraic methods in solving problems. However, they used no mathematical symbols other than primitive numerals. This lack of symbolism in algebra continued for many centuries. Gradually, some of the more common words used in mathematics were abbreviated, which lead to a syncopated algebra. However, it was not until almost 1500 A.D. that a symbolic algebra began to emerge.

The plus, +, and minus, —, signs first appeared in 1489 A.D. and were regularly used by 1544 A.D. The equals sign, =, was first used in 1557 by Robert Recorde in England. The raised dot, ·, and juxtaposition were used for multiplication about 1600, and the symbol × about 1620. We find the division symbol, ÷, appearing in Rahn's *Teutsche Algebra* in 1659.

If one person can be credited with developing symbolic algebra, he is the French mathematician Vieta in about 1600. His book *In Artem* uses vowels for unknown quantities and consonants for known quantities. A few years later, Descartes used x and y for variables and also used exponential notation.

Once symbolism began to appear in mathematics, it was only a few decades before a highly polished language of mathematics was universally adopted. Anyone who has tried to do mathematics both with and without symbolism is bound to conclude that mathematics is much easier to understand and apply with symbols than without.

KEY IDEAS AND KEY WORDS

Numbers are abstract quantities; **numerals** are symbols used to denote numbers.

Decimal notation is our system of numeration, based on the ten **digits** 0, 1, 2, 3, 4, 5, 6, 7, 8, 9.

Exponents are numerals used to indicate **powers** of a number. In each example below, 3 is an exponent.

$$10^3 = 10 \cdot 10 \cdot 10 = 1000$$
$$8^3 = 8 \cdot 8 \cdot 8 = 512$$

Scientific notation is the expression of a number as a product of a number between 1 and 10 and a power of 10. For example, $876 = 8.76 \times 10^2$.

Each **rational number** may be expressed either as a **terminating decimal,** such as $\frac{3}{8} = .375$, or as a **repeating decimal,** such as $\frac{5}{33} = .151515\ldots$

Irrational numbers are real numbers that can *not* be expressed as a quotient of two integers, for example π and $\sqrt{2}$.

A **set** is a collection of objects called **elements.**

The set of all **real numbers** is made up of all rational and irrational numbers.

The **union** of sets A and B is denoted by $A \cup B$ and consists of all elements in either A or B or both.

The **intersection** of A and B is denoted by $A \cap B$ and consists of all elements common to both A and B.

The **empty set** is denoted by $\emptyset$ and is a set with no elements.

If every element of set A is contained in a set B, then A is called a **subset** of B, and we write $A \subset B$.

A **variable** is a symbol used to denote any element of a given set. The given set is called the **domain** of the variable. The elements of the given set are called **values** of the variable.

An **open sentence** is a statement containing one or more variables and having the property that it becomes either true or false when the variables are given specific values from their domains.

The notation $\{x \mid ______\}$ is used for denoting **sets.** For example, the set

$$\{x \mid x \text{ is an integer between 11 and 19}\}$$

consists of the elements 12, 13, 14, 15, 16, 17, and 18.

A line having a real-number scale on it is called a **number line.** The number assigned to a point is called the **coordinate** of the point; the point assigned to a number is called the **graph** of the number.

A **solution** of an open sentence with one variable is a value of the variable for which the statement is true.

The integers 1, 2, 3, 4, and so on are called **positive integers.**

Number theory is the study of the properties of the positive integers.

If a, b, and c are integers such that $a = b \cdot c$, then b and c are called **divisiors,** or **factors,** of a, and a is called a **multiple** of b and c.

A **prime number** is an integer greater than 1 which has 1 and itself as its only positive divisors.

Rule for divisibility of an integer by 3. An integer is divisible by three if, and only if, the sum of its digits is divisible by three.

The **probability of an outcome,** denoted P(outcome), is a measure of the likelihood of that outcome occurring.

An **event** is any set of outcomes of an experiment. P(event) is a measure of the likelihood of some one of the events occurring.

An **algorithm** is a step-by-step procedure for solving a problem which specifically states how to proceed in all circumstances.

A **flow chart** is a picture of an algorithm.

CHAPTER REVIEW

In Exercises 1–11, indicate which of the following statements are true and which are false. Correct the false statements.

1. 5.23 is a numeral for a rational number.
2. $\sqrt{36}$ is a numeral for an irrational number.
3. $\frac{51}{17}$ is a numeral for an integer.
4. $7 + 1$, $\frac{16}{2}$, and 2^3 are different numerals for the same number.
5. In the decimal system we have only nine digits.

6. 243 is a power of 3.

7. 12×10^4 is the representation of 120,000 in scientific notation.

8. $x + 8 \neq 50$ is an open sentence which results in a false statement if x is replaced by 42.

9. 1.2333 . . . represents an irrational number.

10. There is no solution for the open sentence $x + 2 = 2$.

11. If $A = \{4x \mid x$ an integer between 1 and 10$\}$ and if $B = \{x^4 \mid x$ an integer between 1 and 10$\}$, then there is exactly one number in set A which is also in set B.

12. Express as a numeral without exponents.
    (a) $3^2 \cdot 2^3$ (b) 4.5×10^3
    (c) 1^{12} (d) $5 \cdot 7^2$
    (e) 5^2 (f) 2^4
    (g) 3.62×10^5 (h) 3×4^2

13. Express with exponents.
    (a) 27 (b) 81
    (c) $\frac{4}{9}$ (d) .16
    (e) 10,000 (f) 32

14. Give each of the following products in exponential form as a power of one number.
    (a) $2^4 \cdot 2^2$ (b) $3^8 \cdot 3$
    (c) $5^2 \cdot 5^3$

15. (a) If x denotes the number of boys in some class and y the number of girls, write an equation which will give the difference between the number of boys and the number of girls in the class.
    (b) What is the domain of the variables x and y?

16. (a) If there are x boys in a class of 40, what does $40 - x$ stand for?
    (b) What is the domain of the variable x in part (a)?

17. Find the values of $2x^2 + 8$ if x is assigned, in turn, the values 0, 1, 2, 10, and 100.

18. Express each statement in mathematical symbols.
    (a) Three less than the square of the number n
    (b) The number of pennies in r dimes and q quarters
    (c) The number of seconds in h hours
    (d) The distance a plane covers in t hours if the speed of the plane in still air is 500 miles per hour but there is a tail wind of 25 miles per hour
    (e) The sum of the two even integers immediately following the even integer N

19. Use variables to write mathematical expressions for each of the following phrases.
(a) Twelve more than a number
(b) Twice one number decreased by three times another
(c) Seven less than one-half of a number
(d) The value in cents of x nickels and y dimes

20. Classify each of the following statements as either true, false, or open.
(a) $3^2 - 1 \neq 8$
(b) $4(9 - 6) = (4 \cdot 9) - 6$
(c) $5x + 2 = 7x$
(d) $x + 5 = 2x$

21. List the elements in each of the following sets.
(a) $A = \{5n - 1 \mid n \text{ is an integer between 0 and 7}\}$
(b) $B = \{x \mid 6x + 4 = 76\}$
(c) $C = \{x - 3 \mid x \text{ is an integer between 4 and 10}\}$
(d) $D = \{x \mid 7x + 3 = 8x - 1\}$

22. Find the decimal representation of the rational number $\frac{3}{19}$.

23. Find the decimal representation of the rational number $\frac{7}{27}$.

24. If $P = \{5x \mid x \text{ is an integer between 0 and 10}\}$ and
$Q = \{2x \mid x \text{ is an integer between 0 and 10}\}$,
(a) what are the elements of $P \cap Q$?
(b) what are the elements of $P \cup Q$?

25. $P = \{3x \mid x \text{ is an integer between 0 and 10}\}$
$Q = \{x \mid x \text{ is an even integer between 0 and 10}\}$
(a) Find $P \cup Q$.
(b) Find $P \cap Q$.

26. A bag contains 20 marbles of the same size: 10 red, 6 green, and 4 blue. One marble is picked from the bag. What is P(red), the probability that the marble selected is red? P(blue)? P(green)? P(red or green)? P(red, green, or blue)?

27. Two dice are tossed. What is the probability that one die will show 5 and the other 6?

28. A nickel and a dime are tossed. What is the probability that the nickel will show heads and the dime tails?

29. A nickel and a dime are tossed. What is the probability that either one will show heads when the other shows tails?

30. A dice is tossed. What is the probability that a 4 is shown?

The following flow chart illustrates a method for multiplying an integer n by $\frac{5}{6}$ and simplifying your answer.

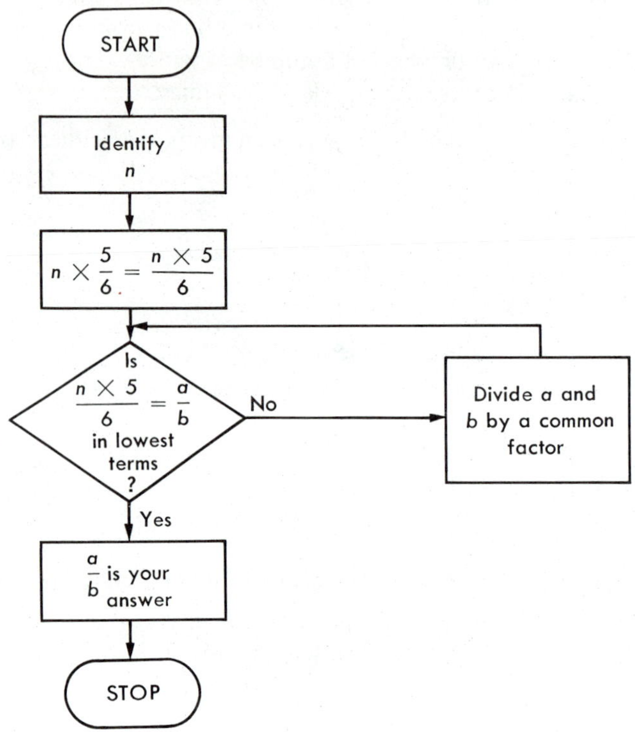

Use the flow chart with the following values of n.

31. (a) $n = 9$ (b) $n = 8$ **32.** (a) $n = 7$ (b) $n = 12$

CHAPTER TEST

1. Give a numeral without exponents for each of the following.
(a) 3^2 (b) $2^3 \times 4^2$ (c) 5.32×10^4

2. Express the following as numerals with exponents.
(a) 8 (b) 100 (c) 64 (d) 1,000,000

3. Give each of the following products in exponential form as a power of one number.
(a) $2^3 \cdot 2^4$ (b) $3 \cdot 3^2$ (c) $7^3 \cdot 7^2 \cdot 7$

4. Find the decimal representation of the rational number $\frac{4}{27}$.

5. List the elements of each of the following sets.

$$A = \{x - 3 \mid x \text{ is an integer between 0 and 5}\}$$
$$B = \{x \mid 2x + 8 = 7x - 2\}$$

6. (a) If x denotes the number of chairs in some classroom and y the number of students, write an equation which will give the difference between the number of chairs and the number of students in the classroom.
(b) What is the domain of the variables x and y?

7. Use variables to write mathematical expressions for each of the following phrases.
(a) Seven more than a number
(b) Twice one number decreased by four times another
(c) Three less than one-third of a number
(d) The value in cents of x quarters and y nickels

8. Classify each of the following statements as either true, false, or open.
(a) $4^2 + 1 \neq 18$ (b) $5(9 - 8) = (5 \cdot 9) - 8$
(c) $3x + 2 = 5x$ (d) $x + 2 = 2x$

9. A nickel and a penny are tossed. What is the probability that one will show heads and the other will show tails?

10. Two dice are tossed. What is the probability that one die will show 2 and the other 5?

CHAPTER 2

Axioms of Algebra

Objectives . . .

- To recognize and use the basic axioms of algebra.
- To apply the order axioms.
- To perform basic operations with positive and negative numbers.
- To use axioms to simplify certain algebraic expressions.

2–1 GENERAL REMARKS

Axioms are the assumptions which are made about operations with numbers. They constitute the basic rules of algebra. Mathematicians began to formulate these axioms in about 1830 and had completed the formulations by 1900.

The basic axioms of addition and multiplication in the set of real numbers will be discussed in this chapter.

The equals sign, =, is used in algebra to indicate the identity of two objects. For example,

$$3 + 4 = 7$$

indicates the $3 + 4$ and 7 are two numerals for the same number. The equals sign is also used in open sentences such as

$$x + 5 = 9$$

which is a true statement if $x = 4$ and false for all other values of x. A sentence of the form $x + 5 = 9$ is also called an *equation.*

Some of the axioms of equality which you use many times without realizing it are:

REFLEXIVE AXIOM

$a = a$ **(*Ref-A*)**

SYMMETRIC AXIOM

If $a = b$***, then*** $b = a$. **(*Sym-A*)**

TRANSITIVE AXIOM

If $a = b$ and $b = c$, then $a = c$. ***(Trans-A)***

ADDITIVE AXIOM

If $a = b$ and $c = d$, then $a + c = b + d$. ***(Add-A)***

MULTIPLICATIVE AXIOM

If $a = b$ and $c = d$, then $ac = bd$. ***(Mult-A)***

You may use these axioms whenever you wish.

In algebra, symbols used for grouping such as parentheses, ( ), and brackets, [ ], indicate the order in which certain operations are to be performed. For example, the parentheses in

$$5 \cdot (3 + 7)$$

indicate that 3 and 7 are to be added. Their sum is then to be multiplied by 5 to yield 50. On the other hand, you would understand

$$(5 \cdot 3) + 7$$

to mean that the product of 5 and 3 is to be added to 7.

$$\begin{aligned}(5 \cdot 3) + 7 &= 15 + 7 \\ &= 22\end{aligned}$$

If there are grouping symbols within grouping symbols in an expression, eliminate them by first doing the arithmetic in the innermost pair and repeating the process until all pairs of parentheses and brackets are gone. The following problem illustrates this process.

$$\begin{aligned}6[7(8 - 5) + 3(6 + 4)] &= 6[(7 \cdot 3) + (3 \cdot 10)] \\ &= 6(21 + 30) \\ &= 6 \cdot 51 \\ &= 306\end{aligned}$$

If there are no parentheses in an expression, it is assumed that you will perform the multiplication and/or division before any addition and/or subtraction.

Exercises

Which of the following statements are true, which are false, and which are open? Why?

1. (a) $(2 + 3) + (8 + 9) = (1 + 4) + (10 + 7)$
(b) $(6 + 5) + (7 + 9) = (10 + 1) + (6 + 10)$

2. (a) $(19 + 12) \times 21 = (13 + 18) \times (7 + 14)$
(b) $(6 + 7) \times (10 + 12) = 13 \times (8 + 14)$

3. (a) $10 + (9 - 6) = 10 + (7 - 5)$
(b) $12 + (13 - 10) = 13 + (12 - 10)$

4. (a) $(7 + 6) \times 52 = (5 + 8) \times 52$
(b) $(8 + 7) \times (3 + 12) = (7 + 8) \times (10 + 5)$

5. (a) $x + 7 = y + (6 + 1)$
(b) $\pi + 3 = (\pi + 1) + 2$

6. (a) $x + r = y + r$
(b) $x \cdot r = y \cdot r$

Simplify each of the following expressions.

7. (a) $6 \cdot (4 + 5)$
(b) $(6 \cdot 4) + 5$

8. (a) $6 + (4 \cdot 5)$
(b) $(6 + 4) \cdot 5$

9. (a) $2[(2 + 9) + (2 + 3)]$
(b) $3[(5 - 2) + (5 + 2)]$

10. (a) $[19 + (5 \cdot 2)] - [19 - (5 \cdot 2)]$
(b) $[16(14 - 4)] - [9(10 - 7)]$

11. (a) $10[5(20 - 10) + 10(15 - 5)]$
(b) $4[6(13 - 11) + 4(7 - 2)]$

12. (a) $2(2[10 + 2(4 - 2)])$
(b) $3(6[8 + 5(12 - 8)])$

13. (a) $2[3(6 - 4) + 4(6 + 4)]$
(b) $5[2(7 + 3) + 3(5 + 2)]$

14. (a) $7(2 + 5[3 + 2(4 + 12)])$
(b) $8(1 + 3[7 + 3(2 + 9)])$

Simplify each expression in Exercises 15–18, and find the value when $a = 2$, $b = 1$, $x = 3$, and $y = 0$.

15. (a) $2(a + 1) + 3(5 + a)$
(b) $3(a + b) + 2(2b + 3a)$

16. (a) $2[3(x + 2y) + 4(2x + y)]$
(b) $3[4(2x + 3y) + 3(x + 4y)]$

17. (a) $5[a(7 + 2b) + 2b(5a)]$
(b) $2[3(4x + 3y) + 5x(2y)]$

18. (a) $4(x + 2[y + 3(x + 2y)])$
(b) $7(a + 5[b + 2(2a + 3b)])$

For what value of x is each of the following statements true?

19. (a) $a + 3 = b + x$ and $a = b$
(b) $a + 7 = 2b + x$ and $a = 2b$

20. (a) $a + x = 15$ and $a = 6$
(b) $a + x = {}^{-}15$ and $a = {}^{-}15$

21. (a) $5x = 15$
(b) $ax = 10$ and $a = 2$

22. (a) $a + 2 = 4 + x$ and $a = 3$
(b) $a + 5 = 3 + x$ and $a = 2$

Name the axiom of equality which is illustrated in each of the following exercises.

23. If $7 + 9 = 16$ and $16 = 2 \cdot 8$, then $7 + 9 = 2 \cdot 8$.

24. If $23 - 12 = 11$, then $11 = 23 - 12$.

25. $3 + 4 = \frac{6}{2} + \frac{100}{25}$

26. $5 \times 3 = \frac{15}{3} \times \frac{21}{7}$

27. If $a = b$, then $a + 3 = b + 3$.

28. If $a = b$, then $a \times 5 = b \times \frac{10}{2}$.

29. $x \times 2 = x \times \frac{4}{2}$

---

30. What value or values of x make the equation

$$a - (b - x) = (a - b) - x$$

true for all values of a and b?

2–2 REARRANGEMENT PROPERTIES

The sum of 30 and 25 is the same as the sum of 25 and 30,

$$30 + 25 = 25 + 30.$$

Similarly, the product of 12 and 7 is the same as the product of 7 and 12,

$$12 \times 7 = 7 \times 12.$$

These examples illustrate the following basic axiom of algebra.

COMMUTATIVE AXIOM

$x + y = y + x$ (*Commutative axiom for addition*) (*C-A*)

$xy = yx$ (*Commutative axiom for multiplication*) (*C-M*)

Each of the equations is true for all numbers x and y.

The phrase "for all numbers x and y" will often be used in place of "for all values of x and y" to emphasize that the domain of each variable is the set of all real numbers.

Addition and multiplication are *binary operations,* that is, they operate on pairs of numbers. Three or more numbers are added or multiplied two at a time. For example, you might add 5, 4, and 8 by adding 5 and 4 and then adding their sum, 9, to 8: $(5 + 4) + 8$. On the other hand, you might add 5 to the sum of 4 and 8: $5 + (4 + 8)$. In either case, the answer is 17.

Similarly, you could multiply 5, 4, and 8 in two different ways: $(5 \times 4) \times 8$ or $5 \times (4 \times 8)$. In either case, the answer is 160.

These examples illustrate another basic axiom of algebra.

ASSOCIATIVE AXIOM

$(x + y) + z = x + (y + z)$
(*Associative axiom for addition*) (*A-A*)

$(xy)z = x(yz)$
(*Associative axiom for multiplication*) (*A-M*)

Each of the equations is true for all numbers x, y, and z.

Because of the associative axiom, you usually write

$5 + 4 + 8$ in place of either $(5 + 4) + 8$ or $5 + (4 + 8)$,

and, similarly, you write

$5 \times 4 \times 8$ in place of either $(5 \times 4) \times 8$ or $5 \times (4 \times 8)$.

In a sum such as $5 + 4 + 8$, the individual numbers 5, 4, and 8 are called *addends.* In a product such as $5 \times 4 \times 8$, the individual numbers 5, 4, and 8 are called *factors.*

You can find a sum of several addends in a variety of ways by repeated use of the commutative and associative axioms. For example, you can find $2 + 9 + 7 + 12$ in each of the following ways.

(a) $2 + [9 + (7 + 12)] = 2 + (9 + 19) = 2 + 28 = 30$

(b) $(2 + 9) + (7 + 12) = 11 + 19 = 30$

(c) $(9 + 2) + (12 + 7) = 11 + 19 = 30$

(d) $9 + [2 + (12 + 7)] = 9 + (2 + 19) = 9 + 21 = 30$

Computations (a) and (b) give the same sum by associativity; (b) and (c) by commutativity; (c) and (d) by associativity.

This combination of the associative and commutative axioms of addition is called the *rearrangement property of addition.* It is not one of the basic axioms of the real number system, but it is derived from the basic axioms and is useful in actual computations. Statements that are derived from the basic axioms are often called *properties.*

REARRANGEMENT PROPERTY OF ADDITION

The addends of a sum may be rearranged in any order. **(*R-A*)**

Since multiplication is also commutative and associative, the rearrangement property holds for multiplication as well.

REARRANGEMENT PROPERTY OF MULTIPLICATION

The factors of a product may be rearranged in any order. **(*R-M*)**

For example, you can use (R-M) to compute $3 \times 4 \times 8$ in any one of the following ways.

(a) $(3 \times 4) \times 8 = 12 \times 8 = 96$

(b) $3 \times (4 \times 8) = 3 \times 32 = 96$

(c) $4 \times (3 \times 8) = 4 \times 24 = 96$

These properties can help simplify your work in solving problems. For example, in finding the sum $3 + 8 + 7 + 2$, you might use the rearrangement property of addition, (R-A), to get

$$(3 + 7) + (8 + 2) = 10 + 10 = 20.$$

Exercises

The commutative axiom states that interchanging two numbers will not alter their sum when you add, or their product when you multiply. Are any of the following pairs of acts commutative? In other words, will the order in which they are performed affect the result?

1. (a) Putting on shoes and putting on socks
(b) Studying your mathematics and taking an examination on it

2. (a) Reading a book and writing a report on it
(b) Buying a paper and buying a loaf of bread

Complete each of the following equations so that it is a correct example of the commutative axiom for addition or multiplication.

3. (a) $18 + 71 = \underline{\ ?\ } + \underline{\ ?\ }$
(b) $18 \cdot 71 = \underline{\ ?\ } \cdot \underline{\ ?\ }$

4. (a) $\underline{\ ?\ } \cdot 23 = \underline{\ ?\ } \cdot 4$
(b) $9 + \underline{\ ?\ } = 7 + \underline{\ ?\ }$

5. (a) $(x + 2) \cdot \underline{\ ?\ } = 3 \cdot (\underline{\ ?\ })$
(b) $ab + \underline{\ ?\ } = (c + d) + \underline{\ ?\ }$

In Exercises 6–13, replace the $*$ by $=$ or $\neq$ to make the statements true. Name the axiom or axioms of which the statements with $=$ are examples.

6. (a) $7 + 5 * 5 + 7$
(b) $17 + 53 * 53 + 17$

7. (a) $2 \cdot 3 * 3 \cdot 2$
(b) $18 \cdot 9 * 9 \cdot 18$

8. (a) $(2 + 3) + 7 * 2 + (3 + 7)$
(b) $2 \cdot (3 + 7) * (2 \cdot 3) + 7$

9. (a) $(12 \cdot 13) \cdot \frac{1}{13} * 12 \cdot (13 \cdot \frac{1}{13})$
(b) $2^{10} + (3^{10} + 4^{10}) * (2^{10} + 3^{10}) + 4^{10}$

10. (a) $9 + (21 \cdot 37) * (21 \cdot 37) + 9$
(b) $3 + (4 \cdot 5) * 4 + (3 \cdot 5)$

11. (a) $6(12 + 4) * (4 + 12)6$
(b) $(8 + 9) \cdot 10 * 10(9 + 8)$

12. (a) $(63 - 18) - 2 * 63 - (18 - 2)$
(b) $(12 \div 6) \div 2 * 12 \div (6 \div 2)$

13. (a) $8 \div 4 * 4 \div 8$
(b) $8 - 4 * 4 - 8$

First use the rearrangement properties so that the computation is made easier. Then find the sums or products of the expressions in Exercises 14–18.

14. (a) $5 \cdot 7 \cdot 2 \cdot 9$
(b) $6 \cdot 39 \cdot 5 \cdot \frac{1}{3}$

15. (a) $36 + 8 + 14 + 12$
(b) $11 + 19 + 22 + 18$

16. (a) $\frac{3}{19} + \frac{4}{13} + \frac{16}{19} + \frac{9}{13}$
(b) $\frac{5}{6} + \frac{4}{5} + \frac{7}{6} + \frac{11}{5}$

17. (a) $69 + 77 + 31 + 23$
(b) $17 + 9 + 13 + 11$

18. (a) $14 \times 2 \times 5 \times 3$
(b) $5 \times 17 \times 4 \times 5$

Classify each statement as true, false, or open. State the axiom or property that justifies each true statement.

19. (a) $19 \cdot (82 \cdot 31) = (19 \cdot 82) \cdot 31$
(b) $(9 \cdot 7) \cdot 4 = (7 \cdot 9) \cdot 4$

20. (a) $(6 + 7) + 8 = (7 + 6) + 8$
(b) $(\frac{17}{19} + \frac{13}{14}) + 86 = \frac{17}{19} + (\frac{13}{14} + 86)$

21. (a) $(31 + 19) + 50 = 19 + (31 + 50)$
(b) $8 + (x + 3) = (8 + 3) + x$

22. (a) $\frac{1}{10} \cdot (x \cdot 20) = \frac{1}{10} \cdot (20 \cdot x) = (\frac{1}{10} \cdot 20)x$, or $2x$
(b) $x + (9 + y) = x + (y + 9) = (x + y) + 9$

23. (a) $(7 + 8) + (9 + 10) = [7 + (9 + 8)] + 10$
(b) $(a + b + c) + d = a + [(c + b) + d]$

24. (a) $4 \div 2 + 3 = \dfrac{4}{2 + 3}$
(b) $9 + (3 \div 3) = 4$

25. (a) $x + 3 = 3x$
(b) $x \cdot 2 = 2x$

26. (a) $20x \cdot 5 = 20 \cdot (x \cdot 5) = 20 \cdot (5x) = 100x$
(b) $(17 + y) + 9 = 17 + (y + 9) = 17 + (9 + y) = 26 + y$

EXTRA!

The imaginary people of Venus add numbers differently than you do. A radio telescope in Venezuela has discovered that 1 plus 1 equals 5 and 3 plus 2 equals 13 on Venus. That is, using the symbol $\bigtriangledown\!\!\!\!+$ for "Venus addition,"

$$1 \bigtriangledown\!\!\!\!\!+\; 1 = 5, \quad 3 \bigtriangledown\!\!\!\!\!+\; 2 = 13.$$

After deciphering many radio messages from Venus, scientists have decided that in terms of Earth's addition and multiplication,

$$x \bigtriangledown\!\!\!\!\!+\; y = 3x + 2y.$$

Thus, for example,

$$3 \bigtriangledown\!\!\!\!\!+\; 2 = 3 \cdot 3 + 2 \cdot 2 = 13.$$

1. Find:
   (a) $3 \mathbin{\bigtriangledown\!\!\!\!\!+} 5$ (b) $5 \mathbin{\bigtriangledown\!\!\!\!\!+} 3$ (c) $\frac{1}{3} \mathbin{\bigtriangledown\!\!\!\!\!+} \frac{1}{2}$
   (d) $\frac{1}{2} \mathbin{\bigtriangledown\!\!\!\!\!+} \frac{1}{3}$ (e) $4 \mathbin{\bigtriangledown\!\!\!\!\!+} 4$
2. Is the operation $\bigtriangledown\!\!\!\!\!+$ commutative? (Refer to your answers in Exercise 1.)
3. For what numbers x and y is $x \mathbin{\bigtriangledown\!\!\!\!\!+} y = y \mathbin{\bigtriangledown\!\!\!\!\!+} x$?
4. Find:
   (a) $(4 \mathbin{\bigtriangledown\!\!\!\!\!+} 5) \mathbin{\bigtriangledown\!\!\!\!\!+} 6$
   (b) $4 \mathbin{\bigtriangledown\!\!\!\!\!+} (5 \mathbin{\bigtriangledown\!\!\!\!\!+} 6)$
   (c) $(1 \mathbin{\bigtriangledown\!\!\!\!\!+} 7) \mathbin{\bigtriangledown\!\!\!\!\!+} 3$
   (d) $1 \mathbin{\bigtriangledown\!\!\!\!\!+} (7 \mathbin{\bigtriangledown\!\!\!\!\!+} 3)$
5. Is the operation $\bigtriangledown\!\!\!\!\!+$ associative? (Refer to your answers in Exercise 4.)
6. For what numbers x, y, and z is

$$(x \mathbin{\bigtriangledown\!\!\!\!\!+} y) \mathbin{\bigtriangledown\!\!\!\!\!+} z = x \mathbin{\bigtriangledown\!\!\!\!\!+} (y \mathbin{\bigtriangledown\!\!\!\!\!+} z)?$$

2-3 DISTRIBUTIVE AXIOM

The method used in arithmetic for finding the product of two integers is illustrated below.

$$\begin{array}{r} 32 \\ \times 4 \\ \hline 128 \end{array} \qquad \begin{array}{r} 30 + 2 \\ \times\ 4 \\ \hline 120 + 8 \end{array}$$

Use is made of the fact that

$$(30 + 2) \times 4 = (30 \times 4) + (2 \times 4).$$

This example illustrates the *distributive axiom.*

DISTRIBUTIVE AXIOM

$$x(y + z) = xy + xz$$

$$(y + z)x = yx + zx \qquad (D)$$

Each of the equations is true for all numbers x, y, and z.

By the distributive axiom, (D), you can find 6×128 as follows:

$$\begin{aligned} 6 \times 128 &= 6 \times (120 + 8) \\ &= (6 \times 120) + (6 \times 8) \\ &= 720 + 48 \\ &= 768. \end{aligned}$$

The other axioms of the real number system are about addition alone or about multiplication alone. The distributive axiom differs from them in that it relates addition and multiplication.

You can use the distributive axiom to simplify your work in solving problems. Some examples are given below.

Problem 1. Find $(21 \cdot 7) + (21 \cdot 13)$.

Solution.

First method: $$\begin{aligned} (21 \cdot 7) + (21 \cdot 13) &= 147 + 273 \\ &= 420. \end{aligned}$$

Second method: $$\begin{aligned} (21 \cdot 7) + (21 \cdot 13) &= 21 \cdot (7 + 13), \\ &= 21 \cdot 20 \\ &= 420. \end{aligned}$$

In the first method, you perform the multiplication first and then the addition. In the second, you use the distributive axiom to perform the addition first, and then the multiplication. For this problem, the second seems to be the easier method.

Problem 2. Find $24 \cdot (\frac{2}{3} + \frac{1}{4})$.

Solution.

First method: $$\begin{aligned} 24 \cdot (\tfrac{2}{3} + \tfrac{1}{4}) &= (24 \cdot \tfrac{2}{3}) + (24 \cdot \tfrac{1}{4}) \\ &= 16 + 6 \\ &= 22. \end{aligned}$$

Second method: $$\begin{aligned} 24 \cdot (\tfrac{2}{3} + \tfrac{1}{4}) &= 24 \cdot (\tfrac{8}{12} + \tfrac{3}{12}) \\ &= 24 \cdot \tfrac{11}{12} \\ &= 22. \end{aligned}$$

In this problem, the first method seems to be easier.

Exercises

Explain the use of the distributive axiom in each of the following multiplications.

1. (a) $\begin{array}{r} 52 \\ \times 3 \\ \hline 156 \end{array}$ (b) $\begin{array}{r} 71 \\ \times 4 \\ \hline 284 \end{array}$

2. (a) $\begin{array}{r} 112 \\ \times 7 \\ \hline 784 \end{array}$ (b) $\begin{array}{r} 136 \\ \times 5 \\ \hline 680 \end{array}$

Complete each of the following examples of the distributive axiom.

3. (a) $8(3 + 17) = (8 \cdot \underline{\ ?\ }) + (8 \cdot \underline{\ ?\ })$
(b) $19(12 + 5) = (19 \cdot \underline{\ ?\ }) + (19 \cdot \underline{\ ?\ })$

4. (a) $43(16 + 14) = (\underline{\ ?\ } \cdot 16) + (43 \cdot \underline{\ ?\ })$
(b) $16(18 + 23) = (\underline{\ ?\ } \cdot 18) + (16 \cdot \underline{\ ?\ })$

5. (a) $(9 + 3) \cdot (8 + 17) = [(9 + 3) \cdot \underline{\ ?\ }] + [(9 + 3) \cdot \underline{\ ?\ }]$
(b) $(a + b) \cdot (c + d) = [(a + b) \cdot \underline{\ ?\ }] + [\underline{\ ?\ } \cdot \underline{\ ?\ }]$

Use the distributive axiom to write each of the products as a sum of two products.

6. (a) $5 \cdot (4 + 7)$ (b) $8 \cdot (6 + 9)$

7. (a) $(8 + 12) \cdot 7$ (b) $(10 + 11) \cdot 12$

8. (a) $18 \cdot (\frac{2}{3} + \frac{5}{6})$ (b) $(21 + 14) \cdot \frac{2}{7}$

Use the distributive axiom to write each of the sums as a product.

9. (a) $(43 \cdot 7) + (43 \cdot 3)$
(b) $(49 \cdot 63) + (49 \cdot 37)$

10. (a) $(24 \cdot \frac{1}{11}) + (31 \cdot \frac{1}{11})$
(b) $(82 \cdot \frac{2}{9}) + (8 \cdot \frac{2}{9})$

Do each computation two ways to check your work.

11. (a) $29(12 + 8)$ (b) $(31 + 19) \cdot 7$

12. (a) $15(\frac{2}{3} + \frac{4}{5})$ (b) $30(\frac{5}{6} + \frac{1}{10})$

13. (a) $\frac{3}{7}(20 + 22)$ (b) $\frac{2}{9}(19 + 17)$

14. (a) $.01(4.1 + .9)$ (b) $.2(.1 + .02)$

In Exercises 15–20, which statements are true and which are false? Justify your answers.

15. (a) $11 \cdot (3 + 4) = (11 \cdot 3) + (11 \cdot 4)$
(b) $11 \cdot (3 + 4) \neq (3 \cdot 11) + (4 \cdot 11)$

16. (a) $(11 \cdot 6) + (5 \cdot 6) \neq 6 \cdot (11 + 5)$
(b) $(31 \cdot 17) + (41 \cdot 17) = (31 + 41) \cdot 17$

17. (a) $(5 \cdot \frac{1}{3}) + (7 \cdot \frac{1}{3}) = (5 + 7) \cdot \frac{1}{3}$
(b) $(22 \cdot \frac{2}{7}) + (83 \cdot \frac{2}{7}) = \frac{2}{7}(22 + 83)$

18. (a) $213 + 84 = 3(71 + 28)$
(b) $52 + 91 = (4 + 7) \cdot 13$

---

19. $(8 + 7)(9 + 6) = [(8 + 7) \cdot 9] + [(8 + 7) \cdot 6]$

20. $(3 + 4)(5 + 6) = [3 \cdot (5 + 6)] + [4 \cdot (5 + 6)]$

21. Find two integral factors greater than one for each of the following integers.

(a) 12
(b) $(3 \cdot 2) + (3 \cdot 3)$
(c) $(7 \cdot 9) + (7 \cdot 2)$
(d) $(17 \cdot 4) + (17 \cdot 9)$
(e) $(6 \cdot 37) + (7 \cdot 37)$
(f) $(4 \cdot 5) + (4 \cdot 5)$

---

22. (a) Give the axiom or property which justifies each numbered equals sign in the following sequence.

$$\begin{aligned} 4 + [2 \cdot (3 + 5)] &\overset{1}{=} 4 + [(2 \cdot 3) + (2 \cdot 5)] \\ &\overset{2}{=} 4 + [(2 \cdot 5) + (2 \cdot 3)] \\ &\overset{3}{=} [4 + (2 \cdot 5)] + (2 \cdot 3) \end{aligned}$$

(b) Using four variables, write an equation that illustrates the pattern of the following:

$$4 + [2 \cdot (3 + 5)] = [4 + (2 \cdot 5)] + (2 \cdot 3).$$

(c) Test the correctness of the pattern you found in (b) by using four numbers of your own choice.

23. If the domains of x and y are restricted to integers, what kind of number is $2x$? $2y$? $2(x + y)$? What does the equation

$$2x + 2y = 2(x + y)$$

tell us about the sum of two even integers?

24. Give the axiom or property which justifies each numbered equals sign.

$$\begin{aligned} (a + b) \cdot (y + z) &\overset{1}{=} [(a + b) \cdot y] + [(a + b) \cdot z] \\ &\overset{2}{=} (a \cdot y) + (b \cdot y) + (a \cdot z) + (b \cdot z) \end{aligned}$$

or

$$\begin{aligned}(a+b)(y+z) &\overset{3}{=} [a\cdot(y+z)] + [b\cdot(y+z)]\\ &\overset{4}{=} [(a\cdot y)+(a\cdot z)] + [(b\cdot y)+(b\cdot z)]\\ &\overset{5}{=} (a\cdot y)+(b\cdot y)+(a\cdot z)+(b\cdot z)\end{aligned}$$

25. Use the formula

$$(a+b)\cdot(y+z) = ay + by + az + bz$$

to find the product $(3+5)(7+9)$.

2–4 IDENTITY AND INVERSE ELEMENTS

The numbers 0 and 1 have special properties. Some examples of these properties are

$$0+3=3,\quad 121+0=121,\quad \tfrac{22}{7}+0=\tfrac{22}{7},$$

and

$$1\times 3=3,\quad 121\times 1=121,\quad \tfrac{22}{7}\times 1=\tfrac{22}{7}.$$

In other words, the sum of any number and zero is the number itself; and the product of any number and one is the number itself. Since the number to which zero is added keeps its identity under this addition, zero is called the *additive identity.* Since the number which is multiplied by one keeps its identity under this multiplication, one is called the *multiplicative identity.*

ADDITIVE IDENTITY AXIOM

There exists a number 0 such that the equations

$$x+0=x \quad \text{and} \quad 0+x=x \qquad (Id\text{-}A)$$

are true for every number x.

MULTIPLICATIVE IDENTITY AXIOM

There exists a number 1 such that the equations

$$x\cdot 1=x \quad \text{and} \quad 1\cdot x=x \qquad (Id\text{-}M)$$

are true for every number x.

A number b is called an *additive inverse* of a number a if $a + b = 0$ and $b + a = 0$. Real numbers have additive inverses according to the axiom below.

ADDITIVE INVERSE AXIOM

Every real number has a unique additive inverse. **(*Inv-A*)**

The additive inverse of a number x is commonly called the *negative* of x and is denoted by ^-x. The equations

$$x + {}^-x = 0, \quad \text{and} \quad {}^-x + x = 0$$

are true for every real number x according to the additive inverse axiom, (Inv-A). For example, $^-3$ is the negative, or additive inverse, of 3 since

$$3 + {}^-3 = 0, \quad {}^-3 + 3 = 0.$$

Since $^-3 + 3 = 0$ and $3 + {}^-3 = 0$, 3 is the negative, or additive inverse, of $^-3$. In symbols,

$$3 = {}^-({}^-3).$$

Similarly, the equation

$$x = {}^-({}^-x)$$

is true for every number x.

Since $0 + 0 = 0$ by the additive identity axiom, (Id-A), 0 is its own additive inverse.

A number b is called a *multiplicative inverse* of a number a if $a \cdot b = 1$ and $b \cdot a = 1$. Real numbers have multiplicative inverses according to the axiom below.

MULTIPLICATIVE INVERSE AXIOM

Every nonzero real number has a unique multiplicative inverse. **(*Inv-M*)**

The multiplicative inverse of a nonzero number x is commonly called the *reciprocal* of x and is denoted by

$$\frac{1}{x} \quad \text{or} \quad 1/x.$$

By the multiplicative inverse axiom, (Inv-M), the equations

$$x \cdot \frac{1}{x} = 1 \quad \text{and} \quad \frac{1}{x} \cdot x = 1$$

are true for every nonzero number x. For example, $\frac{1}{3}$ or $1/3$ is the multiplicative inverse, or reciprocal, of 3 and

$$3 \cdot \tfrac{1}{3} = 1, \quad \tfrac{1}{3} \cdot 3 = 1.$$

The equations above show that 3 is the multiplicative inverse, or reciprocal, of $\frac{1}{3}$:

$$3 = \frac{1}{\frac{1}{3}}.$$

Similarly, the equation

$$x = \frac{1}{\frac{1}{x}}$$

is true for every nonzero number x.

Since $1 \cdot 1 = 1$ by the multiplicative identity axiom, (Id-M), 1 is its own multiplicative inverse,

$$1 = \tfrac{1}{1}.$$

As other examples of multiplicative inverses, $\frac{2}{5}$ is the multiplicative inverse, or reciprocal, of $\frac{5}{2}$ since

$$\tfrac{2}{5} \cdot \tfrac{5}{2} = 1;$$

and 1.25 is the multiplicative inverse, or reciprocal, of .8 since

$$1.25 \times .8 = 1.$$

Exercises

Find the additive inverse of each of the following numbers.

1. (a) 6 (b) 12

2. (a) 1 (b) 0

3. (a) $\frac{1}{2}$ (b) $\frac{1}{4}$

4. (a) $^{-}5$ (b) $^{-}10$

5. (a) $^{-}\frac{1}{3}$ (b) $^{-}\frac{3}{7}$

6. (a) $^{-}(^{-}17)$ (b) $^{-}(^{-}13)$

Find the reciprocals of the following numbers where possible.

7. (a) 15 (b) 100

8. (a) $\frac{1}{6}$ (b) $\frac{2}{3}$

9. (a) .7 (b) .625

10. (a) 0 (b) 1

11. (a) $5\frac{2}{3}$ (b) $6\frac{2}{5}$

12. (a) $\frac{41}{83}$ (b) $\frac{99}{101}$

Which of the following pairs of numbers are additive inverses, or negatives, of each other? Why?

13. (a) $2, {}^{-}2$ (b) $3, {}^{-}({}^{-}3)$

14. (a) ${}^{-}5, {}^{-}({}^{-}5)$ (b) ${}^{-}({}^{-}6), {}^{-}({}^{-}6)$

15. (a) $x, {}^{-}({}^{-}x)$ (b) ${}^{-}x, {}^{-}({}^{-}x)$

Which of the following pairs of numbers are multiplicative inverses, or reciprocals, of each other? Why?

16. (a) $2, \frac{1}{2}$ (b) $\frac{1}{10}, 10$

17. (a) 2, .5 (b) .1, 10

18. (a) .4, 2.5 (b) .6, $1\frac{2}{3}$

19. (a) $\frac{1}{2}, \frac{1}{2}$ (b) 1, .1

Find the value of x for which each of the following equations is true.

20. (a) $12 + x = 12$ (b) $x + 34 = 34$

21. (a) $3 + x + 2 = 5$ (b) $10 = 6 + 4 + x$

22. (a) $4 \cdot x = 4$ (b) $x \cdot 14 = 14$

23. (a) $\frac{2}{5} \cdot x = 1$ (b) $\frac{3}{4} \cdot \frac{4}{3} = x$

24. (a) $x(7 + 8) = 15$ (b) $x(11 + 9) = 20$

25. (a) $.5x = 1$ (b) $.125x = 1$

---

26. Is the reciprocal of an integer ever an integer?

27. If a and b denote nonzero integers, what is the multiplicative inverse of the rational number a/b?

---

28. Fill in the blanks with the reasons for the key steps in the following completed exercises.

(a) $29 + {}^{-}12 = (17 + 12) + {}^{-}12$

$= 17 + [12 + {}^{-}12]$ ___?___

$= 17 + 0$ ___?___

$= 17$

(b) $^{-}36 + 15 = (^{-}21 + {}^{-}15) + 15$

$= {}^{-}21 + (^{-}15 + 15)$ __?__

$= {}^{-}21 + 0$ __?__

$= {}^{-}21$

(c) $^{-}(\frac{12}{35}) + (\frac{10}{21}) = {}^{-}(\frac{36}{105}) + (\frac{50}{105})$

$= {}^{-}(\frac{36}{105}) + [(\frac{36}{105}) + (\frac{14}{105})]$

$= [^{-}(\frac{36}{105}) + (\frac{36}{105})] + (\frac{14}{105})$ __?__

$= 0 + (\frac{14}{105})$ __?__

$= \frac{2}{15}$

2-5 ORDER AXIOMS

The set of real numbers is made up of positive numbers, negative numbers, and zero. Such numbers as 7, 248, 3/2, $\sqrt{2}$, and 4.718 are positive numbers. The numbers $^{-}4$, $^{-}3.16$, $^{-}\sqrt{5}$, and $^{-}12/5$ are negative numbers.

There are three basic axioms, called the order axioms, relating the positive and negative numbers.

ORDER AXIOM 1

The set of positive numbers is closed under addition; that is, the sum of two positive numbers is a positive number.

ORDER AXIOM 2

The set of positive numbers is closed under multiplication; that is, the product of two positive numbers is a positive number.

ORDER AXIOM 3

Every real number is either a positive number, zero, or a negative number.

By Order Axiom 1, you know that if you add two positive numbers, the sum is positive. For example,

$$4 + 2 = 6,$$
$$93 + 56 = 149.$$

By Order Axiom 2, you know that if you multiply two positive numbers, the product is a positive number. For example,

$$2 \times 8 = 16,$$
$$12 \times 14 = 168.$$

These axioms about the set of positive numbers are not true of all subsets of the positive numbers. For example, the set of all odd integers is not closed under addition since

$$3 + 5 = 8, \text{ and 8 is not an odd integer.}$$

The set of all primes, $\{2, 3, 5, 7, 11, 13, 17, \ldots\}$, is not closed under multiplication. For example,

$$3 \times 5 = 15, \text{ and 15 is not a prime number.}$$

Looking more closely at Order Axiom 3 and using your knowledge of set theory brings you to the following conclusions: If R denotes the set of real numbers, P, the positive numbers, and N, the negative numbers, then

$$R = P \cup N \cup \{0\}$$

where $\{0\}$ denotes the set containing the single element 0. Also, P and N have no elements in common:

$$P \cap N = \emptyset.$$

Exercises

Are the following sets closed under the specified operations?

1. (a) All odd integers, multiplication
(b) All even integers, multiplication

2. (a) All even integers, addition
(b) All prime numbers, addition

3. (a) All rational numbers, multiplication
(b) All rational numbers, addition

4. (a) All integers, multiplication
(b) All integers, addition

5. (a) $\{1, 2, 3, 4\}$, addition
(b) $\{1, 2, 3, 4\}$, multiplication

6. (a) $\{0, 1\}$, multiplication
(b) $\{0, 1\}$, addition

7. (a) All positive numbers, subtraction
(b) All rational numbers, subtraction

8. (a) All nonzero rational numbers, division
(b) All positive numbers, division

9. (a) All odd integers, subtraction
(b) All even integers, subtraction

10. (a) All positive even integers, division
(b) All odd integers, division

Review for Sections 2–1 through 2–5

Simplify each of the following expressions.

1. $4 \cdot (2 + 3)$

2. $4 \cdot 2 + 3$

3. $3 + 7 \cdot 4 + 6$

4. $(3 + 7)(4 + 6)$

5. $2[3 + 5 \cdot (2 + 4)]$

6. $10[4(6 + 1) + 2]$

For what value of x is each of the following statements true?

7. $a + 4 = a + x$, for every number a

8. $5 \cdot b = x \cdot b$, for every number b

9. $3 + x = 3$

10. $a = 5$ and $x = a$

11. $n + 4 = x + 4$ and $n = 6$

12. $x + 5 = 0$

13. $7 \cdot x = 1$

Complete each of the following equations to make it a true statement. Name the axiom or property that is illustrated in each.

14. $53 + 42 =$ _?_ $+ 53$

15. $74 \cdot (37 \cdot$ _?_$) = (74 \cdot 37) \cdot 21$

16. $17(12 + 16) =$ _?_ $\cdot 12 + 17 \cdot 16$

17. $\frac{1}{2} \cdot 2 =$ _?_

18. $75 + ($_?_ $+ 27) = ($_?_ $+ 36) + 27$

19. $61 \cdot$ _?_ $= 53 \cdot$ _?_

20. $42 + {}^{-}42 =$ _?_

21. $72 \cdot 46 + 72 \cdot 31 =$ _?_$(46 + 31)$

22. $15 +$ _?_ $= 0$

Which of the statements in Exercises 23–31 are true and which are false? Name the axiom or property which supports each true statement.

23. $53(26 + 4) = 53 \cdot 26 + 53 \cdot 4$

24. $17 \cdot 4 + 17 \cdot 5 = 4 + 5 \cdot 17$

25. $35 + 26 \neq 26 + 35$

26. $(17 + 4) + 3 = (17 + 3) + 4$

27. $13 + 0 = 13$

28. $74 \cdot 1 = 75$

29. $82 + {}^{-}82 = 0$ **30.** $\frac{1}{2} \cdot 2 = 2\frac{1}{2}$ **31.** $15 \cdot \frac{1}{15} = 1$

Find the additive inverse and multiplicative inverse for each of the following numbers.

32. 12 **33.** 5 **34.** 36 **35.** $\frac{1}{2}$ **36.** $\frac{2}{3}$

Are the following sets closed under the specified operations?

37. All even integers, division

38. {2, 4, 6}, addition

39. {0, 1, 2}, multiplication

40. All positive numbers, addition

Answers to Review for Sections 2–1 through 2–5

1. 20	**2.** 11	**3.** 37
4. 100	**5.** 66	**6.** 300
7. 4	**8.** 5	**9.** 0
10. 5	**11.** 6	**12.** $^{-}5$
13. $\frac{1}{7}$	**14.** 42, (C-A)	**15.** 21, (A-M)
16. 17, (D)	**17.** 1, (Inv-M)	**18.** 36, 75, (A-A)
19. 53, 61, (C-M)	**20.** 0, (Inv-A)	**21.** 72, (D)
22. $^{-}15$, (Inv-A)	**23.** True, (D)	**24.** False
25. False	**26.** True, (R-A)	**27.** True, (Id-A)
28. False	**29.** True, (Inv-A)	**30.** False
31. True, (Inv-M)	**32.** $^{-}12$, $\frac{1}{12}$	**33.** $^{-}5$, $\frac{1}{5}$
34. $^{-}36$, $\frac{1}{36}$	**35.** $^{-}\frac{1}{2}$, 2	**36.** $^{-}\frac{2}{3}$, $\frac{3}{2}$ or $\dfrac{1}{\frac{2}{3}}$
37. No	**38.** No	**39.** No
40. Yes		

2–6 ADDITION OF POSITIVE AND NEGATIVE NUMBERS

Addition of positive and negative numbers may be interpreted as a series of directed moves on a number line. For example, the sum

$$2 + {}^{-}3$$

is equivalent to moving a distance of two units to the right of 0 and then three units to the left, finishing one unit to the left of 0. Therefore,

$$2 + {}^{-}3 = {}^{-}1,$$

as shown in Fig. 2–1.

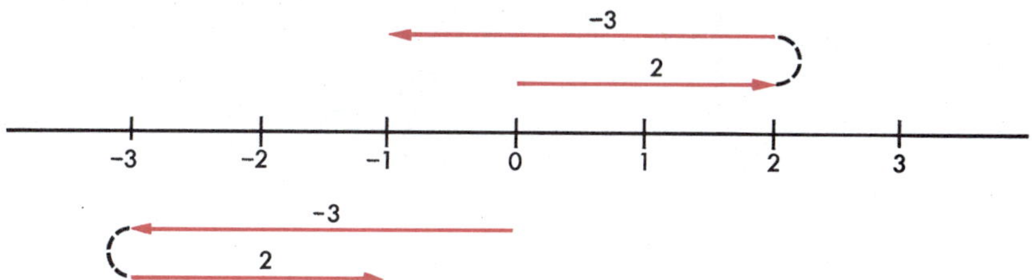

FIGURE 2-1

If you first move a distance of three units to the left of 0 and then two units to the right, you still finish one unit to the left of 0. Thus, $^{-}3 + 2 = {}^{-}1$. Of course, you knew the answer would be the same since

$$2 + {}^{-}3 = {}^{-}3 + 2,$$

by the commutative axiom for addition, (C-A).

The following examples illustrate a more formal way of finding the sum of positive and negative numbers.

Using the additive inverse axiom, (Inv-A), and other properties of addition, you can find the sum of any positive number and any negative number, as follows:

$$\begin{aligned} 7 + {}^{-}3 &= (4 + 3) + {}^{-}3 && (7 = 4 + 3) \\ &= 4 + (3 + {}^{-}3) && \text{(A-A)} \\ &= 4 + 0 && \text{(Inv-A)} \\ &= 4. && \text{(Id-A)} \end{aligned}$$

Thus,

$$7 + {}^{-}3 = 4.$$

Similarly,

$$\begin{aligned} {}^{-}76 + 33 &= ({}^{-}43 + {}^{-}33) + 33 && ({}^{-}76 = {}^{-}43 + {}^{-}33) \\ &= {}^{-}43 + ({}^{-}33 + 33) && \text{(A-A)} \\ &= {}^{-}43 + 0 && \text{(Inv-A)} \\ &= {}^{-}43. && \text{(Id-A)} \end{aligned}$$

Exercises

Find each of the sums in Exercises 1–6 by starting at 0 on a number line and using the addends as successive directed moves. Sketch each problem.

1. (a) (i) $3 + {}^{-}5$
(ii) ${}^{-}5 + 3$

(b) (i) $9 + {}^{-}5$
(ii) ${}^{-}9 + 5$

2. (a) (i) $4 + {}^{-}3 + 7$
(ii) $4 + 7 + {}^{-}3$
(b) (i) $7 + {}^{-}12 + {}^{-}3$
(ii) ${}^{-}12 + 7 + {}^{-}3$

3. (a) (i) $2 + {}^{-}8 + 7$
(ii) $7 + 2 + {}^{-}8$
(b) (i) $3 + {}^{-}6 + 10$
(ii) $10 + 3 + {}^{-}6$

4. (a) $17 + {}^{-}6 + {}^{-}17$
(b) ${}^{-}13 + {}^{-}12 + 12$

5. (a) $1 + 0 + {}^{-}6$
(b) $0 + 1 + {}^{-}6$

6. (a) $2 + {}^{-}2 + 5 + {}^{-}5$
(b) ${}^{-}5 + 4 + {}^{-}3 + 2$

First write the answer to each of the following problems as the indicated sum of positive or negative numbers; then do the addition.

7. (a) A rise in temperature of 5° was followed by a rise of 6°. Express the total rise in temperature.
(b) A hiker is on a mountain 800 feet above sea level and climbs another 300 feet. Express his altitude after the climb.

8. (a) A \$100 decrease in the price of a used car is followed by a second decrease of \$200. Express the total decrease.
(b) A fall of \$$1\frac{3}{8}$ in the price of a particular stock is followed by a fall of \$$2\frac{1}{8}$. Express the total fall in the price.

9. (a) An airplane climbs to 9000 feet after takeoff. Later the pilot descends 2000 feet and levels off. At what altitude does he level off?
(b) A submarine dives to 600 feet below sea level. Later it ascends 100 feet. How many feet below sea level is it now?

10. (a) A football team has these results in four plays: a loss of 4 yards, a gain of 7 yards, a loss of 5 yards, and a gain of 15 yards. Find the total number of yards lost or gained.
(b) On a cold day the temperature dropped two degrees in an hour. The next hour it dropped another three degrees. The third hour it rose one degree, but the fourth hour it fell four degrees. Find the total change in temperature.

In adding several positive and negative numbers, you can either add them in the order in which they appear, or you can use the rearrangement property for addition (R-A) to add first the positive numbers together, then the negative numbers, and finally add the sums of the positive and negative numbers. For example, you have either

$$(2 + {}^{-}3) + 6 + {}^{-}4 = ({}^{-}1 + 6) + {}^{-}4 = 5 + {}^{-}4 = 1,$$

or

$$2 + {}^{-}3 + 6 + {}^{-}4 = (2 + 6) + ({}^{-}3 + {}^{-}4) = 8 + {}^{-}7 = 1.$$

Use both ways described on page 68 to find each of the following sums.

11. $10 + {}^{-}16 + 7 + {}^{-}9$

12. $7 + {}^{-}9 + 3 + {}^{-}2$

13. $24 + {}^{-}19 + {}^{-}5 + 3$

14. ${}^{-}(\frac{21}{4}) + {}^{-}(\frac{7}{3}) + \frac{21}{4}$

15. $\frac{1}{3} + {}^{-}(\frac{2}{3}) + {}^{-}(\frac{4}{3}) + \frac{7}{3}$

16. $3.4 + 2.7 + 1.3 + {}^{-}1.2$

17. $17 + {}^{-}13 + {}^{-}13 + 37$

---

18. Find the sums of the following pairs of numbers. Give a reason for each of your steps.

(a) ${}^{-}26 + 10^2$
(b) ${}^{-}72 + 203$
(c) $\frac{3}{4} + {}^{-}(\frac{3}{6})$
(d) ${}^{-}.37 + 3.7$

19. Multiplication can be thought of as repeated addition. For example,

$$3({}^{-}5) = {}^{-}5 + {}^{-}5 + {}^{-}5 = {}^{-}15.$$

Use this method to find the following products.

(a) $2({}^{-}7)$
(b) $4({}^{-}10)$
(c) $5({}^{-}8)$
(d) $11({}^{-}3)$

2–7 SUBTRACTION OF POSITIVE AND NEGATIVE NUMBERS

You may have noticed that the addition of positive and negative numbers looks suspiciously like subtraction. For example, $7 + {}^{-}3 = 4$ by the preceding section, and $7 - 3 = 4$ according to subtraction. This suggests that the operation of subtraction in the set of real numbers can be defined in the following manner.

Definition of subtraction

The equation

$$x - y = x + {}^{-}y \qquad \textit{(Def-Sub)}$$

is true for all numbers x and y.

For example,

$$6 - 2 = 6 + {}^{-}2, \quad \text{or } 4,$$

and

$$59 - 11 = 59 + {}^{-}11, \quad \text{or } 48.$$

In order to find a difference such as $8 - {}^{-}2$, you must remember that the additive inverse of ${}^{-}2$ is ${}^{-}({}^{-}2)$, or 2, so that

$$8 - {}^{-}2 = 8 + 2, \quad \text{or } 10.$$

By similar reasoning,

$${}^{-}10 - 3 = {}^{-}10 + {}^{-}3, \quad \text{or } {}^{-}13,$$

and

$${}^{-}5 - {}^{-}2 = {}^{-}5 + 2, \quad \text{or } {}^{-}3.$$

Addition and subtraction are opposites in the sense that the equation

$$(a + b) - b = a$$

is true for all real numbers a and b. For example,

$$\begin{aligned} (6 + 9) - 9 &= (6 + 9) + {}^{-}9 && \text{(Def-Sub)} \\ &= 6 + (9 + {}^{-}9) && \text{(A-A)} \\ &= 6 + 0 && \text{(Inv-A)} \\ &= 6. && \text{(Id-A)} \end{aligned}$$

Now that subtraction has been defined, there should be no confusion in writing $-a$ in place of ${}^{-}a$. Then $b + (-a)$ is simply $b - a$. Henceforth, the negative of a number will be denoted by $-a$ instead of ${}^{-}a$.

Exercises

Complete the computation.

1. (a) $8 - 5 = 8 + (-5) =$ __?__
(b) $9 - (-4) = 9 + 4 =$ __?__

2. (a) $-27 - 19 = -27 + (-19) =$ __?__
(b) $-84 - 62 = -84 + (-62) =$ __?__

3. (a) $-65 - (-33) = -65 + 33 = \underline{\ ?\ }$
(b) $-62 - (-49) = -62 + 49 = \underline{\ ?\ }$

Compute the following.

4. (a) $(41 + 9) - 9$
(b) $(17 + 4) - 4$

5. (a) $[27 - (-6)] + (-6)$
(b) $[65 - (-9)] + (-9)$

6. (a) $-762 + (-762)$
(b) $-345 - (-345)$

7. (a) $(6 + 8) - (6 + 8)$
(b) $[7 + (-9)] + [7 + (-9)]$

8. (a) $(-4 - 7) - (-4 - 7)$
(b) $[4 + (-7)] + [4 - (-7)]$

Find the differences.

9. (a) $17 - 10$ (b) $102 - 73$

10. (a) $96 - (-96)$ (b) $102 - (-63)$

11. (a) $-52 - 83$ (b) $-63 - 102$

12. (a) $-67 - (-92)$ (b) $-13 - (-23)$

13. (a) $0 - (-23)$ (b) $0 - (-42)$

14. (a) $\frac{9}{5} - (-\frac{3}{7})$ (b) $\frac{11}{14} - (-\frac{9}{10})$

15. (a) $-.08 - (-1.7)$ (b) $-.11 - (-1.1)$

16. (a) $79 - (-97)$ (b) $53 - (-35)$

By how much does the first number exceed the second? In other words, what number added to the second yields the first?

17. $19, -3$

18. $7, -12$

19. $-8, -15$

20. $0, -24$

21. $24, 0$

22. $-3, -4$

23. $7, 0$

24. $9, -9$

25. $-.16, -.9$

26. $-\frac{1}{3}, -\frac{1}{3}$

27. $-4.2, -6.4$

28. $18, -20$

---

29. Find the value of x for which each equation is true.
(a) $x + 21 = 9$ (b) $x + 15 = 26$
(c) $x + (-45) = -18$ (d) $x + (-35) = -47$

30. (a) Does the commutative axiom hold for subtraction? In other words, is the equation $a - b = b - a$ true for all numbers a and b? Justify your answer.

(b) Does the associative axiom hold for subtraction? In other words, is the equation $a - (b - c) = (a - b) - c$ true for all numbers a, b, and c? Justify your answer.

Preparation for Section 2–8

Which of the following pairs of numbers are additive inverses of each other? Why?

1. $10, -10$

2. $-\frac{1}{2}, \frac{1}{2}$

3. $-(-1), 1$

4. $-3, -(-3)$

Which of the following pairs of numbers are multiplicative inverses of each other? Why?

5. $3, \frac{1}{3}$

6. $\frac{1}{4}, \frac{1}{4}$

7. $\frac{5}{6}, \frac{6}{5}$

8. $.3, \frac{10}{3}$

Classify each of the following statements as true, false, or open. State the axiom or property that justifies each true statement.

9. $7 + 8 \neq 8 + 7$

10. $7 + 8 + 11 + 10 = 7 + 10 + 11 + 8$

11. $18 + 0 = 18$

12. $x \cdot 1 = 30$

13. $\frac{1}{14} \cdot 14 = 1$

14. $18 + (-18) = 0$

2–8 NEGATIVES AND RECIPROCALS

The sum of $8 + 13$ and $-8 + (-13)$ is 0 by the calculations below.

$$\begin{aligned} (8 + 13) + [-8 + (-13)] &= [8 + (-8)] + [13 + (-13)] && \text{(R-A)} \\ &= 0 + 0 && \text{(Inv-A)} \\ &= 0 && \text{(Id-A)} \end{aligned}$$

Whenever the sum of two numbers is zero, one of the numbers is the additive inverse of the other. Therefore, $-8 + (-13)$ is the additive inverse or negative, of $8 + 13$; that is

$$-(8 + 13) = -8 + (-13).$$

In the same way, you can demonstrate the following result.

Negative of a sum

The equation

$$-(x + y) = -x + (-y) \qquad \textit{(Neg-Sum)}$$

is true for all numbers x and y.

This property of the number system is not an axiom. In fact, it is derived from the axioms by reasoning as shown above for $8 + 13$.

The process of arriving at new properties from given ones is called *deductive reasoning.* True statements which are deduced from the axioms or other known properties are often called *theorems* or *laws.* The reasoning used to show the truth of a particular property is called a *proof.* (To *prove* a theorem is to give a proof of it.) The negative of a sum, (Neg-Sum), above is an example of a theorem.

As another example of deductive reasoning, you can use steps similar to those for the negative of a sum, (Neg-Sum), to show that $\frac{1}{4} \cdot \frac{1}{5}$ is the multiplicative inverse, or reciprocal, of $4 \cdot 5$. You need only show that the product of $4 \cdot 5$ and $\frac{1}{4} \cdot \frac{1}{5}$ is 1.

$$\begin{aligned} (4 \cdot 5) \cdot (\tfrac{1}{4} \cdot \tfrac{1}{5}) &= (4 \cdot \tfrac{1}{4}) \cdot (5 \cdot \tfrac{1}{5}) && \text{(R-M)} \\ &= 1 \cdot 1 && \text{(Inv-M)} \\ &= 1 && \text{(Id-M)} \end{aligned}$$

You have proven

$$\frac{1}{4 \cdot 5} = \frac{1}{4} \cdot \frac{1}{5}.$$

Replacing 4 by x and 5 by y in the proof above, you can prove the following theorem.

Reciprocal of a product

The equation

$$\frac{1}{x \cdot y} = \frac{1}{x} \cdot \frac{1}{y} \qquad \textit{(Recip-Prod)}$$

is true for all nonzero numbers x and y.

Exercises

Which statements are true and which are false? Give reasons for your answers.

1. (a) (i) $-(5 + 8) = -5 + (-8)$
(ii) $-(5 + 8) = -5 - 8$
(b) (i) $-(13 + 2) = -13 + (-2)$
(ii) $-(13 + 2) = -13 - 2$

2. (a) (i) $-(5 - 8) = -5 - (-8)$
(ii) $-(5 - 8) = -5 + 8$
(b) (i) $-(13 - 2) = -13 - (-2)$
(ii) $-(13 - 2) = -13 + 2$

3. (a) $\frac{1}{2} \cdot \frac{1}{3} = \frac{1}{6}$
(b) $\dfrac{1}{2} \cdot \dfrac{1}{3} = \dfrac{1}{2 + 3}$

Replace the $*$ by $=$ or $\neq$ to make each statement true.

4. (a) $-(5 + 6) * -5 + (-6)$
(b) $-(5 - 6) * -5 + (-6)$

5. (a) $-[-6 + (-7)] * 6 - 7$
(b) $-[-4 + (-8)] * 4 + 8$

6. (a) $-(4 - 7) * -4 + 7$
(b) $-(-7 + 9) * 7 + (-9)$

7. (a) $9 + [-(-6)] * 9 + 6$
(b) $-(-1) + [-(-2)] * 1 + 2$

8. (a) $-(-6 - 10) * 6 - (-10)$
(b) $-(-7 - 11) * 7 - (-11)$

9. (a) $-[-5 - (-9)] * 5 - 9$
(b) $-[-4 - (-8)] * 4 + (-8)$

10. (a) $-[-11 + (-3)] * 11 - 3$
(b) $-(-8 - 7) * 8 + 7$

11. (a) $\dfrac{1}{12 \cdot 13} * \dfrac{1}{12} \cdot \dfrac{1}{13}$
(b) $\dfrac{1}{2 \cdot 2} * \dfrac{1}{2} + \dfrac{1}{2}$

12. (a) $\dfrac{1}{(-6) \cdot (-5)} * \left(\dfrac{1}{-6}\right) \cdot \left(\dfrac{1}{-5}\right)$
(b) $\dfrac{1}{(-4) \cdot (-3)} * \left(\dfrac{1}{-4}\right) \cdot \left(\dfrac{1}{-3}\right)$

13. (a) $\dfrac{1}{24} * \dfrac{1}{2} \cdot \dfrac{1}{12}$
(b) $\dfrac{1}{36} * \dfrac{1}{4} \cdot \dfrac{1}{9}$

14. (a) $\dfrac{1}{3 + 3} * \dfrac{1}{3} \cdot \dfrac{1}{3}$
(b) $\dfrac{1}{2 + 2} * \dfrac{1}{2} \cdot \dfrac{1}{2}$

What is the additive inverse of each of the following numbers?

15. (a) $-(-2)$ (b) $-(-9)$

16. (a) $7 - 4$ (b) $4 - 7$

17. (a) $-(2 + 3)$ (b) $-2 + (-3)$
18. (a) $-4 - (-5)$ (b) $-[-4 - (-5)]$

What is the multiplicative inverse of each of the following numbers?

19. (a) $3 \cdot 8$ (b) $12 \cdot 2$

20. (a) $\frac{1}{2} \cdot \frac{1}{3}$ (b) $\frac{1}{7} \cdot \frac{1}{2}$

21. (a) $\dfrac{1}{7 \cdot 9}$ (b) $\dfrac{1}{10 \cdot 12}$

22. (a) $3 \cdot 6$ (b) $6 \cdot 7$

Which of the following pairs of numbers are additive inverses of each other? Why?

23. (a) $12 - 15,\ 15 - 12$ (b) $12 + 15,\ 12 - 15$

24. (a) $-12 + 15,\ 12 - 15$ (b) $-8 + 10,\ 8 + 10$

Which of the following pairs of numbers are multiplicative inverses of each other? Why?

25. (a) $8,\ \frac{1}{2} \cdot \frac{1}{4}$ (b) $5,\ \frac{1}{2} + \frac{1}{3}$

26. (a) $.4,\ 2 \cdot \frac{1}{5}$ (b) $.6,\ 5 \cdot \frac{1}{3}$

---

27. Two numbers are additive inverses if their sum is zero. Give the reason for each step of the following proof that $y - x$ is the additive inverse of $x - y$.

$$
\begin{aligned}
(x - y) + (y - x) &= [x + (-y)] + [y + (-x)] && \underline{\ ?\ } \\
&= [x + (-x)] + [(-y) + y] && \underline{\ ?\ } \\
&= 0 + 0 && \underline{\ ?\ } \\
&= 0 && \underline{\ ?\ }
\end{aligned}
$$

28. Give the reason for each step of the following proof that

$$-[-(x - y)] = -(y - x).$$

$$
\begin{aligned}
-[-(x - y)] &= -(-[x + (-y)]) && \underline{\ ?\ } \\
&= -(-x + [-(-y)]) && \underline{\ ?\ } \\
&= -[-x + y] && \underline{\ ?\ } \\
&= -[y + (-x)] && \underline{\ ?\ } \\
&= -(y - x) && \underline{\ ?\ }
\end{aligned}
$$

29. Give the reason for each step of the following proof that $\left(\frac{1}{x \cdot y}\right) \cdot x = \frac{1}{y}$.

$$\left(\frac{1}{x \cdot y}\right) \cdot x = \frac{1}{x} \cdot \frac{1}{y} \cdot x \qquad \text{—?—}$$
$$= \frac{1}{x} \cdot x \cdot \frac{1}{y} \qquad \text{—?—}$$
$$= 1 \cdot \frac{1}{y} \qquad \text{—?—}$$
$$= \frac{1}{y} \qquad \text{—?—}$$

Preparation for Section 2–9

Classify each of the following statements as true or false. State the axiom or property of the real number system that justifies each true statement.

1. $(4 + 3) + 5 = 4 + (3 + 5)$
2. $6 + 12 = 12 + 6$
3. $-(7 + 9) = -7 + (-9)$
4. $-7 + (-9) = -7 - 9$
5. $\frac{1}{6} \cdot 7 \cdot 12 = \frac{1}{6} \cdot 12 \cdot 7$
6. $5(12 + 6) = 5 \cdot 12 + 5 \cdot 6$
7. $8 + 5 + 6 + 4 = 5 + 6 + 8 + 4$
8. $4 \cdot 8 + 4 \cdot 20 = 4(8 + 20)$
9. $1 \cdot (16 + 4) = 20$
10. $(16 + 4) + 0 = 20$

2–9 ALGEBRAIC EXPRESSIONS

Certain algebraic expressions involving numerals, variables, and operations can be made shorter, or simpler, by use of the basic properties of the real number system.

Problem 1. Simplify $(33x + 14y) + (25x + 73y)$.

Solution. You have

$$(33x + 14y) + (25x + 73y) = (33x + 25x) + (14y + 73y) \qquad \text{(R-A)}$$
$$= (33 + 25)x + (14 + 73)y \qquad \text{(D)}$$
$$= 58x + 87y.$$

Problem 2. Simplify $(5xy) \cdot (\frac{4}{5}xy)$, and find its value when $x = 2$ and $y = 3$.

Solution. By the rearrangement property of multiplication, (R-M), $(5xy) \cdot (\frac{4}{5}xy) = (5 \cdot \frac{4}{5}) \cdot (x \cdot x) \cdot (y \cdot y) = 4x^2y^2$. Thus, $4x^2y^2$ is the simplified expression.

When $x = 2$ and $y = 3$,

$$\begin{aligned}(5xy) \cdot (\tfrac{4}{5}xy) &= (5 \cdot 2 \cdot 3) \cdot (\tfrac{4}{5} \cdot 2 \cdot 3) \\ &= 30 \cdot \tfrac{24}{5} \\ &= 144.\end{aligned}$$

The value of the simplified expression must be the same.

$$\begin{aligned}4x^2y^2 &= 4 \cdot 2^2 \cdot 3^2 \\ &= 4 \cdot 4 \cdot 9 \\ &= 144\end{aligned}$$

In each of the examples above, you are replacing an algebraic expression by an equivalent one as defined below.

Definition of equivalent expressions

Two algebraic expressions in the same variables are equivalent if they have the same value for every set of values of the variables.

Problem 3. Simplify $(3x - y) - (2x + 3y)$, and find its value when $x = -5$ and $y = 7$.

Solution. Replace the given expression by equivalent ones until you obtain the simplest expression possible.

$$\begin{aligned}(3x - y) - (2x + 3y) &= [3x + (-y)] + [-(2x + 3y)] && \text{(Def-Sub)} \\ &= [3x + (-y)] + [-2x + (-3y)] && \text{(Neg-Sum)} \\ &= [3x + (-2x)] + [-y + (-3y)] && \text{(R-A)} \\ &= [3 + (-2)]x + [-1 + (-3)]y && \text{(D)} \\ &= x + (-4y) \\ &= x - 4y && \text{(Def-Sub)}\end{aligned}$$

When $x = -5$ and $y = 7$, the expression $x - 4y$ has the value

$$-5 - (4 \cdot 7), \quad \text{or } -33.$$

This is also the value of $(3x - y) - (2x + 3y)$ when $x = -5$ and $y = 7$.

$$\begin{aligned}[3 \cdot (-5) - 7] - [2 \cdot (-5) + 3 \cdot 7] &= [-15 - 7] - [-10 + 21] \\ &= -22 - 11 \\ &= -33\end{aligned}$$

Exercises

Simplify each expression. Find the value of the simplified form when $x = 4$ and $y = 5$. To check your work, find the values of the expressions before simplification.

1. (a) $3x + 7x$ (b) $5x + x$

2. (a) $3x \cdot 7x$ (b) $5x \cdot x$

3. (a) $4xy + 5xy$ (b) $3xy + 6xy$

4. (a) $\frac{7}{x} + \frac{9}{x}$ (b) $\frac{3}{y} + \frac{2}{y}$

5. (a) $\frac{4y}{5} + \frac{3y}{5}$ (b) $\frac{3x}{4} + \frac{7x}{4}$

6. (a) $18xy \cdot \frac{2}{3}xy$ (b) $12xy \cdot \frac{1}{6}xy$

7. (a) $(5x) \cdot (3y^2) \cdot \left(\frac{1}{5x}\right), \quad x \neq 0$

(b) $(3y) \cdot (2x^2) \cdot \left(\frac{1}{3y}\right), \quad y \neq 0$

Simplify.

8. (a) $x + 19x$ (b) $x + x + 1$

9. (a) $5a + ab + 6a$ (b) $4rs + s + 7rs$

10. (a) $x \cdot 7x$ (b) $a \cdot 5a$

11. (a) $5ab \cdot ab \cdot 6ab$ (b) $4rs \cdot rs \cdot 7rs$

12. (a) $\frac{7x}{9} + \frac{2x}{3} + \frac{5x}{6}$ (b) $\frac{3x}{4} + \frac{5x}{2} + \frac{4x}{3}$

13. (a) $(2x) \cdot (9xy) \cdot \left(\frac{1}{2x}\right), \quad x \neq 0$

(b) $(3xy) \cdot (2x^2) \cdot \left(\frac{1}{3xy}\right), \quad x \cdot y \neq 0$

14. (a) $4c - d + 6c + 2d$
(b) $6x - y + 12x - 2y$

15. (a) $8a - (7 + 2a)$
(b) $8x - (4x + 2)$

16. (a) $\frac{1}{2}h + \frac{3}{4}t - 5h + t$
(b) $-\frac{1}{4}x + \frac{1}{5}y - \frac{1}{5}x - \frac{3}{4}y$

17. (a) $4y - (8 - y)$
(b) $7a - (3a - 24)$

18. (a) $(3xy) \cdot (\frac{1}{9}x)$
(b) $(5xy) \cdot (\frac{1}{15}xy)$

19. (a) $(3x) \cdot \left(\frac{1}{2y}\right) \cdot \left(\frac{1}{3x}\right) \cdot (2y), \quad x \neq 0, \quad y \neq 0$

(b) $\left(\frac{1}{4xy}\right) \cdot \left(\frac{1}{2x}\right) \cdot (4xy) \cdot (2x), \quad x \neq 0, \quad y \neq 0$

20. (a) $(8a - 2b) - (8a - 2b)$

(b) $(x + y) - (x - y)$

In Exercises 21–27, tell which pairs of algebraic expressions are equivalent.

21. $5(x + y) + 7(x + y), \quad 12(x + y)$

22. $2x(1 + y) + 3x(1 + y), \quad 5x(1 + y)$

23. $9a(x - y) - 4a(x - y), \quad 5a(x - y)$

24. $3b(x - y) + a(x - y), \quad (3b + a)(x - y)$

25. $5(y + 2) + (y + 2), \quad 5(y + 2)$

26. $5(y + 3) + (y + 3), \quad 6(y + 3)$

27. $a - ay, \quad a(1 - y)$

28. Simplify.

(a) $3(a + b) + 8(a + b)$

(b) $5(x - 2y) + 6(x - 2y)$

(c) $\frac{x}{5} - \frac{y}{3} - \frac{x}{2} - \frac{y}{4}$

(d) $7(x - 3) + 2(5x - 4)$

(e) $5(2y - 3) - 4(3y - 7)$

(f) $3(a - b) - 2(a + b)$

State the axioms or properties of the real number system which make each of the equations below true for all values of the variables.

29. $9 \cdot (\frac{5}{9}y) = (9 \cdot \frac{5}{9})y$

30. $(x \cdot 13) + (x \cdot 17) = x(13 + 17)$

31. $9y + y = (9 + 1)y$

32. $(a + 3) + 9 = a + 12$

33. $\frac{7x}{8} + \frac{5x}{9} = \left(\frac{7x}{8} \cdot \frac{9}{9}\right) + \left(\frac{5x}{9} \cdot \frac{8}{8}\right)$

34. $(3x \cdot 9x) = 27x^2$

35. $(3x + 5y) + 0 = 3x + 5y$

Many times the distributive axiom may be used to write sums as products. For example,

$$2x + 2y = 2(x + y) \text{ and } 5^2 + 5^5 = 5^2(1 + 5^3).$$

In Exercises 36–45, use the distributive axiom to write each of the sums as a product.

36. $rx + sx$ **37.** $c + cd$ **38.** $63xy + 9x^2y^2$ **39.** $121bc + 55b^2$

40. $3(a + x) + 7(a + x)$

41. $2\pi r^2 + 2\pi rh$

42. $2(a + b) + (a + b)$

43. $19^4 + 19^5$

44. $17^2 + 17^3$

45. $3^2 + 3^3 + 3^4$

46. An algebraic expression of the form

$$1 + x + y + xy$$

can be factored (that is, changed so that it becomes the product of two or more expressions) by the following method. Give the reason for each step.

$$\begin{aligned} 1 + x + y + xy &= (1 + x) + (y + xy) \\ &= (1 + x) + [(1 \cdot y) + (xy)] \\ &= (1 + x) + [(1 + x) \cdot y] \\ &= (1 + x) \cdot 1 + (1 + x) \cdot y \\ &= (1 + x) \cdot (1 + y) \end{aligned}$$

47. Use the method shown in Exercise 46 to factor each of the following expressions.

(a) $2m + n + 2m^2 + mn$
(b) $5 + x + 5y + xy$
(c) $x + 1 + xy + y$
(d) $3a + b + 3a^2 + ab$
(e) $6x + y + 18x^2 + 3xy$
(f) $5z + 2w + 5wz + 2w^2$

2–10 CANCELLATION LAWS AND PROPERTIES OF ZERO

If $a = b$, then $a + 3 = b + 3$ by the additive axiom of equality, (Add-A). Conversely, if $a + 3 = b + 3$, then $a = b$ by the following proof.

$$\begin{aligned} a + 3 &= b + 3 & \\ (a + 3) + (-3) &= (b + 3) + (-3) & \text{(Add-A)} \\ a + [3 + (-3)] &= b + [3 + (-3)] & \text{(A-A)} \\ a + 0 &= b + 0 & \text{(Inv-A)} \\ a &= b & \text{(Id-A)} \end{aligned}$$

That is, if $a + 3 = b + 3$, you can "cancel" the 3's,

$$a + \cancel{3} = b + \cancel{3}$$

getting $a = b$.

Using c in place of 3 in the argument above, you can easily prove the following theorem.

CANCELLATION LAW OF ADDITION

If $a + c = b + c$, then $a = b$. (Can-A)

There is also a multiplicative cancellation law as stated below.

CANCELLATION LAW OF MULTIPLICATION

If $ca = cb$ and $c \neq 0$, then $a = b$. (Can-M)

For example, if $3a = 3b$ then $a = b$ by the cancellation law of multiplication, (Can-M).

You may remember from arithmetic that $0 \cdot 2 = 0$, $0 \cdot 17 = 0$, and so forth. Can you prove that $0 \cdot 2 = 0$, using only the axioms and theorems stated previously? See if you can follow the argument below.

$$\begin{aligned} 0 \cdot 2 &= (0 + 0) \cdot 2 && \text{(since } 0 = 0 + 0) \\ 0 \cdot 2 &= 0 \cdot 2 + 0 \cdot 2 && \text{(D)} \\ 0 + 0 \cdot 2 &= 0 \cdot 2 + 0 \cdot 2 && \text{(Id-A)} \\ 0 + \cancel{0 \cdot 2} &= 0 \cdot 2 + \cancel{0 \cdot 2} && \text{(Can-A)} \\ 0 &= 0 \cdot 2 \end{aligned}$$

On replacing 2 by x above, you can prove the following theorem.

MULTIPLICATION BY ZERO

The equation

$$0 \cdot x = 0 \qquad \text{(Zero-M)}$$

is true for every number x.

Exercises

In Exercises 1–13, make each of the statements true by filling in the blank. State the reason for each answer.

1. (a) If $x + 17 = 4 + 17$, then $x =$ __?__.
(b) If $x + 12 = 13 + 12$, then $x =$ __?__.

2. (a) If $76 \cdot x = 76 \cdot 99$, then $x =$ __?__.
(b) If $x \cdot 13 = 18 \cdot 13$, then $x =$ __?__.

3. (a) If $93 + 5x = 93 + 22$, then $5x =$ __?__.
(b) If $8x + 24 = 16 + 24$, then $8x =$ __?__.

4. (a) If $(2x)(36) = (4)(36)$, then $2x =$ __?__.
(b) If $(12x)(14) = (24)(14)$, then $12x =$ __?__.

5. (a) If $\frac{4}{5}x = 30 \cdot \frac{4}{5}$, then $x =$ __?__.
(b) If $\frac{3}{7}x = 14 \cdot \frac{3}{7}$, then $x =$ __?__.

6. (a) $8 \cdot 0 =$ __?__
(b) $0 + 8 =$ __?__

7. (a) $0 + x =$ __?__ for every real number x
(b) $0 \cdot x =$ __?__ for every real number x

8. (a) If $x + 9 = 17$, then $x + 9 =$ __?__ $+ 9$ and $x =$ __?__.
(b) If $x + 12 = 30$, then $x + 12 =$ __?__ $+ 12$ and $x =$ __?__.

9. (a) If $x + 6 = 4$, then $x + 6 =$ __?__ $+ 6$ and $x =$ __?__.
(b) If $x + 10 = 6$, then $x + 10 =$ __?__ $+ 10$ and $x =$ __?__.

10. (a) If $3x = 30$, then $3x = 3 \cdot$ __?__ and $x =$ __?__.
(b) If $5x = 45$, then $5x = 5 \cdot$ __?__ and $x =$ __?__.

11. (a) If $2 + 3x = 5$, then $3x =$ __?__ and $x =$ __?__.
(b) If $4 + 2y = 12$, then $2y =$ __?__ and $y =$ __?__.

12. (a) If $4 + 9y = 4$, then $9y =$ __?__ and $y =$ __?__.
(b) If $2x + 11 = 11$, then $2x =$ __?__ and $x =$ __?__.

13. (a) If $5 + 4x = 9 + 2x$, then $5 + 2x =$ __?__,
$2x =$ __?__, and $x =$ __?__.
(b) If $8y - 3 = 4y + 1$, then $4y - 3 =$ __?__,
$4y =$ __?__, and $y =$ __?__.

Use the cancellation laws to find the value of x for which each equation is true.

14. (a) $x + 5 = 7$ (b) $10 + x = 12$

15. (a) $5 = 10 + x$ (b) $-5 = x + 3$

16. (a) $2x = 20$ (b) $7x = 56$

17. (a) $3x + 4 = 4$ (b) $8x + 25 = 25$

18. (a) $5x + 10 = 15$ (b) $6x + 7 = 25$

Where possible, use the cancellation laws to find the value of x for which each equation is true.

19. $2x - 5 = 10$

20. $3x + 3 = 18 - 15$

21. $6x - 6 = 21 - 27$

22. $5x - 8 = x$

23. $3x + 1 = 2x + 11$

24. $16 + x = x$

25. $7x = 7x + 3$

Review for Sections 2–6 through 2–10

Compute the following.

1. $14 + (-3) = _?_$

2. $6 - (-2) = _?_$

3. $-7 + (-8) = _?_$

4. $-6 + 4 = _?_$

5. $-15 - 5 = _?_$

6. $17 + (-9) = _?_$

7. $-53 - (-12) = _?_$

8. $13 - 5 = _?_$

9. $-24 + 6 = _?_$

Replace the * by = or ≠ to make each statement true.

10. $-7 + (-13) * -(7 + 13)$

11. $\frac{1}{4 \cdot 8} * (-4)(-8)$

12. $-8 + 17 * -(8 + 17)$

13. $7(x + y) + 3(x + y) * 10(x + y)$

14. $-8 + 17 * -(8 - 17)$

15. $\frac{1}{6} \cdot \frac{1}{12} * \frac{1}{6 \cdot 12}$

16. $-(53 + 46) * -53 - 46$

17. $14xy + 7x * 7x(2y + 1)$

18. $5x + 4y * 9xy$

19. $\frac{1}{3 \cdot 5} * \frac{1}{3} \cdot \frac{1}{5}$

What is the additive inverse of each of the following numbers?

20. $-(-16)$

21. -5

22. $4 + 8$

23. $-(6 + 3)$

24. $15 - 9$

25. $-5 - 24$

What is the multiplicative inverse of each of the following numbers?

26. $7 \cdot 52$

27. $\frac{1}{3} \cdot \frac{1}{15}$

28. $\frac{1}{6 \cdot 12}$

29. $8 \cdot 93$

Simplify each of the following expressions and then find the value of each when $x = 2$ and $y = 4$.

30. $x + 3y + 7x$

31. $y + 4x + 3y$

32. $2x \cdot 3y + 4$

33. $3y \cdot 7x + 3$

34. $5xy + 2xy + 14$

35. $3xy + 6x$

Use the cancellation laws to find the value of x for which each of the following equations is true.

36. $x + 10 = 4 + 10$

37. $15 + x = 15 + 73$

38. $x + 2 = 10$

39. $3 \cdot x = 3 \cdot 12$

40. $4x = 16$

41. $2x + 3 = 13$

42. $3x + 5 = 26$

43. $4x + 20 = 28$

Answers to Review for Sections 2–6 through 2–10

1. 11 **2.** 8 **3.** -15 **4.** -2
5. -20 **6.** 8 **7.** -41 **8.** 8
9. -18 **10.** $=$ **11.** $\neq$ **12.** $\neq$
13. $=$ **14.** $=$ **15.** $=$ **16.** $=$
17. $=$ **18.** $\neq$ **19.** $=$ **20.** -16
21. 5 or $-(-5)$ **22.** $-4 + (-8)$ or $-(4 + 8)$ or -12
23. $6 + 3$ **24.** $-15 + 9$ or $-(15 - 9)$
25. $5 + 24$ or $-(-5 - 24)$ **26.** $\dfrac{1}{7 \cdot 52}$
27. $3 \cdot 15$ **28.** $6 \cdot 12$
29. $\dfrac{1}{8 \cdot 93}$
30. $8x + 3y$, 28 **31.** $4(x + y)$, 24
32. $6xy + 4$, 52 **33.** $21xy + 3$, 171
34. $7(xy + 2)$, 70 **35.** $3x(y + 2)$, 36
36. 4 **37.** 73
38. 8 **39.** 12
40. 4 **41.** 5
42. 7 **43.** 2

2–11 ABSOLUTE VALUE

Associated with each number x is a number called the *absolute value* of x, which is denoted by

$$|x|.$$

Perhaps you can guess the definition of $|x|$ from the following examples:

$$|-2| = 2 \quad \text{and} \quad |2| = 2,$$
$$|-\tfrac{3}{7}| = \tfrac{3}{7} \quad \text{and} \quad |\tfrac{3}{7}| = \tfrac{3}{7}.$$

The definition of $|x|$ has three parts.

Definition of absolute value of x, $|x|$.

1. $|x| = x$ if x is a positive number.
2. $|0| = 0$.
3. $|x| = -x$ if x is a negative number.

In the last case do not be misled into thinking that $-x$ is a negative number. Remember if x is a negative number, then $-x$ is a positive number. For example,

$$|-3| = -(-3), \quad \text{or } 3.$$

Remember that if $x \neq 0$, $|x|$ is always a positive number. By definition,

$$|x| = |-x|$$

for every number x.

Exercises

Compute the following.

1. (a) $9 + |-6|$ (b) $|20| + |-38|$
2. (a) $16 - |-7|$ (b) $21 - |-5|$
3. (a) $|-29| - |47|$ (b) $|-51| - |40|$
4. (a) $|-18| - |18|$ (b) $|-51| - |51|$
5. (a) $|14| + |-14|$ (b) $|36| + |-36|$
6. (a) $|-19 - 10|$ (b) $|-18 - 13|$
7. (a) $|-19 - (-11)|$ (b) $|-20 - (-38)|$
8. (a) $-|-36 - 12|$ (b) $-|-17 - 18|$
9. (a) $|-2| + |-3| - |-5|$ (b) $|-7| + |-6| - |-4|$

Tell which of the following statements are true and which are false.

10. (a) $|3| + |-3| = 2 \cdot |3|$ (b) $|-6| + |6| = 2 \cdot |-6|$
11. (a) $|-5| - |5| = 0$ (b) $|-7| + |7| = 0$
12. (a) $|-(\frac{3}{2})| = |1 + \frac{1}{2}|$ (b) $|-(\frac{5}{2})| = |3 - \frac{1}{2}|$
13. (a) $|5 + (-7)| = |5| + |-7|$ (b) $|-8 + 6| = |-8| + |6|$
14. (a) $|-6 + (-9)| = |-6| + |-9|$
    (b) $|-7 - 9| = |-7| + |-9|$
15. (a) $-|7 - 9| = -|7| + |9|$
    (b) $7 + |-9| = |7 - 9|$

In Exercises 16–22, find all possible values of x that will make each statement true.

16. (a) $|x| = 63$ (b) $|x| = 0$
17. (a) $|x + 3| = 4$ (b) $|x + 5| = 7$
18. (a) $|x - 3| = 4$ (b) $|x - 5| = 7$

19. (a) $|x| + 6 = 0$ (b) $|x| + 8 = 2$

20. (a) $|x| + 7 = 23$ (b) $|x| + 6 = 14$

21. (a) $x + |-40| = 0$ (b) $x + |16| = 0$

22. (a) $|x + 18| = |2| + |18|$ (b) $|x + 5| = |3| + |5|$

Which of the statements are true and which are false? Explain your answers.

23. For every real number x, $|x + 5| = |x| + |5|$.

24. For every real number x, $x + |5| = x + |-5|$.

25. For every real number x, $|x + (-5)| = |x| + |-5|$.

26. For every real number x, $|x - 4| = |x| - |4|$.

Give three elements of each of the following sets. Describe each set without using absolute values.

27. $A = \{x \mid |x + 3| = |x| + |3|\}$

28. $B = \{x \mid |x + (-2)| = |x| + |-2|\}$

29. $C = \{x \mid |-5 - x| = 5 + |x|\}$

30. $D = \{x \mid |x - 6| = |x| - |6|\}$

Fill in the blanks so that the statements are true.

31. $|-13| = |$__?__$|$

32. $|$__?__$| = 12$

33. $|21| + |-21| =$ __?__

34. $|15| + |-15| =$ __?__

35. $|19| - |$__?__$| = 0$

36. $|4 + (-2)| =$ __?__

37. $3 + |-3| =$ __?__

38. $-|-10| =$ __?__

2-12 MULTIPLICATION OF POSITIVE AND NEGATIVE NUMBERS

You know how to multiply positive numbers from arithmetic. For example,

$$3 \cdot 4 = 12.$$

To multiply a positive number, such as 3, by a negative number, such as -4, proceed as follows:

$$\begin{aligned} [3 \cdot (-4)] + (3 \cdot 4) &= 3 \cdot (-4 + 4), && \text{(D)} \\ [3 \cdot (-4)] + 12 &= 3 \cdot 0, && \text{(Inv-A)} \\ [3 \cdot (-4)] + 12 &= 0, && \text{(Zero-M)} \\ [3 \cdot (-4)] + 12 &= -12 + 12, && \text{(Inv-A)} \\ 3 \cdot (-4) &= -12. && \text{(Can-A)} \end{aligned}$$

By replacing 3 with x and 4 with y and using the same steps, you can prove the following theorem.

NEGATIVE MULTIPLICATION

The equations

$$x(-y) = -(x \cdot y)$$

and ***(Neg-M)***

$$(-x)y = -(x \cdot y)$$

are true for all numbers x and y.

For example,

$$7 \cdot (-8) = -(7 \cdot 8), \text{ or } -56,$$

by the first equation of (Neg-M), and

$$(-7)8 = -(7 \cdot 8), \text{ or } -56,$$

by the second equation of (Neg-M).

The product of two negative numbers can also be found by (Neg-M). For example,

$-7(-8) = -[7(-8)],$	(by Eq. 1 of (Neg-M))
$= -(-56),$	(by Eq. 2 of (Neg-M))
$= 56$	$(-(-x) = x)$

Therefore,

$$(-7)(-8) = 56.$$

As these examples show, the product of two numbers of OPPOSITE sign is a NEGATIVE number, and the product of two numbers of the SAME sign is a POSITIVE number.

1. $x \cdot y = -|x| \cdot |y|$ ***if x and y are numbers of opposite sign.***

2. $x \cdot y = |x| \cdot |y|$ ***if x and y are numbers of the same sign.***

A product such as $4(5-2)$ can be found using the distributive axiom, (D), and negative multiplication, (Neg-M), as follows:

$$\begin{aligned} 4(5-2) &= 4[5+(-2)] && \text{(Def-Sub)} \\ &= (4 \cdot 5) + 4(-2) && \text{(D)} \\ &= (4 \cdot 5) + [-(4 \cdot 2)] && \text{(Neg-M)} \\ &= (4 \cdot 5) - (4 \cdot 2), \quad \text{or } 12. && \text{(Def-Sub)} \end{aligned}$$

This example illustrates that the distributive axiom holds for subtraction as well as addition. That is, the equation

$$x(y-z) = xy - xz$$

is true for all real numbers x, y, and z.

Exercises

In Exercises 1–3, make each statement true by filling in the blank.

1. (a) (i) $-4 + (-4) =$ __?__ (ii) $2(-4) =$ __?__ (b) (i) $-5 + (-5) + (-5) =$ __?__ (ii) $3(-5) =$ __?__

2. (a) (i) $10 + (-10) =$ __?__ (ii) $2[10 + (-10)] =$ __?__ (b) (i) $5 + (-5) =$ __?__ (ii) $3[5 + (-5)] =$ __?__

3. (a) $-3(-6 + 6) =$ __?__ (b) $-4(-18 + 18) =$ __?__

In Exercises 4–20, find the products.

4. (a) $(-4)(-51)$ (b) $(-8)(-7)$

5. (a) $(12)(-7)$ (b) $9 \cdot (-9)$

6. (a) $(-16) \cdot 2$ (b) $(-12) \cdot 8$

7. (a) $(-.4)(.08)$ (b) $(-.2)(.01)$

8. (a) $(-\frac{1}{2}) \cdot 2$ (b) $(-\frac{3}{4})(\frac{4}{3})$

9. (a) $10 \cdot 31$ (b) $100 \cdot 310$

10. (a) $(-819) \cdot 0$ (b) $0 \cdot (-613)$

11. (a) $(-1) \cdot 1$ (b) $(-1)(-1)$

12. (a) $(-6)(-\frac{13}{3})$ (b) $(-14)(-\frac{15}{7})$

13. (a) $6 \cdot (-5)(-3)$ (b) $(-2)(-4) \cdot 8$

14. (a) $(-5) \cdot 0 \cdot 6$ (b) $7 \cdot (-9) \cdot 0$

15. (a) $5(8 - 9)$ (b) $7(12 - 14)$

16. (a) $5[-8 + (-2)]$ (b) $7[-4 + (-6)]$

17. (a) $[5 \cdot (-8)] + [5 \cdot (-2)]$ (b) $[7 \cdot (-4)] + [7 \cdot (-6)]$

18. (a) $(-\frac{3}{5})(-30)(-\frac{7}{2})$ (b) $(\frac{3}{5})(-\frac{1}{3}) \cdot 5$

19. (a) $(-7)(-9)(-\frac{1}{3})(-2)$ (b) $(-3)(-2)(-\frac{1}{12})(-4)$

20. (a) $(-3) \cdot 2 \cdot 4 \cdot (-5)$ (b) $(-3) \cdot 2 \cdot (-4)(-5)$

---

21. From your observations in the above exercises, formulate a general rule about the following:
(a) Is the product of an odd number of negative numbers positive or negative?
(b) Is the product of an even number of negative numbers positive or negative?

22. Make each statement true by filling in the blank.
(a) If the product of two numbers is a positive number and one of the numbers is positive, then the other number must be __?__.
(b) The product of two negative numbers is a __?__ number.
(c) The product of a __?__ number and a positive number is a negative number.
(d) The product of a __?__ number and a __?__ number is a negative number.

Compute the following.

23. $4[(-3) \cdot (-7)]$

24. $[(4) \cdot (-3)](-7)$

25. $|-5| \cdot |-3| - |(-5) \cdot (-3)|$

26. $|-5 - 3| \cdot 2$

---

27. Simplify each of the following algebraic expressions.
(a) $8(\frac{1}{4}x - \frac{1}{2}y)$ (b) $-6(\frac{2}{3}x - \frac{1}{2}y)$
(c) $-5(6a - 11b) - 7(6a - b)$ (d) $8(5a - 13b) - 12(5a - 13b)$
(e) $3(2x + y) + 4(-x + 3y)$ (f) $-5(y - 3x) - (-7[3x - (-2y)])$

Preparation for Section 2–13

Fill in the blanks.

1. $x^5 = x \cdot x \cdot$ __?__

2. $x^6 \cdot x^2 = (x \cdot$ __?__$) \cdot (x \cdot x)$
$=$ __?__

State the axiom or property of the real number system that justifies each of the following statements.

3. $17 \cdot 9 \cdot 6 \cdot 5 = 5 \cdot 17 \cdot 6 \cdot 9$

4. $\frac{1}{8} \cdot 8 = 1$

5. $12 \cdot 1 = 12$

6. $\frac{1}{12 \cdot 13} = \frac{1}{12} \cdot \frac{1}{13}$

2–13 DIVISION OF POSITIVE AND NEGATIVE NUMBERS

The operation of division is defined as follows.

Definition of division

$$x \div y = x \cdot \frac{1}{y} \quad \text{if } y \neq 0 \qquad \text{(Def-Div)}$$

The notations

$$\frac{x}{y} \text{ and } x/y$$

are also used for $x \div y$.

Every rational number may be considered either as a quotient of two integers or as a product of an integer by the reciprocal of an integer. For example,

$$\tfrac{7}{12} = 7 \div 12 \quad \text{or} \quad \tfrac{7}{12} = 7 \cdot \tfrac{1}{12}.$$

A product such as $(2 \div 3)(5 \div 7)$, or $\frac{2}{3} \cdot \frac{5}{7}$, can be found as follows:

$$\begin{aligned} \tfrac{2}{3} \cdot \tfrac{5}{7} &= (2 \cdot \tfrac{1}{3}) \cdot (5 \cdot \tfrac{1}{7}) && \text{(Def-Div)} \\ &= (2 \cdot 5) \cdot (\tfrac{1}{3} \cdot \tfrac{1}{7}) && \text{(R-M)} \\ &= (2 \cdot 5) \cdot \left(\frac{1}{3 \cdot 7}\right) && \text{(Recip-Prod)} \\ &= \frac{2 \cdot 5}{3 \cdot 7}, \quad \text{or } \frac{10}{21}. && \text{(Def-Div)} \end{aligned}$$

If you replace $\frac{2}{3}$ with a/b and $\frac{5}{7}$ with c/d, then you can use the reasoning above to prove the following theorem.

MULTIPLICATION OF FRACTIONS

The equation

$$\frac{a}{b} \cdot \frac{c}{d} = \frac{a \cdot c}{b \cdot d} \qquad \text{(Fr-M)}$$

is true for all numbers a and c, and all nonzero numbers b and d.

If you let $b = 1$ in the multiplication of fractions theorem, (Fr-M), then $a/1 = a$ and the multiplication of fractions theorem (Fr-M), becomes

$$a \cdot \frac{c}{d} = \frac{a \cdot c}{d}.$$

You observe that division by zero is not allowed in the definition of division. The reason for this is that 0 does not have a reciprocal. That is, there is no number x such that $0 \cdot x = 1$. In fact, the equation

$$0 \cdot x = 1$$

is false for every number x since $0 \cdot x = 0$ by the multiplication by zero theorem, (Zero-M). Consequently,

division by 0 is not defined.

Since $x \cdot 1/x = 1$ and 1 is a positive number, x and $1/x$ necessarily have the same sign. That is, the reciprocal of a positive number is positive; the reciprocal of a negative number is negative.

The sign of x/y can be determined once you realize

$$\frac{x}{y} = x \cdot \frac{1}{y}.$$

For example, if x is positive and y is negative, then $1/y$ is negative and $x \cdot 1/y$ is negative by the negative multiplication theorem, (Neg-M). Reasoning in this way, you can prove the following statements.

$$\frac{x}{y} = -\frac{|x|}{|y|}$$ ***if x and y are numbers of opposite sign.***

$$\frac{x}{y} = \frac{|x|}{|y|}$$ ***if x and y are numbers of the same sign.***

For example,

$$\frac{12}{-4} = \frac{-12}{4} = -\left(\frac{12}{4}\right) = -3,$$

or

$$12 \div (-4) = -12 \div 4 = -(12 \div 4) = -3.$$

Problem. Simplify $(12ab^3) \div (3ab)$.

Solution.

$$\begin{aligned}
(12ab^3) \div (3ab) &= (12ab^3) \cdot \frac{1}{3ab} && \text{(Def-Div)}\\
&= (12ab^3)\left(\frac{1}{3} \cdot \frac{1}{a} \cdot \frac{1}{b}\right) && \text{(Recip-Prod)}\\
&= (12 \cdot \tfrac{1}{3})\left(a \cdot \frac{1}{a}\right)\left(b \cdot b \cdot b \cdot \frac{1}{b}\right) && \text{(R-M)}\\
&= 4 \cdot \left(a \cdot \frac{1}{a}\right)(b \cdot b)\left(b \cdot \frac{1}{b}\right) && \text{(Fr-M)}\\
&= 4 \cdot 1 \cdot b \cdot b \cdot 1 && \text{(Inv-M)}\\
&= 4b^2. && \text{(Id-M)}
\end{aligned}$$

Exercises

Which statements are true and which are false?

1. (a) Since $-5 \cdot \frac{-1}{5} = 1$, the reciprocal of -5 is $\frac{-1}{5}$.

(b) Since $-4 \cdot \frac{1}{-4} = 1$, the reciprocal of -4 is $\frac{1}{-4}$.

2. (a) $\frac{1}{0} = 0$

(b) $\frac{1}{-1} = 1$

3. (a) $\frac{-7}{8} \cdot \frac{-8}{7} = 1$

(b) $\frac{5}{-4} \cdot \frac{-4}{5} = 1$

4. (a) $-\left(\frac{1}{2}\right) = \frac{-1}{2} = \frac{1}{-2} = -\left(\frac{-1}{-2}\right)$

(b) $\frac{1}{-3} = \frac{-1}{3} = -\left(\frac{1}{3}\right) = \frac{-1}{-3}$

Divide as indicated.

5. (a) $\frac{15}{5}$ (b) $\frac{-62}{-1}$

6. (a) $\frac{-248}{-4}$ (b) $\frac{-128}{-4}$

7. (a) $\frac{42}{-7}$ (b) $\frac{84}{-6}$

8. (a) $\frac{0}{-4}$ (b) $\frac{0}{-19}$

9. (a) $\frac{-14.4}{.2}$ (b) $\frac{16}{-.8}$

10. (a) $\frac{-132}{11}$ (b) $\frac{156}{-12}$

Compute the following.

11. (a) $-207 \div 9$ (b) $-105 \div 3$

12. (a) $(6.3) \div (-21)$ (b) $(5.7) \div (-.19)$

13. (a) $-.133 \div (-19)$ (b) $-8.8 \div (-.44)$

14. (a) $6.33 \div (-2.11)$ (b) $4.82 \div (-2.41)$

15. (a) $-1 \div \frac{4}{7}$ (b) $1 \div \frac{-4}{7}$

16. (a) $[(-5)(6)] \div (-3)$ (b) $[(-7)(-9)] \div 3$

17. (a) $-5[6 \div (-3)]$ (b) $-7(-9 \div 3)$

18. (a) $(-20 \div 5) \div 4$ (b) $(-36 \div 12) \div 3$

19. (a) $-20 \div (5 \div 4)$ (b) $-36 \div (12 \div 3)$

20. (a) $|-24| \div (3 - 15)$ (b) $|-40| \div (14 - 10)$

---

21. Compute the following.

(a) $(-10)^3 \div (-10)^2$

(b) $(-3)^5 \div (-3)^2$

(c) $(-1)^{10} \div (-1)^{48}$

(d) $(-2)^{50} \div (-2)^{48}$

(e) $[-13 + (-37)] \div (-5)$

(f) $[-13 \div (-5)] + [-37 \div (-5)]$

22. Complete the statements so that they are true.

(a) If the quotient of two numbers is a positive number, and one of the numbers is positive, then the other must be __?__.

(b) The quotient of two negative numbers is a __?__ number.

(c) The quotient of a __?__ number by a positive number is a negative number.

(d) The quotient of a __?__ number by a __?__ number is a negative number.

(e) If the quotient of two real numbers is negative, their product is __?__.

(f) If the product of two real numbers is positive, their quotient is __?__.

23. Supply the reasons for the key steps in the completed exercises.

(a) $\frac{x}{3} + \frac{x}{5} = \left(\frac{5}{5} \cdot \frac{x}{3}\right) + \left(\frac{3}{3} \cdot \frac{x}{5}\right)$

$= \frac{5x}{15} + \frac{3x}{15}$

$= \left(5x \cdot \frac{1}{15}\right) + \left(3x \cdot \frac{1}{15}\right)$ —?—

$= (5x + 3x) \cdot \frac{1}{15}$ —?—

$= [(5 + 3)x] \cdot \frac{1}{15}$ —?—

$= 8x \cdot \frac{1}{15}$

$= \frac{8x}{15}$ —?—

(b) $36x^4y^2 \div 15x^3y^5 = (36x^4y^2) \cdot \left(\frac{1}{15x^3y^5}\right)$ —?—

$= (36x^4y^2) \cdot \left(\frac{1}{15} \cdot \frac{1}{x^3} \cdot \frac{1}{y^5}\right)$ —?—

$= \left(36 \cdot \frac{1}{15}\right) \cdot \left(x^4 \cdot \frac{1}{x^3}\right)\left(y^2 \cdot \frac{1}{y^5}\right)$ —?—

$= \frac{12}{5} \cdot \left(x \cdot x^3 \cdot \frac{1}{x^3}\right) \cdot \left(y^2 \cdot \frac{1}{y^2} \cdot \frac{1}{y^3}\right)$

$= \frac{12}{5} \cdot x \cdot \left(x^3 \cdot \frac{1}{x^3}\right) \cdot \left(y^2 \cdot \frac{1}{y^2}\right) \cdot \frac{1}{y^3}$

$= \frac{12}{5} \cdot x \cdot 1 \cdot 1 \cdot \frac{1}{y^3}$ —?—

$= \frac{12}{5} \cdot \frac{x}{1} \cdot \frac{1}{y^3}$ —?—

$= \frac{12x}{5y^3}$

Simplify, and give reasons for your key steps.

24. $\left(\frac{17}{c} + \frac{4}{d}\right) + \left(\frac{9}{c} + \frac{16}{d}\right)$

25. $\frac{15}{a^2} + \frac{7}{b^2} + \frac{23}{a^2}$

26. $21x^4y \div 14xy^5$

27. $(3x^2y + 5xy^2) \cdot \frac{1}{15xy}$

28. $18a^2b^2 \div 3ab^3$

29. $-8xy^7 \div 20x^5y^3$

30. $\left(\frac{1}{a} - \frac{7}{b}\right)3 + \left(\frac{2}{b} - \frac{4}{a}\right)6$

31. $\frac{x}{3y} \div \frac{4y^2}{3x}$

*2–14 PROOF IN ALGEBRA I

Some examples of theorems and proofs were given in Sections 8 through 13. A further discussion of proofs is given in this section.

A statement of the form

If Harry goes on a trip, *then* he buys gasoline

or

If I pass the examination, *then* I pass the course

is called an *if-then* statement. Thus, each *if-then* statement has the form

If A, then B

where A and B are themselves statements. Associated with each statement

If A, then B

is its *converse*

If B, then A.

For example,

If Harry buys gasoline, *then* he goes on a trip

is the converse of the first *if-then* statement above.

A statement

If A, then B

is called an *implication* if statement B is true whenever statement A is true (that is, if the truth of A *implies* the truth of B). For example, the statement

$$\text{If } \underbrace{x = 5 \quad \text{and} \quad y = 3}_{A}, \quad \text{then } \underbrace{x + y = 8}_{B}$$

is an implication; if A is true, then B is also true. Incidentally, the converse of this *if-then* statement,

$$\text{If } \underbrace{x + y = 8}_{B}, \quad \text{then } \underbrace{x = 5 \quad \text{and} \quad y = 3}_{A},$$

is *not* an implication; from $x + y = 8$, you cannot conclude that $x = 5$ and $y = 3$ (you might have $x = 4$, $y = 4$).

The additive axiom of equality, (Add-A), is the implication

$$\text{If } a = b, \quad \text{then } a + c = b + c.$$

Its converse

$$\text{If } a + c = b + c, \quad \text{then } a = b$$

is also an implication, which you should recognize to be the cancellation law of addition, (Can-A).

If an *if-then* statement

If A, then B

and its converse

If B, then A

are both implications, then you may write

A if, and only if, B.

For example,

$$a = b \text{ if, and only if, } a + c = b + c.$$

Definitions are often written as implications. When this is done, the converse of the implication is always assumed to be an implication also. For example, the definition of equivalent expressions (page 77) is understood to have the following form:

> Two algebraic expressions in the same variables are equivalent if, and only if, they have the same value for every set of values of the variables.

A theorem which asserts the truth of some statement can often be proved by using a chain of implications along with the axioms and known true statements. Some examples of this procedure are given below.

Problem 1. Prove the cancellation law of multiplication.

Solution. You are asked to prove that the statement

$$\text{if } a \cdot c = b \cdot c \quad \text{and} \quad c \neq 0, \quad \text{then } a = b$$

is true. You can prove this using a series of implications derived from the axioms you already know. Proceed as follows.

$$\text{If } ac = bc \text{ and } c \neq 0, \text{ then } (ac)\,\frac{1}{c} = (bc)\,\frac{1}{c}. \qquad (Q)$$

$$\text{If } (ac)\,\frac{1}{c} = (bc)\,\frac{1}{c}, \text{ then } a\left(c \cdot \frac{1}{c}\right) = b\left(c \cdot \frac{1}{c}\right). \qquad (R)$$

$$\text{If } a\left(c \cdot \frac{1}{c}\right) = b\left(c \cdot \frac{1}{c}\right), \quad \text{then } a \cdot 1 = b \cdot 1. \qquad (S)$$

And finally

$$\text{If } a \cdot 1 = b \cdot 1, \quad \text{then } a = b. \qquad (T)$$

Implication (Q) is derived from the multiplication axiom of equality, (Mult-A); (R) from the associative axiom for multiplication, (A-M); (S) from the multiplicative inverse axiom, (Inv-M), and (T) from the multiplicative identity axiom, (Id-M). Thus, you can say if the statement $ac = bc$ and $c \neq 0$ is true, then so is the statement $a = b$. This is the implication you wished to prove.

Problem 2. Prove that the equation

$$x + (y + z) = y + (x + z)$$

is true for all numbers x, y, and z.

Solution. As you know, this equation is true by the rearrangement property of addition, (R-A). However, (R-A) is not one of the axioms; it can be proved from the axioms. Since this is an equation and not an *if-then* statement, you may prove it in either of the following two ways.

I. With your knowledge of using implications to prove a given statement you could proceed as follows.

$$\text{If } x + (y + z) = (x + y) + z, \text{ then } x + (y + z) = (y + x) + z \qquad (1)$$
$$\text{If } x + (y + z) = (y + x) + z, \text{ then } x + (y + z) = y + (x + z) \qquad (2)$$

Implication (1) is derived from the associative and commutative axioms for addition, (A-A) and (C-A); (2) is derived from the associative axiom for addition (A-A). Therefore the equation

$$x + (y + z) = y + (x + z)$$

is true for all numbers x, y, and z.

II. The following method may also be used to prove that the equation is true. Proceed by obtaining a series of equations which are derived from the basic axioms.

$$x + (y + z) = (x + y) + z \qquad \text{(A-A)}$$
$$(x + y) + z = (y + x) + z \qquad \text{(C-A)}$$
$$(y + x) + z = y + (x + z) \qquad \text{(A-A)}$$

And, by the transitive axiom of equality,

$$x + (y + z) = (x + y) + z.$$

This is what you set out to prove.

Exercises

1. In the following proof, give the reason why equation (Q) is true; then tell why (Q) implies (R), (R) implies (S), (S) implies (T).
Prove:

$$\frac{a}{b} \cdot \frac{c}{d} = \frac{a \cdot c}{b \cdot d}. \qquad \text{(Fr-M)}$$

Assume that a and c are any numbers and b and d any nonzero numbers. Then

$$\frac{a}{b} \cdot \frac{c}{d} = \left(a \cdot \frac{1}{b}\right) \cdot \left(c \cdot \frac{1}{d}\right), \qquad \text{(Q)}$$
$$\left(a \cdot \frac{1}{b}\right) \cdot \left(c \cdot \frac{1}{d}\right) = (a \cdot c) \cdot \left(\frac{1}{b} \cdot \frac{1}{d}\right), \qquad \text{(R)}$$
$$(a \cdot c) \cdot \left(\frac{1}{b} \cdot \frac{1}{d}\right) = (a \cdot c) \cdot \left(\frac{1}{b \cdot d}\right), \qquad \text{(S)}$$
$$(a \cdot c) \cdot \left(\frac{1}{b \cdot d}\right) = \frac{a \cdot c}{b \cdot d}. \qquad \text{(T)}$$

2. In the following proofs give the reason why Equation (A) is true; then tell why (A) implies (B), (B) implies (C), (C) implies (D), (D) implies (E), (E) implies (F).
(a) Prove:

$$-(x + y) = -x + (-y).$$

$$(x + y) + [-(x + y)] = 0 \qquad \text{(A)}$$
$$-x + (-y) + ((x + y) + [-(x + y)]) = -x + (-y) + 0 \qquad \text{(B)}$$
$$(-x + x) + (-y + y) + [-(x + y)] = -x + (-y) + 0 \qquad \text{(C)}$$
$$(-x + x) + (-y + y) + [-(x + y)] = -x + (-y) \qquad \text{(D)}$$
$$0 + 0 + [-(x + y)] = -x + (-y) \qquad \text{(E)}$$
$$-(x + y) = -x + (-y) \qquad \text{(F)}$$

The above equations prove the negative sum property.

(b) If x and y are nonzero numbers, prove the following.

$$\frac{1}{x \cdot y} = \frac{1}{x} \cdot \frac{1}{y}$$

$$(x \cdot y) \cdot \frac{1}{(x \cdot y)} = 1 \qquad \text{(A}')$$

$$\left(\frac{1}{x} \cdot \frac{1}{y}\right)\left[(x \cdot y) \cdot \frac{1}{x \cdot y}\right] = \left(\frac{1}{x} \cdot \frac{1}{y}\right) \cdot 1 \qquad \text{(B}')$$

$$\left(\frac{1}{x} \cdot x\right) \cdot \left(\frac{1}{y} \cdot y\right) \cdot \frac{1}{x \cdot y} = \frac{1}{x} \cdot \frac{1}{y} \cdot 1 \qquad \text{(C}')$$

$$\left(\frac{1}{x} \cdot x\right) \cdot \left(\frac{1}{y} \cdot y\right) \cdot \frac{1}{x \cdot y} = \frac{1}{x} \cdot \frac{1}{y} \qquad \text{(D}')$$

$$1 \cdot 1 \cdot \frac{1}{x \cdot y} = \frac{1}{x} \cdot \frac{1}{y} \qquad \text{(E}')$$

$$\frac{1}{x \cdot y} = \frac{1}{x} \cdot \frac{1}{y} \qquad \text{(F}')$$

The above equations prove the reciprocal product property.

(c) Observe that Equation (A′) of part (b) is obtained from Equation (A) of part (a) by replacing addition by multiplication, negatives by reciprocals, and 0 by 1. Is this also true for every other step?

In the following proofs assume that Equation (1) is true; then tell why (1) implies (2), (2) implies (3), (3) implies (4), (4) implies (5), (5) implies (6), (6) implies (7), and so on.

3. Prove: If $\frac{a}{b} = \frac{c}{d}$ and $bd \neq 0$, then $\frac{a}{c} = \frac{b}{d}$.

Assume that a, b, c, and d are numbers and $bd \neq 0$ such that the equation

$$\frac{a}{b} = \frac{c}{d} \qquad (1)$$

is true. Then

$$a \cdot \frac{1}{b} = c \cdot \frac{1}{d}, \qquad (2)$$

$$\left(a \cdot \frac{1}{b}\right) \cdot \left(b \cdot \frac{1}{c}\right) = \left(c \cdot \frac{1}{d}\right) \cdot \left(b \cdot \frac{1}{c}\right), \qquad (3)$$

$$\left(a \cdot \frac{1}{c}\right) \cdot \left(\frac{1}{b} \cdot b\right) = \left(c \cdot \frac{1}{c}\right) \cdot \left(b \cdot \frac{1}{d}\right), \qquad (4)$$

$$\left(a \cdot \frac{1}{c}\right) \cdot 1 = 1 \cdot \left(b \cdot \frac{1}{d}\right), \qquad (5)$$

$$a \cdot \frac{1}{c} = b \cdot \frac{1}{d}, \qquad (6)$$

$$\frac{a}{c} = \frac{b}{d}. \qquad (7)$$

4. Prove: If $\frac{a}{b} = \frac{c}{d}$ and $bd \neq 0$, then $a \cdot d = b \cdot c$.

Assume that a, b, c, and d are numbers and $bd \neq 0$ such that the equation

$$\frac{a}{b} = \frac{c}{d} \tag{1}$$

is true. Then

$$a \cdot \frac{1}{b} = c \cdot \frac{1}{d}, \tag{2}$$

$$\left(a \cdot \frac{1}{b}\right) \cdot (b \cdot d) = \left(c \cdot \frac{1}{d}\right) \cdot (b \cdot d), \tag{3}$$

$$(a \cdot d) \cdot \left(\frac{1}{b} \cdot b\right) = (b \cdot c) \cdot \left(\frac{1}{d} \cdot d\right), \tag{4}$$

$$(a \cdot d) \cdot 1 = (b \cdot c) \cdot 1, \tag{5}$$

$$a \cdot d = b \cdot c. \tag{6}$$

5. Prove: If $\frac{a}{b} = \frac{c}{d}$ and $bd \neq 0$, then $\frac{b}{a+b} = \frac{d}{c+d}$.

In Exercise 4, we proved that if the equation $\frac{a}{b} = \frac{c}{d}$ is true, then $a \cdot d = b \cdot c$. If

$$a \cdot d = b \cdot c, \tag{1}$$

then

$$ad + bd = bc + bd, \tag{2}$$

$$(a + b) \cdot d = b \cdot (c + d), \tag{3}$$

$$(a + b) \cdot d \cdot \frac{1}{a+b} \cdot \frac{1}{c+d} = b \cdot (c + d) \cdot \frac{1}{a+b} \cdot \frac{1}{c+d}, \tag{4}$$

$$\left[(a + b) \cdot \frac{1}{a+b}\right] \cdot d \cdot \frac{1}{c+d} = b \cdot \frac{1}{a+b} \cdot \left[(c + d) \cdot \frac{1}{c+d}\right], \tag{5}$$

$$1 \cdot d \cdot \frac{1}{c+d} = b \cdot \frac{1}{a+b} \cdot 1, \tag{6}$$

$$d \cdot \frac{1}{c+d} = b \cdot \frac{1}{a+b}, \tag{7}$$

$$\frac{d}{c+d} = \frac{b}{a+b}, \tag{8}$$

$$\frac{b}{a+b} = \frac{d}{c+d}. \tag{9}$$

6. Prove: If $\frac{a}{b} = \frac{c}{d}$ and $bd \neq 0$, then

$$\frac{a+c}{c} = \frac{b+d}{d}.$$

Your steps in the proofs of parts (a) and (b) of Exercises 7–9 should be similar to those in Exercise 2; part (b) should be obtained from part (a) by interchanging addition and multiplication, negatives and reciprocals, subtraction and division, and 0 and 1.

7. (a) Prove that

$$-(x - y) = y - x$$

for all numbers x and y.

(b) Prove that

$$\frac{1}{\frac{x}{y}} = \frac{y}{x}$$

for all nonzero numbers x and y.

8. (a) Prove that

$$(x + y) - y = x$$

for all numbers x and y.

(b) Prove that

$$\frac{x \cdot y}{y} = x$$

for all numbers x and y, $y \neq 0$.

9. (a) Prove the cancellation law of subtraction: If $a - c = b - c$, then $a = b$.

(b) Prove the cancellation law of division: If $a/c = b/c$, and $c \neq 0$, then $a = b$.

10. Prove that $a(b + c + d) = ab + ac + ad$ for all numbers a, b, c, and d.

2-15 FLOW CHARTS

The algorithm for finding the product of two numbers can be written as follows.

1. Pick numbers x and y.
2. Do x and y have the same sign?
   (a) If yes, go to step 3.
   (b) If no, go to step 4.
3. Find $|x| \cdot |y|$. This is your answer.
4. Find $-|x| \cdot |y|$. This is your answer.

The flow chart for this algorithm is pictured in Fig. 2-2.

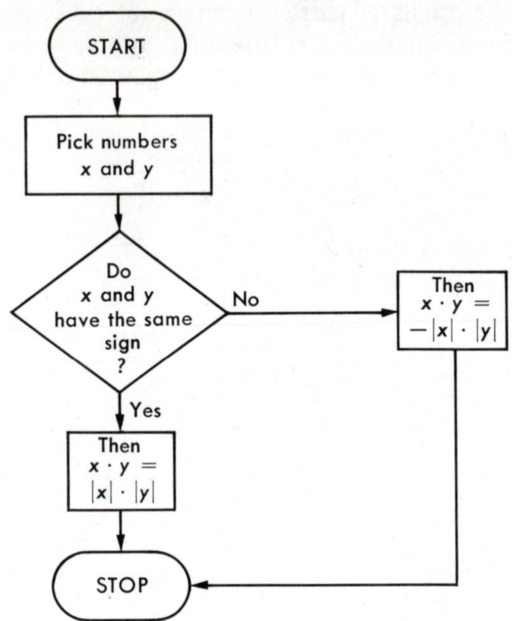

FIGURE 2–2

You know that, in lowest form,

$$38 \div 8 = 4 + \tfrac{6}{8}, \quad \text{or } 4\tfrac{3}{4},$$

and that

$$152 \div 5 = 30 + \tfrac{2}{5}, \quad \text{or } 30\tfrac{2}{5}.$$

An algorithm for carrying out this division, called the *repeated subtraction algorithm,* is as follows.

Your goal is to express the quotient $N \div D$ of two positive integers in lowest form; that is, as a sum of an integer and a non-negative rational number less than one.

1. Let $N =$ dividend, $D =$ divisor, and $A = 0$.
2. Is N greater than or equal to D?
   If it is, go to step 3.
   If it is not, go to step 5.
3. Replace N by $N - D$ and A by $A + 1$.
4. Go back to step 2.
5. Answer is $A + N/D$.

The flow chart for this algorithm is given in Fig. 2–3.

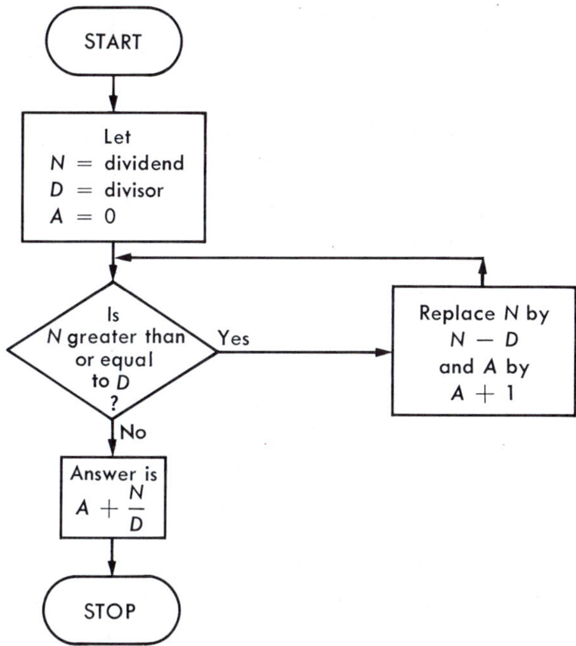

FIGURE 2–3

Example. Find $38 \div 8$.

Solution. Let $N = 38$, $D = 8$, $A = 0$.

Since 38 is more than 8, let $N = 38 - 8 = 30$ and $A = 1$.
Since 30 is more than 8, let $N = 30 - 8 = 22$ and $A = 2$.
Since 22 is more than 8, let $N = 22 - 8 = 14$ and $A = 3$.
Since 14 is more than 8, let $N = 14 - 8 = 6$ and $A = 4$.

Since 6 is less than 8, the answer is $A + N/D = 4 + \frac{6}{8}$ or $4\frac{3}{4}$.

Exercises

Using the flow chart for multiplication, compute the following:

1. (a) $14 \cdot (-3)$ (b) $6 \cdot 73$
2. (a) $(-6) \cdot (-5)$ (b) $(-5) \cdot 26$
3. (a) $42 \cdot 9$ (b) $723 \cdot (-6)$
4. (a) $(-7) \cdot 243$ (b) $(-75)(-2)$
5. (a) $27 \cdot (-31)$ (b) $14 \cdot 73$

Using the flow chart for division of positive numbers, compute the following:

6. (a) $32 \div 4$ (b) $72 \div 8$

7. (a) $23 \div 7$ (b) $16 \div 5$

8. (a) $31 \div 16$ (b) $48 \div 13$

9. (a) $55 \div 7$ (b) $35 \div 6$

10. (a) $46 \div 17$ (b) $53 \div 20$

11. (a) $63 \div 15$ (b) $56 \div 14$

12. (a) $52 \div 12$ (b) $74 \div 21$

13. (a) $45 \div 12$ (b) $48 \div 16$

14. (a) $54 \div 18$ (b) $62 \div 10$

---

15. (a) Write an algorithm for finding the absolute value of a number.
(b) Draw the flow chart for the algorithm in Exercise 15(a).

Using the flow chart you drew for Exercise 15(b), find the following absolute values:

16. (a) $|-15|$ (b) $|-23|$

17. (a) $|16|$ (b) $|27|$

18. (a) $|3 - 16|$ (b) $|14 - 18|$

19. (a) $|17 - 3|$ (b) $|54 - 12|$

---

20. (a) From your knowledge of division, you know there is an easier way to find the answer to $4624 \div 12$ than by repeated subtraction of 12. Can you write an algorithm (or alter the algorithm) for division which allows some shortcuts?
(b) Draw the flow chart for the algorithm you wrote in Exercise 20(a).

KEY IDEAS AND KEY WORDS

AXIOMS OF EQUALITY

Reflexive Axiom
$a = a$ (Ref-A)

Symmetric Axiom
If $a = b$, then $b = a$. (Sym-A)

Transitive Axiom
If $a = b$ and $b = c$, then $a = c$. (Trans-A)

Additive Axiom
If $a = b$ and $c = d$, then $a + c = b + d$. (Add-A)

Multiplicative Axiom
If $a = b$ and $c = d$, then $ac = bd$. (Mult-A)

AXIOMS

Commutative Axioms
$x + y = y + x$ (Commutative Axiom for addition) (C-A)
$xy = yx$ (Commutative Axiom for multiplication) (C-M)
Each of the equations is true for all numbers x and y.

Associative Axioms
$(x + y) + z = x + (y + z)$ (Associative Axiom for addition) (A-A)
$(xy)z = x(yz)$ (Associative Axiom for multiplication) (A-M)
Each of the equations is true for all numbers x, y, and z.

Distributive Axioms
$x(y + z) = xy + xz$
$(y + z)x = yx + zx$ (D)
Each of the equations is true for all numbers x, y, and z.

Additive Identity Axiom
There exists a number 0 such that the equations

$$x + 0 = x \quad \text{and} \quad 0 + x = x \qquad \text{(Id-A)}$$

are true for every number x.

Multiplicative Identity Axiom
There exists a number 1 such that the equations

$$x \cdot 1 = x \quad \text{and} \quad 1 \cdot x = x \qquad \text{(Id-M)}$$

are true for every number x.

Additive Inverse Axiom
Every real number has a unique additive inverse. (Inv-A)

Multiplicative Inverse Axiom
Every nonzero real number has a unique multiplicative inverse. (Inv-M)

Order Axioms

1. The set of positive numbers is closed under addition.
2. The set of positive numbers is closed under multiplication.
3. Every real number is either a positive number, zero, or a negative number.

THEOREMS

Rearrangement Property of Addition
The addends of a sum may be rearranged in any order. (R-A)

Rearrangement Property of Multiplication
The multiplicands of a product may be rearranged in any order. (R-M)

The additive inverse of a negative number is a positive number. That is,

$$-(-x) = x.$$

The multiplicative inverse of a unit fraction is an integer. That is,

$$\frac{1}{\frac{1}{x}} = x.$$

Negative of a Sum
The equation $-(x + y) = -x + (-y)$ is true for all numbers x and y. (Neg-Sum)

Reciprocal of a product
The equation $\frac{1}{x \cdot y} = \frac{1}{x} \cdot \frac{1}{y}$ is true for all nonzero numbers x and y. (Recip-Prod)

Cancellation Law of Addition
If $a + c = b + c$, then $a = b$. (Can-A)

Cancellation Law of Multiplication
If $ca = cb$ and $c \neq 0$, then $a = b$. (Can-M)

Multiplication by Zero
The equation $0 \cdot x = 0$ is true for every number x. (Zero-M)

Negative Multiplication
The equations $x \cdot (-y) = -(xy)$ and $(-x)y = -(xy)$ are true for all numbers x and y. (Neg-M)

DEFINITIONS

Addition and multiplication are **binary** operations. In other words, when two numbers are combined, there is a **unique** number which is their sum or product.

The **additive identity** element is zero, and the **multiplicative identity** element is one.

All **real numbers** have additive **inverses,** commonly called **opposites.** All **real numbers** except zero have multiplicative inverses, commonly called **reciprocals.**

Subtraction can be performed by adding the negative, and **division** can be performed by multiplying by the reciprocal.

Two algebraic expressions in the same variables are called **equivalent** if they have the same value for every set of values of the variables.

The **absolute value** of a nonzero number is always positive.

CHAPTER REVIEW

Fill in a word, phrase, numeral or symbol to make the following statements true.

1. The equation $y + _?_ = 0$ is true for every real number y.
2. The true statement

$$(-3 \cdot 8) \cdot 125 = -3 \cdot (8 \cdot 125)$$

is an example of the _?_ axiom for _?_.
3. To subtract a is the same as to _?_.
4. Another name for the multiplicative inverse of a number is the _?_ of the number.
5. The true statement

$$(-5 + 7) + 20 = 20 + (-5 + 7)$$

is an example of the _?_ axiom for _?_.
6. Another name for the additive inverse of a number is the _?_ of the number.
7. To divide by x is the same as to _?_.
8. The equation $x \cdot (-y) = _?_$ is true for all numbers x and y.
9. The true statement

$$(7 \cdot 27) + (7 \cdot 23) = 7 \cdot 50$$

is an example of the _?_ axiom.

Tell which of the statements in Exercises 10–17 are true and which are false. Justify your answer by citing a definition, a property, or an axiom, whenever possible. Otherwise, cite one or more specific cases to support your answer.

10. $\frac{5}{0} = 0$
11. $\frac{0}{5} = 0$
12. The equation $|x| = x$ is true for all numbers x.

13. The equation $(-a) - b = -a + (-b)$ is true for all numbers a and b.
14. The equation $a - (-b) = a + b$ is true for all numbers a and b.
15. The following has the value 0 if we let x equal 8.

$$\left(\frac{x+4}{6} - \frac{3x-4}{4} + \frac{7x}{8}\right)$$

16. If $5x + 3 = 3x + 5$, then $x + 3 = 3x$.
17. Zero is either positive or negative.

Do the computation.

18. $-8 \cdot [5(-6)]$
19. $[(-8)5] \div (-6)$
20. $-8 + [2(-12)]$
21. $(-8 + 2)(-12)$
22. $-9 - (14 + 5)$
23. $(-1)^{10} - (-2)^3$
24. $-8 + 7 + (-20)$
25. $(18 - 26) - (6 - 10)$
26. $15 - (17 + 8)$
27. $[6 + (-13)]^2(-3)$

Use the properties of the real number system to determine the value of x.

28. $x + (-83) = -83$
29. $\frac{-2}{5}x = 1$
30. $-17x = -17$
31. $3x = 21$
32. $x + (-32) = -15$
33. $-4x = 48$
34. $x + 18 = 0$
35. $3x + 7 = 9x + 7$
36. $51x = 51 \cdot 4$
37. $7 + x = 0$
38. $4x + (-x) = 12$

Simplify, and find the value of each expression when $x = -3$ and $y = 4$.

39. $9x + 5x$
40. $9x \cdot 5x$
41. $3y^2 - 5y^2$
42. $(\frac{1}{4}xy^2)(\frac{3}{2}xy)$
43. $\left(\frac{6}{x} - \frac{5}{y}\right) - \left(\frac{9}{x} + \frac{11}{y}\right)$
44. $(2x^2y) \div \left(\frac{-1}{3}xy\right)$
45. $(7x + y) - (x - 2y)$
46. $9x - 5y - 7x - 8y$
47. Use the distributive axiom to write each sum as a product.
   (a) $2x + xy$
   (b) $x + 5x$
   (c) $7x + 42x^2$
   (d) $x + xy$
   (e) $23^2 + 23^3$
   (f) $13xy^2 + 26x^2y$

Simplify.

48. $8x - 5y - 14x + y$
49. $9x - (2x + 3)$
50. $14x - (6x - 5)$
51. $8(x + 3y) + [-2(5x - y)]$
52. $\frac{x}{9} - \frac{y}{2} - \frac{x}{6} + \frac{y}{3}$
53. $5(x - 2) + 3(4x + 7)$
54. $18x - (3x + 4y) + y$
55. $-\frac{1}{2}h + k + 2h - \frac{4}{3}k$

56. $-2(x - y) - [(2x - 3y) - (5x + 2y)]$

57. $6(a - 2b) + [-4(a - 3b)]$

58. $2x - [x - (1 - 2x)]$

59. $3y - [2y - (3 - y)]$

CHAPTER TEST

Give a reason or reasons why the statement is true.

1. $(12 + 9) + 7 = (9 + 12) + 7$
2. $17(25 \cdot 4) = 17(4 \cdot 25)$
3. $(9 \cdot 52) + (x \cdot 52) = 52(9 + x)$
4. $3(\frac{1}{3}x) = (3 \cdot \frac{1}{3})x$
5. $(3 \cdot \frac{1}{3})x = x$
6. $(3 - 3)(-5) = 0$

Compute the following.

7. $-7 - (-20)$
8. $-30 \div (-6)$
9. $-7 + (-20)$
10. $-23 + 4$
11. $-4 \cdot (-3)^2$
12. $10 - 24$
13. $-(-6 - 2)$

Simplify the following.

14. $(5xy)(-3xy)(\frac{1}{15}xy)$
15. $7x + x + 10$
16. $9x \cdot 3x$
17. $8x - 5y - 26x - y$
18. $3x^2y\left(\frac{-1}{6}xy\right)^2$
19. $-5x^3y^2 \div 10x^4y$
20. $7x - 4(5x - 3)$
21. $\frac{8}{a} - \frac{7}{b} - \frac{10}{a} - \frac{13}{b}$
22. $(3x + 4y) - (8x - y)$
23. $(-5x - y) - (7x + 8y)$

CHAPTER 3

Equations

Objectives . . .

- To decide if two equations are equivalent.
- To solve simple linear equations.
- To find rational expressions for numbers represented as repeating decimals.
- To apply your algebraic skills to the solution of word problems.

3–1 SOLUTION SET OF AN EQUATION AND EQUIVALENT EQUATIONS

A mathematical sentence involving an equals sign, =, is called an *equation.* Thus,

$$12 + 9 = 21$$

is an equation. It is a *true* equation, since the sum of 12 and 9 is 21. The sentence

$$8 + 9 = 21$$

is also an equation. However, it is a *false* equation since the sum of 8 and 9 is not 21. (In an equation such as $12 + 9 = 21$, $12 + 9$ and 21 are called the *sides* of the equation.)

The equation

$$N + 9 = 21$$

involving a variable N is *open;* that is, it is neither true nor false. It becomes either true or false when you give the variable N a value. For example, the equation is true when $N = 12$ and false when $N = 8$. A value for the variable which makes the equation true is called a *solution* of the equation. Hence, 12 is a solution of the equation $N + 9 = 21$.

Definition of solution

A solution of an open equation in one variable is a value of the variable for which the equation is true. The set of all solutions is called the* solution set *of the equation.

The process of finding solutions of an open equation is called *solving the equation.* You will learn methods of solving equations in this chapter.

Problem 1. Solve the equation $k + 4 = 13$.

Solution. By trial and error, giving k various values, you find that, if $k = 9$, then $k + 4 = 13$ is a true equation $(9 + 4 = 13)$. Thus, 9 is a solution.

In Problem 1, you found that 9 is a solution to the equation $k + 4 = 13$. Is 9 the only solution? In other words, is 9 the only value for k that makes $k + 4 = 13$ a true equation? One way to find out is to replace the original equation with simpler equivalent equations by using the properties you have learned.

Definition of equivalent equations

Two equations are said to be equivalent if they have the same solution set.

In using this method, you hope to eventually find an equivalent equation with an obvious solution set.

By the additive axiom of equality, (Add-A), you can say

$$\begin{aligned} (k + 4) + (-4) &= 13 + (-4), && \\ k + [4 + (-4)] &= 13 - 4, && \text{(A-A) and (Def-Sub)} \\ k + 0 &= 9, && \text{(Inv-A)} \\ k &= 9. && \text{(Id-A)} \end{aligned}$$

Since $\{9\}$ is the obvious solution set of the equation $k = 9$, $\{9\}$ is also the solution set of the equivalent equation $k + 4 = 13$.

Problem 2. Solve the equation $x + 14 = 6$.

Solution. Find an equivalent equation having the variable x as one side of the equation and a numeral as the other side. You might proceed as follows. (Each equation is equivalent to the one directly above it for the stated reason.)

$$\begin{aligned} x + 14 &= 6 && \\ (x + 14) + (-14) &= 6 + (-14) && \text{(Add-A)} \\ x + [14 + (-14)] &= -8 && \text{(A-A) and (Def-Sub)} \\ x + 0 &= -8 && \text{(Inv-A)} \\ x &= -8 && \text{(Id-A)} \end{aligned}$$

The last equation has the solution set $\{-8\}$. Since the equations above are all equivalent, the given equation also has the solution set $\{-8\}$.

You should check your answer to be sure an error has not been made.

Check.

$$-8 + 14 \stackrel{?}{=} 6$$
$$6 \stackrel{\checkmark}{=} 6$$

The question mark above the equals sign indicates that you are questioning the truth of the equation. The check mark indicates the equation is true. Thus,

$$\{-8\}$$

is the solution set of the equation $x + 14 = 6$.

Problem 3. Solve the equation $7m = 28$.

Solution. Find an equivalent equation having the variable m on one side of the equals sign and a numeral on the other.

$$\begin{aligned} 7m &= 28 & & \\ (\tfrac{1}{7}) \cdot 7m &= (\tfrac{1}{7}) \cdot 28 & & \text{(Mult-A)} \\ (\tfrac{1}{7} \cdot 7)m &= \tfrac{28}{7} & & \text{(A-M) and (Def-Div)} \\ 1 \cdot m &= 4 & & \text{(Inv-M)} \\ m &= 4 & & \text{(Id-M)} \end{aligned}$$

The solution set of $m = 4$ is $\{4\}$. Is this the solution set of the equation $7m = 28$?

Check.

$$7 \cdot 4 \stackrel{?}{=} 28$$
$$28 \stackrel{\checkmark}{=} 28$$

Exercises

Find the solution set of each equation in Exercises 1–17. Check each solution.

1. (a) $x + 5 = 10$ (b) $x + 9 = 11$

2. (a) $x + 4 = 15$ (b) $x + 7 = 18$

3. (a) $6 + x = 20$ (b) $5 + x = 23$

4. (a) $x + 7 = 3$ (b) $x + 11 = 2$

5. (a) $19 = x + 20$ (b) $20 = x + 22$

6. (a) $3x = 12$ (b) $4y = 20$

7. (a) $6m = 36$ (b) $5x = 25$

8. (a) $2m = 21$ (b) $3m = 10$

9. (a) $10x = 45$ (b) $8m = 52$

10. (a) $7x = -14$ (b) $3x = -12$

11. (a) $x + 23 = 30$ (b) $x + 40 = 51$

12. (a) $x - 12 = 6$ (b) $x - 8 = 2$

13. (a) $x - 15 = 20$ (b) $x - 22 = 22$

14. (a) $x - 11 = -2$ (b) $x - 14 = -1$

15. (a) $11x = 33$ (b) $16x = 32$

16. (a) $x + 4 = 5 + 9$ (b) $x + 6 = 12 + 1$

17. (a) $x - 6 = 12 - 18$ (b) $x - 10 = 10 - 20$

Give the reason for each step in the solution of the following equations.

18.

$$\begin{aligned} x - 14 &= -6 \\ x + (-14) &= -6 && \text{_?_} \\ [x + (-14)] + 14 &= -6 + 14 && \text{_?_} \\ x + (-14 + 14) &= -6 + 14 && \text{_?_} \\ x + 0 &= 8 && \text{_?_} \\ x &= 8 && \text{_?_} \end{aligned}$$

19.

$$\begin{aligned} -13x &= -26 \\ (-\tfrac{1}{13})(-13x) &= (-\tfrac{1}{13})(-26) && \text{_?_} \\ [-\tfrac{1}{13}(-13)]x &= \tfrac{1}{13} \cdot 26 && \text{_?_} \\ 1 \cdot x &= 2 && \text{_?_} \\ x &= 2 && \text{_?_} \end{aligned}$$

20.

$$\begin{aligned} 4x &= 3x + 5 \\ 4x + (-3x) &= (3x + 5) + (-3x) && \text{_?_} \\ 4x + (-3x) &= [3x + (-3x)] + 5 && \text{_?_} \\ [4 + (-3)]x &= [3x + (-3x)] + 5 && \text{_?_} \\ x &= 0 + 5 && \text{_?_} \\ x &= 5 && \text{_?_} \end{aligned}$$

Solve each of the following equations.

21. $4m = -16$

22. $-12x = 36$

23. $-5x = -25$

24. $-7x = -49$

25. $-3m = 24$

26. $3m = -15$

27. $8x = 7x + 1$

28. $2x + 4 = 3x$

29. $7x = 12 + 9$

30. $5x + 1 = 4x$

31. $2x + 3 = x + 1$

32. $4x - 3 = 3x - 2$

3–2 SOLVING EQUATIONS I

You have seen how an equation can be solved by finding an equivalent equation with an obvious solution set (an equation with the variable as one side and a numeral as the other). The solution set of the equivalent equation is the same as that of the equation you are trying to solve.

Problem 1. Solve the equation $4n + 3 = 33$.

Solution. First get an equivalent equation with $4n$ as one side, then proceed to get an equivalent equation with just n as one side.

$$\begin{aligned}
4n + 3 &= 33 & \\
(4n + 3) + (-3) &= 33 + (-3) & \text{(Add-A)} \\
4n + [3 + (-3)] &= 33 - 3 & \text{(A-A) and (Def-Sub)} \\
4n + 0 &= 30 & \text{(Inv-A)} \\
4n &= 30 & \text{(Id-A)} \\
\tfrac{1}{4} \cdot (4n) &= \tfrac{1}{4} \cdot 30 & \text{(Mult-A)} \\
(\tfrac{1}{4} \cdot 4)n &= \tfrac{30}{4} & \text{(A-M) and (Def-Div)} \\
1 \cdot n &= \tfrac{15}{2} & \text{(Inv-M)} \\
n &= \tfrac{15}{2} & \text{(Id-M)}
\end{aligned}$$

The solution set of $n = \frac{15}{2}$ is $\{\frac{15}{2}\}$. Thus, the solution set of $4n + 3 = 33$ is $\{\frac{15}{2}\}$.

Check.

$$\begin{aligned}
4 \cdot \tfrac{15}{2} + 3 &\stackrel{?}{=} 33 \\
\tfrac{60}{2} + 3 &\stackrel{?}{=} 33 \\
30 + 3 &\stackrel{?}{=} 33 \\
33 &\stackrel{\checkmark}{=} 33
\end{aligned}$$

Problem 2. Solve the equation $13y + 5 = 37y - 11$.

Solution. You add $-13y$ to each side of this equation to obtain an equivalent equation having the variable y on only one side.

$$\begin{aligned}
13y + 5 &= 37y - 11 \\
-13y + (13y + 5) &= -13y + (37y - 11) \\
(-13y + 13y) + 5 &= (-13y + 37y) + (-11) \\
0 + 5 &= (-13 + 37)y + (-11) \\
5 &= 24y + (-11)
\end{aligned}$$

Then add 11 to each side of this equation.

$$\begin{aligned}
5 + 11 &= [24y + (-11)] + 11 \\
16 &= 24y + (-11 + 11) \\
16 &= 24y
\end{aligned}$$

Finally, multiply each side by the reciprocal of 24.

$$\frac{1}{24} \cdot 16 = \frac{1}{24} \cdot (24y)$$
$$\frac{16}{24} = (\frac{1}{24} \cdot 24)y$$
$$\frac{2}{3} = y$$

The equations above are all equivalent. Therefore,

$$\{\frac{2}{3}\}$$

is the solution set of the given equation.

Check.

$$13 \cdot \frac{2}{3} + 5 \stackrel{?}{=} 37 \cdot \frac{2}{3} - 11$$
$$\frac{26}{3} + \frac{15}{3} \stackrel{?}{=} \frac{74}{3} - \frac{33}{3}$$
$$\frac{26 + 15}{3} \stackrel{?}{=} \frac{74 - 33}{3}$$
$$\frac{41}{3} \stackrel{\checkmark}{=} \frac{41}{3}$$

Problem 3. Solve the equation $4 - 3N = 7 - 3N$.

Solution. Using the cancellation law, you obtain the equivalent equation

$$4 = 7.$$

This equation is false; yet it is equivalent to the given equation $4 - 3N = 7 - 3N$. Therefore, the given equation is *false* for every value of N and its solution set is the empty set, $\emptyset$.

Exercises

Find the solution set of each equation and check your solution.

1. (a) $4x + 3 = 27$ (b) $3x + 4 = 25$

2. (a) $7x + 9 = 16$ (b) $5x + 2 = 17$

3. (a) $x = 2x + 9$ (b) $3x = 4x + 5$

4. (a) $2x - 5 = 17$ (b) $3x - 14 = 4$

5. (a) $3a + 2a = 25$ (b) $6a - 4a = 10$

6. (a) $7x - 5 = 16$ (b) $14x + 5 = -9$

7. (a) $9x - 2x = 21$ (b) $7n - n = 24$

8. (a) $4y + 528 = 0$ (b) $6x - 126 = 0$

9. (a) $x - 4 = 2x + 7$ (b) $2x - 4 = x + 9$

10. (a) $2x + 3 = 3x - 10$ (b) $4 + x = 15 + 2x$

11. (a) $x - 3 = 2x + 6$ (b) $5y + 7 = 4y - 23$

12. (a) $8y - 11 = 9y + 7$ (b) $12x + 15 = 13x - 4$

13. (a) $15 + 3b = 8b$ (b) $9a = 20 + 4a$

14. (a) $3x + 5 = x - 10$ (b) $8x + 7 = 5x - 9$

15. (a) $5 + y = y + 3$ (b) $3x - 10 = 14 + 3x$

16. (a) $3y - 9 = y + 4 + y$ (b) $6x + 1 + 2x = 4 + 9x$

17. (a) $7x - 3 = 10 - 4x$ (b) $12 - 17x = 3x - 8$

18. (a) $15 + 11x = 17x + 9$ (b) $14 + 5x = 10x + 4$

19. (a) $3g = 42 - 4g$ (b) $13x - 27 = 3x$

20. (a) $m + 3m = 72 - 5m$ (b) $11x + x = 52 - 6x$

21. (a) $36n - 42 = 102 + 12n$ (b) $11x + 3 = 2x - 24$

22. (a) $11p + 2 = 3 + p$ (b) $6s + 3 = 4 + 3s$

23. (a) $6 - 8q = 4 - 4q$ (b) $9 - 2z = 30 - z$

24. (a) $6y + 25 = -2y - 15$ (b) $4x - 15 = -4x - 9$

25. (a) $3y - 17 = 4 + 3y$ (b) $2x - 3 + 5x = 4x + 3x$

---

26. (a) $3(a + 5) + 4(a + 5) = 21$
(b) $2(x + 4) + 5(x + 4) = 14$

27. (a) $7n + 5 + n = 3n - 59 + n$
(b) $12x - 3 + 4x = x + 10 + 2x$

28. (a) $15 - (a - 6) = 6 - (3a + 5)$
(b) $7y - (9y - 15) = 14 - (y + 19)$

29. (a) $(8c - 11) - (-15 - c) = 0$
(b) $(2x - 3) - (-2x + 4) = 0$

30. (a) $18x + (-6)(3x + 1) = 8x - 5(x - 3)$
(b) $2(5x - 4) - 10x = 6x + 3(2x - 5)$

---

31. Solve the equation

$$3(x + 5) = 15 + 3x.$$

How does this solution set differ from those in the previous exercises?

Preparation for Section 3–3

Simplify.

1. $\frac{1}{6} + \frac{1}{18}$

2. $\frac{1}{3} + \frac{4}{5}$

3. $\frac{1}{2} - \frac{1}{6} + \frac{5}{12}$

4. $3(\frac{1}{6} + \frac{2}{3} - \frac{4}{9})$

5. $4(\frac{11}{16} - \frac{3}{4} + \frac{5}{2})$

6. $9(\frac{5}{6} + \frac{2}{3} - \frac{4}{9})$

3–3 SOLVING EQUATIONS II

Not all equations have integral terms like those you solved in Sections 3–1 and 3–2. In fact, you will probably encounter equations with fractional terms quite often.

Problem 1. Solve the equation $\frac{3}{4}x = 6$.

Solution. You could select one of two ways to solve this equation. One way would be to multiply both sides of the equation by the reciprocal of $\frac{3}{4}$. That is, multiply by $\frac{4}{3}$.

$$\begin{aligned} \tfrac{4}{3} \cdot (\tfrac{3}{4}x) &= \tfrac{4}{3} \cdot 6 \\ (\tfrac{4}{3} \cdot \tfrac{3}{4})x &= 8 \\ 1 \cdot x &= 8 \\ x &= 8 \end{aligned}$$

Another way is to *multiply both sides of the equation by the least common denominator of all the fractions in the given equation.* This will give you an equation free of fractions which you already know how to solve.

$$\begin{aligned} 4 \cdot (\tfrac{3}{4}x) &= 4 \cdot 6 \\ (4 \cdot \tfrac{3}{4})x &= 24 \\ 3x &= 24 \\ \tfrac{1}{3} \cdot (3x) &= \tfrac{1}{3} \cdot 24 \\ (\tfrac{1}{3} \cdot 3)x &= 8 \\ 1 \cdot x &= 8 \\ x &= 8 \end{aligned}$$

Since the solution set of $x = 8$ is obviously $\{8\}$, the solution set of $\frac{3}{4}x = 6$ is also $\{8\}$.

Check.

$$\begin{aligned} \tfrac{3}{4} \cdot 8 &\stackrel{?}{=} 6 \\ \tfrac{24}{4} &\stackrel{?}{=} 6 \\ 6 &\stackrel{\checkmark}{=} 6 \end{aligned}$$

Problem 2. Solve the equation $\frac{3}{4}y + \frac{1}{2} = 1$.

Solution. The least common denominator of the fractions in this equation is 4. Using the multiplicative axiom of equality, proceed as follows.

$$
\begin{aligned}
4 \cdot (\tfrac{3}{4}y + \tfrac{1}{2}) &= 4 \cdot 1 \\
4 \cdot (\tfrac{3}{4}y) + 4 \cdot \tfrac{1}{2} &= 4 \\
(4 \cdot \tfrac{3}{4})y + 2 &= 4 \\
3y + 2 &= 4 \\
(3y + 2) + (-2) &= 4 + (-2) \\
3y + [2 + (-2)] &= 2 \\
3y + 0 &= 2 \\
3y &= 2 \\
\tfrac{1}{3} \cdot 3y &= \tfrac{1}{3} \cdot 2 \\
y &= \tfrac{2}{3}
\end{aligned}
$$

Thus, the solution set of $\frac{3}{4}y + \frac{1}{2} = 1$ is $\{\frac{2}{3}\}$.

Check.

$$
\begin{aligned}
\tfrac{3}{4} \cdot \tfrac{2}{3} + \tfrac{1}{2} &\stackrel{?}{=} 1 \\
\tfrac{1}{2} + \tfrac{1}{2} &\stackrel{?}{=} 1 \\
1 &\stackrel{\checkmark}{=} 1
\end{aligned}
$$

Problem 3. Solve the equation $\frac{1}{4}m - \frac{1}{2} = \frac{2}{3}m + \frac{3}{4}$.

Solution. The least common denominator of the fractions in this equation is 12.

$$
\begin{aligned}
12(\tfrac{1}{4}m - \tfrac{1}{2}) &= 12(\tfrac{2}{3}m + \tfrac{3}{4}) \\
12(\tfrac{1}{4}m) + 12 \cdot (-\tfrac{1}{2}) &= 12(\tfrac{2}{3}m) + 12 \cdot \tfrac{3}{4} \\
3m - 6 &= 8m + 9 \\
-8m + (3m - 6) &= -8m + (8m + 9) \\
(-8m + 3m) - 6 &= (-8m + 8m) + 9 \\
-5m - 6 &= 9 \\
(-5m - 6) + 6 &= 9 + 6 \\
-5m + (-6 + 6) &= 15 \\
-5m &= 15 \\
(-\tfrac{1}{5})(-5m) &= (-\tfrac{1}{5}) \cdot 15 \\
m &= -3
\end{aligned}
$$

The solution set of $\frac{1}{4}m - \frac{1}{2} = \frac{2}{3}m + \frac{3}{4}$ is $\{-3\}$.

Check.

$$
\begin{aligned}
\tfrac{1}{4} \cdot (-3) - \tfrac{1}{2} &\stackrel{?}{=} \tfrac{2}{3} \cdot (-3) + \tfrac{3}{4} \\
-\tfrac{3}{4} - \tfrac{1}{2} &\stackrel{?}{=} -2 + \tfrac{3}{4} \\
-\tfrac{3}{4} - \tfrac{2}{4} &\stackrel{?}{=} -\tfrac{8}{4} + \tfrac{3}{4} \\
-\tfrac{5}{4} &\stackrel{\checkmark}{=} -\tfrac{5}{4}
\end{aligned}
$$

Exercises

Find the solution set of each equation and check your solution.

1. (a) $\frac{1}{3}x = 9$ (b) $\frac{1}{4}x = 8$

2. (a) $\frac{2}{5}x = 4$ (b) $\frac{3}{11}x = 9$

3. (a) $\frac{1}{6}x = \frac{1}{2}$ (b) $\frac{1}{4}x = \frac{1}{6}$

4. (a) $\frac{3}{4}x = -2$ (b) $\frac{4}{5}x = -5$

5. (a) $-\frac{3}{7}x = 6$ (b) $-\frac{3}{10}x = 9$

6. (a) $\frac{1}{2}x + \frac{1}{3} = 1$ (b) $\frac{1}{3}x + \frac{1}{2} = 1$

7. (a) $\frac{4}{7} - \frac{1}{2}x = 2$ (b) $\frac{1}{2} - \frac{4}{7}x = 2$

8. (a) $\frac{1}{5}x - \frac{1}{3} = \frac{1}{15}$ (b) $\frac{1}{7}x - \frac{1}{3} = \frac{1}{21}$

9. (a) $\frac{1}{11}x - 2 = 5$ (b) $\frac{1}{13}x - 3 = 4$

10. (a) $-\frac{1}{4} = 8 - \frac{3}{8}x$ (b) $-\frac{1}{6} = 4 - \frac{5}{12}x$

11. (a) $\dfrac{h}{2} = 7 - \dfrac{2h}{3}$ (b) $\dfrac{t}{5} = 17 - \dfrac{3t}{2}$

12. (a) $9 + 4 = \dfrac{x}{3} - 3$ (b) $10 + 13 = \dfrac{x}{2} - 4$

13. (a) $2 - \dfrac{c}{2} = \dfrac{c}{9} + \dfrac{1}{6}$ (b) $\dfrac{3}{10} - \dfrac{x}{5} = \dfrac{x}{4} + 3$

14. (a) $\dfrac{x}{4} - 2 = \dfrac{x}{6} + \dfrac{x}{8}$ (b) $\dfrac{3x}{5} - 1 = \dfrac{2x}{3} + \dfrac{x}{30}$

15. (a) $\dfrac{n}{9} - \dfrac{2}{3} = -\dfrac{2}{9} + n$ (b) $\dfrac{3n}{13} - \dfrac{1}{2} = -\dfrac{n}{26} - \dfrac{3}{13}$

16. (a) $-1 - \frac{1}{2}p = \frac{1}{10}p + 5$ (b) $-4 - \frac{1}{4}x = \frac{1}{16}x + 1$

17. (a) $\dfrac{3y}{5} + \dfrac{5}{2} = -\dfrac{y}{5} - \dfrac{3}{2}$ (b) $\dfrac{4x}{11} - \dfrac{1}{3} = -\dfrac{2x}{3} + \dfrac{1}{11}$

18. (a) $\dfrac{3}{2} - \dfrac{d}{3} = 5 + \dfrac{d}{6}$ (b) $\dfrac{5}{2} + \dfrac{x}{4} = 3 - \dfrac{3x}{8}$

19. (a) $\dfrac{5}{2} - \dfrac{1}{12} = \dfrac{9}{4} - \dfrac{2s}{3}$ (b) $\dfrac{3x}{2} + \dfrac{5}{24} = \dfrac{5x}{8} + \dfrac{13}{12}$

20. (a) $\dfrac{1}{5} - \dfrac{3x}{20} = \dfrac{x}{4} + 2$ (b) $\dfrac{1}{14} - \dfrac{4x}{7} = \dfrac{1}{2} + x$

21. (a) $\dfrac{4x}{5} + \dfrac{2x}{3} - x = \dfrac{1}{2}$ (b) $\dfrac{3x}{15} + \dfrac{7x}{30} - 2x = \dfrac{47}{15}$

22. (a) $\dfrac{x}{2} - \dfrac{x}{3} + \dfrac{1}{12} = \dfrac{5x}{24} - \dfrac{1}{8}$ (b) $\dfrac{x}{6} - \dfrac{x}{9} + \dfrac{1}{3} = \dfrac{3x}{2} - \dfrac{4}{9}$

Find the solution set of each equation and check your solution.

23. $\dfrac{x + 2}{3} = \dfrac{2x - 7}{5}$

24. $\dfrac{5 - 3t}{9} = \dfrac{4t - 9}{2} + \dfrac{7}{18}$

25. $\dfrac{3-x}{2} = \dfrac{-6-5x}{7}$

26. $\dfrac{2h-3}{4} - 1 = \dfrac{h-5}{3}$

27. $\dfrac{9-7}{2} - \dfrac{1}{3} = 1 + \dfrac{a+9}{9}$

28. $\dfrac{5x+3}{7} = \dfrac{3x+13}{3} - 10$

29. $\frac{3}{4}(3e+1) + 11 = \frac{8}{7}(4e-8)$

30. $\dfrac{3y+5}{2} - \dfrac{3y+1}{4} = 3$

31. $\dfrac{7x-3}{10} = \dfrac{x+10}{4} - 1$

32. $\dfrac{5y-9}{8} - \dfrac{3y-12}{4} = 0$

3–4 REPEATING DECIMALS

In Chapter 1, you saw how every rational number could be represented by either a terminating decimal or an infinite repeating decimal. In this section you will see that every terminating decimal and every infinite repeating decimal represents a rational number.

A convenient way of indicating repeating decimals is to underline the shortest repeating block of digits. Thus, you write

$$
\begin{aligned}
.\underline{3} &\quad \text{to indicate} \quad .3333\ldots, \\
.\underline{21} &\quad \text{to indicate} \quad .21212121\ldots, \\
2.1\underline{6} &\quad \text{to indicate} \quad 2.16666\ldots, \\
-1.3\underline{71} &\quad \text{to indicate} \quad -1.3717171\ldots.
\end{aligned}
$$

Note that only the underlined digits are repeated. From now on, this notation will be used to indicate repeating decimals.

Every repeating decimal can be expressed as a quotient of two integers as shown below.

Problem 1. Find the rational number represented by the repeating decimal $.5\underline{4}$.

Solution. If N denotes the rational number so represented,

$$N = .54444\ldots, \quad \text{or } .5\underline{4}.$$

Then,

$$10N = 5.44444\ldots, \quad 5.\underline{4},$$

and

$$100N = 54.44444\ldots, \quad \text{or } 54.\underline{4}.$$

Since $100N$ and $10N$ have the same fractional part, their difference is an integer.

$$
\begin{array}{rrl}
100N = & 54.44444\ldots, & \text{or } 54.\underline{4} \\
-\quad 10N = & 5.44444\ldots, & \text{or } \;\; 5.\underline{4} \\
\hline
90N = & 49.00000\ldots, & \text{or } 49
\end{array}
$$

You can solve this equation for N as follows:

$$
\begin{aligned}
90N &= 49, \\
N &= \frac{49}{90}.
\end{aligned}
$$

Therefore, $\frac{49}{90}$ is the rational number having the repeating decimal $.5\underline{4}$.

Problem 2. Find the rational number represented by the repeating decimal $.4\underline{36}$.

Solution. If you let N denote the number

$$N = .436363636\ldots,$$

then

$$
\begin{aligned}
10N &= 4.36363636\ldots, \\
100N &= 43.6363636\ldots, \\
1000N &= 436.36363636\ldots.
\end{aligned}
$$

Now subtract $10N$ from $1000N$. You know that $1000N - 10N$ is an integer, because $1000N$ and $10N$ have the same fractional part.

$$
\begin{array}{rrl}
1000N = & 436.363636\ldots, & \text{or } 436.\underline{36} \\
-\quad 10N = & 4.363636\ldots, & \text{or } \;\; 4.\underline{36} \\
\hline
990N = & 432.000000\ldots, & \text{or } 432
\end{array}
$$

You solve $990N = 432$ for N as follows:

$$
\begin{aligned}
990N &= 432 \\
N &= \tfrac{432}{990}, \quad \text{or } \tfrac{24}{55}.
\end{aligned}
$$

Hence, $\frac{24}{55}$ is the rational number having the repeating decimal $.4\underline{36}$.

Exercises

Find the rational numbers represented by the repeating decimals.

1. (a) $.1111\ldots$, or $.\underline{1}$ (b) $.2222\ldots$, or $.\underline{2}$

2. (a) $.3333\ldots$, or $.\underline{3}$ (b) $.4444\ldots$, or $.\underline{4}$

3. (a) .5555 . . . , or $.\underline{5}$ (b) .6666 . . . , or $.\underline{6}$
4. (a) .7777 . . . , or $.\underline{7}$ (b) .8888 . . . , or $.\underline{8}$
5. (a) .9999 . . . , or $.\underline{9}$ (b) .49999 . . . , or $.4\underline{9}$
6. (a) 1.6666 . . . , or $1.\underline{6}$ (b) 1.8888 . . . , or $1.\underline{8}$
7. (a) .96666 . . . , or $.9\underline{6}$ (b) .17777 . . . , or $.1\underline{7}$
8. (a) 5.121212 . . . , or $5.\underline{12}$ (b) 6.424242 . . . , or $6.\underline{42}$
9. (a) .2565656 . . . , or $.2\underline{56}$ (b) .3494949 . . . , or $.3\underline{49}$
10. (a) .213213 . . . , or $.\underline{213}$ (b) .314314 . . . , or $.\underline{314}$

Find the rational numbers represented by the repeating decimals.

11. $1.\underline{2}$
12. $.6\underline{8}$
13. $.36\underline{9}$
14. $.\underline{21}$
15. $.2\underline{10}$
16. $.3\underline{78}$
17. $3.\underline{41}$
18. $17.\underline{13}$
19. $.\underline{100}$
20. $1.\underline{123}$
21. $7.\underline{892}$
22. $.\underline{601}$
23. $10.\underline{003}$
24. $.4\underline{321}$

Find the rational numbers represented by the repeating decimals.

25. $.\underline{12345}$
26. $.\underline{54321}$
27. $9.\underline{1122}$
28. $-.\underline{428571}$
29. Compare the denominators of the rational numbers represented by the following repeating decimals.
    (a) $.\underline{4}$ (b) $.\underline{5}$ (c) $.\underline{6}$ (d) $.\underline{7}$

Review for Sections 3–1 through 3–4

Find the solution set of each equation in Exercises 1–30. Check each solution.

1. $x + 14 = 20$
2. $3m = 15$
3. $24 - 7 = k$
4. $2x + 6 = 18$
5. $6r = 24$
6. $5m + 3 = 28$
7. $k + 5 = 2k - 5$
8. $m - 6 = -6$
9. $4t = -16$
10. $3x + 4 + 2x = 54$
11. $\frac{2}{3}y = 6$
12. $2p + 6 = 3p - 6$
13. $x + 7 = 1$
14. $5k = 0$
15. $4r + 3 = 7r - 3$
16. $\frac{1}{2}m + 3 = 7$
17. $5t + 3 = 7t + 3$
18. $\frac{3}{4}k - \frac{1}{2} = 6$
19. $8m + 4 = 6m + 12$
20. $\frac{1}{3}x + \frac{1}{2} = 2$

21. $m - 5 = -2$ **22.** $\frac{3}{5}x + \frac{1}{2}x = 11$

23. $k + 5 = 10 - 7$ **24.** $\frac{1}{4}m + \frac{2}{3} = \frac{7}{12}$

25. $7r + 2 + 3r = 8r + 8$ **26.** $y + 3 = 2$

27. $\frac{3}{4}k - \frac{2}{3} = \frac{1}{2}$ **28.** $p - 4 = 2 + (-6)$

29. $\frac{5}{6}x + \frac{2}{3}x = 15$ **30.** $2m + 3 = 4m - 9$

Find the rational numbers represented by each of the following repeating decimals.

31. .3131 . . . , or $.\underline{31}$ **32.** .4666 . . . , or $.4\underline{6}$

33. .7222 . . . , or $.7\underline{2}$ **34.** .5353 . . . , or $.\underline{53}$

35. .6464 . . . , or $.\underline{64}$ **36.** .2555 . . . , or $.2\underline{5}$

37. .9777 . . . , or $.9\underline{7}$ **38.** .8989 . . . , or $.\underline{89}$

39. .531531 . . . , or $.\underline{531}$ **40.** .65454 . . . , or $.6\underline{54}$

Answers to Review for Sections 3–1 through 3–4

1. $\{6\}$	**2.** $\{5\}$	**3.** $\{17\}$	**4.** $\{6\}$
5. $\{4\}$	**6.** $\{5\}$	**7.** $\{10\}$	**8.** $\{0\}$
9. $\{-4\}$	**10.** $\{10\}$	**11.** $\{9\}$	**12.** $\{12\}$
13. $\{-6\}$	**14.** $\{0\}$	**15.** $\{2\}$	**16.** $\{8\}$
17. $\{0\}$	**18.** $\{\frac{26}{3}\}$	**19.** $\{4\}$	**20.** $\{\frac{9}{2}\}$
21. $\{3\}$	**22.** $\{10\}$	**23.** $\{-2\}$	**24.** $\{-\frac{1}{3}\}$
25. $\{3\}$	**26.** $\{-1\}$	**27.** $\{\frac{14}{9}\}$	**28.** $\{0\}$
29. $\{10\}$	**30.** $\{6\}$	**31.** $\frac{31}{99}$	**32.** $\frac{42}{90}$ or $\frac{7}{15}$
33. $\frac{65}{90}$ or $\frac{13}{18}$	**34.** $\frac{53}{99}$	**35.** $\frac{64}{99}$	**36.** $\frac{23}{90}$
37. $\frac{88}{90}$ or $\frac{44}{45}$	**38.** $\frac{89}{99}$	**39.** $\frac{531}{999}$ or $\frac{59}{111}$	**40.** $\frac{648}{990}$ or $\frac{36}{55}$

3–5 FORMULAS

Formulas are equations which have arisen from mathematical applications. They express known facts about one variable in terms of other variables and, thus, help you solve particular problems. Some examples are given below.

Problem 1. If the distance from New York to Southampton is 2800 nautical miles and the Queen Elizabeth II makes the trip in 80 hours, what is her average speed?

Solution. The distance, d, speed, r, and time, t, are related by the formula

$$d = rt.$$

The problem tells you that $d = 2800$ and $t = 80$. Hence,

$$2800 = 80r,$$

and

$$\tfrac{2800}{80} = r, \quad \text{or } r = 35.$$

Thus, the average speed of the Queen Elizabeth II is 35 nautical miles per hour, or 35 knots.

Problem 2. If you dropped a ball from a fourth-story window, and it took 2 seconds to hit the ground, how high is the window?

Solution. The formula $d = 16t^2$ relates the distance, d (in feet), that an object will fall in t seconds. You know that $t = 2$.

$$\begin{aligned} d &= 16t^2 \\ d &= 16 \cdot (2^2) \\ &= 16 \cdot 4 \\ &= 64. \end{aligned}$$

The window is 64 feet from the ground.

Problem 3. A quadrilateral with two and only two parallel sides is called a *trapezoid*. (See Fig. 3–1.) If the parallel sides have lengths a and b and if the altitude has length h, the formula for the area of a trapezoid is

$$A = \tfrac{1}{2}(a + b)h.$$

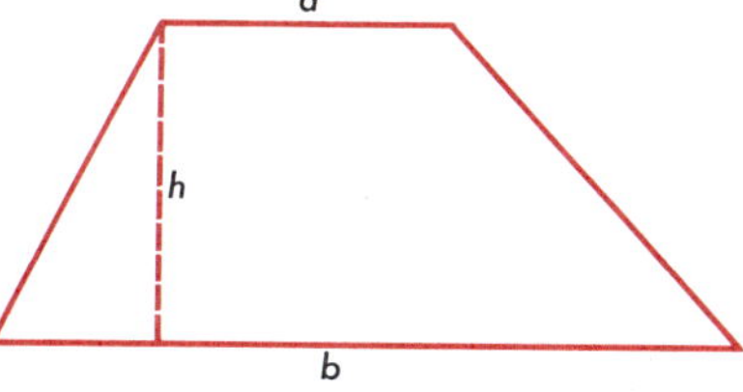

FIGURE 3–1

Find the altitude of a trapezoid with parallel sides 5 and 10 inches long and an area of 45 square inches.

Solution.

$$\begin{aligned} A &= \tfrac{1}{2}(a + b)h \\ 45 &= \tfrac{1}{2}(5 + 10)h \\ 45 &= \tfrac{15}{2}h \\ \tfrac{2}{15} \cdot 45 &= \tfrac{2}{15} \cdot \tfrac{15}{2}h \\ 6 &= h \end{aligned}$$

Thus, the altitude is 6 inches.

Exercises

Use the values which are given, and find the value of the remaining variable in each of the following formulas.

1. $A = P + Pr$
   (a) $A = 1040, P = 1000$ (b) $A = 800, P = 600$
2. $E = mc^2$
   (a) $m = 5, c^2 = 3 \times 10^{10}$ (b) $m = 2, c^2 = 3 \times 10^{10}$
3. $S = \dfrac{W}{W - w}$
   (a) $W = 69, w = 39$ (b) $S = 1.5, W = 120$
4. $T = a + (n - 1) \cdot d$
   (a) $T = 51, a = 3, n = 13$ (b) $T = 38, a = 5, d = 3$
5. $S = \dfrac{a}{1 - r}$
   (a) $S = 6, r = \frac{1}{3}$ (b) $S = 50, r = -\frac{1}{2}$
6. $T = ar^3$
   (a) $T = 375, r = 3$ (b) $T = 1008, r = 2$
7. $x^2 + y = z^2$
   (a) $x = 5, z = 13$ (b) $x = 4, z = -3$
8. The perimeter of a square is given by the formula $p = 4s$. Find:
   (a) the perimeter when $s = 9$ in.
   (b) the length of a side of the square when $p = 20$ in.
9. The perimeter of a rectangle is given by the formula $p = 2l + 2w$ where l and w denote the measures of the length and width respectively. Find:
   (a) the length of the rectangle if the perimeter is 50 in. and the width is 10 in.
   (b) the width of the rectangle if the perimeter is 80 in. and the length is 30 in.
10. The volume of a cube is given by the formula $V = e^3$. Find:
    (a) the volume of the cube if $e = 6$ in.
    (b) the length of one edge (e) if $V = 512$ cu in.
11. (a) How far is Town City from Central City if it takes 3 hours to travel the distance at an average speed of 50 miles per hour?
    (b) How long will it take to travel a distance of 400 miles at an average speed of 60 miles per hour?
12. (a) Find the area of a rectangle with length 12 inches and width 5 inches. (Hint: $A = l \cdot w$).
    (b) Find the width of a rectangle if its area is 100 sq in. and its length is 25 in.

13. Solve each of the following formulas for the specified variable, deriving a new formula for one variable in terms of the others.

(a) $L = 2\pi r + 3s$. Solve for r. (b) $3x + 2y = 12$. Solve for y.

(c) $I = \frac{C}{D^2}$. Solve for C. (d) $R = \frac{kl}{d^2}$. Solve for l.

14. The so-called normal weight of an adult may be found by using the formula $2W = 11(h - 40)$. Here, W represents the weight in pounds, and h represents the height in inches.

(a) What is the normal weight for Miss Brown if she is 5 feet 8 inches tall?

(b) Mr. O'Hara's normal weight is 176 pounds. How tall is he?

---

15. Physicists say that if an object is dropped from a point above the earth, the number of feet it falls will be 16 times the square of the number of seconds it is falling. Formulate this sentence mathematically and use it to answer these questions.

(a) If a boy drops a stone off a bridge and it takes 2 seconds to reach the water, how high above the water is he?

(b) Answer the question in (a) for the following time: 5 seconds.

(c) If an object is dropped from a balloon 400 feet in the air, how long will it take the object to reach the earth?

(d) Answer the question in (c) for the following height: 256 feet.

16. Salesmen for a certain company earn a salary of \$70 weekly and, in addition, a commission that is 5% of their sales for that week. Express the weekly earnings of each salesman by a formula.

(a) What is Mr. Clark's income in a 3-week period during which his weekly sales were as follows: \$1300, \$2500, \$0?

(b) How much would Mr. Clark have to sell in a three-week period in order to earn \$600?

In Exercises 17–21, write a relevant formula and use it to find the required information.

17. One base of a trapezoid is 3 feet 4 inches long, and the other base is 5 feet long. If the area is 30 square feet, what is the length of the altitude of this trapezoid?

18. The perimeter of a rectangular lot is 240 feet. If the width of the lot is 49 feet, find its length.

19. How much money will you have in the bank after one year if you deposit \$2300 in a bank paying $3\frac{3}{4}\%$ interest per year?

20. Find the total surface area of a cube if each edge is 5 inches long.

21. The cost of riding in a taxi is 35¢ for the first quarter of a mile and 10¢ for every quarter of a mile thereafter. Each trunk costs 25¢ extra.

(a) What is your fare to the railroad station seven miles away if you have two trunks?

(b) If your fare is $2.65 and you have no baggage, how far have you ridden?

Preparation for Section 3–6

Solve each of the following equations.

1. $3x + 4 = 19$

2. $5x - 12 = 13$

3. $3x - 7 = 4(x - 5)$

4. $5x + 6 = 3(x + 4)$

5. $x + 2(3x) + 5(2x + 1) = 22$

6. $(x + 7) + 4 = 4(x + 1)$

7. $.2x = .1(400 - x)$

8. $.34x + 1.66(100 - x) = 34$

3–6 WORD PROBLEMS

If the variables appearing in a problem are related by a formula, you can solve the problem using the methods shown in Section 3–5. In some problems, however, the variables are not related by a given formula; they are related only in the way stated in the problem. Then you must develop your own formula for expressing this relationship.

Problem 1. Sue's mother is 35 years old. Three years ago, she was four times as old as Sue was then. How old is Sue now?

Solution.

Denote Sue's age now by	S.
Three years ago, Sue's age was	$S - 3$.
Four times Sue's age three years ago is	$4(S - 3)$.
Mother's age three years ago was	$35 - 3$, or 32.

We have been told that her mother's age three years ago was four times Sue's age three years ago. Therefore,

$$\begin{aligned} 4(S - 3) &= 32, \\ \tfrac{1}{4} \cdot 4(S - 3) &= \tfrac{1}{4} \cdot 32, \\ S - 3 &= 8, \\ S &= 8 + 3, \quad \text{or } 11. \end{aligned}$$

Thus, Sue is now 11 years old.

Check. Three years ago, Sue was 8 and her mother was 32, and

$$32 \stackrel{\checkmark}{=} 4 \cdot 8.$$

Problem 2. Angela was paid \$1264 for working 152 hours last month. She gets \$8 an hour for regular hours and \$12 an hour for overtime. How many regular hours did she work last month?

Solution.

Denote the number of regular hours by	H.
Then the number of overtime hours was	$152 - H$.
Pay for regular hours was	$8H$.
Pay for overtime hours was	$12(152 - H)$.

$$\underbrace{8H}_{\text{Pay for regular hours}} \underbrace{+}_{\text{plus}} \underbrace{12(152 - H)}_{\text{pay for overtime}} \underbrace{=}_{\text{equals}} \underbrace{1264}_{\text{total pay.}}$$

$$\begin{aligned} 8H + 1824 - 12H &= 1264 \\ 1824 - 4H &= 1264 \\ 1824 - 1264 &= 4H \\ 560 &= 4H \\ 140 &= H \end{aligned}$$

Thus, she worked 140 regular hours last month.

Problem 3. Coffee worth 75¢ per pound is mixed with coffee worth 93¢ per pound to produce 10 pounds of a mixture worth 85¢ per pound. How much of each type is used?

Solution.

The number of pounds of the 75¢ coffee is denoted by	x.
Then the number of pounds of the 93¢ coffee is	$10 - x$.
The value in dollars of the 75¢ coffee is	$.75x$.
The value in dollars of the 93¢ coffee is	$.93(10 - x)$.
The value in dollars of the mixture is	$.85 \times 10$.

$$\underbrace{.75x}_{\text{value of 75¢ coffee}} + \underbrace{.93(10 - x)}_{\text{value of 93¢ coffee}} = \underbrace{.85 \times 10}_{\text{value of mixture}}$$

You can solve this equation as follows:

$$\begin{aligned} .75x + 9.3 - .93x &= 8.5, \\ -.18x &= -.8, \\ x &= 4\tfrac{4}{9} \text{ and } 10 - x = 5\tfrac{5}{9}. \end{aligned}$$

Thus, $4\frac{4}{9}$ pounds of the 75¢ coffee and $5\frac{5}{9}$ pounds of the 93¢ coffee are used.

Exercises

An algebraic expression such as $3(x + 1)$ can be interpreted in many ways, depending on the domain of the variable x. If you think of x as an integer, $3(x + 1)$ denotes three times the next integer after x. If you think of x as the number of inches in each side of an equilateral triangle, then $3(x + 1)$ is the number of inches in the perimeter of a second equilateral triangle having each side one inch longer than the side of the first equilateral triangle. If you think of x as Leann's age in years now, then $3(x + 1)$ is three times Leann's age next year.

Give your own interpretation of each of the following algebraic expressions. First state clearly what you intend the variable to denote.

1. (a) $3x$ (b) $3(x - 1)$

2. (a) $x + 2$ (b) $x - 5$

Let x stand for Linda's age in years now, and write an algebraic expression for each of the following phrases.

3. (a) Linda's age three years ago
(b) Her age five years from now

4. (a) Twice her age last year
(b) Three times her age next year

5. (a) Two times her age five years ago
(b) Seven years less than six times her age

6. An open equation may be interpreted in many ways. For example, if we think of x as denoting Barry's present age in years, the equation

$$3x - 5 = 2(x + 3)$$

states that five less than three times Barry's present age is equal to twice Barry's age three years from now.

Give your own interpretation of the following open equations.

(a) $2x + 5 = 39$ (b) $3x - 7 = 89$

Write an algebraic expression for each of the phrases.

7. (a) The total cost of 5 records if some cost \$3.98 each and the rest cost \$4.95 each
(b) The cost of 15 stamps if some cost 5¢ each and the rest 8¢ each

8. (a) The total receipts from tickets sold for a play if main-floor seats were \$2.00 per ticket, balcony seats were \$1.50 per ticket, and 500 people bought tickets
(b) The cost of 8 pounds of candy if it contains bridge mix at 69¢ a pound, twice as much fudge as bridge mix at 89¢ a pound, and the remainder is caramels at 79¢ a pound

For each exercise, write an open equation and solve it. Do not forget to give the meaning of the variable you use.

9. (a) Dick is 18 years older than Harry. If the sum of their ages is 46 years, how old is each?

(b) Joan's age is three times Kathy's age. The sum of their ages is 24 years. Find the present age of each girl.

10. (a) In six years, Mary will be four times as old as she is now. How old is Mary?

(b) Steve is now seven times as old as Bob. In six years, Steve will be five times as old as Bob. Find their present ages.

11. (a) Pat is twice as old as Andrea. Eight years ago she was ten times as old as Andrea. How old is each girl now?

(b) Richard is twice as old as his brother. Four years ago he was four times as old as his brother. Find their present ages.

12. (a) I have stamps worth 87¢. If I have twice as many 3¢ stamps as 1¢ stamps and three more 4¢ stamps than 3¢ stamps, how many of each kind do I have?

(b) Linda has $1.85 in quarters, nickels, and dimes. If she has three times as many dimes as quarters, and five fewer nickels than dimes, how many of each coin does she have?

13. (a) Mr. Jones invests $2500, part of it in a bank paying $4\frac{1}{2}\%$ interest annually and the remainder in a loan association paying 5% interest annually. If his total income for one year is $119, how much has he invested at each rate?

(b) How can $6300 be invested, one part at 4% and the remainder at 5%, so that the income will be the same on each investment?

For each exercise, write an open equation and solve it. Do not forget to give the meaning of the variable you use.

14. Mary is six years older than Sue. In two years Mary will be twice as old as Sue. Find their ages.

15. Johnny has four more nickels than pennies. If he were to spend three of his nickels and one of his pennies, he would have 70¢ left. How many nickels and how many pennies does he have?

16. Pat has $5.77 in her coin bank. If the number of quarters is three more than the number of pennies, the number of dimes is two less than the number of pennies and the number of nickels is one and one-half times the number of pennies, find how many coins of each kind Pat has.

17. Tea worth 85¢ a pound is mixed with tea worth $1.10 a pound to produce 20 pounds of a mixture worth $1.00 a pound. How much of each type is used?

18. The total receipts for 900 tickets were \$950. If students paid 75¢ per ticket and nonstudents paid \$1.25 per ticket, how many student tickets were sold?

19. One solution is 50% sulfuric acid, while a second is 75% sulfuric acid. How many liters of each should be used to make 10 liters of a solution which is 60% sulfuric acid?

Preparation for Section 3–7

1. Write algebraic expressions for the following phrases.

(a) The distance a person can drive in 5 hours if he averages x miles per hour

(b) The distance driven in x hours at an average speed of 40 miles per hour

(c) The time in hours required for a trip of 500 miles at an average speed of x miles per hour

(d) The time in seconds that someone runs x yards at a speed of 30 feet per second

(e) The average speed maintained if a person drives x miles in 7 hours

(f) The average speed a person must maintain to travel 525 miles in x hours

Write algebraic expressions for the phrases.

2. The sum of three consecutive integers if the middle one is denoted by $2n$

3. The next larger even integer if $x + 7$ represents an even integer

4. The product of two consecutive integers if the smaller one is x

3–7 FLOW CHART FOR SOLVING WORD PROBLEMS

A flow chart for solving word problems might be illustrated as in Fig. 3–2.

Problem 1. The sum of three consecutive integers is 69. Find the three integers.

Solution. If you know one of the integers, then you can easily find the other two.

Let N denote the smallest of the three integers.

Then the three integers are N, $N + 1$, and $N + 2$.

The problem is expressed by the equation

$$N + (N + 1) + (N + 2) = 69.$$

START
Read the problem
Identify what is to be found
Denote the unknown number by a variable
Write algebraic expressions for the facts given
Can you write an equation relating the algebraic expressions ?
No
Yes
Reread the problem
Look for a different unknown
Write and solve the equation
Solve the problem
Does your solution check?
No
Yes
STOP

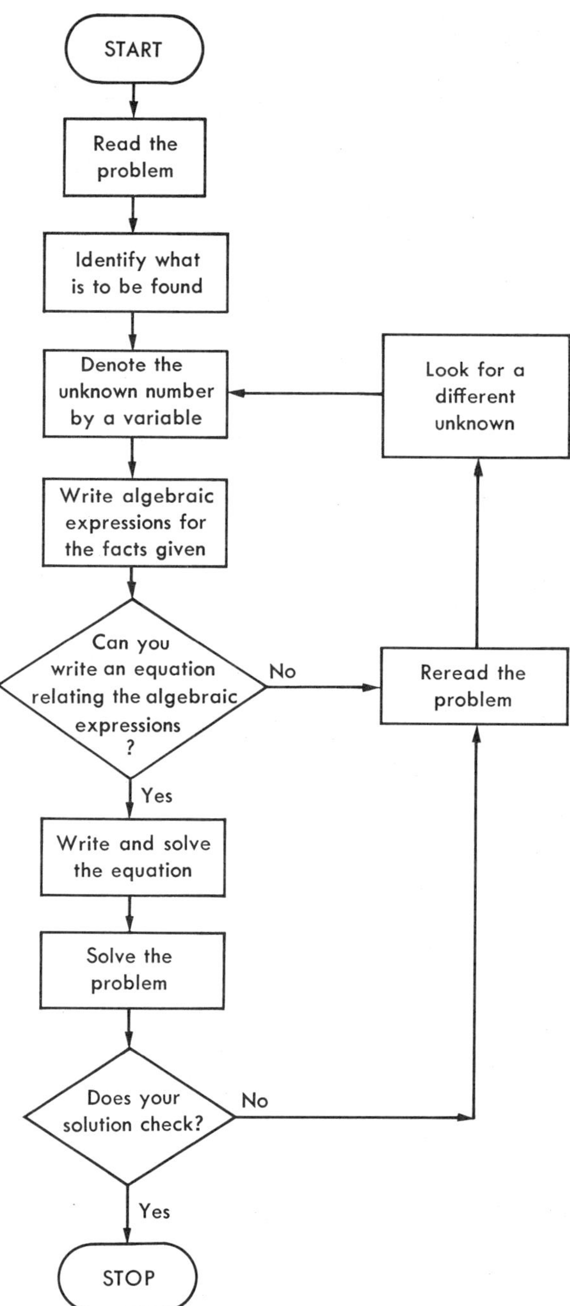

FIGURE 3-2

You can solve this equation as follows:

$$\begin{aligned}(N + N + N) + (1 + 2) &= 69,\\ 3N + 3 &= 69,\\ 3N &= 66,\\ N &= 22,\ N + 1 = 23,\ N + 2 = 24.\end{aligned}$$

The three integers are 22, 23, 24.

Check. $22 + 23 + 24 \stackrel{\checkmark}{=} 69$

(You could have solved this problem in an alternate way by observing that the first integer is one less and the third is one more than the middle integer. Therefore, the sum of the three integers is three times the middle one. Hence, the middle integer is one-third of 69, or 23.)

Problem 2. Mr. Chiu leaves Stratford at 8 A.M. and drives north on the highway at an average speed of 48 miles per hour. Ms. Garcia leaves Stratford at 8:30 A.M. and drives north on the same highway at an average speed of 56 miles per hour. At what time will Ms. Garcia overtake Mr. Chiu?

Solution. You want the time at which Ms. Garcia overtakes Mr. Chiu.

Let h denote the time at which Ms. Garcia overtakes Mr. Chiu. It is difficult to express the problem by an equation involving h. So reread the problem. If you know the number of hours Ms. Garcia travels before she overtakes Mr. Chiu, then you can find the *time* at which Ms. Garcia overtakes Mr. Chiu.

Let t denote the number of hours Ms. Garcia travels before overtaking Mr. Chiu.

Then Mr. Chiu travels $t + \frac{1}{2}$ hours before being overtaken by Ms. Garcia.

Ms. Garcia travels $56 \cdot t$ miles.

Mr. Chiu travels $48 \cdot (t + \frac{1}{2})$ miles.

However, each travels the same distance:

$$56t = 48(t + \tfrac{1}{2}).$$

You can solve this equation as follows:

$$\begin{aligned}56t &= 48t + 48 \cdot \tfrac{1}{2},\\ 56t - 48t &= 24,\\ 8t &= 24,\\ t &= 3.\end{aligned}$$

The time at which Ms. Garcia overtakes Mr. Chiu is 3 hours after Ms. Garcia leaves, or 11:30 A.M.

Check. At 11:30 A.M. Mr. Chiu has traveled for $3\frac{1}{2}$ hours at 48 mph, and therefore has gone a distance of $\frac{7}{2} \cdot 48$, or 168, miles. By 11:30 A.M., Ms. Garcia has traveled for 3 hours at 56 mph. Thus, she has gone a distance of $3 \cdot 56$, or 168, miles also.

Problem 3. Eight girls want to rent a gym for basketball practice. Each girl will pay one-eighth of the monthly rent. If they could find two more girls to share the cost of the monthly rent, each of them would pay $2.50 less. What is the monthly rent?

Solution. You want to find the monthly rent. Let R denote the monthly rent, in dollars. Then $R/8$ is the cost to each girl if eight share equally, and $R/10$ is the cost to each girl if ten share equally.

$$\underbrace{\text{The cost for each of ten girls}}_{\frac{R}{10}} \quad \underbrace{\text{equals}}_{=} \quad \underbrace{\text{\$2.50 less than the cost for each of eight girls.}}_{\frac{R}{8} - 2.50}$$

You can solve this equation as follows.

$$\begin{aligned} 2.50 &= \frac{R}{8} - \frac{R}{10} \\ 2.50 &= (\tfrac{1}{8} - \tfrac{1}{10})R \\ 2.50 &= \tfrac{1}{40}R \\ 40 \times 2.50 &= 40 \times \tfrac{1}{40}R \\ 100 &= R \end{aligned}$$

Therefore, the monthly rent is $100.

Check. With eight girls sharing, each pays $\frac{100}{8}$, or $12.50. With ten girls sharing, each pays $\frac{100}{10}$, or $10. The difference between $12.50 and and $10 is $2.50.

Exercises

In Exercises 1–3, write algebraic expressions for the phrases given.

1. (a) The distance between a car and a train x hours after the car, which averaged 55 miles per hour, passed the train, which averaged 48 miles per hour (Assume the car and train are traveling in the same direction.)

(b) The distance between the car and the train in Exercise 1(a), assuming that they were traveling in opposite directions

2. (a) The distance traveled in 9 hours if an average speed of 42 miles per hour was maintained for x hours and then slowed to 36 miles per hour
   (b) The distance traveled by Mrs. Smith if she drove for x hours on a tollway at 60 miles per hour and 3 hours on a side road at 40 miles per hour
3. (a) The sum of three integers if x denotes the first one and it is known that the second one is four more than one-half the first one and the third one is three times the second one
   (b) The sum of three integers related in the same way as those described in Exercise 3(a) with x denoting the second of the three

Find solutions for the following problems.

4. (a) The sum of three consecutive even integers is 942. What are the integers?
   (b) The sum of three consecutive odd integers is 1503. What are the integers?
5. (a) Two cars start from the same town at the same time and travel in opposite directions. One car averages 50 miles an hour and the other 60 miles an hour. In how many hours will they be 550 miles apart?
   (b) A westbound jet leaves an airport traveling 600 miles an hour. At the same time, an eastbound plane departs at 350 miles an hour. In how many hours will the planes be 1900 miles apart?
6. (a) A train left a town at 8 A.M. traveling 50 miles per hour. At noon a plane also left the same town and traveled in the same direction at 300 miles per hour. At what time did the plane overtake the train?
   (b) A freight train left a station at 1 P.M. traveling 30 miles per hour. At 7 P.M. an express train traveling 60 miles per hour left the same station. When did the express overtake the freight?
7. (a) Find three consecutive integers such that twice the smallest is 57 less than three times the largest.
   (b) Find three consecutive integers such that four times the largest is 22 more than three times the second one.
8. (a) The sum of two numbers is 25. If $\frac{2}{3}$ of the smaller one is two more than $\frac{1}{4}$ of the larger one, find the two numbers.
   (b) The sum of two numbers is 31. If $\frac{1}{5}$ of the smaller one is five less than $\frac{1}{2}$ of the larger one, find the two numbers.
9. (a) Find three consecutive odd integers such that the sum of the last two is 85 more than the first.
   (b) Find three consecutive even integers such that the sum of the first two is 90 more than the last.
10. (a) John took a trip of 1910 miles. He traveled by train at 55 miles an hour and three times that number of hours by plane at 300 miles an hour. How many hours did the trip take?

(b) A man crossed a lake 18 miles wide in 5 hours. He set out in a motorboat at 6 miles an hour, and when the motor broke down, he rowed the remaining distance at 2 miles an hour. How far did he row?

---

11. Are there three consecutive integers whose sum is 100? Explain your answer.

12. The perimeter of a triangle is 488 centimeters. The length of the longest side is four times that of the shortest side, and the length of the third side is 15 centimeters less than that of the longest side. Find the length of each side.

13. The length of a rectangle is 1.2 meters more than the width. If the length were decreased by 1.5 meters, and the width were increased by .6 meter, the perimeter would be 6.2 meters. Find the length and width of the original rectangle.

14. The length of a rectangle is three times its width. If the length is decreased by 4 inches and the width is doubled, the perimeter remains the same. What were the original width and length?

15. John sets out from camp at 9:00 A.M. and walks 4 miles an hour. Jim leaves camp at 9:30 A.M. and walks 5 miles an hour. At what time will Jim overtake John? How far will each have walked?

16. A freight train averaging 45 miles per hour and an express train averaging 60 miles per hour leave the station at the same time. How long will it take the express train to get 100 miles ahead of the freight train?

17. Two cars traveling in opposite directions pass each other. One averages 45 miles per hour, the other 30 miles per hour. How many hours later will they be 100 miles apart?

18. A man drives out into the country at an average rate of 40 miles an hour. In the evening traffic, he drives back at 30 miles an hour. If the total traveling time is 7 hours, what is the total distance he traveled?

19. The average speed of a train is 20 miles per hour faster than that of a car. In 8 hours and 40 minutes the train covers the same distance that the car covers in 13 hours. Find the average speed of each.

---

20. (a) A man leaves Alexandria, driving toward a resort at 40 miles per hour. At the same time, a second man leaves the same resort, driving toward Alexandria at 50 miles per hour. If the distance from Alexandria to the resort is 185 miles, how long will it be before they meet? (Assume that the men are on the same highway.)

(b) If the man at the resort leaves at 9:00 A.M. and the man at Alexandria leaves at 9:45 A.M., at what time will they meet?

21. Mr. Smith leaves Andover at noon, traveling west at 50 miles per hour. His neighbor leaves at 1 P.M. with the brief case that Mr. Smith forgot. How fast must the neighbor drive if he is to catch up with Mr. Smith by 5 P.M.?

22. The ratio of the speed v of an object to the speed V of sound in the surrounding air is called the *Mach number*. Thus, if the ballistic missile *Corporal* travels at a speed greater than Mach 3, it is traveling more than three times as fast as sound. Suppose that a guided missile, traveling at Mach 2, is shot out to sea. Two minutes later an antimissile, averaging Mach 5, is shot along the same path toward the first one. In how many minutes will the two missiles collide?

23. James has \$40 more than Charles. He gives Charles \$10 and then finds that he has $\frac{4}{3}$ as much money as Charles. How much does Charles then have? How much does James have?

*3–8 PROOF IN ALGEBRA II

This section is a continuation of a study of proof in algebra started in Chapter 2 (Section 2–14).

Problem 1. Prove that the equation

$$x \cdot (-y) = -(x \cdot y)$$

is true for all numbers x and y.

Solution. To prove this you must use the additive inverse axiom, (Inv-A), which says

$$y + (-y) = 0.$$

Then the equations

$$x \cdot [y + (-y)] = x \cdot 0 \qquad \text{(Mult-A)}$$

and

$$x \cdot y + x \cdot (-y) = 0 \qquad \text{(D), (Zero-M)}$$

are true for all numbers x and y. Again using the additive inverse axiom, (Inv-A),

$$0 = x \cdot y + [-(x \cdot y)].$$

Hence, the equations,

$$x \cdot y + x \cdot (-y) = x \cdot y + -(x \cdot y)$$

and

$$x \cdot (-y) = -(x \cdot y) \qquad \text{(Can-A)}$$

are true for all numbers x and y. This completes the proof.

Problem 2. Prove the following implication:

$$\text{If } x \cdot y = 0 \quad \text{and} \quad x \neq 0, \quad \text{then } y = 0.$$

Solution. If the statement

$$x \cdot y = 0 \quad \text{and} \quad x \neq 0$$

is true, then each of the statements below is also true:

$$\frac{1}{x} \cdot (x \cdot y) = \frac{1}{x} \cdot 0 \qquad \text{(Mult-A)}$$

$$\left(\frac{1}{x} \cdot x\right) \cdot y = \frac{1}{x} \cdot 0 \qquad \text{(A-M)}$$

$$1 \cdot y = 0 \qquad \text{(Inv-M), (Zero-M)}$$

$$y = 0. \qquad \text{(Id-M)}$$

Exercises

1. In the following proof that multiplication is distributive with respect to subtraction, give the reason why equation (1) is true, why (1) implies (2), and so on.

   (a) $x \cdot (y - z) = x[y + (-z)]$ (1)
   $x \cdot (y - z) = (x \cdot y) + [x \cdot (-z)]$ (2)
   $x \cdot (y - z) = (x \cdot y) + [-(x \cdot z)]$ (3)
   $x \cdot (y - z) = (x \cdot y) - (x \cdot z)$ (4)

   (b) Is $x \cdot (y - z) = (y - z) \cdot x$ true for all numbers x, y, and z? Why?

2. Provide the reasons for each step in the proof of the following theorem:

   Every real number x has *exactly one* additive inverse, namely $-x$.

   Proof: Assume that x has another additive inverse, say n, such that $x + n = 0$. Then

   $-x + (x + n) = -x + 0$ __?__
   $(-x + x) + n = -x$ __?__
   $0 + n = -x$ __?__
   $n = -x$ __?__

   Hence, since n is only another name for $-x$, every real number x has exactly one additive inverse.

3. Prove that every nonzero real number x has exactly one multiplicative inverse, namely $\frac{1}{x}$.

4. Supply the reason for each step of the proof that

$$\frac{x}{y} + \frac{r}{s} = \frac{xs + ry}{ys}$$

Proof: $\frac{x}{y} + \frac{r}{s} = \frac{xs}{ys} + \frac{ry}{sy}$ —?—

$= \frac{xs}{ys} + \frac{ry}{ys}$ —?—

$= (xs)\left(\frac{1}{ys}\right) + (ry)\left(\frac{1}{ys}\right)$ —?—

$= (xs + ry)\left(\frac{1}{ys}\right)$ —?—

$= \frac{xs + ry}{ys}$ —?—

5. Prove that $\frac{x}{y} - \frac{r}{s} = \frac{xs - ry}{ys}$ for all real numbers x and r and all nonzero real numbers y and s.

6. Prove that

$$\frac{\frac{x}{y}}{\frac{r}{s}} = \frac{x}{y} \cdot \frac{s}{r}$$

for all real numbers x and all nonzero real number y, r, and s.

7. Prove that $x \cdot (-1) = -x$.

8. Prove that if $x = 0$ or $y = 0$, then $xy = 0$.

9. Prove that $-0 = 0$.

10. Prove that if $\frac{a}{b} = c + \frac{d}{b}$, then $a = bc + d$.

11. Prove $a \cdot 0 = 0$ for every number a, by starting with the true equation $a(1 + 0) = a$.

12. (a) Prove $x - (y - z) = (x - y) + z$ for all numbers x, y, and z.
(b) Prove

$$\frac{x}{\frac{y}{z}} = \frac{x}{y} \cdot z$$

for all nonzero numbers x, y, and z.

13. (a) Prove $\frac{0}{x} = 0$ for all nonzero numbers x.

(b) Prove $(a - b) + (c - d) = (a + c) - (b + d)$ for all numbers a, b, c, and d.

14. (a) Prove $-(-x) = x$ for all x.

(b) Prove

$$\frac{1}{\frac{1}{x}} = x$$

if $x \neq 0$.

KEY IDEAS AND KEY WORDS

The **solution set** of an equation is the set of all solutions of the equation. Two equations are called **equivalent** if they have the same solution set.

Every rational number can be represented by either a **terminating** or an infinite **repeating decimal,** and every infinite repeating decimal represents a rational number. We denote repeating decimals by underscoring:

$.\underline{6}$ indicates .666 . . . ,
$3.\underline{89}$ indicates 3.898989 . . . .

CHAPTER REVIEW

Solve each of the following equations. Check your answers.

1. $5 - 8x = 40 - 13x$

2. $12 - 11x = 2 - 9x$

3. $4x + 17 - 11 = 8x$

4. $5 - 7(2 - x) = 31 - 4(2x - 5)$

5. $\frac{1 + 12}{6} = \frac{9 - x}{3} - 1$

6. $7 - \frac{x}{4} = \frac{x}{2} + 2 - \frac{x}{3}$

7. $\frac{x - 6}{5} + 2 = \frac{x + 9}{10}$

8. $\frac{x + 4}{3} - \frac{x}{5} = \frac{1}{2} + \frac{x}{6}$

9. Describe the solution set of each of the following equations.

(a) $3(y + 4) + 7(2 - y) = 4(6 - y) + 2$

(b) $15 + 4x = 18 + 4x$

Find the rational number represented by each of the repeating decimals.

10. $.\underline{153}$ **11.** $.2\underline{57}$ **12.** $.3\underline{28}$ **13.** $2.\underline{73}$

14. $1.\underline{403}$

15. Solve the formulas for the variable y.

(a) $x = 2h + 3y$

(b) $s = yt - 16t^2$

(c) $(a \cdot b) + (b \cdot y) + c = 0$

16. $A = P + Pr$. Find r if $A = 2250$ and $P = 1500$.

17. $V = \frac{1}{3}bh$. Find h if $V = 50$ and $b = 25$.

18. The sum of three consecutive odd positive integers is 381. Find the integers.

19. Jack's present age is three times his brother's age two years ago. If Jack is 18 years old, how old is his brother now?

20. The difference between two-thirds of a number and one-sixth of the same number is 78. Find the number.

21. A grocer wishes to combine one grade of coffee worth $1.05 per pound with another grade worth 85¢ per pound to obtain 20 pounds of a blend which he can sell at 90¢ per pound. How many pounds of each grade should he use?

22. A clerk sells a dozen more $4 blouses than $5 blouses. How many of each did she sell to take in $210?

23. The total receipts for 120 tickets were $140.80. If student tickets sold for $1.00 each, and tickets for adults sold for $1.40, how many tickets of each type were sold?

24. (a) Tea worth 90¢ a pound is mixed with tea worth 75¢ a pound to produce 300 pounds of a mixture worth 80¢ a pound. How much of each type should be used?
    (b) If we use these two types of tea, can we produce a mixture worth 70¢ a pound? 95¢ a pound?

25. Sue is two years older than Mary. Eleven years ago Sue was twice as old as Mary. How old is each girl now?

26. (a) The sum of the first and third of three consecutive integers is 48 more than half the middle one. Find the largest of the three.
    (b) What type of integer is the sum of three consecutive integers? Explain your answer.

27. (a) Two planes leave the same airport at 12:00 noon. One flies east at 400 miles per hour, and the other flies west at 550 miles per hour. At what time will they be 1425 miles apart?
    (b) What assumption did you make in Exercise 27(a)?

CHAPTER TEST

Solve each of the following equations. Check your answers.

1. $7x - 5 = 2x + 15$

2. $\dfrac{3x + 4}{2} = 11$

3. Find the rational number represented by the repeating decimal $.4\underline{28}$.

4. Solve the following formula for a.

$$s = at - 16t^2$$

5. The sum of three consecutive odd positive integers is 327. Find the integers.

6. Jack's present age is four times his brother's age three years ago. If Jack is 20 years old, how old is his brother now?

7. The difference between two times a number and one-sixth of the same number is 11. Find the number.

8. A grocer wishes to combine one grade of coffee worth \$1.10 per pound with another grade worth 85¢ per pound to obtain 20 pounds of a blend which he can sell at 90¢ per pound. How many pounds of each grade should he use?

9. $A = P + Pr$. Find r if $A = 2500$ and $P = 2000$.

10. $V = \frac{1}{3}Bh$. Find h if $V = 75$ and $B = 25$.

CHAPTER 4

Inequalities

Objectives . . .

- To simplify inequalities by applying the basic properties of inequalities.
- To find solution sets of linear inequalities.
- To use number lines to graph the solution sets of inequalities.

4-1 LESS-THAN AND GREATER-THAN RELATIONS

The solution to a practical problem is not always a single number. You often find a solution that is any number less than a certain number or any number between two numbers. In mathematics, it is, therefore, important to study not only equations but also *less-than* and *greater-than* relationships. Statements involving less-than or greater-than relations are called *inequalities.*

The order axioms, given in Section 2–5, are restated below for convenience. (P denotes the set of positive real numbers.)

ORDER AXIOM 1

If x and y are in P, then $x + y$ is in P. (*Closure-A*)

ORDER AXIOM 2

If x and y are in P, then xy is in P. (*Closure-M*)

ORDER AXIOM 3

For every real number x, either x is in P, $x = 0$, or $-x$ is in P. (*Tri*)

Because of the three possibilities it offers, Axiom 3 is often called the *trichotomy axiom.*

When you say "9 is greater than 4," or, equivalently, "4 is less than 9," you mean that the difference $9 - 4$, or 5, is a positive number. This relationship between 9 and 4 is expressed by either

$$9 > 4 \quad \text{or } 4 < 9.$$

The symbol $>$ denotes *is greater than* and $<$ denotes *is less than.* Each of these symbols "points" toward the smaller number and "opens up" toward the larger number.

As suggested by the example above, the following definition holds for all numbers x and y.

Definition of greater than

$x > y$ if $x - y$ is a positive number. ***(Def $>$)***

Less than is opposite to greater than.

Definition of less than

$x < y$ if $x - y$ is a negative number. ***(Def $<$)***

According to these definitions,

$$x > y \quad \text{if, and only if,} \quad y < x.$$

$6 > 1$ because $6 - 1 = 5$, and 5 is a positive number.
$2 > -10$ because $2 - (-10) = 12$, and 12 is a positive number.
$0 < 5$ because $0 - 5 = -5$, and -5 is a negative number.
$-9 < -2$ because $-9 - (-2) = -7$, and -7 is a negative number.

If it seems strange that 2 is greater than -10, remember that a temperature of 2° above zero is warmer, or higher, than a temperature of 10° below zero (-10° on the thermometer). Similarly, -9 is less than -2, and a temperature of -9° is colder, or lower, than one of -2°.

By the trichotomy axiom, for all numbers x and y, either

$$x - y \text{ is positive}, \quad x - y = 0, \quad \text{or} \quad -(x - y) \text{ is positive}.$$

Since $-(x - y) = -x + y$, or $y - x$, you may express the trichotomy axiom in the following way.

TRICHOTOMY AXIOM

For all numbers x and y, either

$x > y$, $x = y$, or $x < y$. ***(Tri)***

Problem. Which number is larger, $\frac{2}{5}$ or $\frac{3}{7}$?

Solution. Because these numbers are unequal, one is larger than the other by the trichotomy axiom. This problem may be solved by subtracting one number from the other and observing whether the difference is positive or negative. Thus,

$$\begin{aligned}\tfrac{2}{5} - \tfrac{3}{7} &= (\tfrac{2}{5} \cdot \tfrac{7}{7}) - (\tfrac{3}{7} \cdot \tfrac{5}{5}) \\ &= \tfrac{14}{35} - \tfrac{15}{35}, \quad \text{or } -\tfrac{1}{35}.\end{aligned}$$

Since $-\frac{1}{35}$ is a negative number, $\frac{2}{5} < \frac{3}{7}$ or, conversely,

$$\tfrac{3}{7} > \tfrac{2}{5}.$$

Exercises

Use the words *positive* or *negative* and the symbols $>$ or $<$ to complete the following statements.

1. (a) $18 - 12 = 6$, a _?_ number, and 18 _?_ 12.
(b) $27 - 15 = 12$, a _?_ number, and 27 _?_ 15.

2. (a) $15 - 20 = -5$, a _?_ number, and 15 _?_ 20.
(b) $13 - 17 = -4$, a _?_ number, and 13 _?_ 17.

3. (a) $-7 - (-20) = 13$, a _?_ number, and -7 _?_ -20.
(b) $-11 - (-15) = 4$, a _?_ number, and -11 _?_ -15.

4. (a) $-3 - (-2) = -1$, a _?_ number, and -3 _?_ -2.
(b) $-9 - (-5) = -4$, a _?_ number, and -9 _?_ -5.

Use the symbols $>$ or $<$ to make a correct comparison between the first and second number in the given pairs.

5. (a) 5, 21 (b) 7, 4 **6.** (a) $-5, -21$ (b) $-9, -18$

7. (a) 0, 3 (b) 2, 0 **8.** (a) $0, -3$ (b) $-9, 0$

9. (a) $\frac{5}{8}, -\frac{7}{8}$ (b) $-\frac{7}{9}, \frac{5}{9}$ **10.** (a) $-\frac{5}{9}, -\frac{6}{11}$ (b) $-\frac{3}{8}, -\frac{4}{9}$

In Exercises 11–18, fill in the blanks with numerals to make the statements true.

11. (a) $5 <$ _?_ (b) $27 <$ _?_

12. (a) _?_ < 5 (b) _?_ < 27

13. (a) $-9 <$ _?_ (b) $-7 <$ _?_

14. (a) _?_ < -9 (b) _?_ < -7

15. (a) $-20 >$ _?_ (b) $-2 >$ _?_

16. (a) $-20 <$ _?_ (b) $-12 <$ _?_

17. (a) $0 >$ _?_ (b) $0 <$ _?_

18. (a) $|-5| <$ _?_ (b) $|-5| >$ _?_

19. (a) Which of the blanks in Exercises 11–18 can be filled with only positive numbers?
(b) Which of the blanks in Exercises 11–18 can be filled with only negative numbers?

20. (a) Which of the blanks in Exercises 11–18 can be filled with any positive number?
(b) Which of the blanks in Exercises 11–18 can be filled with any negative number?
(c) Which of the blanks in Exercises 11–18 can be filled with zero?

Preparation for Section 4–2

Do the computation.

1. $(-6 + 15) + (-7 + 3)$

2. $-7 \cdot 12$

3. $13 \cdot (-3)$

4. $-10 \cdot 4 \cdot (-5)$

5. $-6 \cdot (-2) + (-17) \cdot 0$

6. $|4 - 7| + |13 - 20|$

7. $(7 - 11) \cdot |7 - 11|$

8. $-6 \cdot 4 + |3 \cdot (-4)|$

Find the solution set for each of the following equations.

9. $|x| = 3$

10. $|x - 5| = 6$

4–2 PROPERTIES OF INEQUALITIES

A number line such as the one shown in Fig. 4–1 provides a convenient way of describing the relationships greater than and less than. Recall that every real number (positive, negative, or zero) is represented by a point on the line. The figure is incomplete, but imagine that the line extends indefinitely far in both directions. Thus, 10^6 and -10^6 are assigned to points on the line, although these points are not shown in the figure.

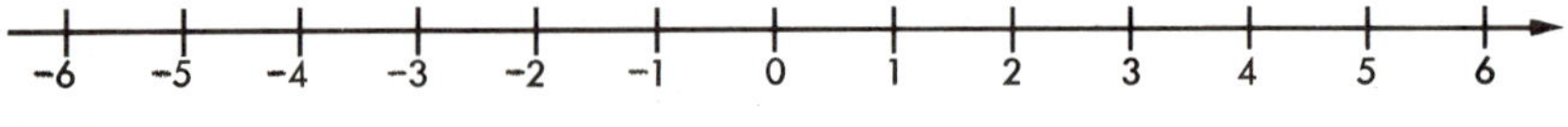

FIGURE 4–1

Numbers are usually assigned to points on a number line in such a way that the numbers become larger as you move from left to right. Thus, one number is *larger* than another number if it appears to the *right* of it; and one number is *smaller* than another number if it appears to the *left* of it. For example,

$6 > 1$ because the graph of 6 appears to the right of 1.
$-3 < 2$ because the graph of -3 appears to the left of 2.

Each number line has a *direction* assigned to it, namely, the direction of increasing numbers. The direction is indicated by an arrowhead, as shown in Fig. 4–1 and 4–2.

If you compare three numbers and find that the first number is greater than the second and the second greater than the third, you know that the first number is greater than the third. For example,

$23 > 19$ because $23 - 19 = 4$, and 4 is a positive number,

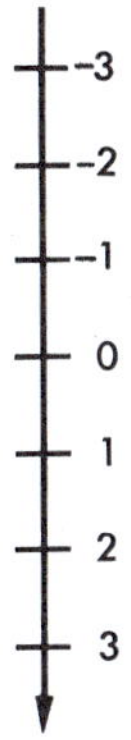

FIGURE 4–2

and

$19 > -3$ because $19 - (-3) = 22$, and 22 is a positive number.

Does it follow that $23 > (-3)$? Consider the following:

$$(23 - 19) + [19 - (-3)] = 4 + 22,$$
$$[23 - (-3)] + (19 - 19) = 26,$$

and

$$23 - (-3) = 26.$$

Note that 26 is a positive number. Therefore, $23 > -3$.

The property illustrated above can be stated in the following way.

TRANSITIVE LAW FOR GREATER THAN

If $x > y$ and $y > z$, then $x > z$. $(T >)$

The transitive law is also true for the less-than relationship.

TRANSITIVE LAW FOR LESS THAN

If $x < y$ and $y < z$, then $x < z$. $(T <)$

Note that $x > 0$ if, and only if, x is a positive number; and $y < 0$ if, and only if, y is a negative number. Thus,

$\{x \mid x > 0\}$ is the set P of all positive numbers.
$\{y \mid y < 0\}$ is the set of all negative numbers.

It is convenient to write

$$7 > 4 > 2 \quad \text{or} \quad 2 < 4 < 7$$

to indicate that 4 is less than 7 but greater than 2, that is, to indicate that 4 is *between* 2 and 7. As other examples, 0 is between -3 and 10,

$$-3 < 0 < 10,$$

and 27 is between 30 and 20,

$$30 > 27 > 20.$$

You should never write $7 > 4 < 9$, although this is a true statement if it means that 7 is greater than 4 and 4 is less than 9. It is not true that 4 is between 7 and 9, and statements written in this form can be confusing. Similarly, you should never write an expression such as $x < y > z$.

The set of all integers between 0 and 100 can be indicated in terms of inequality signs as follows:

$$\{x \mid 0 < x < 100, x \text{ an integer}\}.$$

If the domain of the variable is not given, it is understood to be the set of all real numbers. Thus,

$$\{x \mid 0 < x < 1\}$$

is the set of all real numbers between 0 and 1. Included in this set are

$$\frac{1}{2}, \quad \frac{2}{3}, \quad .001, \quad \frac{\sqrt{2}}{2}, \quad \frac{\pi}{4},$$

and infinitely many other numbers.

The set

$$\{5x \mid 0 < x < 21, x \text{ an integer}\}$$

consists of the positive multiples of 5 up to and including 100.

Exercises

1. (a) In the figure, on which side of -1000 would -999 appear? Which number is smaller?
(b) On which side of -500 would -501 appear? Which number is smaller?

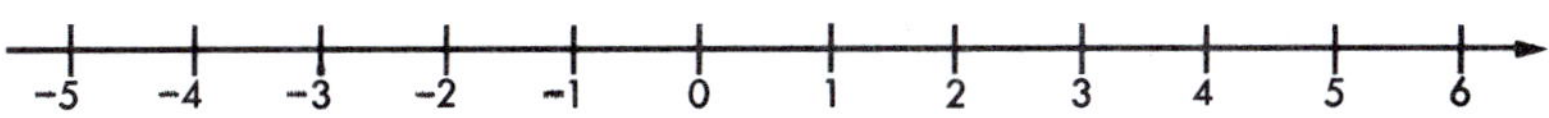

In Exercises 2–8 tell which statements are true and which are false.

2. (a) $(8 - 9) < (9 - 8)$
(b) $(8 - 5) > (5 - 8)$

3. (a) $-6 < -5 < 0$
(b) $-12 < -11 < -10$

4. (a) $(-7 + 1) < (3 + 1) < (10 + 1)$
(b) $(-17 + 3) > (-20 + 3) > (-19 + 3)$

5. (a) $-2 > |3| > -4$ (b) $2 < |-3| < 4$

6. (a) $-4 \cdot 3 < -3 \cdot 4$ (b) $-2 \cdot 5 < -5 \cdot 2$

7. (a) $3 \cdot (-6) < 3 \cdot 0 < 3 \cdot 2$ (b) $-2 \cdot 4 < -2 \cdot 5 < -2 \cdot 6$

8. (a) $|9 - 15| > 7$ (b) $|7 - 12| > 2$

In Exercises 9–11 list the elements of each of the sets. Which sets are equal?

9. (a) $A = \{x \mid x > -4 \text{ and } x < 4, x \text{ an integer}\}$
(b) $B = \{n \mid -5 < n < 15, n \text{ an integral multiple of } 3\}$

10. (a) $C = \{\frac{1}{2}y \mid -8 < y < 8, y \text{ an even integer}\}$
(b) $D = \{3m \mid -2 < m < 5, m \text{ an integer}\}$

11. (a) $E = \{x|\ |x| < 6, x \text{ an integer}\}$
(b) $F = \{y|\ |y| < 5 \text{ or } |y| = 5, y \text{ an integer}\}$

---

12. Explain why each of the following statements is true.
(a) If $x > 0$, then $x - 0$ is positive.
(b) If $x - 6$ is positive, then $x > 6$.
(c) If $x < 5$, then $5 - x$ is positive.

13. If possible, list all the elements in each of the following sets. If it is not possible to list all the elements, list any five.

(a) $A = \{x \mid |x| < 10, x \text{ an integer}\}$
(b) $B = \{-3y \mid y > 3, y \text{ an even integer}\}$
(c) $C = \{2x \mid x < -3, x \text{ an odd integer}\}$
(d) $D = \left\{\frac{1}{x} \,\middle|\, 0 < x < 11, x \text{ an integer}\right\}$

14. For each integer x, which number is larger: $x + 6$ or $x + 8$?

15. For each integer x, which number is larger: $x - 6$ or $x - 8$?

---

16. If $a + 3 < b + 3$, how are a and b related?

17. If $a - 3 < b - 3$, how are a and b related?

18. Tell whether the inequality

$$8k > 6k$$

is true or false when k is given, in turn, the values 5, 3, 1, 0, -1, -2, -6, and -10.

Preparation for Section 4–3

Which of the following pairs of numbers are additive inverses of each other, and which are multiplicative inverses? Which are neither?

1. $-4, 4$

2. $\frac{1}{7}, 7$

3. $-\frac{1}{11}, -11$

4. $-\frac{1}{20}, \frac{1}{20}$

5. $-\frac{1}{3}, 3$

6. $-(2x - 1), 2x - 1$

7. $4x - 3, 3 - 4x$

8. $11x - 3, -11x - 3$

4–3 SOLUTION SETS OF INEQUALITIES

Mathematical statements such as

$$2x + 1 > 5 \quad \text{and} \quad 3 - N < 7$$

are called inequalities. An *inequality* is a statement consisting of two algebraic expressions called the *sides,* connected by an inequality sign such as $>$ or $<$.

An inequality involving one variable is called an *inequality in one variable.* A *solution* of an inequality in one variable is a value of the variable which makes the inequality *true.*

For example, 3 is a solution of $2x + 1 > 5$, because

$$(2 \cdot 3) + 1 > 5, \quad \text{or } 7 > 5,$$

is a true statement. Note that 2 is not a solution, because the inequality

$$(2 \cdot 2) + 1 > 5, \quad \text{or } 5 > 5,$$

is false.

The set of all solutions of an inequality is called the *solution set* of the inequality. When you *solve* an inequality, you find its solution set. Inequalities in one variable are solved in much the same way equations are solved. In other words, you replace any given inequality by a succession of equivalent ones until you obtain an inequality with an obvious solution set.

Definition of equivalent inequalities

Two inequalities are equivalent if they have the same solution set.

Knowing that $9 > 4$, you also know that $9 + 3 > 4 + 3$ and $9 + (-3) > 4 + (-3)$. In each case, the difference between the left side and the right side is 5. This illustrates the following property of inequalities.

ADDITION PROPERTY OF INEQUALITIES

$a > b$ if, and only if, $a + c > b + c$ ***($A >$)***

The property above consists of 2 parts:

If $a + c > b + c$, then $a > b$.
If $a > b$, then $a + c > b + c$.

Knowing that $9 > 4$, you also know that $9 \cdot 3 > 4 \cdot 3$; that is, $(9 \cdot 3) - (4 \cdot 3) = (9 - 4) \cdot 3$ is positive by Order Axiom 2 since both $9 - 4$ and 3 are positive. This illustrates the following property of inequalities.

POSITIVE MULTIPLICATION PROPERTY OF INEQUALITIES

If $c > 0$, then $a > b$ if, and only if, $ac > bc$. ***($M >$)***

Similar properties hold for less than.

$$a < b \text{, if, and only if, } a + c < b + c. \qquad (A <)$$
$$\text{If } c > 0 \text{, then } a < b \text{ if, and only if, } ac < bc. \qquad (M <)$$

These four properties may be used to obtain equivalent inequalities from a given one, as shown in the following problems.

Problem 1. Solve the inequality $2x + 1 > 5$.

Solution. Each of the following inequalities is equivalent to the given one.

$$(2x + 1) + (-1) > 5 + (-1)$$
$$2x > 4$$

Next, multiply each side by $\frac{1}{2}$, which is a positive number:

$$\tfrac{1}{2} \cdot 2x > \tfrac{1}{2} \cdot 4,$$
$$x > 2.$$

Thus,

$$\{x \mid x > 2\}$$

is the solution set of the given inequality, $2x + 1 > 5$.

As this problem shows, the solution set of an inequality is frequently an infinite set. An equation, on the other hand, usually has a finite solution set.

Problem 2. Solve the inequality $3 - N < 7$.

Solution. By adding -3 to each side you have the following equivalent inequalities.

$$3 - N + (-3) < 7 + (-3)$$
$$-N < 4$$

The multiplication property of inequalities does not allow you to multiply each side by a negative number, but you can proceed as follows:

$$-N + (N - 4) < 4 + (N - 4),$$
$$-4 < N,$$

or

$$N > -4.$$

Therefore, the solution set of the given inequality is

$$\{N \mid N > -4\}.$$

The following are a few examples of the many possible elements of this solution set:

$$-3, \quad -\tfrac{1}{2}, \quad 0, \quad \text{and} \quad 10^6.$$

You can shorten the solution of Problem 2 by using the following property of inequalities.

NEGATIVE MULTIPLICATION PROPERTY OF INEQUALITIES

If $c < 0$, then $a > b$ if, and only if, $ac < bc$. $(-M >)$

A similar statement is valid for less than.

If $c < 0$, then $a < b$ if, and only if, $ac > bc$. $(-M <)$

Using the property of multiplication of each side by a negative number, you can start with the inequality $-N < 4$ in Problem 2 and proceed as follows:

$$\begin{aligned} -N &< 4, \\ (-1)\cdot(-N) &> (-1)\cdot 4, \\ N &> -4. \end{aligned}$$

Exercises

1. (a) If $S = \{0, 5, 2, -1, -10, 3, 4\}$, which elements of S are solutions of the inequality $3x - 5 < 13 - 3x$? Which are not?
   (b) If $S = \{0, -7, 2, 6, -6, -1, 1, 10, -10\}$, which elements of S are solutions of the inequality $3x - 4 < 4x + 1$? Which are not?
2. (a) Starting with the known true inequality $8 > -5$, tell why each of the inequalities is true or false.
   (i) $8 + 12 > -5 + 12$
   (ii) $8 + (-7) > -5 + (-7)$
   (iii) $8 \cdot 4 > -5 \cdot 4$
   (iv) $-3 \cdot 8 > -3 \cdot (-5)$
   (v) $-10 \cdot 8 < -10 \cdot (-5)$

(b) Starting with the known true inequality $-10 > -15$, tell why each of the inequalities is true or false.

(i) $-10 + 3 > -15 + 3$
(ii) $-10 + (-4) > -15 + (-4)$
(iii) $-10 \cdot (-4) > -15 \cdot (-4)$
(iv) $-10 \cdot 17 > -15 \cdot 17$
(v) $-10 \cdot (-1) < 15 \cdot (-1)$

3. (a) If $a < 6$ is true, which inequalities are also true? Why?

(i) $-4 \cdot a < -4 \cdot 6$
(ii) $a + (-2) < 6 + (-2)$
(iii) $2 \cdot a < 2 \cdot 6$
(iv) $a + 3 < 6 + 3$
(v) $-1 \cdot a > -1 \cdot 6$

(b) If $b < 5$ is true, which inequalities are also true? Why?

(i) $b + (-3) < 5 + (-3)$
(ii) $b + 7 < 5 + 7$
(iii) $b \cdot 7 < 5 \cdot 7$
(iv) $b \cdot (-1) < 5 \cdot (-1)$
(v) $b \cdot (-5) > 5 \cdot (-5)$

Tell whether the following statements are true or false, and correct the statements that you marked false so that they become true statements.

4. (a) $2 < 3$ and $-2 < -3$ (b) $5 > -6$ and $-5 < 6$

5. (a) $-10 > -3$ and $10 > 3$ (b) $-1 > -3$ and $1 < 3$

For what values of x are the following statements true?

6. (a) $x > -5$ and $-x < 5$ (b) $2x > 4$ and $-2x < -4$

7. (a) $x - 3 < 5$ (b) $x + 4 > -3$

8. (a) $2x - 3 < x + 3$ (b) $2x - 4 > x + 1$

9. (a) $5 - x < 9$ (b) $6 - x < -1$

10. (a) $5 + x < 9$ (b) $6 + x < -1$

11. (a) $10 - 2x < 2$ (b) $12 < 3 - 9x$

Solve each of the following inequalities.

12. $8x - 10 > 7x - 4$

13. $11 - 5x < 11 - 4x$

14. $8x - 13 < -13$

15. $\frac{16}{3}x - 7 < 5x - 8$

16. $\frac{29}{14}x + 7 > 2x + \frac{13}{2}$

17. $8 + 2x < 28 - 3x$

18. $9 - x > 4 - 7x$

19. $6 + 8x < 4 + 5x$

20. $7 - 10x > 12 - 12x$

21. $13x - 8 > 17x - 11$

4-4 SOLVING INEQUALITIES

The relationship *greater than or equal to* is denoted by $\geqq$ and is defined in the following way.

$$a \geqq b \quad \textit{if either} \quad a > b \quad \textit{or} \quad a = b.$$

Similarly, *less than or equal to* is denoted by $\leqq$ and defined in the following way.

$$a \leqq b \quad \textit{if either} \quad a < b \quad \textit{or} \quad a = b.$$

For example,

$$12 \geqq 2 \quad \text{because} \quad 12 > 2,$$
$$7 \leqq 7 \quad \text{because} \quad 7 = 7.$$

The properties stated in the preceding section for solving inequalities involving $<$ or $>$ also hold for inequalities involving $\leqq$ or $\geqq$. Some examples are given below.

Problem 1. Solve the inequality $3x - 5 \leqq 13 - 3x$.

Solution. Each of the following inequalities is equivalent to the original one for the stated reason.

$$\begin{array}{rll} 3x - 5 + 5 \leqq & 13 - 3x + 5 & (\text{A} <) \\ 3x + (-5 + 5) \leqq & -3x + (13 + 5) & (\text{A-A}) \text{ and } (\text{C-A}) \\ 3x \leqq & -3x + 18 & (\text{Inv-A}) \\ 3x + 3x \leqq & 3x + (-3x + 18) & (\text{A} <) \\ 6x \leqq & (3x - 3x) + 18 & (\text{A-A}) \\ 6x \leqq & 18 & (\text{Inv-A}) \\ \frac{1}{6} \cdot 6x \leqq & \frac{1}{6} \cdot 18 & (\text{M} <) \\ x \leqq & 3 & (\text{Inv-M}) \end{array}$$

Thus, $\{x \mid x = 3 \text{ or } x < 3\}$ is the solution set of the original inequality.

Problem 2. Solve the inequality $7 - 5a \geqq 19 - 13a$.

Solution. Each of the following inequalities is equivalent to the given one.

$$\begin{aligned} 7 - 5a + 13a &\geqq 19 - 13a + 13a \\ 7 + (-5a + 13a) &\geqq 19 + (-13a + 13a) \\ 7 + 8a &\geqq 19 \\ -7 + 7 + 8a &\geqq -7 + 19 \\ (-7 + 7) + 8a &\geqq 12 \\ 8a &\geqq 12 \\ \tfrac{1}{8} \cdot 8a &\geqq \tfrac{1}{8} \cdot 12 \\ a &\geqq \tfrac{3}{2} \end{aligned}$$

Therefore,

$$\{a \mid a = \tfrac{3}{2} \text{ or } a > \tfrac{3}{2}\} \quad \text{or} \quad \{a \mid a \geqq \tfrac{3}{2}\}$$

is the solution set.

Problem 3. Solve the inequality $5 - 8n \leqq 21$.

Solution.

$$\begin{aligned} 5 - 8n &\leqq 21 \\ -5 + 5 - 8n &\leqq -5 + 21 \\ (-5 + 5) - 8n &\leqq 16 \\ -8n &\leqq 16 \\ -\tfrac{1}{8} \cdot (-8n) &\geqq -\tfrac{1}{8} \cdot 16 \\ [-\tfrac{1}{8} \cdot (-8)]n &\geqq -2 \\ 1 \cdot n &\geqq -2 \\ n &\geqq -2 \end{aligned}$$

Thus, $\{n \mid n \geqq -2\}$ is the solution set.

Exercises

Which statements are true and which are false?

1. (a) $9 \geqq 9$ (b) $-8 \geqq -9$

2. (a) $7 \leqq 6$ (b) $7 \geqq 6$

3. (a) $0 \geqq x$ for every negative number x.
(b) $0 \leqq x$ for every positive number x.

4. Set A consists of all real numbers greater than 5. This can be expressed in set notation as follows:

$$A = \{x \mid x > 5\}.$$

In like manner, write each of the following in set notation.

(a) Set B consists of 6 and all real numbers greater than 6.
(b) Set C consists of -3, 2, and all real numbers between -3 and 2.

If the set S is defined as

$$\{x \mid x \leqq 2\},$$

you may make the following observations about the elements of S:

(i) S has a *largest* element, 2, but it has no smallest element.
(ii) S contains an infinite number of positive elements, all but one of which are less than 2.
(iii) Every negative number is an element of S.
(iv) S contains the element 0.

Use this method to characterize the elements of the following sets. Also, list six elements of each set.

5. (a) $A = \{x \mid x \geqq -5\}$
(b) $B = \{x \mid x \leqq -6\}$

6. (a) $C = \{x \mid x > -10 \text{ and } x \leqq 17\}$
(b) $D = \{x \mid x \geqq 17 \text{ or } x < -10\}$

Solve each of the following inequalities.

7. (a) $6x + 3 \leqq 5x - 1$ (b) $11x - 2 \geqq 10x + 1$

8. (a) $7 - 2x \geqq 2 - 3x$ (b) $2 - 3x \geqq 5 - 4x$

9. (a) $5x - 2 \leqq 2x + 10$ (b) $4x - 3 \geqq 6x + 3$

10. (a) $18 - 9x \geqq 12 - 2x$ (b) $4x - 3 \geqq 6x - 5$

11. (a) $-2x + 3 \leqq -4x - 5$ (b) $-7x + 13 \geqq -2x + 9$

12. (a) $x - 2 \leqq 10 + \frac{1}{2}x$ (b) $\frac{3}{2}x + 3 \geqq x - 2$

13. (a) $\frac{1}{5}x - 3 \geqq 17$ (b) $-12 \geqq \frac{1}{3}x - 1$

14. (a) $3x - \frac{2}{5} \leqq 5x + \frac{3}{5}$ (b) $\frac{9}{7}x + 2 \leqq \frac{2}{7}x + 5$

---

15. (a) Give three positive and three negative solutions of the inequality $|x| \leqq 1$.
(b) What is the largest number in the solution set of $|x| \leqq 1$? What is the smallest number?

16. (a) Give three positive and three negative solutions of the inequality $|x| \geqq 4$.
(b) Does the solution set of $|x| \geqq 4$ contain a smallest element? a smallest positive element?
(c) Does the solution set of $|x| \geqq 4$ contain a largest element? a largest negative element?

Review for Sections 4–1 through 4–4

Use the symbols $<$ or $>$ to make the following statements true. Justify your answers.

1. 16 __?__ 14 **2.** -13 __?__ 21 **3.** 13 __?__ -21
4. $\frac{2}{3}$ __?__ $\frac{1}{4}$ **5.** 0 __?__ 7 **6.** 0 __?__ -2
7. -5 __?__ 2 **8.** -15 __?__ -10 **9.** 4 __?__ -10

Make the following statements true by filling in the blanks with an appropriate numeral.

10. $12 >$ __?__ **11.** $-13 <$ __?__ **12.** $15 <$ __?__
13. $-6 >$ __?__ **14.** $0 >$ __?__ **15.** $-4 <$ __?__

Which of the following statements are true and which are false?

16. $15 < -18$ **17.** $24 > 13 + 7$ **18.** $3 < |-6| < 9$
19. $15 \cdot (-3) < 3 \cdot (-5)$ **20.** $13 + 6 > 16 + 2$ **21.** $-6 > -2$
22. $17 \cdot 4 < 14 \cdot 7$ **23.** $|4 - 9| > -4$ **24.** $-17 > |10 - 7|$

List the elements of each of the following sets. Are any of the sets equal? Which ones?

25. $K = \{k \mid 0 < k < 1, k \text{ an integer}\}$
26. $R = \{r \mid 2 < r < 10, r \text{ an even integer}\}$
27. $F = \{f \mid 3 < f < 6, f \text{ an integral multiple of } 3\}$
28. $P = \{p \mid 7 < p < 16, p \text{ an integral multiple of } 3\}$
29. $S = \{3s \mid 2 < s < 6, s \text{ an integer}\}$
30. $M = \{2m \mid 1 < m < 5, m \text{ an integer}\}$

For what values of x are each of the following inequalities true?

31. $8 + x < 4$ **32.** $x - 10 > 6$ **33.** $5 + 2x < 3x$
34. $x + 5 > 3x$ **35.** $2x + 1 > 15$ **36.** $4 - x < -1$

Solve the following inequalities.

37. $x + 5 \leqq 57 - x$ **38.** $5x + 3 \geqq 12 - 3x$
39. $\frac{1}{2}x + \frac{2}{3} \leqq 3$ **40.** $8 - 2x \leqq 5 - 3x$
41. $x + 26 \geqq 18 - x$ **42.** $\frac{3}{4} - \frac{1}{2}x \geqq \frac{1}{2}$
43. $13 + 5x \geqq 35 - 6x$ **44.** $\frac{2}{5}x \leqq 10 - x$
45. $2x + 13 \leqq 27$

Answers to Review for Sections 4–1 through 4–4

1. $>$ **2.** $<$ **3.** $>$ **4.** $>$ **5.** $<$
6. $>$ **7.** $<$ **8.** $<$ **9.** $>$

10. Any number less than 12; for example, 11

11. Any number greater than -13; for example, -12

12. Any number greater than 15; for example, 16

13. Any number less than -6; for example, -7

14. Any negative number; for example, -1

15. Any number greater than -4; for example, -3

16. False **17.** True **18.** True **19.** True **20.** True

21. False **22.** True **23.** True **24.** False

25. $K = \emptyset$ **26.** $R = \{4, 6, 8\}$ **27.** $F = \emptyset$

28. $P = \{9, 12, 15\}$ **29.** $S = \{9, 12, 15\}$ **30.** $M = \{4, 6, 8\}$

Equal sets: $K = F$, $R = M$, $P = S$

31. $x < -4$ **32.** $x > 16$ **33.** $x > 5$

34. $x < \frac{5}{2}$ **35.** $x > 7$ **36.** $x > 5$

37. $\{x \mid x \leqq 26\}$ **38.** $\{x \mid x \geqq \frac{9}{8}\}$ **39.** $\{x \mid x \leqq \frac{14}{3}\}$

40. $\{x \mid x \leqq -3\}$ **41.** $\{x \mid x \geqq -4\}$ **42.** $\{x \mid x \leqq \frac{1}{2}\}$

43. $\{x \mid x \geqq 2\}$ **44.** $\{x \mid x \leqq \frac{50}{7}\}$ **45.** $\{x \mid x \leqq 7\}$

4–5 THE GRAPH OF A SET OF NUMBERS

Integers can be represented as equally spaced points on a line, as indicated in Fig. 4–3. The three dots between 4 and 10^6 indicate that many integers have been omitted between 4 and 10^6 (in this case, 999,995 of them). Nevertheless, you assume that these integers are on the number line. The three dots to the right of 10^6 remind you that there are infinitely many integers greater than 10^6 whose presence on the line is also assumed. The other sets of dots on the line indicate omissions the same way.

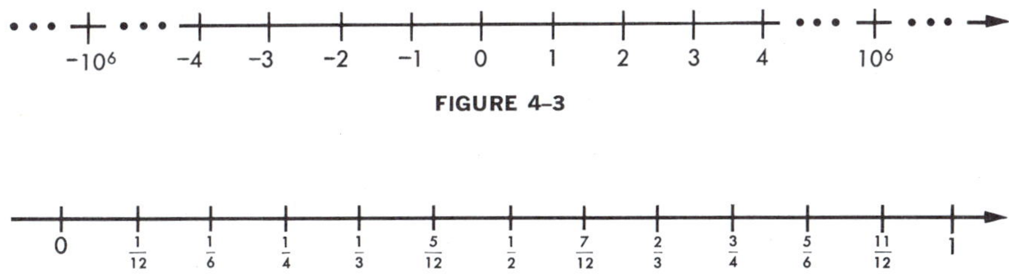

FIGURE 4–3

FIGURE 4–4

Real numbers are represented by proportionately spaced points on the line. For example, a part of the line shown in Fig. 4–3 is magnified in

Fig. 4–4 to show some of the real numbers between 0 and 1. There are, however, an infinite number of real numbers between 0 and 1 which are not shown in Fig. 4–4.

Definition of graph

The point on a number line associated with each number is called the graph of the number. The graph of a set S of numbers is the set of the graphs of the individual numbers of S.

For example, the graph of $\{x \mid x \text{ an integer}\}$ is indicated in Fig. 4–3.

Problem 1. Graph the set $S = \{x \mid -3 \leqq x \leqq 1, x \text{ an integer}\}$.

Solution. The graph of S is the set of five points shown in Fig. 4–5.

If the domain of a variable is unspecified, you can assume it is the set of all real numbers. Now consider some graphs of special sets.

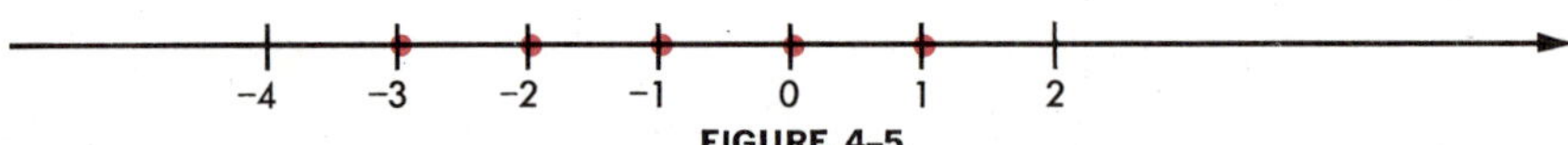

FIGURE 4–5

Problem 2. Graph the set $S = \{x \mid x \geqq 1\}$.

Solution. Since set S contains all real numbers greater than or equal to 1, the graph of S is the set of all points at and to the right of 1 on the number line. (See Fig. 4–6)

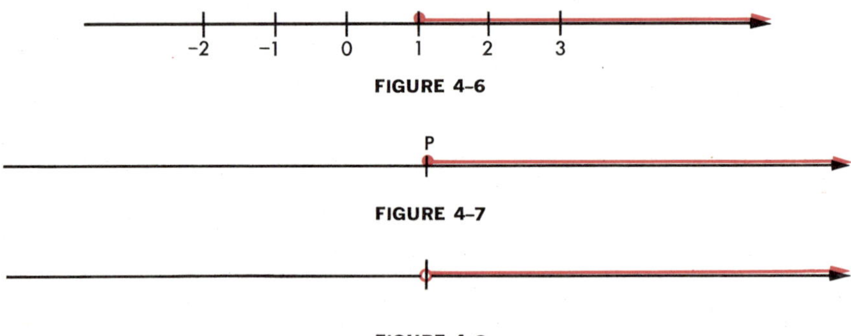

FIGURE 4–6

FIGURE 4–7

FIGURE 4–8

Each point P on a line divides the line into two parts. One of these parts together with point P is called a *ray.* Figure 4–6 is an example of a ray. Figure 4–7 also shows a ray. Point P is called the *endpoint* of the ray. Occasionally, you may wish to consider a ray without its endpoint; such a ray is called an *open ray* or a *half-line.* A hollow dot is always put at the end of an open ray to distinguish it from a ray. (See Fig. 4–8.)

Problem 3. Graph the set $\{x \mid x < 3\}$.

Solution. The graph consists of all points to the left of 3 on the number line in Fig. 4–9. Because the point at 3 is not included, the graph is an open ray.

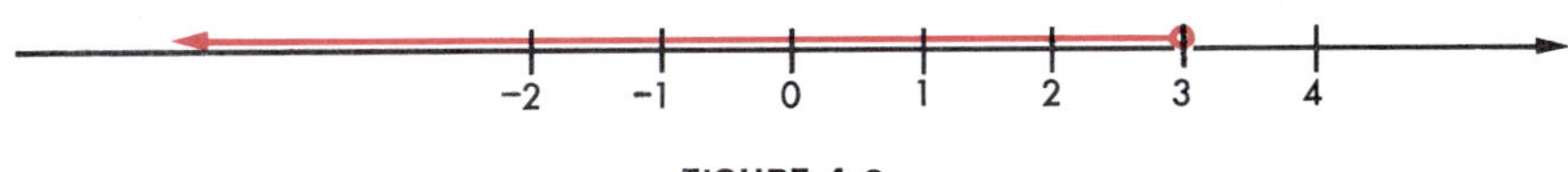

FIGURE 4–9

Exercises

1. Define the sets indicated by the graphs and describe the graphs.

(a)

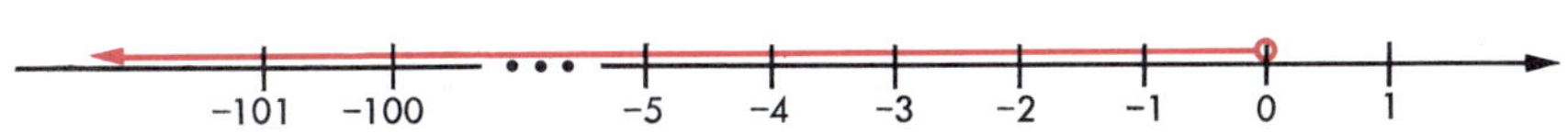

(b)

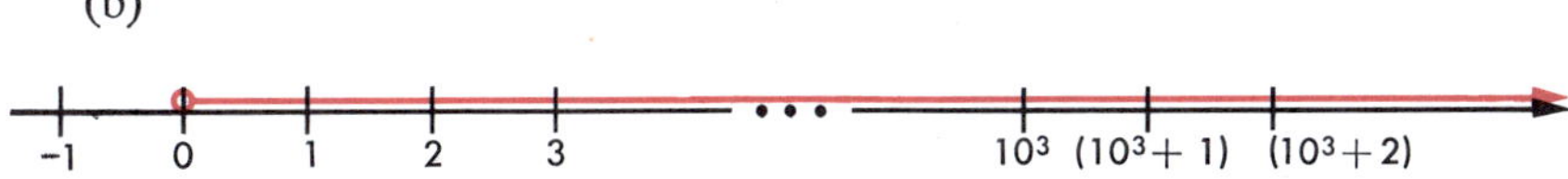

Graph the following sets and list three elements in each. Describe each of the graphs.

2. (a) $A = \{x \mid x < 5\}$ (b) $B = \{x \mid x < 1\}$
3. (a) $C = \{x \mid x \geqq 7\}$ (b) $D = \{x \mid x \geqq 4\}$
4. (a) $E = \{x \mid x < 0\}$ (b) $F = \{x \mid x > -4\}$
5. (a) $G = \{x \mid -x > 5\}$ (b) $H = \{x \mid -x > -4\}$
6. (a) $I = \{x \mid -x \leqq -1\}$ (b) $J = \{x \mid -x \leqq \frac{3}{2}\}$

Simplify the inequality which describes each of the following sets. Then graph each set.

7. (a) $A = \{x \mid 4x + 5 > 3x + 8\}$
   (b) $B = \{x \mid 3x - 5 < 2x + 10\}$
8. (a) $C = \{x \mid 7x + 2 \geqq 5x + 12\}$
   (b) $D = \{x \mid 4 + 2x \geqq 11 - 5x\}$

Graph each of the following sets. If possible, give the number of elements each set contains.

9. (a) $A = \{x \mid -6 \leqq x \leqq 6, x \text{ an integer}\}$
(b) $B = \{x \mid 0 \leqq x \leqq 5, x \text{ an integer}\}$

10. (a) $C = \{x \mid x < 6, x \text{ a positive integer}\}$
(b) $D = \{x \mid x > -6, x \text{ a nonpositive integer}\}$

11. (a) $E = \{x \mid x > 3, x \text{ an even integer}\}$
(b) $F = \{x \mid x < 6, x \text{ an odd integer}\}$

12. (a) $G = \{x \mid -7 < x < 7, x \text{ an even integer}\}$
(b) $H = \{x \mid -7 \leqq x \leqq 7, x \text{ an odd integer}\}$

---

13. If $x > 2$, is x positive or negative?

14. If $x > 2$, is $-x$ positive or negative?

15. Name three elements of the set $\{x \mid x > 2\}$ and then graph the set.

16. Name three elements of the set $\{x \mid -x > 2\}$ and then graph the set.

17. Do the sets in Exercises 15 and 16 have any elements in common?

18. Name three elements of the set $\{x \mid x < -2\}$ and then graph the set.

19. Does the set in Exercise 18 have any elements in common with either of the sets in Exercises 15 and 16?

Simplify the inequality which describes each of the following sets.

20. (a) $R = \{r \mid 3r - 2 > r - 1\}$
(b) $S = \{s \mid 2s + 3 < 3s - 2\}$
(c) $T = \{t \mid -3t + 7 \geqq -2t + 5\}$
(d) $U = \{u \mid 8u - 2(u + 1) \leqq 10\}$
(e) $W = \left\{w \,\middle|\, \frac{w}{5} + \frac{3}{10} > \frac{w}{2} - \frac{3}{5}\right\}$

21. Graph the sets in Exercise 20.

In Exercises 22–29, find four pairs of equivalent inequalities.

22. $x > 3$ **23.** $-x > 3$ **24.** $x < 3$ **25.** $-x > -3$

26. $-x < 3$ **27.** $-x < -3$ **28.** $x > -3$ **29.** $x < -3$

Graph each of the following sets. If possible, state the number of elements the set contains. If not, list five elements of the set.

30. $B = \{x \mid x \geqq -16, x \text{ an integral multiple of } 4\}$

31. $C = \{x \mid x \leqq 10, x \text{ an integer}\}$

32. $D = \{x \mid -6 \leqq x \leqq 6, x \text{ an even integer}\}$

33. $E = \{x \mid x > -17, x \text{ an integral multiple of } 4\}$

34. $F = \{x \mid x < 11, x \text{ an integer}\}$

35. $G = \{x \mid x < -1, x \text{ an odd integer}\}$

4-6 INTERVALS

In this section, special graphs called *intervals* are studied.

Problem 1. Graph the set $T = \{x \mid -2 \leqq x \leqq 1\}$.

Solution. The graph of T consists of the points at -2 and 1 and all points between -2 and 1. (See Fig. 4-10.) Such a graph is called a *closed interval* since both endpoints are included.

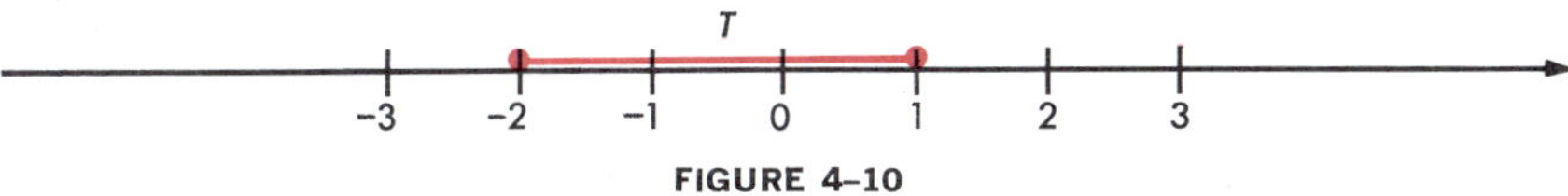

FIGURE 4-10

The closed interval of Fig. 4-10 is closely related to the graphs of the sets

$$\{x \mid x \leqq 1\} \quad \text{and} \quad \{x \mid x \geqq -2\}.$$

In fact, if you look at Fig. 4-11, you can see that the interval T in Fig. 4-10 is simply the intersection, or overlap, of two rays:

$$T = \{x \mid x \leqq 1\} \cap \{x \mid x \geqq -2\}.$$

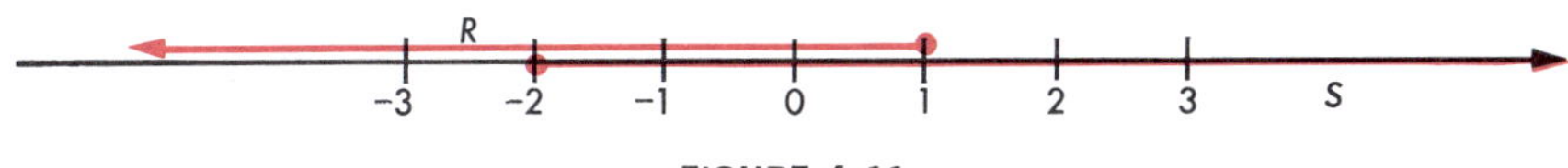

FIGURE 4-11

Problem 2. Graph the set $C = \{x \mid 0 < x < 3.5\}$.

Solution. If you let

$$A = \{x \mid x < 3.5\} \quad \text{and} \quad B = \{x \mid x > 0\},$$

then

$$C = A \cap B.$$

The graph of C, then, is the intersection of the open rays which are the graphs of sets A and B. (See Fig. 4-12.) The graph of set C is shown

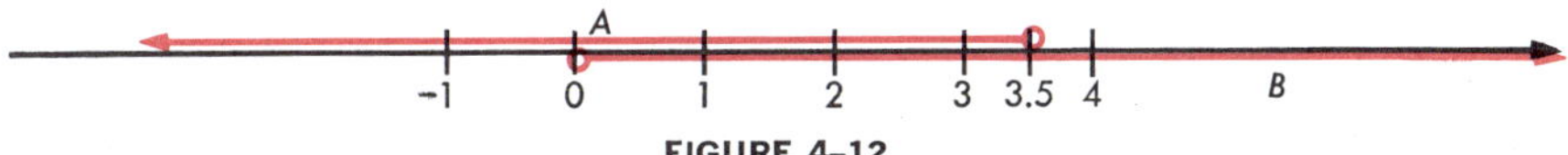

FIGURE 4-12

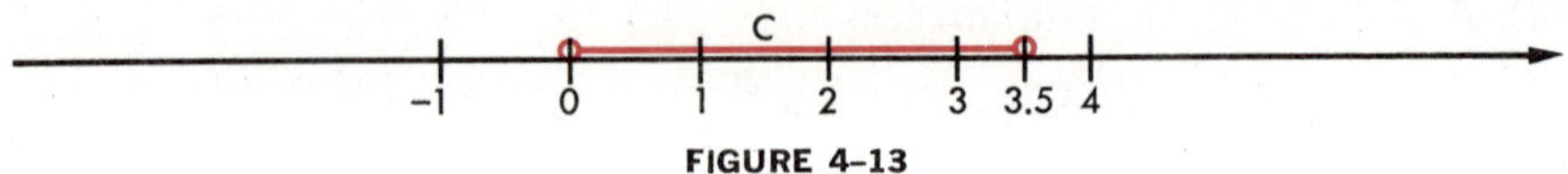

FIGURE 4–13

in Fig. 4–13. Remember that the hollow dots at 0 and 3.5 indicate that these points are not part of the graph.

The interval, C, in Fig. 4–13 is called an *open interval.* Every open interval is the intersection of two open rays.

Problem 3. Graph the set $\{x \mid -2 \leqq x < 2\}$.

Solution. Because $x \geqq -2$ and $x < 2$, you can write:

$$\{x \mid -2 \leqq x < 2\} = \{x \mid x < 2\} \cap \{x \mid x \geqq -2\}.$$

The graph of the given set is the intersection of an open ray and a ray, as shown in Fig. 4–14. The actual graph of $\{x \mid -2 \leqq x < 2\}$ is shown in Fig. 4–15.

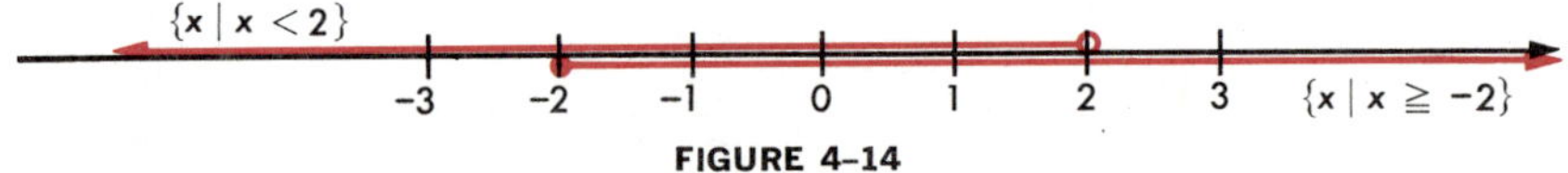

FIGURE 4–14

The interval in Fig. 4–15 contains only one of its endpoints. Therefore, you call it a *half-open interval.* Every half-open interval is the intersection of a ray and an open ray.

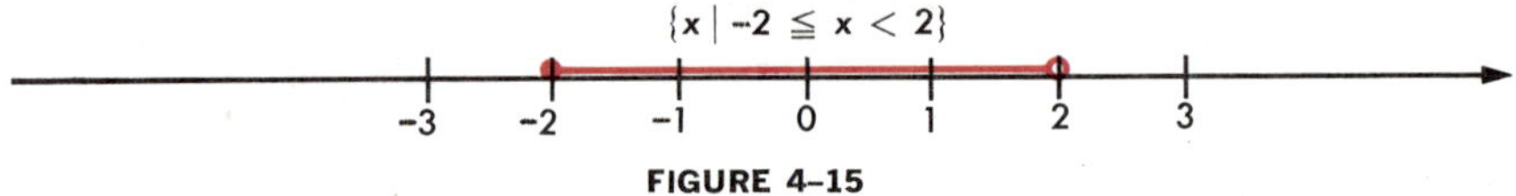

FIGURE 4–15

Problem 4. Graph the set $T = \{x \mid |x| \leqq 5\}$.

Solution. Divide set T into two parts, $\{x \mid x \text{ in } T, x \geqq 0\}$ and $\{x \mid x \text{ in } T, x < 0\}$. If $x \geqq 0$, then $|x| = x$ by definition, and the first part of T is the set

$$R = \{x \mid 0 \leqq x \leqq 5\}.$$

If $x < 0$, then $-x > 0$ and $|x| = -x$ by definition. Therefore, the second part of T is the set

$$S = \{x \mid 0 < -x \leqq 5\}.$$

You may also write S in the following way.

$$S = \{x \mid -5 \leqq x < 0\}$$

The graphs of R and S are intervals, as shown in Fig. 4–16. Set T consists of all the numbers in either R or S. Therefore, you may express T as the union of R and S.

$$T = R \cup S \quad \text{or} \quad T = \{x \mid -5 \leqq x \leqq 5\}$$

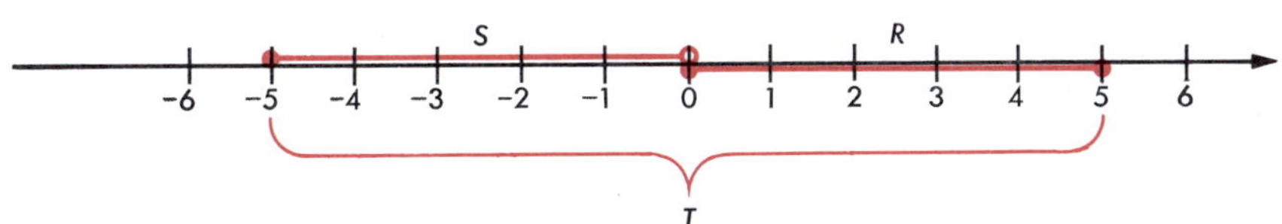

FIGURE 4–16

The graph of $R \cup S$ is the closed interval shown in Fig. 4–16. Notice also that the intersection of R and S is the empty set,

$$R \cap S = \emptyset.$$

Exercises

1. (a) If $D = \{x \mid -5 < x < 12\}$, represent D as the intersection of two sets. Describe the graph of D.
   (b) If $P = \{x \mid -3 \leqq x \leqq 3\}$, represent P as the union of two sets. Describe the graph of P.

Graph each of the following sets. In each case, describe the graph.

2. (a) $A = \{x \mid x \leqq 7\}$ (b) $B = \{x \mid x \leqq \frac{7}{2}\}$
3. (a) $C = \{x \mid |x| \leqq 6\}$ (b) $D = \{x \mid |x| \leq 3\}$
4. (a) $E = A \cap C$ (b) $F = B \cap D$
5. (a) $G = \{x \mid -\frac{9}{2} < x\}$ (b) $H = \{x \mid x > \frac{1}{2}\}$
6. (a) $I = \{x \mid 3 \geqq x\}$ (b) $J = \{x \mid 9 \geqq x\}$
7. (a) $K = G \cap I$ (b) $L = H \cap J$
8. (a) $M = \{x \mid -10 < x < 2\}$ (b) $N = \{x \mid 1 < x < 4.5\}$
9. (a) (i) If $S = \{x \mid 0 < -x \leqq 4\}$ and $T = \{x \mid -4 \leqq x < 0\}$, which of the following are elements of $S \cap T$?

   $$-3, -2, 4, 1, 8, 0, -6, -5, -1$$

   (ii) Graph S and T.
   (iii) In each case, describe the graph.
   (iv) Are S and T the same set?

(b) (i) If $M = \{x \mid -3 < -x < 4\}$ and $N = \{x \mid -4 < x < 3\}$, which of the following are elements of $M \cap N$?

$$-4, -1, 0, 1, 2, 3, -3, -3\tfrac{1}{2}, -2$$

(ii) Graph M and N.
(iii) In each case, describe the graph.
(iv) Are M and N the same set?

Graph each of the following sets and describe the graph.

10. (a) $0 = \{x \mid |x| < 7\}$
(b) $P = \{x \mid -2 < -x < 3\}$

11. (a) $Q = \{x \mid -13 \leqq -x < 0\}$
(b) $R = \{x \mid 0 < -x \leqq 15\}$

In Exercises 12–14, graph each of the sets and describe the graph.

12. (a) $A = \{x \mid |x| \leqq 7\}$
(b) $B = \{x \mid |x| < 2\}$

13. (a) $C = \{x \mid |x| \geqq 6\}$
(b) $D = \{x \mid |x| > 2\}$

14. (a) $E = \{x \mid |x| < 3\}$
(b) $F = \{x \mid |x| \geqq 3\}$

15. (a) Write a description of each set in Exercises 12(a), 13(a), and 14(a) without using absolute value signs.
(b) Write a description of each set in Exercises 12(b), 13(b), and 14(b) without using absolute value signs.

---

16. (a) Sets A, B, and C are defined in the following manner.

$$A = \{x \mid |x| \leqq 3\}, \quad B = \{x \mid 0 \leqq x \leqq 3\}, \quad C = \{x \mid 0 < -x \leqq 3\}$$

To which of the above sets, if any, does each of the following numbers belong?

$$-4, -3, -2, -1, 0, \tfrac{1}{2}, \tfrac{3}{2}, 3, 4$$

(b) Describe $B \cap C$.
(c) Describe $B \cup C$.
(d) Graph set A and describe the graph that you obtain.
(e) Describe $B \cap A$.
(f) Describe $C \cap A$.

4–7 FLOW CHARTS

Flow charts can be used to help list all objects of a set. For example, if you wanted to list all multiples of 2 between 0 and 99, that is, the set $\{x \mid 0 < x < 99$ and x is a multiple of $2\}$, you might use the following.

1. Let $x = 1$.
2. Multiply $2 \cdot x$.
3. Is $2 \cdot x < 99$?
   If yes, go to step 4.
   If no, go to step 7.
4. Write down $2x$.
5. Replace x by $x + 1$.
   Call this new number x.
6. Go back to step 2.
7. You have finished.

The flow chart would look like this:

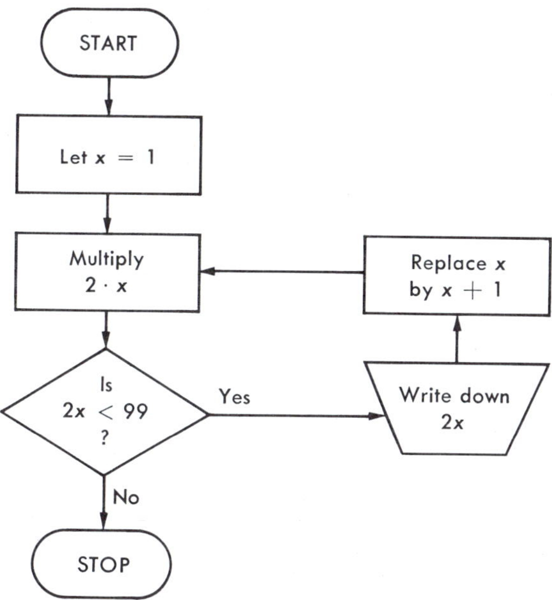

FIGURE 4–17

An alternative method is shown in Fig. 4–18. Will this flow chart give all multiples of 2 between 0 and 99?

When you write a flow chart which gives the procedure for producing the list of elements in a set, you can say it is a flow chart for *generating*

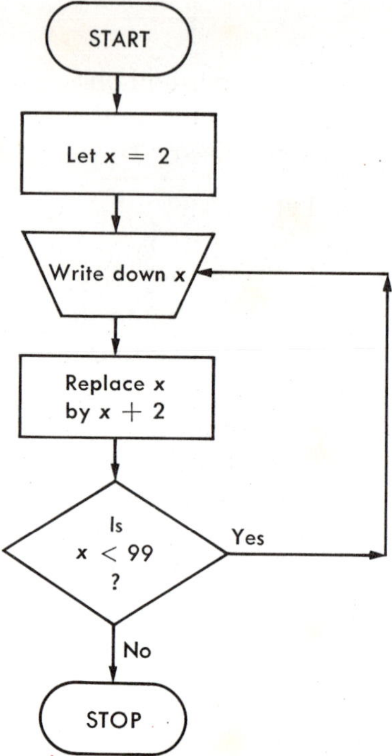

FIGURE 4-18

these elements. In the examples above, you would say the flow charts generate all multiples of 2 between 0 and 99.

Exercises

1. (a) Write an algorithm and draw a flow chart for generating all elements of the set

$$R = \{x \mid 0 < x < 75 \text{ and } x \text{ is a multiple of } 5\}.$$

(b) Write an algorithm and draw a flow chart for generating all elements of the set

$$T = \{y \mid 0 < y < 50 \text{ and } y \text{ is a multiple of } 3\}.$$

2. (a) Using the flow chart you drew in 1(a), list all elements of R.
(b) Using the flow chart you drew in 1(b), list all elements of T.

3. (a) Write an algorithm and draw a flow chart for generating all elements of the set

$$A = \{n^2 \mid 0 < n < 10\}.$$

(b) Write an algorithm and draw a flow chart for generating all elements of the set

$$L = \{k + 2 \mid 8 \leqq k \leqq 15\}.$$

4. (a) Using the flow chart you drew in 3(a), list all elements of A.
(b) Using the flow chart you drew in 3(b), list all elements of L.

5. (a) Write an algorithm and draw a flow chart for generating all elements of the set

$$D = \{2m \mid 0 < m < 5\}.$$

(b) Write an algorithm and draw a flow chart for generating all elements of the set

$$K = \{5a \mid 2 < a < 9\}.$$

6. (a) Using the flow chart you drew in 5(a), list all elements of D.
(b) Using the flow chart you drew in 5(b), list all elements of K.

In the following exercises, describe the set which would be generated by the flow charts.

7. (a) (b)

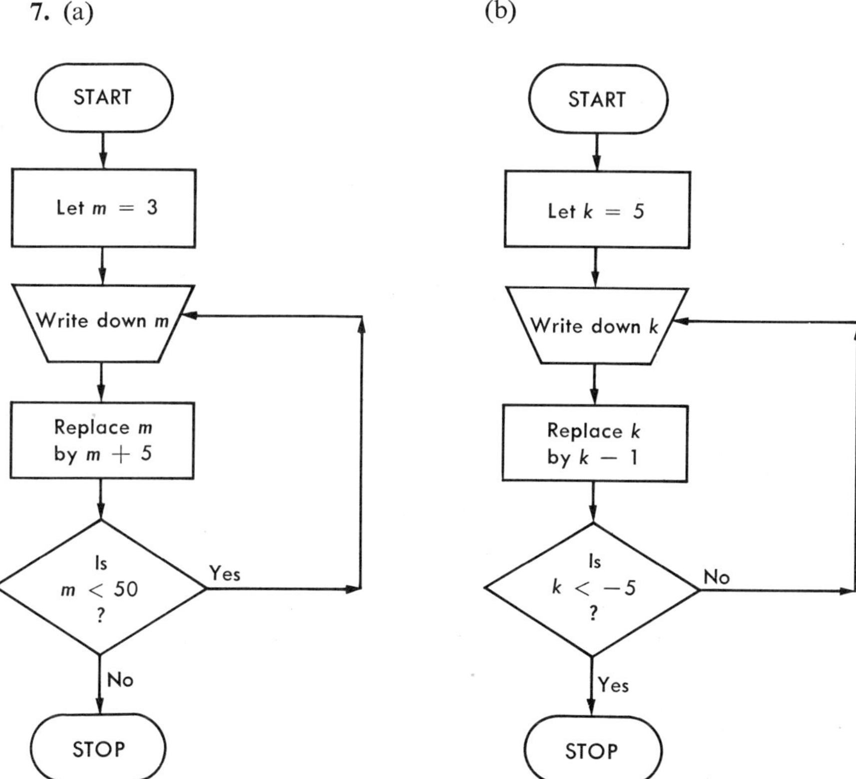

8. (a) (b)

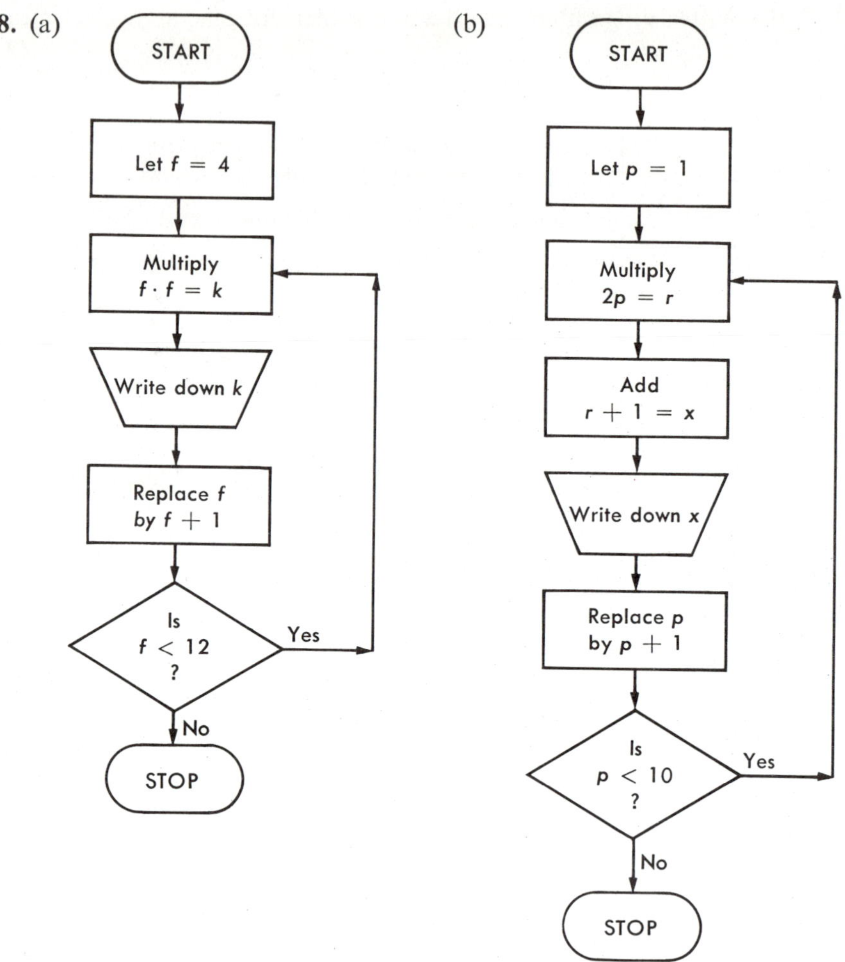

KEY IDEAS AND KEY WORDS

Greater than is denoted by $>$ and **less than** is denoted by $<$.

By definition of **greater than,** $x > y$ if, and only if, $x - y$ is positive. By definition of **less than,** $x < y$ if, and only if, $x - y$ is negative.

The **trichotomy axiom** states that $x > y$, $x = y$, or $x < y$.

The **transitive law** states that if $x > y$ and $y > z$, then $x > z$.

A **solution** of an inequality in one variable is a value of the variable for which the inequality is true. The **solution set** of an inequality is the set of all its solutions. Two **inequalities** are called **equivalent** if they have the same solution set.

The **graph of a number** is the point on a number line associated with the number. The **graph of a set of numbers** is the set of graphs of the individual numbers in the set.

A point P divides a line into two parts. One of these parts together with the point P is called a **ray.** An **open ray** is a ray without its endpoint.

If P and Q are two points on a line, the set of all points on the line between P and Q is called an **open interval.** The set consisting of P and Q and all points between P and Q is a **closed interval.** The open interval together with one, but not both, of the points P and Q constitutes a **half-open interval.**

A **flow chart** gives a method for generating all elements of a set if it gives a procedure for producing the list of all elements in the set.

CHAPTER REVIEW

1. Use the symbols $>$ and $<$ to make a correct comparison between the first and second numbers in the given pairs.

(a) $-1, 5$ (b) $-3, -10$

(c) $0, -6$ (d) $112{,}000; 121{,}000$

(e) $.5, \frac{1}{3}$ (f) $.070, .007$

(g) $\frac{5}{6}, \frac{5}{7}$ (h) $-.61, 6$

2. List five elements of each of the following sets.

(a) $S = \{x \mid x < -2\}$

(b) $S = \{x \mid 0 < x < 1, x \text{ a real number}\}$

(c) $S = \{x \mid x < 100\}$

(d) $S = \{x \mid -3 < x < 1000\}$

3. Is $x + 10 > x + 4$ for all integers x?

4. Is $4x > 3x$ for all integers x?

In Exercises 5–11, tell which of the statements are true and which are false.

5. The graph on a number line of the set $\{x \mid x > -5\}$ is an open ray.

6. The graph on a number line of the set $\{x \mid |x| \geqq 5\}$ is two rays.

7. The set $\{x \mid x \geqq 0\}$ has no least element.

8. The set $\{x \mid x \leqq -2\}$ has no greatest element.

9. The set $\{x \mid |x| \leqq 1\}$ has for its graph a closed interval from -1 to 1.

10. The intersection of the graphs of the sets in Exercises 5 and 6 is one ray.

11. The union of the set $\{x \mid x > 2\}$ and the set $\{x \mid -x > 2\}$ is the same as the set $\{x \mid |x| > 2\}$.

12. Graph each of the following sets on a number line.

(a) $A = \{x \mid x > -4, x \text{ an integer}\}$
(b) $B = \{x \mid x \leqq 5, x \text{ a real number}\}$
(c) $C = \{x \mid |x| < 6, x \text{ a real number}\}$
(d) $D = \{x \mid |x| \geqq 6, x \text{ a real number}\}$
(e) $E = B \cap D$
(f) $F = C \cup D$

13. (a) Name three elements in each of the following sets.

$$A = \{x \mid -x > 5\}$$
$$B = \{x \mid 0 < -x < 3\}$$

(b) Is there a least integer in set A? If so, what is it? Is there a greatest integer in set A? If so, what is it?

(c) Is there a greatest integer in set B? If so, what is it? Is there a least integer in set B? If so, what is it?

Solve the following inequalities. If the solution sets have least or greatest elements, tell what they are.

14. $x + 5 > 7$

15. $x - 2 > 4$

16. $2x + 6 > 10$

17. $4x - 8 > 12$

18. $7x + 10 > 3x + 20$

19. $9x - 3 > 4x + 7$

20. $5x + 2 \geqq 3x + 8$

21. $8x - 5 \leqq 2x + 3$

22. $8x > 12x - 2$

23. $6(x + 2) > 18$

24. $2(x + 1) + 7 \leqq 14$

25. $5(x - 1) \geqq 2(3x - 4)$

26. $\frac{x}{2} + 5 \leqq \frac{x}{3} + 6$

27. $\frac{x}{7} - 3 \geqq \frac{x}{3} + 4$

Write an algorithm and draw a flow chart for generating all numbers of the following sets.

28. $A = \{m \mid 0 < m < 50 \text{ and } m \text{ is a multiple of } 4\}$

29. $B = \{k \mid 0 < k < 100 \text{ and } k \text{ is divisible by } 6\}$

CHAPTER TEST

1. Which of the following sentences are true?

(a) $-1 < -2$ (b) $|x| > 0$ (c) $\frac{3}{8} > \frac{6}{4}$
(d) $|-3| < |-5|$ (e) $-|5| < |-4|$ (f) $5 - 2 > 2 - 5$

2. List two elements in each of the following sets.

(a) $S = \{x \mid 5 < x < 10\}$
(b) $S = \{x \mid -1 < x < 1, x \text{ a real number}\}$

In Exercises 3–11, solve the inequalities.

3. $3 + x > 4$ **4.** $4 - x > 5$ **5.** $2 - 3x > 4 - x$

6. $12x - 1 < 3x - 37$ **7.** $6x - 3 < 42 - 3x$

8. $3x - 8 < 5(2x - 3)$ **9.** $8x - 9 < 21 - 7x$

10. $3x - 7 > 5x + 9$ **11.** $\frac{x}{6} + \frac{1}{2} < \frac{1}{3} - \frac{x}{2}$

12. Find the solution set for the following inequality. List three members of the solution set.

$$\frac{x}{2} + \frac{x}{5} + 6 \geqq x - \frac{x}{10}$$

CUMULATIVE REVIEW I

1. Write the following numbers in scientific notation.
(a) 100 (b) 78
(c) 1100 (d) 8957

2. Find the products.
(a) $8 \cdot 10^4$ (b) 3.68×10^3 (c) 7.94321×10^2

3. Give the products in exponential form.
(a) $3^2 \cdot 3^5$ (b) $2^3 \cdot 2^3$ (c) $5^3 \cdot 5^4$

4. Find the decimal representation of the following rational numbers.
(a) $\frac{4}{5}$ (b) $\frac{7}{8}$
(c) $\frac{5}{6}$ (d) $\frac{1}{12}$

5. Set A consists of the numbers 1, 2, 3, 4, 5, 6, 7; set B consists of the numbers 2, 7, 8, 9, 10, 4. List the elements of the following sets.
(a) $A \cap B$ (b) $B \cap A$
(c) $A \cup B$ (d) $B \cup A$

6. In Exercise 5, is A a subset of B?

7. In Exercise 5, is B a subset of A?

8. In Exercise 5, list all the subsets of A that contain three elements.

9. Write an expression with variables for each of the following phrases.
(a) The difference of two numbers
(b) The product of three numbers
(c) The sum of two consecutive integers
(d) The product of a number and the sum of two other numbers
(e) Eight less than five times a number
(f) The sum of a number and one-half the same number

10. Let x denote Jeff's present age in years. Write mathematical expressions for the following phrases.

(a) Twice his present age
(b) His age three years ago
(c) His age in seven years
(d) Two years less than three times his present age

11. A coin collection contains p pennies, n nickels, and d dimes. Give a mathematical expression for each of the following phrases.

(a) The number of coins
(b) The value in cents of the pennies and nickels
(c) The value in dollars of the pennies and nickels
(d) The value in dollars of the nickels and dimes
(e) The value in cents of the entire collection

12. Let u be the units' digit, t the tens' digit, and h the hundreds' digit for an integer between 100 and 1000. Write a mathematical expression for

(a) the integer.
(b) the sum of its digits.
(c) a new integer which is the original integer with the digits reversed.

13. There are five ribbons of equal length in a paper bag. Two are red, one is green, one is white, and one is blue. If, without looking, you were to draw one out of the bag, what is the probability that

(a) it would be red?
(b) it would be white?
(c) it would be either green or blue?
(d) it would be red or green?
(e) it would be orange?

14. Write an algorithm and draw a flow chart for:

(a) adding three 1-digit integers (that is $a + b + c$).
(b) generating the elements of the set

$$\{n + 3 \mid 0 < n \leqq 10, \text{ and } n \text{ is an integer}\}.$$

15. Are each of the following statements true, false, or open?

(a) $7 + 5 = 1$ (b) $7 + 4 = x$ (c) $7 + 4 = 1 + 10$
(d) $5 + 6 = 12$ (e) $x + 5x = 0$ (f) $3^2 + 2^2 = 5^2$

16. Find the following sums.

(a) $5 + 6$
(b) $-7 - 8$
(c) $7 - 8$
(d) $-7 + 8$
(e) $2 - 3 - 4 + 5$

What property or properties of real numbers make the following statements true?

17. If $x = y$, then $x + 3 = y + 3$.

18. $wz = zw$

19. $(5 + 6) - 7 = 5 + (6 - 7)$

20. $3x + 5 - x + 7 = 3x - x + 5 + 7 = 2x + 12$

21. $3(2 + 4) = 6 + 12$

22. $5 - 5 = 0$

23. $10 \cdot \frac{1}{10} = 1$

24. If $x + y = x + z$, then $y = z$.

25. If $x > y$ and $y > z$, then $x > z$.

Perform the indicated operations.

26. $5(x + y)$

27. $2(x - y) + 7(y - x) + 3x$

28. $(4xy)(5x + y)(\frac{1}{2}xy)$

29. $(9x^2y)(\frac{1}{3}xy)^2(2y)$

30. $(2x - 3y) - (x + y)$

31. $(2.1x + .21y) + (.02x + 3.2y)$

32. $(\frac{2}{3}ab + \frac{4}{7}x) + (\frac{7}{10}ab + \frac{5}{3}x)$

33. $(4xy^2) \cdot (\frac{3}{8}x^2y) \cdot (\frac{10}{21}x^3y^3)$

34. $(7x^2 + 8y^2) + (x^2 + y^2)$

35. $(9x + 7y + 8) + (13x + 14y + 9)$

36. $\left(\frac{3}{2x} + \frac{5}{3y}\right) + \left(\frac{7}{x} \cdot \frac{2}{y}\right)$

Classify each statement as true or false for all values of the variables.

37. $(8 + 80) \cdot a = a \cdot (80 + 8)$

38. $9 + (b \cdot 37) = (b \cdot 37) + 9$

39. $12 \cdot (a + b) = (a + b) \cdot 12$

40. $(x - 5) - 2 = x - (5 - 2)$

41. $(19 + x) \cdot 8 = 8 \cdot (19 + x)$

42. $(19 - x) - 4 = 19 - (x - 4)$

Replace the $*$ by an $=$ or $\neq$ to make each of the statements true.

43. $82 + (9 \cdot 13) * (13 \cdot 9) + 82$

44. $7 - 2 * 2 - 7$

45. $21^3 \cdot 4^3 * 4^3 \cdot 21^3$

46. $47 + 64 * 56 + 65$

47. $15 \cdot 91 * 35 \cdot 39$

48. $31 \cdot 0 * 0 \cdot 29$

49. $0 + 16 * 16 - 0$

50. $(.7 + .07) \cdot 9 * (.07 + .7) \cdot 9$

51. $(3 + \frac{1}{5}) \cdot 5 * 5 \cdot (\frac{1}{5} + 3)$

In Exercises 52–63, perform the indicated operations.

52. $12x - 9y - 20x + 3y$

53. $9(x - 3y) - 4(2x - 5y)$

54. $13a - (7a - 3a)$

55. $30b - (16b - 34b)$

56. $19xy - (7xy - 15xy)$

57. $(3a - 4c) - (2a + 5c)$

58. $2(4x - 3y) + 5(4y - 3x)$

59. $2(b - a) + 3(a - b)$

60. $4(2x - 3y) - (3x - 2y)$

61. $-3[7(a - 2b) - 4(b - 2a)]$

62. $24x^3 \cdot y^2 \div 30xy^3$

63. $\left(\frac{9}{x} \cdot \frac{4}{y}\right) + \left(\frac{6}{x} \cdot \frac{5}{y}\right)$

64. Let $x = 7$ and $y = 9$, and find the value of each expression.

(a) $(8x + 3y) + (x + y)$

(b) $(\frac{2}{3}x + \frac{3}{4}y) + (\frac{4}{3}x + \frac{5}{4}y)$

(c) $(27xy^2) \div (9xy)$

(d) $(\frac{3}{7}xy) \cdot (\frac{4}{9}xy)$

(e) $(9x^3y^2) \div (7x^2y)$

(f) $11(x + 2y) + 9(x + 2y)$

65. If x is the multiplicative inverse of y, what does $x \cdot y$ equal?

66. Describe the set of rational numbers whose reciprocals are integers.

67. (a) Is the reciprocal of the sum of two numbers the same as the sum of the reciprocals? Give three examples that confirm your opinion.

(b) Do you think that the equation

$$\frac{1}{a + b} = \frac{1}{a} + \frac{1}{b}$$

is true for all real numbers a and b? Explain your answer.

Find the solution set of each of the equations.

68. $3x = 9$

69. $5x + 10 = 15$

70. $2x - 3 = 5$

71. $4x - 5 = 4 - 2x$

72. $3 - 2x = 7 - 4x$

73. $4 - 3x = 6 - 4x$

74. $\frac{x}{2} = 7$

75. $\frac{x}{3} = 7 - \frac{x}{6}$

76. $\frac{x + 7}{5} + 4 = \frac{x - 3}{10}$

77. $3(x + 4) = 24$

78. $2(x - 6) + 5 = 9$

79. $4(x - 7) + 2(3 - x) = 4$

80. $3(7 - x) = 5(x + 3)$

81. Solve each of the following inequalities. If the solution set has a largest or smallest element, tell what it is.

(a) $x + 5 > 10$

(b) $6 - x > 2$

(c) $x - 7 \geqq 8 - 2x$

(d) $(x + 5) - (3 - x) \leqq 7 - x$

82. Solve the following equations for the indicated variable.

(a) $A = LW$. Solve for W.

(b) $A = p + prt$. Solve for t.

(c) $3x + 6 = 5y + 2$. Solve for y.

(d) $4w + 2 = 2x - 2$. Solve for x.

83. The sum of Jeff's age and Jim's age is 12. Two years ago Jeff was three times as old as Jim. What is Jeff's age now?

84. The sum of three consecutive positive integers is 36. What are the integers?

85. The difference between one-half a number and one-fourth a number is four. What is the number?

86. A merchant wishes to combine one grade of tea worth 50¢ a pound with another grade of tea worth \$1.50 a pound to obtain 10 pounds of tea worth 75¢ a pound. How many pounds of the \$1.50 tea can he use?

87. Dick has \$4.85 in coins. If he has twice as many quarters as dimes and six fewer dimes than nickels, how many coins of each type does he have?

88. A plane averaging 600 miles per hour can make a trip from Miami to Seattle in $52\frac{1}{4}$ hours less time than it takes a train averaging 50 miles per hour. How long does the flight take?

89. Find I if $I = Prt$ and $P = 3000$, $r = .05$, $t = 2$.

90. Find T if $T = I/Pr$ and $I = 40$, $P = 2000$, $r = .01$.

91. Find F if $\text{F} = \frac{9}{5}\text{C} + 32$ and $\text{C} = 55$.

92. Find S if $S = \frac{1}{2}gt^2$ and $g = 32$, $t = 8$.

93. Find g if $S = \frac{1}{2}gt^2$ and $S = 800$, $t = 10$.

Graph each of the following sets on a number line, and describe each graph.

94. $A = \{x \mid x \leqq 5, x \text{ a real number}\}$

95. $B = \{x \mid x > -5, x \text{ a real number}\}$

96. $C = A \cap B$

97. $D = \{x \mid |x| \geqq 3\}$

CHAPTER **5**

Polynominals in One Variable

Objectives . . .

- To explain why a given expression is or is not a polynomial.
- To apply the basic arithmetic operation to polynomials.
- To factor polynomial expressions.
- To find quotients of polynomials by the long division method.
- To solve polynomials equations.

5–1 POLYNOMIALS IN ONE VARIABLE

The word *polynomial* was originally used in mathematics to describe any algebraic expression which consisted of a sum of other algebraic expressions.

In modern mathematics, the word *polynomial* is used in a more specialized way. The expressions which are now called polynomials are considered to be basic objects in a new algebraic system. In this system, polynomials are added and multiplied just as numbers are added and multiplied in the real number system. Furthermore, the fundamental properties of addition and multiplication of real numbers are valid in the system of polynomials.

Each of the algebraic expressions

$$3x + 4 \quad \text{and} \quad x^3 + \tfrac{3}{2}x^2 - 7x - \tfrac{5}{3}$$

is an example of a polynomial with x as its variable.

The polynomial $3x + 4$ is a sum of two *terms,*

$$3x \quad \text{and} \quad 4.$$

The number 3 in the term $3x$ is called the *coefficient* of x; the term 4 which does not contain x is called the *constant term.*

The polynomial

$$x^3 + \tfrac{3}{2}x^2 - 7x - \tfrac{5}{3},$$

or

$$x^3 + \tfrac{3}{2}x^2 + (-7)x + (-\tfrac{5}{3}),$$

is a sum of four *terms:*

$$x^3, \quad \tfrac{3}{2}x^2, \quad -7x, \quad -\tfrac{5}{3}.$$

The first term, x^3, is of the form $x \cdot x \cdot x$, or the third power of x. Since $x^3 = 1 \cdot x^3$, the *coefficient* of x^3 is 1. The second term, $\frac{3}{2}x^2$, is of the form $\frac{3}{2} \cdot x \cdot x$, and its *coefficient* is $\frac{3}{2}$. The third term, $(-7)x$, has coefficient -7, and the fourth term, $-\frac{5}{3}$, is the *constant term.*

You may recall from Chapter 1 that the expressions

$$x^1 = x, \quad x^2 = x \cdot x, \quad x^3 = x \cdot x \cdot x, \quad x^4 = x \cdot x \cdot x \cdot x,$$

and so on, are called *positive integral powers of x.*

A polynomial, then, is defined in the following manner.

Definition of polynomial

A polynomial in x is either a term or a sum of terms, and each term is either a number or a product of a number by a positive integral power of x.

An example of an algebraic expression that is not a polynomial is

$$x^2 + 1 + \frac{3}{x^2}.$$

Like a polynomial, it is a sum of terms. However, one of the terms,

$$\frac{3}{x^2},$$

is not in the proper form; it is not a number multiplied by a positive integral power of x. Instead, it is a number divided by a power of x.

The variable in a polynomial need not be x. For example,

$$a^2 - 5a + 3$$

is a polynomial in a, and

$$-6y + 4$$

is a polynomial in y.

The sum or difference of two polynomials in x is also a polynomial in x. For example, using the rearrangement property for addition and the distributive axiom, you find that

$$\begin{aligned}(x^2 - 3x + 5) + (2x + 8) &= x^2 + [(-3)x + 2x] + (5 + 8)\\ &= x^2 + (-3 + 2)x + 13\\ &= x^2 - x + 13.\end{aligned}$$

Also,

$$\begin{aligned}(5x^2 - 3x + 9) - (7x^2 + 4x - 5)\\ &= (5x^2 - 3x + 9) + [-(7x^2 + 4x - 5)]\\ &= 5x^2 - 3x + 9 - 7x^2 - 4x + 5\\ &= (5x^2 - 7x^2) + (-3x - 4x) + (9 + 5)\\ &= (5 - 7)x^2 + (-3 - 4)x + 14\\ &= -2x^2 - 7x + 14.\end{aligned}$$

Polynomials are useful in mathematics partly because of the ease with which their values can be computed. For example, the polynomial

$$x^2 - 3x + 2$$

has the value

$$\begin{aligned}3^2 - (3 \cdot 3) + 2 &= 9 - 9 + 2, \text{ or } 2, \quad \text{when } x = 3;\\ 1^2 - (3 \cdot 1) + 2 &= 1 - 3 + 2, \text{ or } 0, \quad \text{when } x = 1;\\ \left(\frac{3}{4}\right)^2 - \left(3 \cdot \frac{3}{4}\right) + 2 &= \frac{9}{16} - \frac{9}{4} + 2\\ &= \frac{9 - 36 + 32}{16}, \quad \text{or } \frac{5}{16}, \text{ when } x = \frac{3}{4}.\end{aligned}$$

Exercises

Some of the following algebraic expressions are polynomials. For each expression that is a polynomial, give the constant term and the coefficients of the powers of x.

1. (a) $2x^3 - x^2 + x - \frac{5}{4}$ (b) $3x^2 - 5x - \frac{1}{3}$

2. (a) $3x + \dfrac{5}{x}$ (b) $\dfrac{7}{x^2 - x + 5}$

3. (a) $7x^4 - \sqrt{2}\,x^3$ (b) $\sqrt{3}\,x^3 - 5x^2$

Find the value of the polynomials for each of the given values of the variable.

4. (a) $x^2 - 5x + 6$, for $x = -2, -1, 0, 1, 2, \frac{5}{2}, 3, 4$
(b) $4 - 9y^2$, for $y = -3, -2, -1, -(\frac{2}{3}), 0, \frac{2}{3}, 1, 2, 3$

5. (a) $4x^2 + 8x - 5$, for $x = -3, -(\frac{5}{2}), -2, -1, 0, \frac{1}{2}, 1$
(b) $\frac{1}{3}x^3 - 2x^2 + 3x + 1$, for $x = -2, -1, 0, 1, 2, 3, 4, 5$

In Exercises 6–9, add the pairs of polynomials.

6. (a) $\begin{array}{l} 3t^2 + 2t + 1 \\ \underline{2t^2 - 6t - 3} \end{array}$ (b) $\begin{array}{l} a^2 + 5a - 9 \\ \underline{3a^2 - 20a - 15} \end{array}$

7. (a) $\begin{array}{l} x^2 - y^2 \\ \underline{\frac{1}{2}x^2 + \frac{1}{2}y^2} \end{array}$ (b) $\begin{array}{l} x^3 + y^3 \\ \underline{\frac{1}{3}x^3 - \frac{1}{3}y^3} \end{array}$

8. (a) $\begin{array}{r} 7v^2 \qquad\quad + 1 \\ -6v^2 + 3v - 1 \\ \hline \end{array}$ (b) $\begin{array}{l} 1 \qquad\quad - x^3 \\ 2 + x + x^3 \\ \hline \end{array}$

9. (a) $\begin{array}{r} 3r^4 + 2r^2 + 1 \\ 5r^2 - 7 \\ \hline \end{array}$ (b) $\begin{array}{l} 6 - 5y - 7y^3 \\ -7 + 8y \\ \hline \end{array}$

In each of the exercises, subtract the lower polynomial from the one above.

10. (a) $\begin{array}{l} 1 + 64x^6 \\ 1 - 64x^6 \\ \hline \end{array}$ (b) $\begin{array}{l} 3 - 4x^5 \\ 3 + 4x^5 \\ \hline \end{array}$

11. (a) $\begin{array}{r} -4u^3 + 2u^2 - u + 5 \\ 2u^3 - u^2 + 3u - 5 \\ \hline \end{array}$ (b) $\begin{array}{r} y^3 - 2y^2 + y - 6 \\ 3y^3 - y^2 + y - 6 \\ \hline \end{array}$

12. (a) $\begin{array}{l} 2n^6 - 3n^4 + n^2 \\ n^6 \qquad\quad + n^2 \\ \hline \end{array}$ (b) $\begin{array}{l} m^3 \qquad\qquad - 1 \\ m^3 - 2m^2 + 1 \\ \hline \end{array}$

13. (a) $\begin{array}{l} x^3 \qquad\qquad\quad - 27 \\ 2x^3 - 3x^2 + x - 5 \\ \hline \end{array}$ (b) $\begin{array}{l} x^3 \qquad\qquad + 2x \\ 3x^3 - 5x^2 + x - 6 \\ \hline \end{array}$

---

14. Find the value of each polynomial when $x = 0$, 1, and -1, respectively.
(a) $3x^2 - 7x + 4$
(b) $3x^8 - 7x^7 + 4$

15. (a) What general statement can be made about the value of a polynomial when the variable is given the value zero?
(b) Describe a quick way to evaluate a polynomial when the variable is given the value 1. (Hint: Compare the value of $30x^2 - 17x + 9$ for $x = 1$ with the number $30 - 17 + 9$.)
(c) State a rule to use for the evaluation of a polynomial in x if $x = -1$.

16. If the sum of two polynomials is zero, each is called the additive inverse, or negative, of the other. Find the additive inverse of each of the polynomials.
(a) $x^2 - 3x + 1$
(b) $-5x^2 + 7x + 10$

17. Fill the blank to make each statement true. What could you call the addend you are supplying?
(a) $(3x + 5) + \underline{\ ?\ } = 0$
(b) $(7x - 2) + \underline{\ ?\ } = 0$
(c) $\underline{\ ?\ } + (6 - 2x - 3x^2) = 0$

18. Find the additive inverse of each polynomial given below.
(a) $a^3 - 2a^2 + 6$
(b) $x^2 + x + 1$
(c) $2b^4 - 3b^2 - 2$
(d) $-3x^3 - x^2 - 5$
(e) $y^7 + 10^7$
(f) $10^6 - y^6$

19. Give a reason or reasons for the first two steps in the following addition of two polynomials.

$$\begin{aligned}(3x^2 - 5x - 7) + (-x^2 + 9x + 3)\\ &= [3x^2 + (-x^2)] + (-5x + 9x) + (-7 + 3)\\ &= [3 + (-1)]x^2 + [(-5) + 9]x + (-4)\\ &= 2x^2 + 4x - 4\end{aligned}$$

20. (a) Which axiom justifies the following statement?

$$3(2x^2 - 9x - 12) = 6x^2 - 27x - 36$$

(b) If a polynomial is multiplied by a nonzero constant, what kind of algebraic expression is the result? Do the powers of the variable change? What does change?

Find the sums or differences of the following polynomials.

21. $(3a^2 - 2a + 5) + (-2a^2 + 5a - 1)$

22. $(6a^2 - 2a + 5) - (10a^2 - 3a + 2)$

23. $(\frac{1}{3}y^2 - \frac{3}{4}y + 6) - (-\frac{2}{3}y^2 + \frac{5}{8}y + 1)$

24. $(7b^5 - 3b^3 + b^2) + (-7b^5 + 3b^3 - b^2)$

25. $(3x^2 - 5x + 7) - (8x - 9)$

26. $0 - (9 - 2x + x^5)$

27. $(y + 3) + (-3y - 1) - (y - 1)$

28. $(2x^2 + 3x - 1) - (x^2 - x - 1) + (4x^2 - 3x + 5)$

29. $2(-3y^2 + 2y - 5) + 5(3y^2 - 2y + 5)$

30. $3(6m^2 + 12m - 11) - 2(m^2 + m + 1)$

31. The sides of a triangle are designated by $2x - 5$, $3x - 4$, and $3x + 4$. Write an expression for the perimenter.

32. The length of a rectangle is indicated by $5x + 7$ and the width by $3x - 2$. Give an expression for the perimeter.

33. Write an expression for the perimeter of a square if each side is designated by $\frac{1}{4}s - 5$.

34. By how much does the perimeter of the rectangle in Exercise 32 exceed the perimeter of the triangle in Exercise 31?

---

35. (a) If the domain of x is the set of positive real numbers, what can you predict about the values of each of the following polynomials?

$$3x^4 + x^2 + 1 \qquad -4x^3 - 3x - 5$$

(b) If the domain of x is the set of negative real numbers, what can you predict about the values of the polynomials in part (a)?

36. Suppose the sides of a triangle have lengths $2x + 5$ inches, $3x - 4$ inches, and $3x + 4$ inches.

(a) What is the domain of x, assuming that it is the set of all numbers for which all three sides have positive lengths?

(b) Find an expression for the perimeter of the triangle.

(c) If $x = 5$, what is the perimeter?

(d) Which two sides of this triangle could have the same length? If they were the same length, what value would x have? What is the length of the third side if two sides have equal length?

(e) Get a further restriction on the domain of x by using this fact: The sum of the lengths of any two sides of a triangle is greater than the length of the third side.

5–2 FIRST LAW OF EXPONENTS

A polynomial in x that consists of only one term is called a *monomial.* For example,

$$7x^2 \quad \text{and} \quad -x^5$$

are monomials. The number 7 in $7x^2$ is called the *coefficient,* and 2 is called the *exponent.* The monomial $-x^5$, or $(-1)x^5$, has a coefficient of -1 and an exponent of 5.

Every polynomial is either a monomial or a sum of monomials. If a polynomial is a sum of two monomials, it is called a *binomial;* if it is a sum of three monomials, it is called a *trinomial.*

The product of two monomials can be found by using the rearrangement property of multiplication. For example, you can find $(3x^2) \cdot (2x^5)$ as follows:

$$\begin{aligned}(3x^2) \cdot (2x^5) &= (3 \cdot x \cdot x) \cdot (2 \cdot x \cdot x \cdot x \cdot x \cdot x)\\ &= 3 \cdot 2 \cdot x \cdot x \cdot x \cdot x \cdot x \cdot x \cdot x\\ &= 6x^7.\end{aligned}$$

As this example indicates, the product of two or more monomials in x is also a monomial in x.

You may recall from Chapter 1 that an exponent indicates how many times x is to be used as a factor. For example,

$$x^3 = x \cdot x \cdot x \quad \text{and} \quad x^{25} = x \cdot \ldots \cdot x.$$

The three dots appearing between the two multiplication dots in the last expression indicate that many more factors of x belong there (in this case, 23).

It is not necessary to write two monomials in factored form in order to find their product. For example, you can find

$$x^{121} \cdot x^{73}$$

without writing one hundred twenty-one factors of x and then writing seventy-three more factors of x. You would reason in the following manner.

$$\begin{aligned} x^{121} \cdot x^{73} &= \overbrace{x \cdot \ldots \cdot x}^{121} \cdot \overbrace{x \cdot \ldots \cdot x}^{73} \\ &= \overbrace{x \cdot \ldots \cdot x}^{121+73} \\ &= x^{194} \end{aligned}$$

This example illustrates a general rule, which is stated below.

FIRST LAW OF EXPONENTS

The equation

$$x^m \cdot x^n = x^{m+n} \qquad \text{(LE-1)}$$

is true for all positive integers m and n and every number x.

The following problem shows how the first law of exponents may be used to simplify products of monomials.

Problem 1. Find each of the products.
(a) $(3x)(7x^5)$
(b) $(3x^2)(\frac{1}{6}x^4)$
(c) $(-2x^{23})(-5x^{79})$

Solution. You may perform the computations in the following manner.
(a) $(3x)(7x^5) = (3 \cdot 7)x^{1+5}$, or $21x^6$
(b) $(3x^2)(\frac{1}{6}x^4) = (3 \cdot \frac{1}{6})x^{2+4}$, or $\frac{1}{2}x^6$
(c) $(-2x^{23})(-5x^{79}) = (-2) \cdot (-5)x^{23+79}$, or $10x^{102}$

The distributive axiom tells you that the product of a monomial and a polynomial is a sum of the products of the monomials. This is illustrated in the following problem.

Problem 2. Perform the operations and simplify your answers.

(a) $5x^2(3x - 2)$ (b) $3x^3(7 - 2x) + 5x(x^2 - 3x + 2)$

Solution. Using the distributive axiom and the rearrangement property, proceed in the following manner.

(a) $5x^2(3x - 2) = (5x^2 \cdot 3x) + [5x^2 \cdot (-2)]$
$= 15x^3 - 10x^2$

(b) $3x^3(7 - 2x) + 5x(x^2 - 3x + 2)$
$= [3x^3 \cdot 7 + 3x^3 \cdot (-2x)] + [5x \cdot x^2 + 5x \cdot (-3x) + 5x \cdot 2]$
$= 21x^3 - 6x^4 + 5x^3 - 15x^2 + 10x$
$= -6x^4 + (21 + 5)x^3 - 15x^2 + 10x$
$= -6x^4 + 26x^3 - 15x^2 + 10x$

Exercises

Use the first law of exponents to find each of the products. Express your answer in exponential form.

1. (a) $x^3 \cdot x^5$ (b) $y^4 \cdot y^7$
2. (a) $y^{10} \cdot y^2$ (b) $r^{17} \cdot r^{19}$
3. (a) $r^4 \cdot r$ (b) $x^{1969} \cdot x$
4. (a) $5^6 \cdot 5^9$ (b) $10^{20} \cdot 10^{30}$
5. (a) $-2^{80} \cdot (-2^{90})$ (b) $-3^{21} \cdot (-3^{22})$
6. (a) $3^8 \cdot 3$ (b) $12^{21} \cdot 12$
7. (a) $t^{13} \cdot t^{15} \cdot t^{17}$ (b) $s^{17} \cdot s^{18} \cdot s^{19}$
8. (a) $u^{20} \cdot u^{22} \cdot u^{24}$ (b) $v^{30} \cdot v^{31} \cdot v^{32}$

Find the products.

9. (a) $x^2 \cdot x^3 \cdot x$ (b) $y^3 \cdot y \cdot y^4$
10. (a) $(-x^{12}) \cdot (x^{14})$ (b) $(-x) \cdot (x^{10})$
11. (a) $(-y^7)(-y^8)$ (b) $(-y^{17})(-y^{32})$
12. (a) $(3a^{20})(a^{50})$ (b) $(x^{17})(5x^{14})$
13. (a) $(4x^{11})(-3x^{21})$ (b) $(10x^{14})(-\frac{1}{5}x^{92})$
14. (a) $(-17y^{25})(-8y^{26})$ (b) $(-12x^{18})(-6x^{32})$

---

15. Give x the value 3 in the following monomials, and compare the resulting values.
(a) x^2 (b) $-x^2$ (c) $(-x)^2$ (d) $-(-x)^2$

16. Give x the value 2 in the following monomials, and compare the resulting values.
(a) x^3 (b) $-x^3$ (c) $(-x)^3$ (d) $-(-x)^3$

17. Compare the resulting values of the four monomials in Exercise 15 when x has the value -4.

18. Compare the resulting values of the four monomials in Exercise 16 when x has the value -5.

Use the distributive axiom, the first law of exponents, and the rearrangement properties to simplify the following expressions.

19. $-3(2x^2 - 5)$ **20.** $3x^2(4x - 5)$

21. $-11y^2(6y - 7)$ **22.** $7y(3y^2 - 2y + 5)$

23. $100x^2(.01x^4 - .1x^2 + .03)$ **24.** $-y^4(7y^6 - 14y^5 - 21y^4)$

25. $-a(1 - 3a - 9a^2)$ **26.** $-a^6(70 - 30a^2 - 20a^4)$

27. $4x(2x - 1) + x(3x - 2)$ **28.** $3x(4x - 5) + 2x(6 - 5x)$

29. $2x^2(x + 3) - 5x(x^2 + 3x)$ **30.** $15x^2 - 6x(x - 2) - x(5 + 10x)$

31. $y(3y - 2) - 2y(4 - y) - 3y^2$

32. $x(x - 1) - x(2 - x) - 3(x + 5) - 2x^2$

Find a simpler form for each of the following.

33. $(2x^4)^3$ **34.** $(3y^5)^2$

35. $(3^6)^2$ **36.** $(2^4)^3$

37. $(5a^3)(-4a)^3$ **38.** $(-10b^5)(-3b^2)^3$

39. $(2y^3)(-5y)^2$ **40.** $7^9 \cdot 2^9$

Find each of the products.

41. $(-4a^3)(3a^5)(-\frac{1}{4})$ **42.** $(-2)(-y^3)(-\frac{1}{2}y)$

43. $(-64x)(\frac{1}{16}x^3)(-\frac{1}{2}x^7)$ **44.** $(-\frac{1}{15}x^4)(-3x^5)(-2x^7)(10x)$

45. $(-y^5)(-\frac{1}{4}y)(-8y^6)(-\frac{1}{3}y^9)$ **46.** $(-\frac{1}{5}b^{20})(10b^{21})(\frac{1}{2}b^{22})$

47. $(-.04x^5)(.10x^6)(-1000x^7)$ **48.** $(-3y^2)(4y^5)(4y^5)$

Use the distributive axiom, the first law of exponents, and the rearrangement properties to simplify the expressions in Exercises 49–54.

49. $3y(y^3 - 3y^2 + 3y - 1) - 5y(2y^3 + 9y^2 - 6y + 8)$

50. $2b(4b^2 - 2b + 1) + (4b^2 - 2b + 1) - (8b^3 + 1)$

51. $x(x^3 + x^2 + x + 1) - (x^3 + x^2 + x + 1) - 2(x^4 - 1)$

52. $3y^2(y^6 - y^4 + y^2 - 1) + 3(y^6 - y^4 + y^2 - 1) - 2(y^8 - 1)$

53. $6y^2(6y^2 - 7) - 7(6y^2 - 7)$

54. $5b^6(25b^{12} - 5b^6 + 1) + (25b^{12} - 5b^6 + 1)$

55. Does the first law of exponents apply directly to the product $2^7 \cdot 6^8$? Express this product as a power of 2 times a power of 3.

5-3 PRODUCTS OF POLYNOMIALS

The product of two polynomials in x is also a polynomial in x. For example,

$$(2x - 4) \cdot (3x + 5)$$

is a polynomial. If you think of $2x - 4$ as a single algebraic expression, you can use the distributive axiom to multiply these two polynomials as follows:

$$(2x - 4)(3x + 5) = (2x - 4) \cdot 3x + (2x - 4) \cdot 5.$$

Use the distributive axiom again to find that

$$(2x - 4) \cdot 3x = 2x \cdot 3x + (-4) \cdot 3x$$

and

$$(2x - 4) \cdot 5 = 2x \cdot 5 + (-4) \cdot 5.$$

Therefore,

$$\begin{aligned}(2x - 4)(3x + 5) &= 2x \cdot 3x + (-4) \cdot 3x + 2x \cdot 5 + (-4) \cdot 5,\\ &= 6x^2 - 12x + 10x - 20,\\ &= 6x^2 + (-12 + 10)x - 20,\\ &= 6x^2 - 2x - 20.\end{aligned}$$

Thus, the product of these two polynomials in x is also a polynomial in x. As a second example, consider

$$(3a + 7) \cdot (4a^2 - a + 3).$$

The distributive axiom is used repeatedly to compute this product.

$$\begin{aligned}&(3a + 7)(4a^2 - a + 3)\\ &\quad = (3a + 7) \cdot 4a^2 + (3a + 7)(-a) + (3a + 7) \cdot 3\\ &\quad = 3a \cdot 4a^2 + 7 \cdot 4a^2 + 3a \cdot (-a) + 7 \cdot (-a) + 3a \cdot 3 + 7 \cdot 3 \quad *\\ &\quad = 12a^3 + 28a^2 - 3a^2 - 7a + 9a + 21\\ &\quad = 12a^3 + (28 - 3)a^2 + (-7 + 9)a + 21\\ &\quad = 12a^3 + 25a^2 + 2a + 21\end{aligned}$$

Note that the product is again a polynomial.

You might have found this product in another way.

$$\begin{aligned}&(3a + 7)(4a^2 - a + 3)\\ &\quad = 3a(4a^2 - a + 3) + 7(4a^2 - a + 3)\\ &\quad = 3a \cdot 4a^2 + 3a \cdot (-a) + 3a \cdot 3 + 7 \cdot 4a^2 + 7 \cdot (-a) + 7 \cdot 3 \quad *\\ &\quad = 12a^3 - 3a^2 + 9a + 28a^2 - 7a + 21\\ &\quad = 12a^3 + 25a^2 + 2a + 21\end{aligned}$$

Naturally, you get the same answer.

If you look closely at both examples of $(3a + 7)(4a^2 - a + 3)$, you will see that two equations are marked by an asterisk (*). After studying these two equations, you should be able to describe the multiplication of one polynomial by another. Your description should be similar to the one below.

To find the product of two polynomials, multiply each monomial of one of the polynomials by each monomial of the other polynomial, and add the resulting monomials.

If one polynomial has two terms and the other has three terms, their product has $2 \cdot 3$, or 6, terms. Of course, it might be possible to combine some of the monomials by addition and thereby reduce the number of terms in the product. Terms were combined in this manner in both of the preceding examples.

The nonzero polynomial $4a^2 - a + 3$ is said to have *degree* 2. The polynomial $3x^5 - 7x + 4$ is said to have degree 5.

$$
\begin{array}{rl}
x^3 + 1 & \text{has degree 3.} \\
4a - a^4 + 12 & \text{has degree 4.} \\
4 - 3y & \text{has degree 1.}
\end{array}
$$

A *constant polynomial* has a constant term only. For convenience, you may say the following.

A nonzero constant polynomial has degree 0. The polynomial 0 does not have degree.

The degree of a *nonconstant polynomial* is defined in the following way.

The degree of a nonconstant polynomial in x is the largest power of x appearing in the polynomial.†

If the term of highest degree in one polynomial is multiplied by the term of highest degree in another polynomial, the resulting monomial

† It is assumed, of course, that different terms have different powers of x and that each term which appears has a nonzero coefficient.

is the term of highest degree in the product of the two polynomials. By the first law of exponents, the degree of the resulting monomial is the sum of the degrees of the two terms.

The degree of the product of two nonzero polynomials is the sum of the degrees of the polynomials.

For example, the degree of $3a + 7$ is 1, and the degree of $4a^2 - a + 3$ is 2. Therefore, the degree of their product is $1 + 2$, or 3. This agrees with our statement above.

Exercises

Find the products and check your work by using 2 as the value for the variable. In each case, state the degrees of the factors and the degree of the product.

1. (a) $x(2x + 1)$ (b) $2x(x + 1)$

2. (a) $(5x + 3)(4x + 7) = (5x + 3) \cdot 4x + (5x + 3) \cdot$ __?__ $=$ __?__
(b) $(5x + 2)(7x + 3) = (5x + 2) \cdot 7x + (5x + 2) \cdot$ __?__ $=$ __?__

3. (a) $(3x + 1)(2x - 5) = (3x + 1) \cdot$ __?__ $+ (3x + 1) \cdot$ __?__ $=$ __?__
(b) $(3x + 7)(2x - 9) =$ __?__ $\cdot 2x + (3x + 7)(-9) =$ __?__

4. (a) $(7x - 4)(x - 6) = (7x - 4) \cdot$ __?__ $+ (7x - 4) \cdot$ __?__ $=$ __?__
(b) $(x - 2)(2x - 3) = (x - 2) \cdot$ __?__ $+ (x - 2) \cdot$ __?__ $=$ __?__

5. (a) $(8x - 5)(x - 3)$ (b) $(5x - 3)(4x + 7)$

6. (a) $(4x + 7)(2x + 9)$ (b) $(3x + 4)(2x + 5)$

7. (a) $(3x + 7)(7x - 3)$ (b) $(2x - 5)(5x - 2)$

8. (a) $(2x + 3)(4x^2 - 6x + 9) =$ __?__ $\cdot (4x^2 - 6x + 9)$
$+$ __?__ $\cdot (4x^2 - 6x + 9) =$ __?__
(b) $(2x - 3)(4x^2 + 6x + 9) = (2x - 3) \cdot$ __?__ $+ (2x - 3) \cdot$ __?__
$+(2x - 3) \cdot$ __?__ $=$ __?__

9. (a) $(3y^2 - 2y + 4)(2y + 5)$ (b) $(3y^2 - 2y + 4)(2y - 5)$

10. (a) $(2x + 11)^2$ (b) $(3x + 4)^2$

11. (a) $(3x - 10)^2$ (b) $(4x - 5)^2$

12. (a) $(x + 2)(x^2 - 2x + 4)$ (b) $(x - 2)(x^2 - 2x + 4)$

13. (a) $(3a - 5)(a^2 - 3a + 7)$ (b) $(b^2 - 1)(b^4 - 5b^2 - 8)$

14. (a) $(2x^3 + 3x - 1)(2x + 1)$ (b) $(4 - 4x + x^2)(3x - x^3)$

15. (a) $(2x^3 - x + 1)(x^3 + 2x^2 + 1)$
(b) $(x^2 - x - 3)(x^2 + 3x - 1)$

16. When checking the arithmetic involved in finding the product of two polynomials, would it be wise to choose 0 or 1 for a value of the variable? Why, or why not?

17. (a) What is the degree of the polynomial $x^2 - 2x + 3$?
(b) What is the degree of each term of the polynomial $x^2 - 2x + 3$?
(c) Is each term of the polynomial $x^2 - 2x + 3$ also a polynomial?

18. The product of the polynomial 0 and any other polynomial is always 0. For example,

$$0(x^2 - 2x + 3) = 0,$$
$$0(x^3 + 2x^2 + x - 2) = 0.$$

Can a degree be assigned to the polynomial 0 so that the rule about the degree of the product of two polynomials is true?

5–4 A METHOD FOR MULTIPLYING POLYNOMIALS

It is possible to devise a method for multiplying polynomials that is quite similar to the one for multiplying integers. The method for multiplying polynomials is illustrated by the three examples below.

$$\begin{array}{rrrrl} & 2x & - & 4 & \\ \times & 3x & + & 5 & \\ \hline & 6x^2 & - & 12x & = (2x - 4)\cdot 3x \\ & & & 10x - 20 & = (2x - 4)\cdot 5 \\ \hline & 6x^2 & - & 2x - 20 & \end{array}$$

$$\begin{array}{rl} 4a^2 - a + 3 & \\ \times \qquad 3a + 7 & \\ \hline 12a^3 - 3a^2 + 9a & = (4a^2 - a + 3)\cdot 3a \\ 28a^2 - 7a + 21 & = (4a^2 - a + 3)\cdot 7 \\ \hline 12a^3 + 25a^2 + 2a + 21 & \end{array}$$

You will notice that the polynomials have their terms arranged in descending order of degrees. They could just as well be arranged in ascending order of degrees. In the product, terms of like power are placed one under the other. This helps us to add the terms of like power. The third example is given below.

$$\begin{array}{rrrrr} y^2 - & 5y & + \; 3 & & \\ \times \quad 2y^2 + & 7y & - \; 4 & & \\ \hline 2y^4 - & 10y^3 + & 6y^2 & & \\ & 7y^3 - & 35y^2 + & 21y & \\ & - & 4y^2 + & 20y - & 12 \\ \hline 2y^4 - & 3y^3 - & 33y^2 + & 41y - & 12 \end{array}$$

Exercises

Find the following products and check your answers by using -1 as the value for the variable.

1. (a) $(5x + 7)(4x - 3)$ (b) $(3x - 7)(11x + 1)$
2. (a) $(4x^2 + 7)(5x^2 - 3)$ (b) $(3x^2 - 5)(6x^2 + 1)$
3. (a) $(8x + 5)(8x - 5)$ (b) $(13x - 2)(13x + 2)$
4. (a) $(8x + 3)^2$ (b) $(5x + 6)^2$
5. (a) $(4x^2 - 5x + 7)(2x - 3)$ (b) $(a + 5)(3a^2 - a - 1)$
6. (a) $(a - 1)(a^3 - 2a^2 + a + 1)$ (b) $(c^2 + c + 1)(c - 1)$

Find the products and check your work by using 2 as a value for the variable.

7. (a) $(16y^2 - 4y + 1)(4y + 1)$ (b) $(25x^2 + 5x + 1)(5x - 1)$
8. (a) $(9x^4 - 3x^2 + 1)(3x^2 - 1)$ (b) $(5x^2 - 4x - 3)(2x^2 - 1)$
9. (a) $(2x^2 - x + 1)(2x^2 + x - 1)$ (b) $(x^2 - x - 1)(x^2 - x + 1)$
10. (a) $(3b^3 - 2b^2 + b)(2b - 3)$ (b) $(2b^5 - b^3 + 3b)(2b + 1)$
11. (a) $(7y^2 - 3y + 1)(6y^2 - y - 2)$
    (b) $(3y^2 + 4y + 5)(y^2 + y - 2)$
12. (a) $(3x^2 - 22)(4x^4 - x^3 + 3x^2 - x + 5)$
    (b) $(3x^2 + 1)(2x^4 + 3x^3 - 2x^2 + x - 2)$

---

13. Suppose that the dimensions of a room are $(x - 2)$ feet by x feet by $(x + 2)$ feet. Find a polynomial which represents
    (a) the volume of the room,
    (b) the total area of the walls, ceiling, and floor.

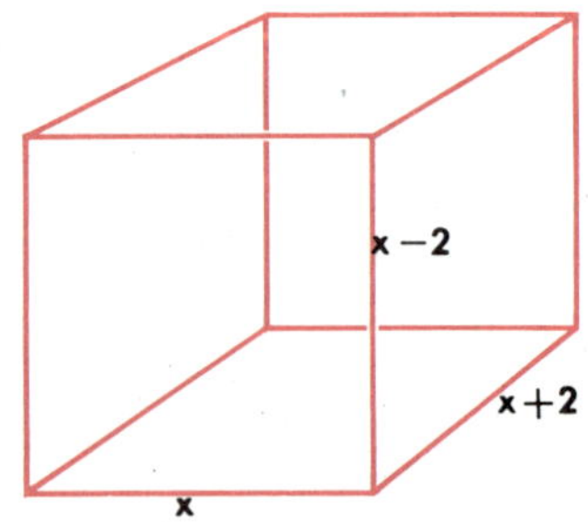

14. Find an expression for the area of a triangle if its base is $(3x + 5)$ inches and its altitude $(4x - 8)$ inches.

15. Find the volume of a cube whose edges are $x - 1$ units long. (Volume of a cube $= e^3$.)

16. Perform the indicated operations and simplify your answers.

(a) $(3x + 5)(2x - 7) - (7x - 4)(x - 1)$
(b) $(3y - 2)(3y + 2) - (3y - 2)^2$
(c) $(x + 2)(x^2 - 2x + 4) - (x - 2)(x^2 + 2x + 4)$
(d) $(x + 3)(x^2 + 6x + 9) - (x + 3)(x^2 - 3x + 9)$

---

17. Find each of the products.

(a) $(x - 1)(x^2 + x + 1)$
(b) $(x - 1)(x^3 + x^2 + x + 1)$
(c) $(x - 1)(x^4 + x^3 + x^2 + x + 1)$

Now write a binomial and another polynomial whose product you believe to be $x^6 - 1$. Multiply to determine whether your guess is right. Write a formula covering this type of problem.

18. Find the products. Then write a formula covering this type of problem.

(a) $(x + 1)(x^2 - x + 1)$
(b) $(x + 1)(x^4 - x^3 + x^2 - x + 1)$
(c) $(x + 1)(x^6 - x^5 + x^4 - x^3 + x^2 - x + 1)$

19. You have seen that numbers written in the decimal system are polynomials in ten. Thus, you could use the methods of this section to multiply two of them. For example,

$$\begin{array}{rrrrr}
314 = & (3 \cdot 10^2) + & (1 \cdot 10) & + 4 & \\
21 = & & (2 \cdot 10) & + 1 & \\
\hline
& (6 \cdot 10^3) + & (2 \cdot 10^2) + & (8 \cdot 10) & \\
& & (3 \cdot 10^2) + & (1 \cdot 10) & + 4 \\
\hline
& (6 \cdot 10^3) + & (5 \cdot 10^2) + & (9 \cdot 10) & + 4
\end{array}$$

and $314 \times 21 = 6594$. Follow this example in the exercises below.

(a) 231×12
(b) 103×33
(c) 203×22
(d) 302×32

20. Find the area of a circle whose diameter is $2x + 10$ units long. (Area of a circle $= \pi r^2$.)

21. Find the volume of a cube if each side is $(5x - 1)$ inches.

5-5 PRODUCTS OF BINOMIALS

With a little practice, you can write the product of two binomials without including each step.

Problem. Find each of the products.

(a) $(x + 2)(x - 3)$ (b) $(2u + 5)(3u - 7)$
(c) $(y - 6)^2$ (d) $(x + 6)(x - 6)$

Solution. The following diagrams indicate the thought processes involved in carrying out each multiplication.

(a) $(x + 2)(x - 3) = x^2 - x - 6$

x^2; -6; $+2x$; $-3x$

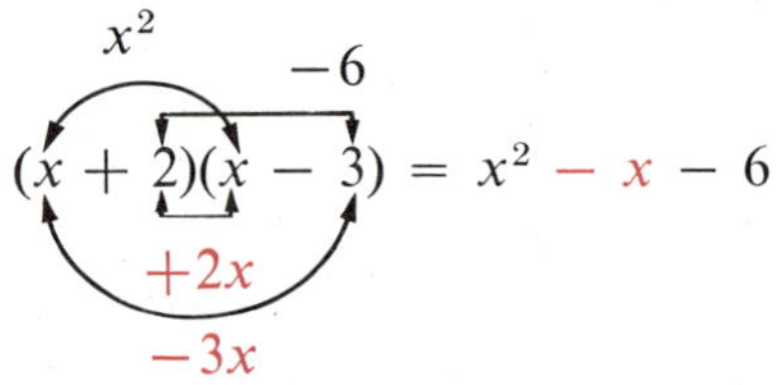

(b) $(2u + 5)(3u - 7) = 6u^2 + u - 35$

$6u^2$; -35; $+15u$; $-14u$

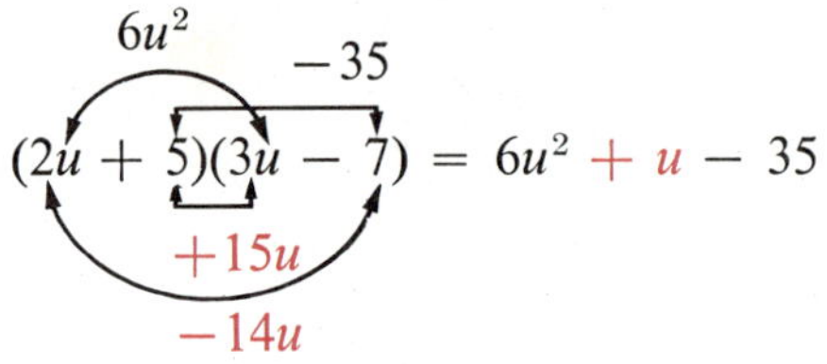

(c) $(y - 6)(y - 6) = y^2 - 12y + 36$

y^2; $+36$; $-6y$; $-6y$

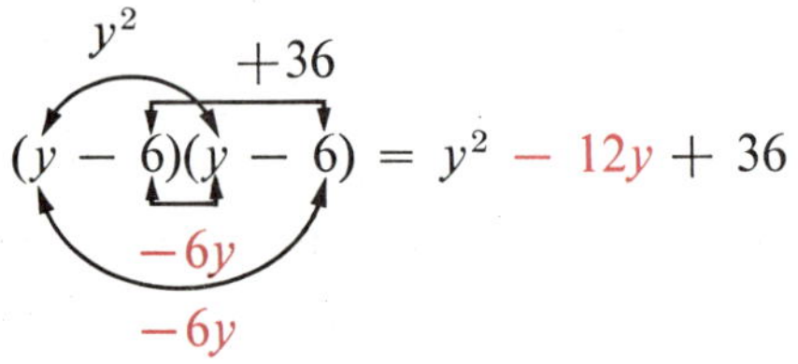

(d) $(x + 6)(x - 6) = x^2 + 0x - 36 = x^2 - 36$

x^2; -36; $+6x$; $-6x$

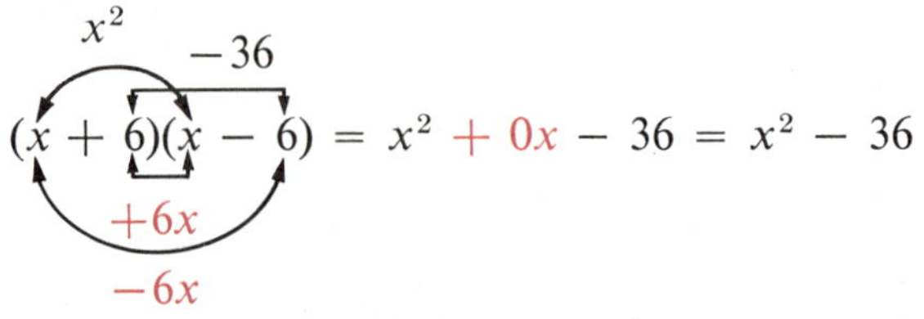

With practice, you will be able to mentally perform the operations that are indicated on the left side of each equation above.

Exercises

1. Continue each exercise in multiplication until you have completely simplified the right side.
   (a) $(3x + 2)(5x + 7) = (3x \cdot 5x) + [(2 \cdot 5x) + (7 \cdot 3x)] + (2 \cdot 7)$
   (b) $(5x - 2)(3x + 4) = (5x \cdot 3x) + [(-2 \cdot 3x) + (4 \cdot 5x)] - (2 \cdot 4)$

Find the following products.

2. (a) $(x + 5)(x + 7)$ (b) $(x + 7)(x + 8)$
3. (a) $(x + 4)(x - 3)$ (b) $(x - 5)(x + 6)$
4. (a) $(x - 7)(x - 1)$ (b) $(y - 9)(y - 3)$
5. (a) $(2x + 3)(5x + 7)$ (b) $(3x + 4)(6x + 1)$
6. (a) $(2y - 5)(3y - 4)$ (b) $(5y - 7)(y - 1)$
7. (a) $(x + 7)^2$ (b) $(y - 6)^2$
8. (a) $(2x - 5)(3x - 10)$ (b) $(3x - 7)(4x - 10)$
9. (a) $(4x - 7)(3x + 2)$ (b) $(x - 3)(2x + 9)$
10. (a) $(7x + 3)(12x - 5)$ (b) $(2y + 1)(7y - 5)$
11. (a) $(2x + 5)^2$ (b) $(9x + 1)^2$
12. (a) $(3y - 10)^2$ (b) $(8y - 1)^2$
13. (a) $(y - 12)(2y + 15)$ (b) $(11y - 9)(y + 2)$
14. (a) $(x + 11)(x - 11)$ (b) $(10 + x)(10 - x)$
15. (a) $(2x + 5)(2x - 5)$ (b) $(7x + 3)(7x - 3)$
16. (a) $(5 - a)(3 + a)$ (b) $(7 - 2b)(9 + b)$

In Exercises 17–20, perform the operations and give your answers in simplest form.

17. (a) $(2x - 3)(5x + 7) + (4x - 1)(4x + 1)$
    (b) $(3x - 2)(7x + 5) + (2x - 1)(2x + 1)$
18. (a) $(x + 4)(3x - 2) - (5x + 2)(3x - 1)$
    (b) $12 - (3x - 5)(3x + 2)$
19. (a) $(2x - 9)^2 - (3x + 4)^2$
    (b) $(2y - 3)^2 - (2y + 5)^2$
20. (a) $5(3x + 1)(3x - 1) - 4(7x + 2)(7x - 2)$
    (b) $(4x + 1)(2x - 5) - 2(6x - 1)(6x + 1)$

---

21. (a) Find $(x - 5)^2$ and $(5 - x)^2$. Are the products in any way related?
    (b) Find $(y + 2)^2$ and $(-y - 2)^2$. How are these products related?

22. Find each of the products.

(a) $(x + 4)(x - 4)$
(b) $(2x + 3)(2x - 3)$
(c) $(\frac{1}{3}x + 7)(\frac{1}{3}x - 7)$
(d) $(8 + x)(8 - x)$
(e) $(6x - 5)(6x + 5)$
(f) $(\frac{1}{3}x - 5)(\frac{1}{3}x + 5)$

23. How many terms did you get for each product in Exercise 20? In each case, study the pair of factors to determine why you are "short" one term in your answer. Write a description of the factors that lead to this two-term product.

24. Fill in the blanks to make the resulting equations true.

(a) $(x - 7)(x - \underline{\ ?\ }) = x^2 - 14x + 49$
(b) $(\underline{\ ?\ }x + 5)(3x + 5) = 9x^2 + 30x + 25$
(c) $(4x + \underline{\ ?\ })(4x - 9) = 16x^2 - 81$
(d) $(x + \underline{\ ?\ })(x + 7) = x^2 + 12x + 35$
(e) $(x - 10)(x + \underline{\ ?\ }) = x^2 - 6x - 40$
(f) $(\underline{\ ?\ } + 1)(3x - 5) = 6x^2 - 7x - 5$

---

25. Study your answers in Exercises 7, 11, and 12 until you can describe what terms the product contains when a binomial is squared. Give the number of terms to be expected, and make a statement about their signs.

Review for Sections 5–1 through 5–5

Find the value of each of the polynomials when m has the values 2, 3, and -1 in turn.

1. $m^2 + 3m - 4$
2. $m^3 + 3m^2 + m - 6$
3. $2m^2 + 4m - 5$
4. $6m^2 - 5m$
5. $\mathrm{m}^3 - 15m^2 + 2$
6. $5m^2 + 3m + 3$

Use the first law of exponents to find each of the following products.

7. $m^3 \cdot m^6$
8. $k^{13} \cdot k^6$
9. $y \cdot y^5$
10. $(3a^6)(2a)$
11. $(4x^2)(5x)$
12. $(3f^2)(3f^2)$

Find the indicated products and check your work by using 2 as a value for the variable.

13. $3k(k + 4)$
14. $7y(2y + 3)$
15. $(8a + 5)(2a + 1)$
16. $(2x + 4)(x + 1)$
17. $(f + 6)(f^2 + 3f - 4)$
18. $3m(2m + 3)(m + 1)$
19. $(n + 4)(3n^2 + 2n + 7)$
20. $(4b - 3)(3b - 6)$
21. $(2k - 4)(2k + 4)$
22. $4t(3t - 6)(2t + 1)$

23. $(y - 4)^2$
24. $(s + 4)^2$
25. $(3a + 2)^2$
26. $(z^2 - 3z + 2)(2z^2 + 5z - 6)$
27. $(x^3 + 5)(x^2 + 2x)$
28. $(f + 1)(f - 1)$
29. $(m + 6)(m - 6)$
30. $(3n + 4)^2$

Answers to Review for Sections 5–1 through 5–5

1. 6, 14, −6
2. 16, 51, −5
3. 11, 25, −7
4. 14, 39, 11
5. −50, −106, −14
6. 29, 57, 5
7. m^9
8. k^{19}
9. y^6
10. $6a^7$
11. $20x^3$
12. $9f^4$
13. $3k^2 + 12k$
14. $14y^2 + 21y$
15. $16a^2 + 18a + 5$
16. $2x^2 + 6x + 4$
17. $f^3 + 9f^2 + 14f - 24$
18. $6m^3 + 15m^2 + 9m$
19. $3n^3 + 14n^2 + 15n + 28$
20. $12b^2 - 33b + 18$
21. $4k^2 - 16$
22. $24t^3 - 36t^2 - 24t$
23. $y^2 - 8y + 16$
24. $s^2 + 8s + 16$
25. $9a^2 + 12a + 4$
26. $2z^4 - z^3 - 17z^2 + 28z - 12$
27. $x^5 + 2x^4 + 5x^2 + 10x$
28. $f^2 - 1$
29. $m^2 - 36$
30. $9n^2 + 24n + 16$

5–6 MONOMIAL FACTORS

You have just seen that if you start with two polynomials, you can find another polynomial which is their product. Now you will learn to do the opposite; that is, start with a polynomial and find two polynomials whose product is the given one. In other words, you will *factor* the given polynomial.

In arithmetic, you learned multiplication facts for the first twelve positive integers. Because you know these facts, you are able to factor certain integers at sight. Thus, you know that

$$24 = 4 \times 6 \text{ or } 2 \times 12 \text{ or } 3 \times 8.$$

In algebra, there are no such multiplication facts to help factor polynomials. Instead, you should look for *patterns* in the coefficients and exponents of the terms of a polynomial.

If every term of a given polynomial in a variable is of degree 1 or more, then the polynomial has a *monomial factor.* For example, the terms of the polynomial

$$3x^2 - 6x$$

are of degree 2 and 1. You can factor a first-degree monomial from this polynomial by using the distributive axiom. For example,

$$\begin{aligned} 3x^2 - 6x &= 3x \cdot x - 3x \cdot 2 \\ &= 3x(x - 2). \end{aligned}$$

The given polynomial has $3x$ and $x - 2$ as factors.

Problem. Find the monomial factor of highest degree of each of the following polynomials. Then express the polynomial as a product of this monomial and a polynomial.

(a) $4x^4 - 14x^3 + 26x^2$ (b) $y^5 - 7y^4 + 13y^3$ (c) $5a^2 + a$

Solution.

(a) A factor of each term is x^2. Therefore,

$$\begin{aligned} 4x^4 - 14x^3 + 26x^2 &= (4x^2 \cdot x^2) - (14x \cdot x^2) + (26 \cdot x^2) \\ &= (4x^2 - 14x + 26)x^2. \qquad \text{(D)} \end{aligned}$$

You may, if you wish, also factor out a 2 and get

$$4x^4 - 14x^3 + 26x^2 = 2x^2(2x^2 - 7x + 13).$$

(b) The lowest degree of any term is 3, and therefore, y^3 is the highest power of y which can be factored from each term.

$$\begin{aligned} y^5 - 7y^4 + 13y^3 &= (y^3 \cdot y^2) - (y^3 \cdot 7y) + (y^3 \cdot 13) \\ &= y^3(y^2 - 7 \;\; + 13) \end{aligned}$$

(c) You have

$$\begin{aligned} 5a^2 + a &= (a \cdot 5a) + (a \cdot 1) \\ &= a(5a + 1). \end{aligned}$$

A polynomial can always be factored in a trivial way. For example,

$$\begin{aligned} x + 2 &= \tfrac{1}{3}(3x + 6), \\ x^2 - 2x + 7 &= 7(\tfrac{1}{7}x^2 - \tfrac{2}{7}x + 1). \end{aligned}$$

Clearly, not much is gained by this kind of factoring. Your primary interest should be to find factors of degree 1 or more. However, occasionally you may find your arithmetic simplified if you factor an expression such as $5x^2 + 5$ in the following manner.

$$5x^2 + 5 = 5(x^2 + 1)$$

Exercises

Use the distributive axiom to factor the following polynomials. One factor should be the monomial of highest possible degree that has the largest possible integer as its coefficient.

1. (a) $8x^2 - 72x$ (b) $2x + x^2$
2. (a) $6b^2 + 6$ (b) $y + y^6$
3. (a) $9a^4 + a$ (b) $13x^2 + x$
4. (a) $3x^5 - 6x^3 + 12x^2$ (b) $3x^3 + 6x^2 + 9x$
5. (a) $15x^4 + 60x^2$ (b) $5x - 10x^2$
6. (a) $x^2 - 7x^4 + x^6$ (b) $x^9 + x^8 + x^7$
7. (a) $5x^4 - 10x^3 + 15x^2$ (b) $8x^4 - 4x^3 + 10x^2$
8. (a) $15x^3 - 21x^2$ (b) $-25a^3 - 15a$
9. (a) $x^5 - 7x^4 + 14x^3$ (b) $x^6 + 8x^8 - 16x^{10}$
10. (a) $9x^2 + 6$ (b) $x^2 + 11x$
11. (a) $18y^4 + 30y^2 - 42y$ (b) $20x^4 - 25x^6 + 15x$
12. (a) $10x^4 + x$ (b) $5b^2 + 5$
13. (a) $z^3 + z^6$ (b) $x^{100} + x^{101}$
14. (a) $5a^2 - 70a^3$ (b) $8y^6 - 12y^3$
15. (a) $5y^3 - y^2$ (b) $6x^4 + x^5$
16. (a) $6r + 2r^2$ (b) $9x + 3x^3$
17. (a) $3x^2 + 12x^5 + 15x^{20}$ (b) $4x^4 + 12x^5 + 40x^{10}$
18. (a) $3x^2 + 9x - 15x^3$ (b) $12y^5 - 8y^3 + 2y$
19. (a) $4x^4 + 3x^3 + 6x^2$ (b) $4x^4 - 10x^3 + 8x^2$
20. (a) $21x^{10} - 35x^9 + 28x^8 - 14x^7$
(b) $12x^7 - 20x^{10} + 8x^4 - 8x^8$

Preparation for Section 5–7

Find the products.

1. $(x + 5)(x - 5)$ **2.** $(6 + x)(6 - x)$
3. $(4x + 7)(4x - 7)$ **4.** $(\frac{1}{2}x^2 + 9)(\frac{1}{2}x^2 - 9)$
5. $(10x^3 + 3)(10x^3 - 3)$ **6.** $(1 - 9x^4)(1 + 9x^4)$

5–7 THE DIFFERENCE OF TWO SQUARES

A polynomial with a nonzero constant term, such as

$$x^2 + 6x + 9 \quad \text{or} \quad 4y^2 - 1,$$

does not have a monomial factor. However, it still might be a product of two polynomials. In fact, each of the second-degree polynomials appearing above is a product of two first-degree polynomials.

One of the easiest polynomials to factor is one which is the *difference of two squares,* that is, a binomial written as one perfect square subtracted from another perfect square.

For example, $4y^2 - 1$ is such a polynomial. The term $4y^2$ is the square of $2y$, and 1 is the square of 1. If you multiply two binomials such as $2y + 1$ and $2y - 1$, you will discover that their product is also a binomial:

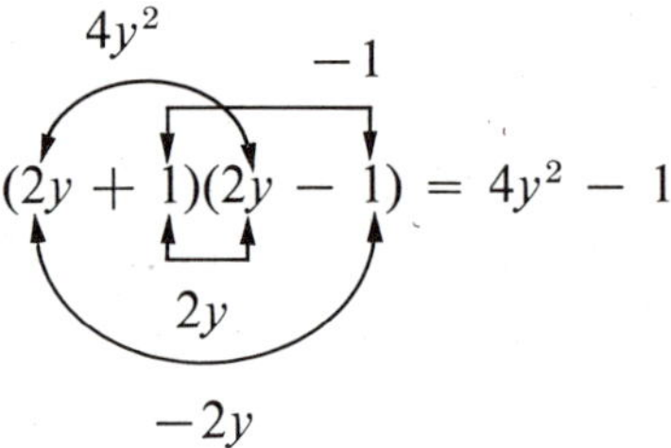

In fact, their product is the *difference of two squares:* $4y^2 - 1$.

This example illustrates the pattern for factoring the difference of any two squares.

(First term)^2 − (Second term)^2
= (First term + Second term)(First term − Second term)

As another example,

$$9y^2 - 49 = (3y)^2 - 7^2.$$

Therefore,

$$9y^2 - 49 = (3y + 7)(3y - 7).$$

You have expressed the second-degree polynomial $9y^2 - 49$ as a product of two first-degree polynomials.

Exercises

Which of the following polynomials is the difference of two squares?

1. (a) $4x^2 + 9$ (b) $169 - x^{47}$

2. (a) $25x^2 - 121$ (b) $625 - x^2$

3. (a) $x^2 - 1$ (b) $4a^2 - 36$

Factor each of the following differences of squares.

4. (a) $x^2 - 25$ (b) $x^2 - 64$

5. (a) $25 - x^2$ (b) $64 - x^2$

6. (a) $x^2 - 100$ (b) $x^2 - 144$

7. (a) $196x^2 - 1$ (b) $169x^2 - 1$

8. (a) $4a^2 - 121$ (b) $196x^2 - 9$

9. (a) $\frac{1}{4} - x^2$ (b) $\frac{1}{36} - x^2$

10. (a) $9y^2 - 16$ (b) $49x^2 - 144$

11. (a) $1 - 100x^2$ (b) $4 - 81x^2$

12. (a) $.01x^2 - 49$ (b) $.04x^2 - 36$

13. (a) $m^6 - 25$ (b) $16t^4 - 9$

14. (a) $16x^4 - 225$ (b) $25x^4 - 169$

15. (a) $x^8 - 36$ (b) $x^8 - 49$

16. (a) $36 - 49x^4$ (b) $121 - 81x^4$

17. (a) $\frac{1}{9}b^2 - \frac{1}{100}$ (b) $\frac{1}{64}y^2 - \frac{1}{25}$

18. (a) $1.44a^4 - .01$ (b) $1.69x^4 - .04$

19. The method of factoring the difference of two squares may be applied to numbers as well as to polynomials. Fill in the blanks to make the given statements true, and the method should become clear to you.

(a) $13^2 - 1 = 13^2 - 1^2$
$= (__?__ + 1)(13 - __?__)$
$= 14 \cdot __?__$

(b) $9991 = 10{,}000 - 9$
$= 100^2 - 3^2$
$= (100 + __?__)(__?__ - 3)$
$= __?__ \cdot 97$

Use the method shown in Exercise 19 to factor the following integers.

20. (a) $50^2 - 3^2$ (b) $12^2 - 1$

21. (a) 3599 (b) 3591

22. (a) 9951 (b) 2484

23. (a) 1591 (b) 4819

24. (a) 7979 (b) 6396

25. You may reverse the method of factoring the difference of two squares and use it to mentally multiply certain pairs of integers. Fill in the blanks to see how this is done.

(a) $99 \times 101 = (\underline{\ ?\ } - 1)(\underline{\ ?\ } + 1)$
$= 100^2 - \underline{\ ?\ }$
$= 10{,}000 - 1$
$= \underline{\ ?\ }$

(b) $(58 \times 62) = (\underline{\ ?\ } - 2)(\underline{\ ?\ } + 2)$
$= 3600 - \underline{\ ?\ }$
$= 3596$

Use the method shown in Exercise 25 to find the following products.

26. (a) $49 \cdot 51$ (b) $68 \cdot 72$

27. (a) $17 \cdot 23$ (b) $29 \cdot 31$

28. (a) $97 \cdot 103$ (b) $65 \cdot 75$

29. (a) $44 \cdot 56$ (b) $32 \cdot 48$

Factor each of the following. First find any monomial factor that exists.

30. (a) $256x^4 - 81$ (b) $3x^2 - 75$

31. (a) $147y^2 - 27$ (b) $72y^3 - 50y$

5-8 PERFECT SQUARES

Certain trinomials can be expressed as squares of binomials. To determine the form of such trinomials you might square some binomials and look for the pattern which distinguishes them.

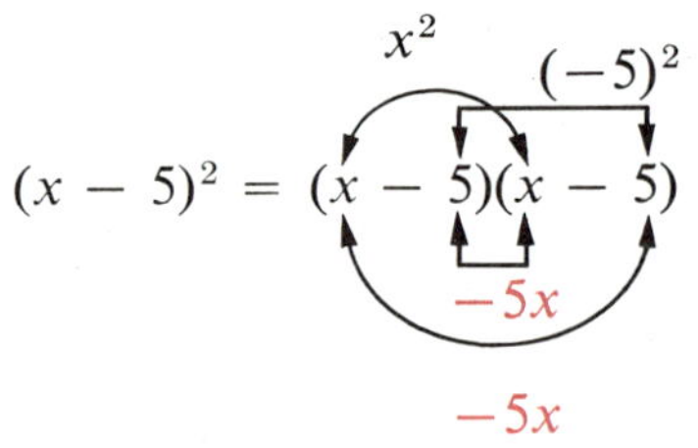

x^2 $(-5)^2$

$$(x - 5)^2 = (x - 5)(x - 5)$$

$-5x$

$-5x$

$$= x^2 + 2 \cdot (-5x) + (-5)^2$$
$$= x^2 - 10x + 25$$

$$(3x + 7)^2 = (3x + 7)(3x + 7)$$

(3x)² 7² +21x +21x

$$= (3x)^2 + 2 \cdot 21x + 7^2$$
$$= 9x^2 + 42x + 49$$

These examples illustrate the following pattern for squaring binomials.

(First term + Second term)^2
= (First term)^2 + 2(First term)(Second term) + (Second term)^2,

or

(First term − Second term)^2
= (First term)^2 − 2(First term)(Second term) + (Second term)^2.

A trinomial is called a *perfect square* if it is the square of some binomial.

Problem. Which of the following trinomials are perfect squares?
(a) $x^2 + 6x + 9$ (b) $25a^2 - 100a + 16$ (c) $y^2 - 12y - 36$

Solution.

(a) Since $x^2 + 6x + 9 = x^2 + 2(x)(3) + 3^2$, it is a perfect square:

$$x^2 + 6x + 9 = (x + 3)^2.$$

(b) If $25a^2 - 100a + 16$ is the square of some binomial, then it must be the square of $5a - 4$. However, the middle term of $(5a - 4)^2$ is $-40a$, not $-100a$. Therefore, the given trinomial is not a perfect square.

(c) You see that

$$y^2 - 12y - 36 = (y)^2 + 2(y)(-6) - (-6)^2.$$

However, $y^2 - 12y - 36$ is not a perfect square, because the constant term is negative. On the other hand, $y^2 - 12y + 36$ is a perfect square:

$$y^2 - 12y + 36 = (y - 6)^2.$$

A knowledge of the pattern for squaring a binomial makes it possible for you to square integers mentally if they are close to multiples of 10. For example,

$$21^2 = (20 + 1)^2 = 20^2 + (2 \cdot 20 \cdot 1) + 1, \quad \text{or } 441.$$
$$18^2 = (20 - 2)^2 = 20^2 - (2 \cdot 20 \cdot 2) + 4, \quad \text{or } 324.$$

Exercises

Tell which of the following trinomials are perfect squares, and factor those which are.

1. (a) $x^2 + 4x + 4$ (b) $x^2 + 14x + 49$
2. (a) $x^2 + 18x + 81$ (b) $x^2 + 8x + 16$
3. (a) $y^2 - 10y + 25$ (b) $b^2 - 12b + 36$
4. (a) $x^2 - 14x + 49$ (b) $x^2 - 18x + 81$
5. (a) $y^2 + 7y + 49$ (b) $y^2 + 6y + 36$
6. (a) $25x^2 + 60x + 36$ (b) $4x^2 + 4x + 1$
7. (a) $64y^2 - 48y + 9$ (b) $16x^2 - 8x + 1$
8. (a) $25y^2 - 20y + 1$ (b) $4a^2 - 12a - 9$
9. (a) $9 + 24x + 16x^2$ (b) $25 + 20x + 4x^2$
10. (a) $4 - 28y + 49y^2$ (b) $49 - 42x + 9x^2$
11. (a) $9x^2 - 12x + 4$ (b) $4x^2 - 12x + 9$
12. (a) $x^4 + 22x^2 + 121$ (b) $y^6 - 2y^3 + 1$

Factor each of the polynomials by the method shown in this or previous sections.

13. (a) $x^2 - 16x + 64$ (b) $x^2 - 20x + 100$
14. (a) $6x^3 + 10x^2 + 2x$ (b) $12x^3 + 6x^2 - 9x$
15. (a) $36a^2 + 132a + 121$ (b) $4a^2 + 36a + 81$
16. (a) $7x^5 - 14x^4 + 21x^3$ (b) $2x^6 - 6x^5 + 12x^4$
17. (a) $100y^2 - 20y + 1$ (b) $64x^2 - 16x + 1$
18. (a) $16a^2 - 72a + 81$ (b) $9x^2 - 42x + 49$

---

19. The monomial factor of the polynomial $48x^3 - 24x^2 + 3x$ is $3x$. Thus,

$$48x^3 - 24x^2 + 3x = 3x(16x^2 - 8x + 1).$$

Furthermore, $16x^2 - 8x + 1$ is a perfect square trinomial. Therefore, the final factorization is

$$48x^3 - 24x^2 + 3x = 3x(4x - 1)(4x - 1),$$

or

$$48x^3 - 24x^2 + 3x = 3x(4x - 1)^2.$$

Use this procedure to factor each of the following polynomials into a monomial multiplied by the square of a binomial.

(a) $5x^3 + 10x^4 + 5x^5$ (b) $7y^{12} + 252y^{10} + 84y^{11}$
(c) $12y^2 + 60y + 75$ (d) $-18x^3 + x^4 + 81x^2$

20. Find the following squares by first writing each integer as a sum or a difference.

(a) $31^2 = (30 + 1)^2$ (b) $29^2 = (30 - 1)^2$
(c) 42^2 (d) 68^2
(e) 58^2 (f) 101^2

---

21. (a) Study the following examples of factoring until you see their relation to the three types of factoring we have just been doing.

(i) $x(x + 2) + 3(x + 2) = (x + 3)(x + 2)$ (D)

(ii) $(y + 1)^2 - 20(y + 1) + 100 = [(y + 1) - 10][(y + 1) - 10]$
$= (y - 9)(y - 9)$
$= (y - 9)^2$

(iii) $(2a - 1)^2 + 8(2a - 1) + 16 = [(2a - 1) + 4][(2a - 1) + 4]$
$= (2a + 3)(2a + 3)$
$= (2a + 3)^2$

(iv) $(3b + 1)^2 - (b - 2)^2$
$= [(3b + 1) + (b - 2)][(3b + 1) - (b - 2)]$
$= (4b - 1)(2b + 3)$

(b) Complete the factorizations.

(i) $2a(3a - 1) - 5(3a - 1) = (_?_)(3a - 1)$

(ii) $6b(2 - b) - 7(2 - b) = (6b - 7)(_?_)$

(iii) $(c - 5)^2 + 18(c - 5) + 81 = [(c - 5) + 9][_?_]$
$= _?_$

(iv) $25(2d + 1)^2 - 30(2d + 1) + 9 = [_?_][5(2d + 1) - 3]$
$= _?_$

(v) $9(4a + 1)^2 - 25(a - 2)^2$
$= [3(4a + 1) - 5(a - 2)][_?_ + _?_]$
$= (_?_)(12a + 3 + 5a - 10)$
$= (7a + 13)(_?_)$

Use the methods developed in Exercise 21 to factor each of the following without first removing parentheses.

22. (a) $x(x + 1) - 3(x + 1)$ (b) $y(7y - 2) - 5(7y - 2)$

23. (a) $x(x + 5) + 3(x + 5)$ (b) $8(3 - 5x) - x(3 - 5x)$

24. (a) $(x + 1)^2 - 16$ (b) $25 - (y + 2)^2$

25. (a) $(x + 3)^2 + 2(x + 3) + 1$ (b) $9(a + 1)^2 + 6(a + 1) + 1$

26. (a) $(2y + 1)^2 - 8(2y + 1) + 16$
(b) $(x + 1)^2 - (2x + 3)^2$

27. In much the same manner as you found the pattern for squaring a binomial, you can find a pattern for cubing a binomial as follows.

$$(A + B)^3 = (A + B)(A + B)^2 = (A + B)(A^2 + 2AB + B^2)$$

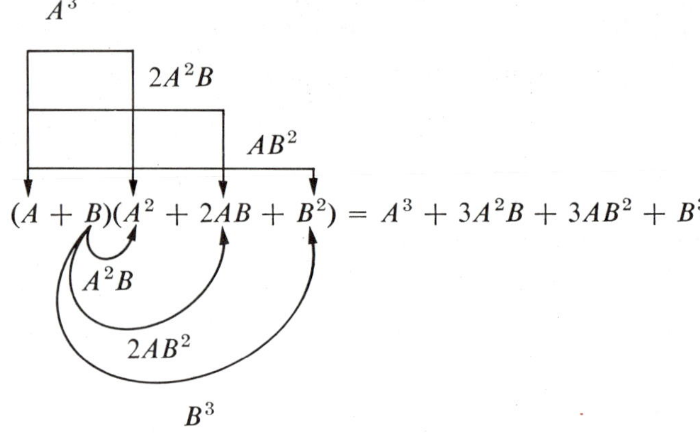

Find:

(a) $(x + 1)^3$ (b) $(x - 1)^3$ (c) $(x + 2)^3$
(d) $(2x - 3)^3$ (e) 21^3 (f) 99^3

EXTRA!

1. A 3-inch cube is painted red and then cut up into 1-inch cubes. Find the number of 1-inch cubes which have
   (a) no red sides
   (b) one red side
   (c) two red sides
   (d) three red sides
2. Answer the same questions for a 2-inch cube.
3. Answer the same questions for a 4-inch cube.
4. Answer the same questions for an n-inch cube where n is any positive integer. (Parts (a) and (d) are the easiest to find.)
5. Do you see any relationship between your answer to Exercise 4 and the following equation?

$$\begin{aligned} n^3 &= [(n - 2) + 2]^3 \\ &= (n - 2)^3 + 3 \cdot (n - 2)^2 \cdot 2 + 3 \cdot (n - 2) \cdot 2^2 + 2^3 \end{aligned}$$

Preparation for Section 5–9

Find the products.

1. $(x + 7)(x + 9)$

2. $(x - 7)(x - 9)$

3. $(x - 7)(x + 9)$

4. $(x + 7)(x - 9)$

5. $(y - 5)(y - 12)$

6. $(x + 4)(x - 10)$

7. $(x + 6)(x + 8)$

8. $(x - 5)(x + 11)$

5–9 FACTORING SECOND-DEGREE POLYNOMIALS

In this section, you will factor polynomials which have integers for coefficients. An example of this kind of polynomial is

$$x^2 + 6x - 7.$$

When factoring this polynomial, you should look only for factors of the first degree which also have integral coefficients.

If $x^2 + 6x - 7$ can be factored, it must have the form

$$x^2 + 6x - 7 = (x + M)(x + N)$$

for some integers M and N. Since

$$(x + M)(x + N) = x^2 + (M + N)x + MN, \tag{1}$$

the integers M and N must be chosen so that

$$M + N = 6 \quad \text{and} \quad MN = -7.$$

Because 7 is a prime number, its only positive factors are 1 and 7. Thus, you must select either of the following in order that $MN = -7$.

$$M = 1 \quad \text{and} \quad N = -7 \quad \text{or} \quad M = -1 \quad \text{and} \quad N = 7$$

However, $M + N = 6$ in only one of these cases: $M = -1$ and $N = 7$. Consequently,

$$x^2 + 6x - 7 = (x - 1)(x + 7).$$

Problem 1. If possible, factor $y^2 + 7y + 12$.

Solution. You must have

$$y^2 + 7y + 12 = (y + M)(y + N)$$

for some integers M and N. Using (1) above,

$$M + N = 7 \quad \text{and} \quad MN = 12.$$

In other words, find two numbers whose sum, $(M + N)$, is 7 and whose product, (MN), is 12.

By trial and error, you can find that $M = 3$ and $N = 4$ are the correct choices:

$$y^2 + 7y + 12 = (y + 3)(y + 4).$$

Problem 2. If possible, factor $a^2 - 25a + 26$.

Solution. If $a^2 - 25a + 26$ can be factored, then

$$a^2 - 25a + 26 = (a + M)(a + N)$$

for some integers M and N. If this is true,

$$M + N = -25 \quad \text{and} \quad MN = 26.$$

Do you see that M and N must both be negative? Hence, the only possible choices are -1 and -26, or -2 and -13. Since $-1 - 26 \neq -25$ and $-2 - 13 \neq -25$, integers M and N do not exist. Hence, $a^2 - 25a + 26$ cannot be factored as a product of two first-degree polynomials having integral coefficients.

This process for factoring second-degree polynomials can be clearly illustrated with a flow chart. To make a flow chart that would apply to the factoring of all polynomials of the form presented in this section, a general polynomial must be used. Let k be the coefficient of the term with degree 1 and let c be the constant term. Any such polynomial which can be factored as a product of two first-degree polynomials can then be expressed as

$$x^2 + kx + c = (x + M)(x + N).$$

The flow chart for factoring a polynomial of the form $x^2 + kx + c$ can be illustrated as in Fig. 5–1. The "failure box" is a provision for the case where a polynomial cannot be factored as a product of two first-degree polynomials having integral coefficients.

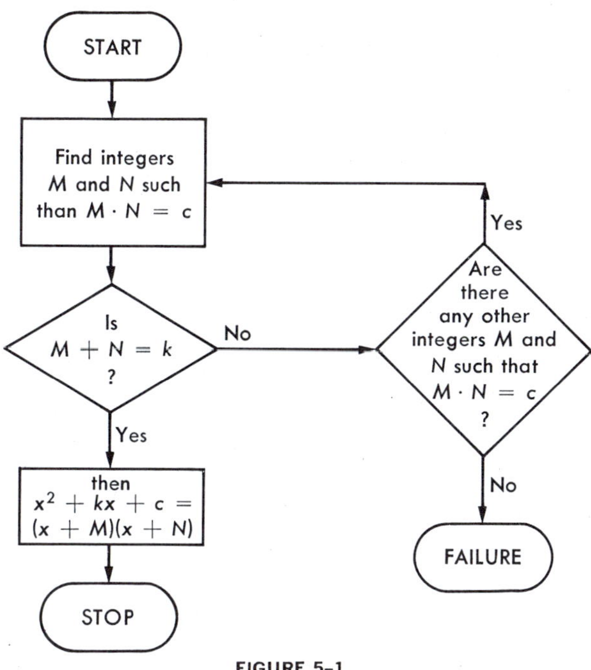

FIGURE 5-1

Exercises

Select the correct factorization for the polynomials in Exercises 1–3.

1. (a) $x^2 + 11x + 24 = \underline{\ ?\ }$ $\begin{cases}(x+1)(x+24)\\(x+6)(x+4)\\(x+12)(x+2)\\(x+8)(x+3)\end{cases}$

(b) $x^2 + 9x + 18 = \underline{\ ?\ }$ $\begin{cases}(x+9)(x+2)\\(x+1)(x+18)\\(x+6)(x+3)\end{cases}$

2. (a) $y^2 - 12y + 32 = \underline{\ ?\ }$ $\begin{cases}(y-1)(y-32)\\(y-2)(y-16)\\(y-4)(y-8)\end{cases}$

(b) $y^2 - 7y + 12 = \underline{\ ?\ }$ $\begin{cases}(y-6)(y-2)\\(y-1)(y-12)\\(y-4)(y-3)\end{cases}$

3. (a) $a^2 - 26a - 56 = \underline{\ ?\ }$ $\begin{cases}(a-1)(a+56)\\(a-4)(a+14)\\(a-7)(a+8)\\(a+2)(a-28)\\(a-2)(a+28)\end{cases}$

(b) $x^2 + 11x - 60 = \underline{\ ?\ }$ $\begin{cases}(x-60)(x+1)\\(x+12)(x-5)\\(x-4)(x+15)\\(x-6)(x+10)\\(x+4)(x-15)\end{cases}$

In each of the following problems, substitute the correct symbol for each *.

4. (a) $y^2 + 7y + 12 = (y * 3)(y * 4)$
   (b) $y^2 - 9y + 20 = (y * 4)(y * 5)$
5. (a) $y^2 + 2y - 15 = (y * 3)(y * 5)$
   (b) $y^2 - 3y - 18 = (y * 6)(y * 3)$

If possible, factor each of the following as a product of polynomials with integral coefficients. (You may find it helpful to refer to the flow chart shown as Fig. 5–1.)

6. (a) $x^2 + 10x + 21$ (b) $x^2 + 31x + 30$
7. (a) $y^2 + 3y + 2$ (b) $y^2 + 62y + 61$
8. (a) $x^2 - 6x + 8$ (b) $x^2 - 7x + 6$
9. (a) $a^2 - 5a - 14$ (b) $a^2 + 6a - 27$
10. (a) $x^2 - 5x + 4$ (b) $a^2 - 6a + 12$
11. (a) $x^2 - 10x + 16$ (b) $x^2 - 7x + 10$
12. (a) $y^2 - 11y + 30$ (b) $y^2 - 26y + 48$
13. (a) $y^2 + 20y + 80$ (b) $x^2 + 12x + 10$
14. (a) $y^2 - y - 56$ (b) $x^2 - 11x + 10$
15. (a) $b^2 + 9b + 14$ (b) $y^2 + y - 42$
16. (a) $a^2 - 8a + 12$ (b) $b^2 - 15b + 14$
17. (a) $y^2 + 8y - 15$ (b) $x^2 - 13x + 12$
18. (a) $b^2 - b - 20$ (b) $y^2 + 3y - 54$
19. (a) $x^2 - x - 72$ (b) $a^2 - 6a - 40$
20. (a) $a^2 - 10a + 21$ (b) $y^2 + 9y + 5$
21. (a) $x^2 + 15x + 54$ (b) $y^2 - 17y + 72$
22. (a) $a^2 + 12a + 35$ (b) $b^2 + 12b - 35$
23. (a) $b^2 - 13b + 42$ (b) $x^2 - 47x + 90$
24. (a) $y^2 + 15y + 26$ (b) $x^2 - 36x + 35$
25. (a) $a^2 - 4a - 21$ (b) $x^2 - 19x - 42$

26. (a) $a^2 + a - 110$ (b) $x^2 + 2x - 195$

27. (a) $y^2 + 18y - 77$ (b) $x^2 + 2x - 42$

If possible, factor each of the following as a product of polynomials with integral coefficients. (Remember to look for monomial factors first.)

28. $y^3 - 6y^2 - 55y$

29. $5x^2 - 30x - 80$

30. $3y^2 - 30y - 225$

31. $y^4 - 26y^3 + 160y^2$

32. $3x^3 + 57x^2 - 102x$

33. $7a^2 + 28a - 147$

34. $5y^3 - 40y^2 + 165y$

35. $36x^2 - 225x$

36. $-75x^3 + 60x^2 - 12x$

37. $7x^2 - 7$

Preparation for Section 5–10

Find the products.

1. $(2x + 3)(4x + 1)$

2. $(2x - 3)(4x - 1)$

3. $(2x + 3)(4x - 1)$

4. $(2x - 3)(4x + 1)$

5. $(3x - 7)(6x - 5)$

6. $(3x + 7)(6x + 5)$

7. $(3x - 7)(6x + 5)$

8. $(3x + 7)(6x - 5)$

5–10 MORE FACTORING

If the coefficient of the second-degree term of a trinomial is other than 1 or -1, the problem of factoring the trinomial is more complicated. For example, consider

$$6x^2 + 7x - 20.$$

Is the following true for some integers A, B, C, and D?

$$6x^2 + 7x - 20 = (Ax + B)(Cx + D)$$

You know that

$$(Ax + B)(Cx + D) = ACx^2 + (AD + BC)x + BD \qquad (1)$$

is true. If the given expression is to be factored, you must have

$$6x^2 + 7x - 20 = ACx^2 + (AD + BC)x + BD$$

and

$$AC = 6, \quad AD + BC = 7, \quad BD = -20.$$

Possible values for A and C are 1 and 6, 2 and 3. Thus, the factored form of $6x^2 + 7x - 20$ might be

$$(x + B)(6x + D), \quad \text{or} \quad (2x + B)(3x + D).$$

In each of these expressions, you can try various values of B and D. For example, if $B = 2$ and $D = -10$, then

$$(x + 2)(6x - 10) = 6x^2 + 2x - 20.$$

This is not the given expression; therefore, these are not the desired values for B and D. The following are other possible values for B and D: 1 and -20, -4 and 5, 5 and -4, and so on. By trial and error, you can discard the wrong choices and find

$$A = 2, \quad C = 3, \quad B = 5, \quad D = -4.$$

These values yield the given expression,

$$6x^2 + 7x - 20 = (2x + 5)(3x - 4).$$

Problem 1. If possible, factor $2x^2 + 11x + 12$.

Solution. Since $2x^2$ can be factored in only one way, you must have

$$2x^2 + 11x + 12 = (2x + M)(x + N)$$

for some integers M and N. Notice that both M and N must be positive. Thus, the possible choices of M and N are as follows:

M	1	2	3	4	6	12
N	12	6	4	3	2	1

Using (1) above, you must also have

$$AD + BC = M + 2N = 11.$$

Hence, $M = 3$ and $N = 4$, and

$$2x^2 + 11x + 12 = (2x + 3)(x + 4).$$

Problem 2. If possible, factor $7a^2 - 19a + 8$.

Solution. Observe that if $7a^2 - 19a + 8$ can be factored, then

$$7a^2 - 19a + 8 = (7a + M)(a + N)$$

for some negative integers M and N. Also observe that the middle term, $AD + BC$, is

$$-19 = M + 7N,$$

so not both M and N can be even. Thus, either $M = -1$ and $N = -8$ or $M = -8$ and $N = -1$. However,

$$-1 + [7 \cdot (-8)] \neq -19,$$
$$-8 + [7 \cdot (-1)] \neq -19.$$

Therefore, the polynomial cannot be factored.

Problem 3. If possible, factor $12x^2 + 44x + 35$.

Solution. To factor this polynomial, you must correctly fill the four blank spaces below.

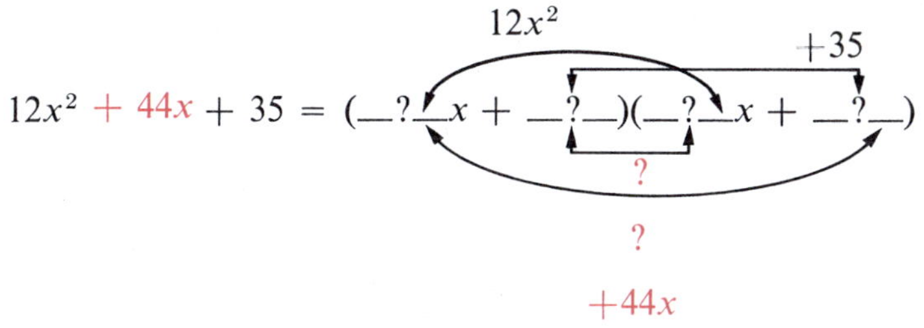

$12x^2$

$+35$

$$12x^2 + 44x + 35 = (_?_x + _?_)(_?_x + _?_)$$

?

?

$+44x$

All four spaces must be filled with positive integers. The spaces in front of the x's must be filled with one of the following pairs of factors of 12:

1 and 12, 2 and 6, 3 and 4.

The other two spaces must be filled with one of the pairs of factors of 35:

1 and 35 or 5 and 7.

By trial and error, you will see that $2 \cdot 7 + 6 \cdot 5 = 44$ and hence,

$$(2x + 5)(6x + 7) = 12x^2 + 44x + 35.$$

Exercises

For Exercises 1–3, substitute the correct sign for each $*$.

1. (a) $24x^2 + 25x + 6 = (Ax * B)(Cx * D)$
   (b) $6x^2 + 29x + 9 = (Ax * B)(Cx * D)$
2. (a) $7x^2 - 20x + 12 = (Ax * B)(Cx * D)$
   (b) $15x^2 - 38x + 24 = (Ax * B)(Cx * D)$

3. (a) $12x^2 + 32x - 35 = (Ax * B)(Cx * D)$
(b) $11x^2 - 41x - 12 = (Ax * B)(Cx * D)$

Complete the following factorizations.

4. (a) $5x^2 + 18x + 9 = (5x_?_)(x_?_)$
(b) $5x^2 + (-18x) + 9 = (5x_?_)(x_?_)$

5. (a) $3y^2 + 14y - 5 = (3y - _?_)(_?_ + 5)$
(b) $8x^2 - 26x - 7 = (4x + _?_)(_?_)$

6. (a) $4x^2 - 25x + 6 = (4x_?_)(_?_ - 6)$
(b) $12x^2 - 11x + 2 = (3x_?_)(_?_ - 1)$

If possible, factor each of the polynomials into a product of polynomials with integral coefficients.

7. (a) $2x^2 + 5x + 2$ (b) $2x^2 + 5x - 1$
8. (a) $2y^2 + 5y - 12$ (b) $3y^2 + 19y - 40$
9. (a) $9x^2 - 42x + 49$ (b) $12y^2 - 25y + 7$
10. (a) $16x^2 + 88x + 121$ (b) $3x^2 + x + 1$
11. (a) $2y^2 - y - 6$ (b) $11x^2 - 12x - 20$
12. (a) $3x^2 - 16x + 16$ (b) $12y^2 - 23y + 5$
13. (a) $7x^2 - 40x - 12$ (b) $24y^2 - 13y - 2$
14. (a) $13x^2 - 16x - 20$ (b) $6x^2 - x - 12$
15. (a) $81x^2 - 169$ (b) $121 - 196a^2$
16. (a) $8x^2 - 34x + 33$ (b) $8x^2 - 34x + 21$
17. (a) $12y^2 + 8y - 7$ (b) $6a^2 + 7a - 3$
18. (a) $28 + 3y - 18y^2$ (b) $6 - 11a - 10a^2$
19. (a) $8y^2 + 2y - 15$ (b) $30x^2 + 37x + 10$
20. (a) $6x^2 + 17x + 12$ (b) $6y^2 + 5y + 1$
21. (a) $33y^2 - y - 14$ (b) $35a^2 - 3a - 2$
22. (a) $12x^2 - 11x + 2$ (b) $x^2 + 4$

First look for monomial factors. Then write each polynomial as a product of as many factors as possible.

23. (a) $6x^2 - 216$ (b) $y^3 - 16y$
24. (a) $48y^3 + 147y$ (b) $2x^4 + 18x^2$
25. (a) $-8x^2 - 56x + 98$ (b) $-10x^2 - 10x + 35$
26. (a) $x^5 - x$ (b) $x^5 - 9x$
27. (a) $30x^2 - 38x + 12$ (b) $6x^2 - 54x + 84$
28. (a) $6x^3 + 15x^2 + 6x$ (b) $4x^3 + 14x^2 + 6x$

Factor each of the following without removing the parentheses.

29. $x^2(x + 1) - 4(x + 1)$

30. $(y + 7)^2 - 5(y + 7) + 6$

31. $(2x - 1)^2 + 9(2x - 1) + 14$

32. $6(a - 5)^2 - 13(a - 5) + 6$

33. $20(4 - x)^2 + 3(4 - x) - 2$

34. $3(x - 2)^2 - 6(x - 2) + 3$

35. Select the correct factorization.

$$3y^2 - 4y - 15 = \begin{cases} (3y - 1)(y + 15) \\ (3y - 5)(y + 3) \\ (3y + 5)(y - 3) \end{cases}$$

Why is the choice of $(3y - 3)(y + 5)$ obviously a poor one?

36. Explain why M and N cannot both be even in Problem 2 on page 214. Which pair of factors for 8 does this fact rule out?

Review for Sections 5–6 through 5–10

Use the distributive axiom to factor the following polynomials. One factor should be the monomial of highest possible degree that has the largest possible integer as its coefficient.

1. $3x^2 + 6x$

2. $5m^3 + 30m$

3. $7y + 21y^2$

4. $4b^5 + 8b^3 + 2b^2$

5. $15a^2 + 3a^5$

6. $6n^2 - 8n^3$

7. $3z^4 - 6z^3 + 9z^2$

8. $16t^3 + 8t^2 - 12t$

9. $10g^3 + 12g^2 - 8g^4$

Factor each of the following differences of squares.

10. $r^2 - 9$

11. $4 - x^2$

12. $25d^2 - 36$

13. $49m^4 - 16$

14. $81k^4 - 1$

15. $16y^8 - 81$

16. $n^8 - b^4$

17. $64a^2 - 25$

18. $9z^2 - 4$

Factor each of the following perfect squares.

19. $t^2 - 6t + 9$

20. $x^4 + 8x^2 + 16$

21. $4g^2 + 20g + 25$

22. $m^2 - 18m + 81$

23. $r^2 + 4rs + 4s^2$

24. $9y^2 + 12y + 4$

25. $16k^2 + 8k + 1$

26. $a^4 + 10a^2b + 25b^2$

Factor each of the polynomials in Exercises 27–34 into a product of two binomials with integral coefficients.

27. $m^2 + 5m + 6$

28. $k^2 + 8k + 15$

29. $x^2 - 7x + 10$

30. $y^2 + 2y - 24$

31. $n^2 + n - 12$

32. $2g^2 + 3g + 1$

33. $3f^2 + 14f + 8$

34. $4t^2 - 6t - 18$

Write each of the following polynomials as a product of as many factors as possible. (Hint: Remember to look for monomial factors.)

35. $2k^3 - 18k$

36. $2r^2 + r - 10$

37. $405x^5 - 80x$

38. $2y^3 + 20y^2 + 50y$

39. $3m^2 + 8m + 5$

40. $8t^2 - 32t - 168$

41. $2a^9 - 2a$

42. $a^4 - 13a^2 + 36$

43. $4p^3 - 24p^2 + 36p$

44. $k^3 - 5k^2 - 24k$

45. $3h^4 - 48h^2$

46. $16r^4 + 4r^2$

47. $5x^3 - 20x$

48. $4y^2 + 14y + 6$

49. $2t^4 - 32$

50. $3m^3 - 12m^2 - 36m$

Answers to Review for Sections 5–6 through 5–10

1. $3x(x + 2)$

2. $5m(m^2 + 6)$

3. $7y(1 + 3y)$

4. $2b^2(2b^3 + 4b + 1)$

5. $3a^2(5 + a^3)$

6. $2n^2(3 - 4n)$

7. $3z^2(z^2 - 2z + 3)$

8. $4t(4t^2 + 2t - 3)$

9. $2g^2(5g + 6 - 4g^2)$

10. $(r + 3)(r - 3)$

11. $(2 + x)(2 - x)$

12. $(5d + 6)(5d - 6)$

13. $(7m^2 + 4)(7m^2 - 4)$

14. $(9k^2 + 1)(3k + 1)(3k - 1)$

15. $(4y^4 + 9)(2y^2 + 3)(2y^2 - 3)$

16. $(n^4 + b^2)(n^2 + b)(n^2 - b)$

17. $(8a + 5)(8a - 5)$

18. $(3z + 2)(3z - 2)$

19. $(t - 3)^2$

20. $(x^2 + 4)^2$

21. $(2g + 5)^2$

22. $(m - 9)^2$

23. $(r + 2s)^2$

24. $(3y + 2)^2$

25. $(4k + 1)^2$

26. $(a^2 + 5b)^2$

27. $(m + 2)(m + 3)$

28. $(k + 5)(k + 3)$

29. $(x - 2)(x - 5)$

30. $(y - 4)(y + 6)$

31. $(n + 4)(n - 3)$

32. $(2g + 1)(g + 1)$

33. $(3f + 2)(f + 4)$

34. $(2t + 3)(2t - 6)$

35. $2k(k + 3)(k - 3)$

36. $(2r + 5)(r - 2)$

37. $5x(9x^2 + 4)(3x + 2)(3x - 2)$

38. $2y(y + 5)^2$

39. $(3m + 5)(m + 1)$

40. $8(t + 3)(t - 7)$

41. $2a(a^4 + 1)(a^2 + 1)(a + 1)(a - 1)$

42. $(a + 2)(a - 2)(a + 3)(a - 3)$

43. $4p(p-3)^2$
44. $k(k+3)(k-8)$
45. $3h^2(h+4)(h-4)$
46. $4r^2(4r^2+1)$
47. $5x(x+2)(x-2)$
48. $2(2y+1)(y+3)$
49. $2(t^2+4)(t+2)(t-2)$
50. $3m(m+2)(m-6)$

5–11 DIVISION BY A MONOMIAL

The properties of algebra may be used to divide two monomials, as the following examples show.

$$\frac{6x^7}{3x} = 6x^6 \cdot x \cdot \frac{1}{3} \cdot \frac{1}{x} = \left(6 \cdot \frac{1}{3}\right) \cdot x^6 \cdot \left(x \cdot \frac{1}{x}\right) = 2x^6$$

$$\frac{2x^9}{9x^2} = 2x^7 \cdot x^2 \cdot \frac{1}{9} \cdot \frac{1}{x^2} = 2 \cdot \frac{1}{9} \cdot x^7 \cdot x^2 \cdot \frac{1}{x^2} = \frac{2}{9}x^7$$

$$\frac{-12x^2}{4x^2} = -12x^2 \cdot \frac{1}{4} \cdot \frac{1}{x^2} = -12 \cdot \frac{1}{4} \cdot x^2 \cdot \frac{1}{x^2} = -3$$

The results of these examples can be used to describe a method for finding the quotient of two monomials. First, you must realize that the degree of the denominator cannot exceed the degree of the numerator if the quotient is to be a monomial. Second, you have to find a relationship between the degree of the quotient and the degrees of the numerator and denominator. Looking at the first two examples above, you see that

$$\frac{6x^7}{3x^1} = 2x^{7-1}, \quad \text{or } 2x^6,$$

$$\frac{2x^9}{9x^2} = \tfrac{2}{9}x^{9-2}, \quad \text{or } \tfrac{2}{9}x^7.$$

In each case, the degree of the quotient equals the degree of the numerator minus the degree of the denominator. If you continue to use this rule, you get

$$\frac{-12x^2}{4x^2} = -3x^{2-2}, \quad \text{or } -3x^0.$$

Until now, you have not had occasion to use the *zero power* of a number. You know that $x^2 \div x^2 = 1$ if $x \neq 0$. Therefore, if $x^2 \div x^2$ is to be x^{2-2}, or x^0, then the *zero power* of a number must be defined in the following manner.

$$\mathbf{x^0 = 1 \quad \textit{if} \quad x \neq 0}$$

Your observations on the degree of a quotient should lead you to state the following.

SECOND LAW OF EXPONENTS

The equation

$$\frac{x^m}{x^n} = x^{m-n}, \quad m \geqq n \qquad \text{(LE-2)}$$

is true for every nonzero number x and all positive integers m and n, subject to the stated condition.

In dividing one monomial by another, you get the coefficient of the quotient by dividing the coefficients of the two monomials. This fact and the second law of exponents can be used to find the following quotients.

Problem 1. Find each of the quotients.

(a) $\frac{36x^3}{9x^2}$ (b) $\frac{8x^7}{-12x^3}$ (c) $\frac{-(\frac{2}{3}x^{12})}{\frac{4}{9}x^3}$ (d) $\frac{x^4}{3x^4}$

Solution. You have

(a) $\frac{36x^3}{9x^2} = \frac{36}{9}x^{3-2} = 4x^1$, or $4x$.

(b) $\frac{8x^7}{-12x^3} = \frac{8}{-12}x^{7-3} = -\frac{2}{3}x^4$.

(c) $\frac{-(\frac{2}{3}x^{12})}{\frac{4}{9}x^3} = -(\frac{2}{3}) \cdot \frac{9}{4}x^{12-3} = -\frac{3}{2}x^9$.

(d) $\frac{x^4}{3x^4} = \frac{1}{3}x^{4-4} = \frac{1}{3}x^0$, or $\frac{1}{3}$.

To divide a polynomial by a monomial, divide each term of the polynomial by the monomial, as illustrated below.

Problem 2. Find each of the quotients.

(a) $(7x^2 - 4x) \div x$

(b) $\frac{x^5 - 4x^4 + 7x^3}{2x^2}$

Solution.

(a) By definition,

$$(7x^2 - 4x) \div x = (7x^2 - 4x) \cdot \frac{1}{x} \cdot$$

Therefore, by the distributive axiom, (D),

$$\begin{aligned}(7x^2 - 4x) \div x &= \left(7x^2 \cdot \frac{1}{x}\right) + \left(-4x \cdot \frac{1}{x}\right) \\ &= \frac{7x^2}{x} - \frac{4x}{x} \\ &= 7x^{2-1} - 4x^{1-1}, \quad \text{or } 7x - 4.\end{aligned}$$

(b) By the distributive axiom, (D),

$$\frac{x^5 - 4x^4 + 7x^3}{2x^2} = \frac{x^5}{2x^2} + \frac{-4x^4}{2x^2} + \frac{7x^3}{2x^2} \cdot$$

Therefore,

$$\begin{aligned}\frac{x^5 - 4x^4 + 7x^3}{2x^2} &= \tfrac{1}{2}x^{5-2} - 2x^{4-2} + \tfrac{7}{2}x^{3-2} \\ &= \tfrac{1}{2}x^3 - 2x^2 + \tfrac{7}{2}x.\end{aligned}$$

Exercises

In Exercises 1–30, find each of the quotients.

1. (a) $24x^9 \div 6x^3$ (b) $9x^{12} \div 3x^6$

2. (a) $-27x^{10} \div 18x^5$ (b) $-30x^8 \div 6x^3$

3. (a) $(-15y^{10}) \div (-3y^5)$ (b) $(-25x^{12}) \div (-5x^7)$

4. (a) $5^{20} \div 5^{17}$ (b) $3^{18} \div 3^{12}$

5. (a) $17x^6 \div 34x$ (b) $16x^8 \div 80x^2$

6. (a) $x^{12} \div 5x^{12}$ (b) $13y^4 \div 26y^4$

7. (a) $4^{21} \div 4^{21}$ (b) $3x^{30} \div 3x^{30}$

8. (a) $-(17x^6) \div (34x^5)$ (b) $-(34x^6) \div (-17x^5)$

9. (a) $-\frac{2}{3}x^8 \div (-\frac{6}{7}x^8)$ (b) $-\frac{4}{5}x^{12} \div (-\frac{16}{25}x^{12})$

10. (a) $\dfrac{12x^8}{-15x^7}$ (b) $\dfrac{80b^{13}}{-18b^{12}}$

11. (a) $\dfrac{-75a^{12}}{15a^4}$ (b) $\dfrac{-56x^{30}}{28x^{10}}$

12. (a) $-\left(\frac{27y^{21}}{54y^{15}}\right)$ (b) $-\left(\frac{24a^9}{18a^3}\right)$

13. (a) $-\left(\frac{19x^6}{57x^2}\right)$ (b) $-\left(\frac{48x^6}{16x^5}\right)$

14. (a) $\frac{4y^7}{-28y^7}$ (b) $\frac{-14x^{11}}{14x^{11}}$

15. (a) $\frac{-54y^{24}}{21y^{23}}$ (b) $-\left(\frac{15x^{33}}{50x^{22}}\right)$

16. (a) $\frac{2x^4 - 3x^3 - 4x^2}{x}$ (b) $\frac{x^3 + x}{x}$

17. (a) $\frac{3x^3 + 6x^2}{3x^2}$ (b) $\frac{4x^7 - 12x^5}{4x^4}$

18. (a) $\frac{56x^8 - 14x^6 - 3x^4}{7x^4}$ (b) $\frac{20x^5 - 8x^4 - 5x^3}{4x^3}$

19. (a) $\frac{63y^9 - 28y^8 - 14y^7}{-7y^6}$ (b) $\frac{65x^5 - 26x^4 + 52x^3}{-13x^3}$

20. (a) $\frac{6a^{15} - 5a^{10} - 2a^5}{30a^5}$ (b) $\frac{22x^{12} - 4x^{11} - 11x^{10}}{44x}$

21. (a) $\frac{3a^{12} - 4a^{10} + 6a^8}{3a^8}$ (b) $\frac{14x^8 - 10x^6 + 5x^4}{10x^4}$

22. (a) $(3x^3 + 15x^2 - 21x) \div 3x$
(b) $(6x^3 + 16x^2 - 20x) \div 2x$

23. (a) $(8x^3 - 4x^2 - 2x) \div (-2x)$
(b) $(20y^6 - 3y^4 - 5y^2) \div (-5y^2)$

24. (a) $(3.5x^9 - .28x^8 - .056x^7) \div (-7x^5)$
(b) $(4.2x^7 - 1.2x^6 - .018x^5) \div (-6x^3)$

25. (a) $(12x^5 + 18x^3 - 6x^2) \div (21x^2)$
(b) $(15x^6 + 25x^5 - 10x^2) \div (5x^2)$

26. (a) $\frac{30x^4 - 45x^2 + 15x}{-75x}$ (b) $\frac{40x^3 + 60x^5 - 80x^7}{-20x^2}$

27. (a) $\frac{16x^4 - \frac{1}{4}x^3 - \frac{1}{3}x^2}{\frac{1}{2}x^2}$ (b) $\frac{18y^{13} - 5y^{12} - 7y^{11}}{\frac{2}{3}y}$

28. (a) $\frac{\frac{3}{4}y^2 + \frac{5}{6}y^3}{\frac{1}{4}y^2}$ (b) $\frac{\frac{1}{2}x^2 - \frac{1}{6}x^3}{\frac{1}{4}x^2}$

29. (a) $\frac{4.2x^2 - .49x}{.07x}$ (b) $\frac{.36y^4 - 4.8y^2}{.06y}$

30. (a) $\frac{\frac{3}{4}y^3 - \frac{1}{4}y^2}{\frac{3}{4}y^2}$ (b) $\frac{\frac{5}{2}x^2 - 5x}{\frac{5}{2}x}$

31. Give the reasons for the steps in the following simplification.

$$\frac{(30x^{10} - 35x^9 - 40x^8)}{20x^5} = (30x^{10} - 35x^9 - 40x^8)\cdot\frac{1}{20x^5} \qquad \underline{\quad ? \quad}$$

$$= \frac{30x^{10}}{20x^5} - \frac{35x^9}{20x^5} - \frac{40x^8}{20x^5} \qquad \underline{\quad ? \quad}$$

$$= \tfrac{3}{2}x^{10-5} - \tfrac{7}{4}x^{9-5} - 2x^{8-5} \qquad \underline{\quad ? \quad}$$

$$= \tfrac{3}{2}x^5 - \tfrac{7}{4}x^4 - 2x^3$$

Perform the operations and give your answers in simplest form.

32. $\dfrac{(-5a^2)(-7a^3)}{35a^5}$

33. $\dfrac{(\frac{1}{2}x^4)(\frac{2}{3}x^6)}{-(\frac{4}{3}x^8)}$

34. $(3x^5)^2 \div (-2x^4)$

35. $100x^9 \div (-2x^3)^2$

36. $[4x(3x - 5) - 7x(6x + 7)] \div (-3x)$

37. $[2y^2(y^3 - 1) - 3y^2(2y^3 + 4)] \div 2y^2$

5–12 DIVISION OF POLYNOMIALS

There is a division process for polynomials similar to the division process for integers. This process is illustrated below.

Problem 1. Use the division process to show that $x + 4$ is a factor of $2x^2 + 3x - 20$.

Solution. First, divide $2x^2$, the term of highest degree in $2x^2 + 3x - 20$, by x, the term of highest degree in $x + 4$, obtaining $2x$.

$$\begin{array}{r} 2x \\ x + 4\overline{)2x^2 + 3x - 20} \end{array}$$

Then multiply $x + 4$ by $2x$, and subtract the resulting polynomial from $2x^2 + 3x - 20$.

$$\begin{array}{r} 2x \\ x + 4\overline{)2x^2 + 3x - 20} \\ \underline{2x^2 + 8x } \\ -5x - 20 \end{array} \qquad = 2x(x + 4)$$

Now repeat the process, this time using polynomial $-5x - 20$ in place of $2x^2 + 3x - 20$. Thus, divide $-5x$, the term of highest degree in $-5x - 20$, by x, the term of highest degree in $x + 4$, obtaining -5.

$$
\begin{array}{r}
2x \quad - 5 \\
x + 4 \overline{)2x^2 + 3x - 20} \\
\underline{2x^2 + 8x \quad\quad} \\
-5x - 20
\end{array}
$$

Then multiply $x + 4$ by -5, and subtract the resulting polynomial from $-5x - 20$.

$$
\begin{array}{rl}
2x \quad - 5 & \\
x + 4 \overline{)2x^2 + 3x - 20} & \\
\underline{2x^2 + 8x \quad\quad} & \\
-5x - 20 & \\
\underline{-5x - 20} & = -5(x + 4) \\
0 &
\end{array}
$$

Since the remainder (that is, the final polynomial) is 0, the polynomial $2x^2 + 3x - 20$ is divisible by $x + 4$, and

$$2x^2 + 3x - 20 = (x + 4)(2x - 5).$$

$2x - 5$ is called the *quotient* on dividing $2x^2 + 3x - 20$ by $x + 4$.

Problem 2. Is $2x + 5$ a factor of $6x^3 + 11x^2 + 4x + 35$?

Solution. If it is a factor, the term of highest degree in the quotient must be $6x^3 \div 2x$, or $3x^2$. Proceed by division as in Problem 1.

$$
\begin{array}{rl}
3x^2 - 2x + 7 & \\
2x + 5 \overline{)6x^3 + 11x^2 + 4x + 35} & \\
\underline{6x^3 + 15x^2 \quad\quad\quad\quad} & = 3x^2(2x + 5) \\
-4x^2 + 4x + 35 & \\
\underline{-4x^2 - 10x \quad\quad} & = -2x(2x + 5) \\
14x + 35 & \\
\underline{14x + 35} & = 7(2x + 5) \\
0 &
\end{array}
$$

Since the remainder is 0, $6x^3 + 11x^2 + 4x + 35$ is divisible by $2x + 5$, and the quotient is $3x^2 - 2x + 7$.

$$6x^3 + 11x^2 + 4x + 35 = (2x + 5)(3x^2 - 2x + 7)$$

Problem 3. Is $x + 5$ a factor of $x^4 + 4x^3 + 28x + 15$?

Solution. Notice that the second polynomial has no term of second degree. Proceed by division.

$$
\begin{array}{r|l}
 & x^3 - x^2 + 5x + 3 \\
x + 5 & x^4 + 4x^3 + 28x + 15 \\
 & x^4 + 5x^3 \qquad = x^3(x+5) \\
 & -x^3 + 28x + 15 \\
 & -x^3 - 5x^2 \qquad = -x^2(x+5) \\
 & 5x^2 + 28x + 15 \\
 & 5x^2 + 25x \qquad = 5x(x+5) \\
 & 3x + 15 \\
 & 3x + 15 \qquad = 3(x+5) \\
 & 0
\end{array}
$$

Since the remainder is 0, $x + 5$ is a factor of $x^4 + 4x^3 + 28x + 15$, and

$$x^4 + 4x^3 + 28x + 15 = (x + 5)(x^3 - x^2 + 5x + 3).$$

Check.

$$
\begin{array}{r}
x^3 - x^2 + 5x + 3 \\
x + 5 \\
\hline
x^4 - x^3 + 5x^2 + 3x \\
5x^3 - 5x^2 + 25x + 15 \\
\hline
x^4 + 4x^3 + 28x + 15
\end{array}
$$

Problem 4. Is $a^2 + 1$ a factor of $a^5 - a^4 + 3a^3 + 4a^2 + 3a + 4$?

Solution. Use the division process to obtain the following.

$$
\begin{array}{r|l}
 & a^3 - a^2 + 2a + 5 \\
a^2 + 1 & a^5 - a^4 + 3a^3 + 4a^2 + 3a + 4 \\
 & a^5 + a^3 \qquad = a^3(a^2+1) \\
 & -a^4 + 2a^3 + 4a^2 + 3a + 4 \\
 & -a^4 - a^2 \qquad = -a^2(a^2+1) \\
 & 2a^3 + 5a^2 + 3a + 4 \\
 & 2a^3 + 2a \qquad = 2a(a^2+1) \\
 & 5a^2 + a + 4 \\
 & 5a^2 + 5 \qquad = 5(a^2+1) \\
 & a - 1
\end{array}
$$

This is as far as you can go, since a is not divisible by a^2. Thus the remainder, $a - 1$, is not zero and you can conclude that $a^2 + 1$ is not a factor of $a^5 - a^4 + 3a^2 + 4a^2 + 3a + 4$.

In the problem on page 225,

$$a^3 - a^2 + 2a + 5$$

is called the *quotient* and $a - 1$ is called the *remainder* on dividing $a^5 - a^5 + 3a^3 + 4a^2 + 3a + 4$ by $a^2 + 1$. The quotient and remainder are related to the given polynomials as follows:

$$a^5 - a^4 + 3a^3 + 4a^2 + 3a + 4$$
$$= (a^2 + 1)(a^3 - a^2 + 2a + 5) + (a - 1).$$

A similar situation arises when you divide two integers. For example, on dividing 443 by 19, the quotient is 23 and the remainder is 6.

$$\begin{array}{rl} & 23 \\ 19\overline{)443} & \\ 380 & = 20 \times 19 \\ \hline 63 & \\ 57 & = 3 \times 19 \\ \hline 6 & \end{array}$$

This means that

$$443 = 19 \cdot 23 + 6.$$

Exercises

1. (a) Fill in the blanks to complete the division.

$$\begin{array}{l} \qquad\quad 4x^2 - 6x \;\; + 9 \\ 2x + 3\overline{)8x^3 + 0x^2 + 0x + 27} \end{array}$$

$$___?___ \qquad = 4x^2(2x + 3)$$

$$___?___ \qquad = -6x(2x + 3)$$

$$___?___ \qquad = 9(2x + 3)$$

Therefore,

$$8x^3 + 27 = (_?_)(_?_).$$

(b) Fill in the blanks to complete the division.

$$\begin{array}{r}4x^2 - 5x + 2 \\ 3x - 2\overline{)12x^3 - 23x^2 + 16x - 4}\end{array}$$

$$\underline{\quad ? \quad} = 4x^2(3x - 2)$$

$$\underline{\quad ? \quad} = -5x(3x - 2)$$

$$\underline{\quad ? \quad} = 2(3x - 2)$$

Therefore,

$$12x^3 - 23x^2 + 16x - 4 = (\underline{\ ?\ })(\underline{\ ?\ }).$$

2. (a) Check the division of Exercise 1(a) by multiplication.
(b) Check the division of Exercise 1(b) by multiplication.

3. (a) Complete the division.

$$\begin{array}{r}2x^2 - 2x - 6 \\ x^2 + x + 4\overline{)2x^4 + 0x^3 + 0x^2 - 3x + 7}\end{array}$$

$$\underline{\quad ? \quad} = 2x^2(x^2 + x + 4)$$

$$\underline{\quad ? \quad} = -2x(x^2 + x + 4)$$

$$\underline{\quad ? \quad} = -6(x^2 + x + 4)$$

Therefore, $2x^4 - 3x + 7 = (\underline{\ ?\ })(\underline{\ ?\ }) + \underline{\ ?\ }.$

(b) Complete the division.

$$\begin{array}{r}2x^2 + 2x - 8 \\ x^2 + 2x + 4\overline{)2x^4 + 6x^3 + 4x^2 - 4x + 1}\end{array}$$

$$\underline{\quad ? \quad} = 2x^2(x^2 + 2x + 4)$$

$$\underline{\quad ? \quad} = 2x(x^2 + 2x + 4)$$

$$\underline{\quad ? \quad} = -8(x^2 + 2x + 4)$$

Therefore,

$$2x^4 + 6x^3 + 4x^2 - 4x + 1 = (\underline{\ ?\ })(\underline{\ ?\ }) + \underline{\ ?\ }.$$

In Exercises 4–12, perform the division, and check the first five by multiplication. In each case, write an equation which relates the quotient and remainder (if one exists) to the given polynomial.

4. (a) $(y^4 + 9y^2 + 20) \div (y^2 + 4)$
   (b) $(16x^4 - 26x^2 - 35) \div (8x^2 + 7)$
5. (a) $(8x^3 - 27) \div (2x - 3)$ (b) $(8x^3 - 27) \div (2x + 3)$
6. (a) $(125y^6 + 64) \div (5y^2 + 4)$ (b) $(125y^6 + 64) \div (5y^2 - 4)$
7. (a) $(27a^3 - 8) \div (9a^2 + 6a + 4)$ (b) $(27a^3 + 8) \div (9a^2 - 6a + 4)$
8. (a) $(b^4 - 16) \div (b^2 - 4)$ (b) $(b^4 - 16) \div (b^2 + 4)$
9. (a) $(b^4 - 16) \div (b + 2)$ (b) $(b^4 - 16) \div (b - 2)$
10. (a) $(x^5 - 32) \div (x - 2)$ (b) $(x^5 - 32) \div (x + 2)$
11. (a) $(x^2 + 36) \div (x + 6)$ (b) $(x^2 + 64) \div (x + 8)$
12. (a) $(y^2 + 25) \div (y - 5)$ (b) $(y^2 + 100) \div (y - 10)$

---

13. $(x^4 - 2x^3 - 1 - 4x^5) \div (2x^3 - 2x^2 + 1)$
14. $(x^4 + x^2 + 1) \div (x^2 - x + 1)$
15. $(y^4 + y^2 + 1) \div (y^2 + y + 1)$
16. $(x^4 - x^2 + 16) \div (x^2 + 3x + 4)$
17. $(x^5 + 243) \div (x + 3)$

Preparation for Section 5–13

Factor each of the following polynomials.

1. $x^2 + 14x + 49$
2. $x^2 - 22x + 121$
3. $x^2 - x - 30$
4. $2x^2 + x - 6$
5. $x^3 - 16x$
6. $2x^3 + 6x^2 + 4x$

Solve each of the following equations.

7. $4x - 9 = 0$
8. $6x + 7 = 1$
9. $12x + 13 = 13$
10. $8x - 1 = 0$

5–13 POLYNOMIAL EQUATIONS

If each side of an equation is a polynomial in x (or in some other variable), the equation is called a *polynomial equation.* In this section you will learn to solve some simple polynomial equations.

Problem 1. Solve the polynomial equation $x^2 + 6x - 7 = 0$.

Solution. This equation is called a *second-degree equation.* You know that $x^2 + 6x - 7 = (x + 7)(x - 1)$; therefore the given equation is equivalent to the equation

$$(x + 7)(x - 1) = 0.$$

The solution set S of this equation is given by

$$S = \{x \mid (x + 7)(x - 1) = 0\}.$$

Before finishing Problem 1, you need to know for what numbers a and b is $ab = 0$. You know that the product of two positive numbers is positive, the product of two negative numbers is positive, the product of a positive and a negative number is negative, and the product of any number and zero is zero. Therefore, you can conclude that the product of two nonzero numbers is not zero. Hence, the product of two numbers is zero if, and only if, one or the other is zero.

FACTORS OF ZERO

ab = 0 if, and only if, a = 0 or b = 0 **(F–0)**

Consequently,

$$(x + 7)(x - 1) = 0 \quad \text{if, and only if,} \quad x + 7 = 0 \text{ or } x - 1 = 0.$$

Therefore,

$$(x + 7)(x - 1) = 0 \quad \text{if, and only if,} \quad x = -7 \text{ or } x = 1.$$

You can conclude that

$$S = \{-7, 1\}$$

is the solution set of equation $x^2 + 6x - 7 = 0$.

Check. $(-7)^2 + [6 \cdot (-7)] - 7 \stackrel{?}{=} 0$ $\quad 1^2 + (6 \cdot 1) - 7 \stackrel{?}{=} 0$

$49 - 42 - 7 \stackrel{\checkmark}{=} 0$ $\quad 1 + 6 - 7 \stackrel{\checkmark}{=} 0$

Problem 2. Solve the second-degree equation $(x - 5)(x - 4) = 42$.

Solution. The polynomial on the left side of the equation is already factored. However, this does not help to solve the equation since the right side is not zero. You know that the product of two numbers is 42, but this tells you nothing about the numbers except that neither is zero.

In order to solve this equation you must find an equivalent equation which has the form of a second-degree polynomial set equal to zero. Proceed as follows:

$$\begin{aligned} x^2 - 9x + 20 &= 42, \\ x^2 - 9x + 20 - 42 &= 42 - 42, \\ x^2 - 9x - 22 &= 0, \\ (x - 11)(x + 2) &= 0. \end{aligned}$$

Thus,

$$\begin{aligned} \{x \mid (x - 5)(x - 4) = 42\} &= \{x \mid (x - 11)(x + 2) = 0\} \\ &= \{x \mid x - 11 = 0\} \cup \{x \mid x + 2 = 0\} \\ &= \{11, -2\}. \end{aligned}$$

Check. $(11 - 5)(11 - 4) \stackrel{?}{=} 42$ $\quad (-2 - 5)(-2 - 4) \stackrel{?}{=} 42$

$6 \cdot 7 \stackrel{\checkmark}{=} 42$ $\quad (-7) \cdot (-6) \stackrel{\checkmark}{=} 42$

Problem 3. Solve the equation $9x^2 - 42x + 49 = 0$.

Solution. You should recognize the given polynomial to be the square of a binomial.

$$(3x - 7)^2 = 0$$

The square of a number is zero if, and only if, the number is zero. Thus, the solution set S of the given equation is

$$\begin{aligned} S &= \{x \mid 3x - 7 = 0\}, \\ S &= \{\tfrac{7}{3}\}. \end{aligned}$$

Check.

$$\begin{aligned} 9 \cdot (\tfrac{7}{3})^2 - (42 \cdot \tfrac{7}{3}) + 49 &\stackrel{?}{=} 0 \\ 9 \cdot \tfrac{49}{9} - (14 \cdot 7) + 49 &\stackrel{?}{=} 0 \\ 49 - 98 + 49 &\stackrel{\checkmark}{=} 0 \end{aligned}$$

Problem 4. Solve the third-degree equation $6x^3 + 9x^2 - 27x = 0$.

Solution. When you factor the polynomial

$$6x^3 + 9x^2 - 27x$$

you find that $3x$ is a monomial factor. Thus,

$$6x^3 + 9x^2 - 27x = 3x(2x^2 + 3x - 9).$$

You may further factor the polynomial as follows:

$$6x^3 + 9x^2 - 27x = 3x(2x - 3)(x + 3).$$

Thus, the given equation is equivalent to the equation

$$3x(2x - 3)(x + 3) = 0.$$

The equation now tells you that the product of these four numbers is zero. You know, then, that one number must be zero. It is not the first factor, 3, so it must be one of the other factors.

$$x = 0 \quad \text{or} \quad 2x - 3 = 0 \quad \text{or} \quad x + 3 = 0$$

Thus,

$$x = 0 \quad \text{or} \quad x = \tfrac{3}{2} \quad \text{or} \quad x = -3.$$

Therefore, the solution set S of the given equation has three elements:

$$S = \{0, \tfrac{3}{2}, -3\}.$$

Check.

$$(6 \cdot 0^3) + (9 \cdot 0^2) - (27 \cdot 0) \overset{\checkmark}{=} 0$$

$$[6 \cdot (\tfrac{3}{2})^3] + [9 \cdot (\tfrac{3}{2})^2] - (27 \cdot \tfrac{3}{2}) \overset{?}{=} 0$$
$$(6 \cdot \tfrac{27}{8}) + (9 \cdot \tfrac{9}{4}) - (27 \cdot \tfrac{3}{2}) \overset{?}{=} 0$$
$$\tfrac{81}{4} + \tfrac{81}{4} - \tfrac{81}{2} \overset{\checkmark}{=} 0$$

$$[6 \cdot (-3)^3] + [9 \cdot (-3)^2] - [27 \cdot (-3)] \overset{?}{=} 0$$
$$[6 \cdot (-27)] + (9 \cdot 9) + 81 \overset{?}{=} 0$$
$$-162 + 81 + 81 \overset{\checkmark}{=} 0$$

Note that this third-degree equation has three solutions.

Exercises

Find the solution set for each of the polynomial equations in Exercises 1–17. Check each solution.

1. (a) $(x + 2)(x + 5) = 0$ (b) $(x + 2)(x - 5) = 0$

2. (a) $(x + 5)(x - 5) = 0$ (b) $(x + 6)(x - 6) = 0$

3. (a) $(3x - 2)(2x + 1) = 0$ (b) $(x - 1)(x + 4)(2x + 3) = 0$

4. (a) $x(x - 4) = 0$ (b) $x(2x + 1) = 0$

5. (a) $(5x - 4)^2 = 0$ (b) $(3x + 7)^2 = 0$

6. (a) $x^2 + 7x + 12 = 0$ (b) $x^2 + 4x + 3 = 0$

7. (a) $x^2 - x - 2 = 0$ (b) $x^2 + 2x - 24 = 0$

8. (a) $y^2 + 16y + 64 = 0$ (b) $x^2 + 24x + 144 = 0$

9. (a) $9x^2 - 6x = 0$ (b) $5x^2 = 4x$

10. (a) $x^2 = 1$ (b) $25x^2 - 16 = 0$

11. (a) $x^2 - 16x + 39 = 0$ (b) $x^2 - 13x + 36 = 0$

12. (a) $3x^2 - 16x + 16 = 0$ (b) $3x^2 + 12x = 2x - 8$

13. (a) $25x^2 + 4 = 20x$ (b) $9x^2 - 66x + 121 = 0$

14. (a) $x^2 = 6x + 7$ (b) $y^2 = 29y - 100$

15. (a) $x^2 - 169 = 0$ (b) $x^2 - 225 = 0$

16. (a) $2x^2 - 5x = 0$ (b) $7x^2 - 4x = 0$

17. (a) $0 = 9 - 16x - 4x^2$ (b) $0 = 15 - 11x + 2x^2$

---

18. (a) Why does the second-degree equation in Problem 1, pages 228 and 229, have two solutions while the second-degree equation in Problem 3, page 230, has only one solution?

(b) Describe the set of second-degree polynomial equations having only one solution.

(c) How many first-degree factors can you get for a second-degree polynomial? What is the maximum number of solutions a second-degree polynomial equation can have?

Simplify each of the polynomial equations, and find the solution set.

19. $(y - 8)^2 = 100$

20. $x(x + 4) = 12(3 - x)$

21. $(x - 3)^2 - 7(x - 3) + 6 = 0$

22. $y(y + 2) = 8$

23. $x^2 = 14x - 49$

24. $y^2 + 20y + 10 = 12y - 2$

25. $(x + 2)^2 = 4$

26. $x(x - 7) = -12$

27. $11x = 10x^2 + 3$

28. $3x(x + 2) = 144$

29. $7y(y - 5) = 168$

30. $y^3 + 3y = 4y^2$

31. $\frac{y^2}{2} - \frac{5y}{6} = \frac{1}{3}$

32. $\frac{1}{2}(y + 7) = \frac{2}{3}y(y + 2)$

33. $3(b - 1)^2 + (b + 2)^2 = 9$

34. $5(a^2 - 5) - 2(3a + 5) = 3a(a - 1)$

35. $2(x - 3)^2 + (x + 4)^2 = 33$

Factor each polynomial on the left side of the equation *without removing parentheses*, and then find the solution set of the equation.

36. $(3x + 7)^2 - 49 = 0$

37. $81 - (2x + 1)^2 = 0$

38. $(2x - 3)^2 - (x + 2)^2 = 0$

39. $64x^2 - 9(x + 1)^2 = 0$

40. $(x^2 - 1)(3x^2 - 4x - 15) = 0$

41. $(5y - 2)^2 - (2y + 1)^2 = 0$

42. $(x - 1)^3 + 5(x - 1)^2 - 14(x - 1) = 0$

43. $16(2x + 1)^2 - 24(2x + 1) + 9 = 0$

KEY IDEAS AND KEY WORDS

Expressions such as

$$x,\ x^2,\ x^3,\ x^4,$$

and so on, are called **powers** of the variable x. Thus, x, or x^1, is called the **first power** of x; x^2 the **second power,** or **square,** of x; x^3 the **third power,** or **cube,** of x; x^4 the **fourth power;** and so on.

A number, or a product of a number by a power of x, is called a **monomial in x.** For example,

$$3x^4,\quad -2x^3,\quad 7x,\quad -9$$

are monomials in x. The number multiplied by a power of x is called the **coefficient** of the power of x. Thus, 3 is the coefficient of x^4 in $3x^4$ and -2 is the coefficient of x^3 in $-2x^3$. We call -9 a **constant monomial.** The superscript 4 in $3x^4$ is called the **exponent** of x in $3x^4$.

A sum of monomials in a variable is called a **polynomial** in that variable. For example,

$$3x^4 - 2x^3 + 7x - 9$$

is a polynomial in x and

$$-\tfrac{2}{3}y^3 + 11y^2 - 31y + \tfrac{4}{7}$$

is a polynomial in y. The individual monomials are called the **terms** of the polynomial, and the constant monomial (if one exists) is called the **constant term** of the polynomial. In the examples above, -9 and $\frac{4}{7}$ are the constant terms. Polynomials are added and multiplied by use of the properties of numbers given in Chapter 2.

The highest power of the variable occurring in a polynomial is called the **degree** of the polynomial. For example,

$$3x^4 - 2x^3 + 7x - 9 \quad \text{has degree 4,}$$
$$4 + 3a \quad \text{has degree 1.}$$

A constant polynomial, consisting of a constant term only, is said to have **degree zero** if the constant term is nonzero. The polynomial 0 does not have degree.

The product of two nonzero polynomials is a polynomial whose degree is the sum of the degrees of the two polynomials.

The pattern for **squaring a binomial** is as follows:

(First term + Second term)^2
= (First term)^2 + 2(First term)(Second term) + (Second term)^2,

or

(First term − Second term)^2
= (First term)^2 − 2(First term)(Second term) + (Second term).^{2}

We call the polynomial $4x^2 - 20x + 25$ a **perfect square** since it is the square of the binomial $2x - 5$.

A **difference of two squares** can be factored as follows:

(First term)^2 − (Second term)^2
= (First term + Second term)(First term − Second term).

The product of two polynomials is a third polynomial, and each of the first two polynomials is a **factor** of the third. The process of expressing a polynomial as a product of factors is called **factoring.** For example, the polynomial $6x^2 - 11x - 10$ is factored as follows:

$$6x^2 - 11x - 10 = (3x + 2)(2x - 5).$$

We define

$$\mathbf{x^0 = 1 \quad \text{if } x \neq 0.}$$

The **laws of exponents**

$$x^m \cdot x^n = x^{m+n} \qquad \text{(LE-1)}$$

$$\frac{x^m}{x^n} = x^{m-n} \quad \text{if} \quad x \neq 0 \quad \text{and} \quad m \geqq n, \qquad \text{(LE-2)}$$

are true for every real number x and all non-negative integers m and n, subject to the stated conditions.

Any polynomial can be divided by another nonzero polynomial to yield a **quotient** and a **remainder** of lower degree than the divisor.

An equation having a second-degree polynomial on one side of the equals sign and zero on the other is called a **second-degree equation.** If a second-degree polynomial can be factored into first-degree polynomials, then the second-degree equation can be solved by the rule for **factors of zero.**

$$\mathbf{a \cdot b = 0 \quad \text{if, and only if,} \quad a = 0 \quad \text{or} \quad b = 0} \qquad \textbf{(F-0)}$$

For example, the equation

$$(x - 3)(x + 5) = 0$$

has solution set

$$\{3, -5\}.$$

CHAPTER REVIEW

For Exercises 1–8 tell which of the statements are true and which are false.

1. The equation $289x^2 - 34x + 1 = 0$ has only one solution.
2. The solutions of the equation $(x - 3)(x + 7) = 24$ are 3 and -7.
3. The additive inverse of a polynomial of the third degree is also of the third degree.
4. The difference of two fifth-degree polynomials is always a fifth-degree polynomial.
5. $2^0 = 2$
6. The monomials $3x^0$ and $5x^0$ have a constant product.
7. The following have a trinomial of the second degree for their product: $(19x + 87)$ and $(19x - 87)$.
8. The equation $x^3 - 9x^2 = 0$ has solution set $\{3, -3\}$.
9. Explain why you think each of the statements you labeled false in Exercises 1–8 is a false statement.
10. Find the value of each polynomial for the given values of the variable.
    (a) $2 + x - x^2$; $x = -3, -2, -1, 0, \frac{1}{2}, 1, 2, 3, 4$
    (b) $y^4 - 5y^2 + 4$; $y = -3, -2, -\frac{3}{2}, -1, 0, 1, \frac{3}{2}, 2, 3$
    (c) $2x^2 - 3x - 8$; $x = -4, -1, 0, 1, 2$

Do the operations and simplify your results.

11. $2(3x^2 - 5x - 2) - 3(x^2 - 6x - 8)$
12. $(-6x^2)(\frac{1}{15}x^{12})$
13. $(-\frac{3}{5}x^{12}) \div (\frac{4}{15}x^3)$
14. $x^8 \cdot x^0$
15. $(9x^5 - 12x^4 + 15x^3) \div 3x^3$
16. $(10x^3 - 21x^2 + 14x - 3) \div (5x - 3)$
17. $2(3t^2 + 2t + 1) - 3(2t^2 - 6t - 3)$
18. $(-4x^8)(\frac{1}{12}x^7)(-5x^9)$
19. $(-2x^2)^3(-3x)^2$
20. $-3x^2(5x^4 - 9x^2 - 7)$
21. $(-7x^{20}) \div (-21x^5)$
22. $(12x^4 - 24x^3 - 6x^2) \div (-36x^2)$
23. $(2x - 5)(3x^2 + 8x + 11)$
24. $(3x^3 + 11x^2 + 11x + 15) \div (x + 3)$
25. $(27y^3 + 125) \div (3y + 5)$
26. $(3x - 1)^2 - 4(2x - 5)^2$

Find the products.

27. $(3x + 7)(3x - 7)$
28. $(9x - 1)^2$
29. $(2x + 5)(7x - 6)$
30. $(2x - 9)(x - 3)$
31. $(5x + 1)(4x - 1)$
32. $(12x^2 + 5)(12x^2 - 5)$
33. $(7x - 5)^2$
34. $(3x + 1)^2$
35. $(6x - 7)(11x + 2)$

Factor each polynomial. (Hint: Look for monomial factors first.)

36. $7x^2 - 21x$
37. $x^2 + 6x - 9$
38. $3x^5 - 48x^3$
39. $17x^2 + 34x + 17$
40. $x^3 - 7x^2 - 18x$
41. $30y^2 + 10y - 100$
42. $15x^5 - 20x^3 + 45x$
43. $3x^7 + x$
44. $121x^2 - 49$
45. $x^2 - 30x + 225$
46. $9x^2 + 30x + 25$
47. $x^2 + x - 42$
48. $5n^2 - 64n + 48$
49. $18y^2 - 5y - 2$
50. $-3x^3 - 27x$
51. $36x^4 - 120x^2 + 100$
52. $y^4 - 81$
53. $4x^4 + 88x^3 + 484x^2$
54. $x(2 + x) - 3(2 + x)$
55. $(x + 17)^2 - 7(x + 17) + 6$

Solve each equation and check the solution.

56. $x(3x + 2) = 0$
57. $x^2 - x = 90$
58. $x^2 - 121 = 0$
59. $3x^2 - 23x - 36 = 0$
60. $(2x + 3)(2x - 5) = 0$
61. $5y^2 + 22y + 24 = 0$
62. $(x - 7)(10x^2 + 19x + 6) = 0$
63. $x(3x + 2) = 8$
64. $9x^2 - 25 = 0$
65. $7x^2 - 5x = 0$

CHAPTER TEST

1. Find the value of $2x^2 - 3x - 8$ for

$$x = -4, -1, 0, 1, 2.$$

2. Simplify.
   (a) $2(3x^2 - 5x - 2) - 3(x^2 - 6x - 8)$
   (b) $(-6x^2)(\frac{1}{15}x^{12})$
   (c) $-(\frac{3}{5}x^{12}) \div (\frac{4}{15}x^3)$
   (d) $x^8 \cdot x^0$
3. Do the computation.
   (a) $(9x^5 - 12x^4 + 15x^3) \div 3x^3$
   (b) $(10x^3 - 21x^2 + 14x - 3) \div (5x - 3)$

4. Give the products.

(a) $(3x + 7)(3x - 7)$ (b) $(9x - 1)^2$
(c) $(2x + 5)(7x - 6)$

5. Factor into the product of as many factors as possible.

(a) $7x^2 - 21x$ (b) $x^2 + 6x - 9$
(c) $3x^5 - 48x^3$ (d) $17x^2 + 34x + 17$
(e) $x^3 - 7x^2 - 18x$ (f) $30y^2 + 10y - 100$

6. Find the solution set for each equation.

(a) $x(3x + 2) = 0$ (b) $x^2 - x = 90$
(c) $x^2 - 121 = 0$ (d) $3x^2 - 23x - 36 = 0$

CHAPTER 6

Functions

Objectives . . .

- To use functions to show the relationship between variables.
- To graph ordered pairs on a cartesian coordinate system.
- To deal with polynomials functions and their application.

6–1 RELATION

To show relationships between variables, you may use equations, formulas, and graphs. If the relationship defined is one in which there is a unique value of one variable for each value of the other variable, the relationship, or correspondence is called a *function.* A function is a special kind of *relation.*

The basketball team scored 64 points in a recent game; 28 points in the first half and 36 in the second half. You could indicate these facts by using the *ordered pair* of numbers (28, 36). The first number in the pair shows the first-half score, the second number shows the second-half score.

Definition of Ordered Pair

Any pair of elements, (x, y), having a first element x and a second element y is called an ordered pair.

The elements x and y need not be different in an ordered pair. For example, the ordered pair (27, 27) could indicate that the basketball team scored 27 points in each half.

Two ordered pairs are *equal* if they have *the same first elements and the same second elements,* and are unequal otherwise. Thus, $(2, 8) = (2, 2^3)$ whereas $(2, 8) \neq (8, 2)$.

Definition of Cartesian Product

If A and B are sets of elements, then the set of all ordered pairs (x, y), where x is an element of A and y is an element of B, is called the cartesian product of A and B. It is denoted by

$$A \times B.$$

For example, if $A = \{\text{Ann, Alice}\}$ and $B = \{\text{Bob, Bill}\}$, then

$$A \times B = \{(\text{Ann, Bob}), (\text{Ann, Bill}), (\text{Alice, Bob}), (\text{Alice, Bill})\}.$$

As another example, if $N = \{1, 2, 3, \ldots\}$ is the set of all positive integers, then $N \times N$ is the set of all ordered pairs of positive integers. Thus, $(12, 13)$, $(10^6, 1001)$, and $(3, 27)$ are elements of $N \times N$.

Everyone has relatives such as brothers, sisters, aunts, parents, and so on. If S is the set of all students in your school, then an ordered pair (Mary, Bob) of elements of S is in the relation "sister" if Mary is a sister of Bob. The set of all ordered pairs (A, B) of elements of S such that A is a sister of B is an example of a *relation.*

Definition of relation

If S is a set, a subset of $S \times S$ is called a relation in S.

For example, if $S = \{1, 2, 3, 4\}$, then the relation "less than" in S consists of all ordered pairs (x, y) such that $x < y$. To be specific, the relation "less than" in S is the following subset of $S \times S$:

$$\{(1, 2), (1, 3), (1, 4), (2, 3), (2, 4), (3, 4)\}.$$

As another example, if N is the set of positive integers, then the relation "is a factor of" in N consists of all ordered pairs (a, b) such that a is a factor of b. The ordered pairs $(3, 12)$, $(8, 32)$, $(1, 7)$, $(10^2, 10^6)$, and $(11, 1001)$ are in this relation.

Exercises

1. Tell whether or not the following ordered pairs are equal.

(a) $(3, 15)$ and $(15, 3)$; $(6, \frac{18}{3})$ and $(6, 6)$; $(-2, 7)$ and $(-2, -7)$; $(-4, 12)$ and $(\frac{24}{2}, -\frac{12}{3})$.

(b) $(-7, 4)$ and $(-\frac{28}{4}, 4)$; $(4, 10)$ and $(\frac{40}{10}, \frac{40}{4})$; $(3, -\frac{1}{3})$ and $(-3, -\frac{1}{3})$; $(6, 11)$ and $(\frac{22}{2}, \sqrt{36})$

In Exercises 2–4, find the cartesian product of R and S.

2. (a) $R = \{\text{John, Bill}\}$, $S = \{\text{Mary, Sue, Ann}\}$
(b) $R = \{\text{A}\}$, $S = \{\text{B, C, D, E}\}$

3. (a) $R = \{2, 4, 6\}$, $S = \{1, 3, 5, 7\}$
(b) $R = \{-4, 0, 4\}$, $S = \{-3, -5, -7, -9\}$

4. (a) $R = \{0, 1\}$, $S = \{1\}$
(b) $R = \{0\}$, $S = \{-1, 0, 1\}$

5. (a) If $A = \{3, 6, 9\}$ and $B = \{2, 4\}$, find $A \times B$. Does $A \times B = B \times A$? (Hint: List the elements of $B \times A$.)
(b) If $R = \{-2, 0\}$ and $S = \{1, 2, 3\}$, find $R \times S$ and $S \times R$. Does $R \times S = S \times R$?

6. (a) If $R = \{-2, 4\}$ and $S = \{3, 10, 11\}$ find $R \times S$ and $S \times R$.
(b) If $R = \{-2, 0, 2\}$ and $S = \{1, 5\}$, find $R \times S$ and $S \times R$.

In exercises 7–9, list the elements of $A \times A$. (If $A \times A$ is an infinite set, list 6 elements of it.)

7. (a) $A = \{3, 9, 10\}$ (b) $A = \{2, 12, 14\}$

8. (a) A the set of all negative integers
(b) A the set of all positive numbers

9. (a) A the set of all even numbers
(b) A the set of all odd numbers

In Exercises 10–12, a set U of real numbers and a relation in U are stated. List the elements of the subset of $U \times U$ which is the given relation.

10. (a) $U = \{-5, 0, 5, 10\}$, "greater than"
(b) $U = \{4, 3, 2, 1\}$, "less than"

11. (a) $U = \{1, 2, 4, 5, 25\}$, "is the square of"
(b) $U = \{1, 2, 3, 8, 27\}$, "is the cube of"

12. (a) $U = \{1, 2, 4, 5, 6\}$, "is a multiple of"
(b) $U = \{\frac{1}{2}, 1, \frac{2}{4}, 2, \frac{4}{2}\}$, "is equal to"

In each of the following, set S and set U are such that S is a relation in U. List the members of the relation.

13. (a) $S = \{(x, y) \mid y \geqq x\}$, where $U = \{2, 4, 6\}$
(b) $S = \{(x, y) \mid y \leqq x\}$, where $U = \{-2, -4, -6\}$

14. (a) $S = \{(x, y) \mid x = 5\}$, where $U = \{4, 5, 6\}$
(b) $S = \{(x, y) \mid y = 5\}$, where $U = \{4, 5, 6\}$

15. (a) $S = \{(x, y) \mid y = x^2\}$, where $U = \{-1, 0, 1, 2\}$
(b) $S = \{(x, y) \mid x = y^2\}$, where $U = \{-1, 0, 1, 2\}$

16. (a) $S = \{(x, y) \mid x + y = 2\}$, where $U = \{-1, 0, 1, 2, 3\}$
(b) $S = \{(x, y) \mid x - y = 2\}$, where $U = \{-1, 0, 1, 2, 3\}$

---

17. (a) If R is the set of positive integers and S is the set of nonpositive integers, describe $R \times S$ and $S \times R$, and list four elements from each set.
(b) If R is the set of positive even integers and S is the set of positive odd integers, describe $R \times S$ and $S \times R$, and list four elements from each set.

18. (a) If R is the set of digits less than 3, and S is the set of positive multiples of 5 less than 20, list the elements of $R \times S$ and $S \times R$.
(b) If R is the set of digits greater than 7 and S is the set of negative integers greater than -4, list the elements of $R \times S$ and $S \times R$.

19. Let $U = \{-2, -1, 0, 1, 2\}$. List the elements of $U \times U$ and list the members of each of the following relations in U.
(a) $\{(x, y) \mid y = 2x\}$ (b) $\{(x, y) \mid y = \frac{1}{2}x\}$
How do the relations differ?

20. Let $U = \{-2, -1, 0, 1, 2\}$. List the members of each of the following relations in U.
(a) $\{(x, y) \mid y = |x|\}$ (b) $\{(x, y) \mid y = x^2\}$
(c) $\{(x, y) \mid y = x^3\}$
How do the relations differ?

Given R, the set of positive integers, find an open sentence defining each of the following relations in $R \times R$.

21. (a) $\{(1, 3), (2, 5), (3, 7), (4, 9), \ldots\}$
(b) $\{(1, 2), (2, 4), (3, 8), (4, 16), \ldots\}$

22. (a) $\{(1, 1), (2, 4), (3, 9), (4, 16), \ldots\}$
(b) $\{(1, 1), (8, 4), (27, 9), (64, 16), \ldots\}$

6–2 FUNCTION

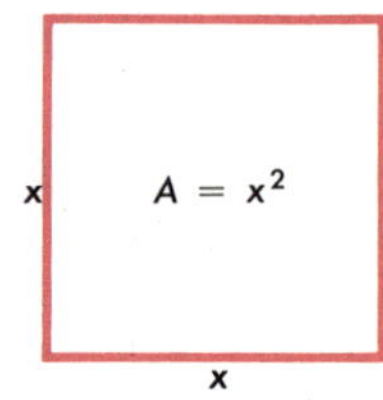

FIGURE 6–1

The formula for finding the area of a square,

$$A = x^2,$$

defines a relation in P, the set of positive numbers. Thus, (x, A) is in the "square" relation if $A = x^2$. This relation has the special property that for each positive number x there is a *unique* (that is, one and only one) ordered pair (x, x^2) in the relation having first element x.

A similar example is given by the equation

$$F = \tfrac{9}{5}C + 32,$$

which relates the temperature C in degrees centigrade to the temperature F in degrees Fahrenheit. Thus,

$$\text{if } C = 0, \text{ then } F = 32;$$
$$\text{if } C = -20, \text{ then } F = -4.$$

The relation made up of all ordered pairs (C, F), related by the equation above, also has the property that to each number C there is a *unique* ordered pair (C, F) in the relation having first element C.

Each of the relations above is an example of a *function.*

Definition of a function

A relation in a set S is called a function if no two different ordered pairs in the relation have the same first element.

For a function in a set S, the set of all numbers x such that some (x, y) is in the function is called the *domain* of the function. The set of all numbers y such that some ordered pair (x, y) is in the function is called the *range* of the function. That is, the *domain* is the set of values for x and the *range* is the set of values for y.

For example, the function above defined by the formula $A = x^2$ has the set of all positive numbers as its domain and its range. Since a temperature cannot go below absolute zero, which is approximately $-273°$ centigrade or $-459.4°$ Fahrenheit, the domain of the temperature function is $\{C \mid C \geqq -273\}$ and its range is $\{F \mid F \geqq -459.4\}$.

Problem 1. Which of the following relations in the set Z of integers are functions?

(a) $\{(1, 3), (2, 5), (3, 7), (4, 5), (5, 3)\}$
(b) $\{(-3, 3), (-2, 2), (-1, 1), (1, 1), (2, 2), (3, 3)\}$
(c) $\{(3, -3), (2, -2), (1, -1), (1, 1), (2, 2), (3, 3)\}$
(d) $\{(x - 1, 2x) \mid x \text{ in } Z\}$
(e) $\{(|x|, 3x \dot{-} 1) \mid x \text{ in } Z\}$

Solution.

(a) No two ordered pairs have the same first element, so this relation is a function.
(b) This is a function.
(c) The ordered pairs $(3, -3)$ and $(3, 3)$ have the same first elements [as do $(2, -2)$ and $(2, 2)$; and $(1, -1)$ and $(1, 1)$]. Therefore the relation is not a function.
(d) If $(x - 1, 2x)$ and $(y - 1, 2y)$ are two ordered pairs in the relation having the same first element, then $x - 1 = y - 1$ and $x = y$. Also, $2x = 2y$ and the ordered pairs $(x - 1, 2x)$ and $(y - 1, 2y)$ are the same. Thus, different ordered pairs have different first elements. The relation is a function.

(e) You know $|1| = |-1|$, $|2| = |-2|$, and so on. If $x = 1$, then $3x - 1 = 2$; if $x = -1$, then $3x - 1 = -4$. Thus, the different ordered pairs (1, 2) and (1, −4) in the relation have the same first element. The relation is not a function.

Exercises

Which of the following relations in the set Z of integers are functions? State the domain and range of each function.

1. (a) $\{(-4, -3), (1, 0), (0, 0)\}$
(b) $\{(-10, -5), (-3, -2), (0, -5)\}$

2. (a) $\{(-2, -4), (-7, -1), (-2, 0), (3, 4)\}$
(b) $\{(-3, 6), (2, 5), (-3, 2), (4, 7)\}$

3. (a) $\{(2, 3), (3, 2)\}$
(b) $\{(2, 2), (3, 3)\}$

4. (a) $\{(2, 2), (2, 3)\}$
(b) $\{(3, 2), (3, 3)\}$

5. (a) $\{(-4, 4), (-2, 2), (0, 0), (2, 2), (4, 4)\}$
(b) $\{(4, -4), (2, -2), (0, 0), (2, 2), (4, 4)\}$

6. (a) $\{(1, 0), (0, 1), (1, 1)\}$
(b) $\{(0, 0), (1, 1)\}$

7. (a) $\{(-3, 3), (-2, 3), (-1, 3), (0, 3), (1, 3)\}$
(b) $\{(-3, 3), (-2, 2), (-1, 1), (1, -1), (2, -2), (3, -3)\}$

8. (a) $\{(x, x + 1) \mid x \text{ in } Z\}$
(b) $\{(x, |x - 1|) \mid x \text{ in } Z\}$

9. (a) $\{(3x, 2x - 1) \mid x \text{ in } Z\}$
(b) $\{(2x + 5, 2x - 5) \mid x \text{ in } Z\}$

10. (a) $\{(x^2, x + 2) \mid x \text{ in } Z\}$
(b) $\{(x^3, x + 2) \mid x \text{ in } Z\}$

11. (a) Ice cream cones are selling for 15¢ each. If n is the number of cones we buy and C is the amount we pay in cents, then $C = 15n$. This equation defines a "cost" function, $\{(n, C) \mid C = 15n\}$. What are the domain and range of the cost function?
(b) A stadium has 600 $2 seats. If r is the amount of money received determined by n, the number of seats sold for any game, then $r = 2n$. This equation defines a function $\{(n, r) \mid r = 2n\}$. What are the domain and range of the function?

Write a formula defining each of the following functions. State the domain and range of each function.

12. (a) The perimeter, P, of a square is four times the side, s.
(b) The volume, V, of a cube is the third power of its edge, e.

13. (a) A trip of D miles requires t hours in a vehicle traveling at the average rate of 45 miles per hour.
(b) The cost, C, of riding a cab is 35¢ plus 10¢ for each one-fifth mile, f, traveled.

Each table below defines a relation consisting of ordered pairs (x, y). Which relations are functions? If possible, express y in terms of x by an equation. (State each rule as an equation in x and y.)

14. (a)

x	-6	1	0	2
y	-2	5	4	6

($y = x +$ _?_)

(b)

x	-4	2	5	8
y	-6	0	3	6

($y = (x +$ _?_)

15. (a)

x	-1	0	1	2	3
y	1	0	1	4	9

(b)

x	-1	0	1	2	3
y	2	1	2	5	10

16. (a)

x	-2	0	2	4	6	7
y	-3	1	5	9	13	15

(b)

x	-2	-1	0	1	3	5
y	-5	-3	-1	1	5	9

17. (a)

x	-3	-1	$\frac{3}{4}$	1	2	3.5
y	$-\frac{1}{3}$	-1	$\frac{4}{3}$	1	$\frac{1}{2}$	$\frac{2}{7}$

(b)

x	-4	-3	0	2
y	-2	$-1\frac{1}{2}$	0	1

---

18. (a) Solve the equation $F = \frac{9}{5}C + 32$ for C in terms of F.
(b) Is centigrade temperature a function of Fahrenheit temperature?

19. (a) Solve the equation $A = x^2$ for x in terms of A. (Remember: $A = x^2$ is the formula for finding the area, A, of a square with side x.)
(b) Is the length of a side of a square a function of the area?

6–3 FUNCTION AS A MAPPING

The function $\{(1, 3), (2, 5), (3, 7), (4, 5), (5, 3)\}$ has domain $\{1, 2, 3, 4, 5\}$ and range $\{3, 5, 7\}$. You can picture this function as shown in Fig. 6–2. The arrows go from the first to the second element of each ordered pair. You can think of the function as *mapping* the domain into the range in this way. Each element of the domain is mapped into

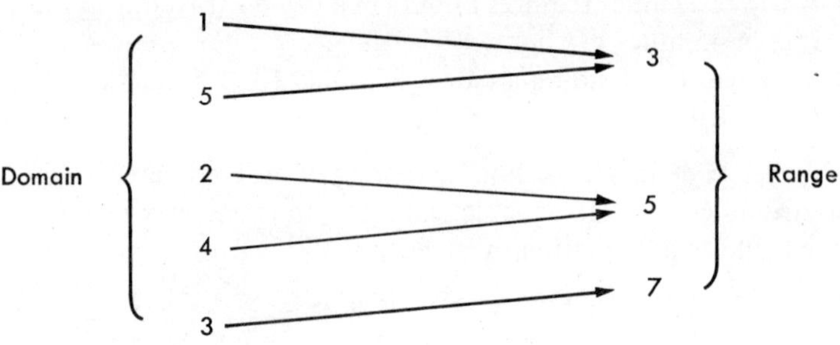

FIGURE 6-2

a unique element of the range. (Observe that different elements of the domain can be mapped into the same element of the range.)

It is common practice to denote functions by symbols such as letters of the alphabet. If f denotes some function, then for each number x in the domain of f, the corresponding number in the range of f is denoted by

$$f(x) \qquad \text{(read "}f\text{ of }x\text{")}.$$

Thus, the function f is the set

$$\{(x, f(x)) \mid x \text{ in domain } f\}.$$

For example, if you denote the function pictured above by f, then

$$f(1) = 3, f(2) = 5, f(3) = 7,$$
$$f(4) = 5, f(5) = 3.$$

If you denote the "area of a square" function by A, then

$$A(x) = x^2$$

and

$$A(5) = 25, A(10) = 100, \quad \text{and} \quad A(\tfrac{3}{4}) = \tfrac{9}{16}.$$

Considering A as a mapping, it maps each positive number x into x^2.

An equation in which one variable, y, is expressed in terms of another variable, x, defines a function consisting of all ordered pairs (x, y). For example, the equation

$$y = 3x + 1$$

defines the function $f = \{(x, 3x + 1) \mid x \text{ a real number}\}$. Thus,

$$f(x) = 3x + 1$$

for every number x: that is, f maps each number x into $3x + 1$.

A somewhat different example of a function is the first-class postal rate. According to postal regulations, the postage on all first-class parcels is 6¢ per ounce or fraction of an ounce. The maximum weight of a parcel that can be sent by first-class mail is 13 ounces. (For a parcel over 13 ounces, air parcel post rates apply.) Thus, when w is the weight in ounces and p is the amount of postage in cents, you have:

$$\begin{aligned} &\text{if } 0 < w \leqq 1, \text{ than } p = 6; \\ &\text{if } 1 < w \leqq 2, \text{ then } p = 12; \\ &\text{if } 2 < w \leqq 3, \text{ then } p = 18; \text{ etc.} \end{aligned}$$

This postage function, consisting of ordered pairs (w, p), has domain $\{w \mid 0 < w \leqq 13, w \text{ a real number}\}$ and range $\{6n \mid 0 < n \leqq 13, n \text{ an integer}\}$.

Exercises

The following ordered pairs are the elements of a function f:

$$\{(-2, -1), (-1, 0), (0, 1), (1, 2)\}.$$

1. (a) Find $f(-1)$ and $f(1)$. (b) Find $f(-2)$ and $f(0)$.

2. (a) Find x if $f(x) = 1$. (b) Find x if $f(x) = 0$.

The following ordered pairs define a function g:

$$\{(2, 5), (3, -4), (4, 5), (5, -1)\}.$$

3. (a) Find $g(2)$ and $g(5)$. (b) Find x if $g(x) = 5$.

4. (a) What is the domain of g? (b) What is the range of g?

The following ordered pairs define a function of f:

$$\{(-4, 10), (-3, 7), (-2, 10), (7, 7), (10, 10)\}.$$

5. (a) Find x if $f(x) = 10$. (b) Find $f(-3)$ and $f(10)$.

6. (a) What is the range of f? (b) What is the domain of f?

The following ordered pairs define a function f:

$$\{(-2, 4), (-1, 1), (0, 4), (1, -2), (2, \tfrac{1}{3})\}.$$

7. (a) Find $f(-2)$ and $f(1)$. (b) Find $f(-1)$ and $f(0)$.

8. (a) Find u if $f(u) = \frac{1}{3}$. (b) Find u if $f(u) = 4$.

9. (a) Can you find $f(4)$? (b) Is there a u such that $f(u) = 0$?

If the function f maps each number x into $x - 4$, find each of the following.

10. (a) $f(10)$ (b) $f(7)$

11. (a) $f(0)$ (b) $f(4)$

If the function g maps each number x into $2x + 5$, find each of the following.

12. (a) $g(0)$ (b) $g(\frac{1}{2})$

13. (a) $g(1)$ (b) $g(-\frac{5}{2})$

14. (a) $g(-1)$ (b) $g(-2)$

If the function h maps each number x into $x^2 - 2$, find each of the following.

15. (a) $h(1)$ (b) $h(0)$

16. (a) $h(2)$ (b) $h(-1)$

17. (a) $h(a)$ (b) $h(w)$

If $f(x) = x + 3$ defines a function, find the number x such that:

18. (a) $f(x) = 5$. (b) $f(x) = 4$.

19. (a) $f(x) = 10$. (b) $f(x) = 6$.

20. (a) $f(x) = 1$. (b) $f(x) = 0$.

21. (a) $f(x) = -3$. (b) $f(x) = -2$.

If the function f maps each number x into $3x - 4$, find the number x such that:

22. (a) $f(x) = 2$. (b) $f(x) = 5$.

23. (a) $f(x) = -1$. (b) $f(x) = -7$.

24. (a) $f(x) = 0$. (b) $f(x) = -4$.

Find the range of each of the following functions with the domain as given.

25. (a) $f(x) = 4x$, $\{-1, 0, 1, 2\}$
(b) $f(x) = -4x$, $\{-1, 0, 1, 2\}$

26. (a) $f(x) = x^2$, $\{-3, -2, -1, 0, 1\}$
(b) $f(x) = x^2 + 1$, $\{-3, -2, -1, 0, 1\}$

27. (a) $f(x) = \sqrt{x}$, $\{25, 16, 9, 1, 0\}$
(b) $f(x) = \sqrt{x - 1}$, $\{26, 17, 10, 2, 1\}$

28. (a) Given that $f(x) = x^2 + 1$, x a real number.
(i) Find $f(0), f(3), f(5), f(-3), f(-2), f(1), f(-1), f(\frac{1}{2})$, and $f(-\frac{1}{2})$.
(ii) What is the range of the function?
(b) Given that $f(x) = x^2 - 2$, x a real number.
(i) Find $f(0), f(1), f(2), f(3), f(-1), f(-2), f(-3), f(\frac{1}{2})$, and $f(-\frac{1}{2})$.
(ii) What is the range of the function?

State whether or not the relation $\{(x, y) \mid x \text{ a real number}\}$ defined by the equation is a function.

29. (a) $y = x + 11$ (b) $y = 2x - 10$

30. (a) $2x = |y|$ (b) $3x + 1 = |y|$

31. (a) $y + x^2 = 1$ (b) $y - 2x^2 = 0$

If $g(x) = x^2 - 1$, find every number x in the domain of g such that:

32. $g(x) = 0$ **33.** $g(x) = 3$. **34.** $g(x) = -1$.

35. (a) If $f(x) = \dfrac{x+2}{x-1}$, find $f(-4), f(0), f(5), f(10)$, and $f(-2)$.
(b) What is the domain of the function? (Hint: Is $f(1)$ defined by the equation?)
(c) Find the value of x for which $f(x) = 3$.
(d) Is there an x for which $f(x) = 1$?
(e) Describe the range of the function.

What element(s) in the domain of each of the functions $\{(x, y) \mid x \text{ a real number}\}$ defined by the following equations is (are) associated with the element 4 in the range?

36. (a) $y = x + 4$ (b) $y = 2x$

37. (a) $y = x^2$ (b) $y = 4$

38. State whether or not the given equation defines a function $y = f(x)$.

(a) $y = \dfrac{1}{x}$ (b) $x^2 + y^2 = 1$

(c) $y + x^2 = 1$ (d) $y^2 + x = 1$

39. Which, if any, of the equations in Exercise 38 define a function $x = g(y)$?

Review for Sections 6–1 through 6–3

Which of the following ordered pairs are equal?

1. (7, 14) and (14, 28) **2.** $(2^2, 5)$ and (4, 5)

3. (3, 5) and (5, −3) **4.** $(\frac{1}{2}, \frac{2}{3})$ and $(\frac{4}{6}, \frac{2}{4})$

5. (3, 9) and $(\frac{9}{3}, \frac{81}{9})$ **6.** (−4, −5) and (4, 5)

What is the cartesian product of M and R?

7. $M = \{\text{Dog, Cat}\}$; $R = \{\text{Fight, Food}\}$

8. $M = \{1, 3, 5\}$; $R = \{2, 4, 6\}$

9. $M = \{4\}$; $R = \{7\}$

10. $M = \{3, 4\}$; $R = \{0\}$

List the elements of $K \times K$. (If K is an infinite set, list four elements of $K \times K$.)

11. $K = \{2, 4, 8\}$ **12.** $K = \{0, 1\}$

13. $K = \{x \mid 0 < x < 1\}$ **14.** $K = \{\text{All positive integers}\}$

In each of the following, set P and set U are given such that P is a relation in U. List all members of set P.

15. $P = \{(a, b) \mid a < b\}$, where $U = \{1, 3, 5\}$

16. $P = \{(a, b) \mid a \leqq b\}$, where $U = \{1, 3, 5\}$

17. $P = \{(a, b) \mid a + 1 = b\}$, where $U = \{4, 5, 7, 8, 9\}$

What is the domain and what is the range in each of the following relations in U? Which relations are functions?

18. $\{(0, 0), (1, 1), (2, 4), (3, 9), (4, 16)\}$ where U is the set of integers

19. $\{(-6, 6), (-4, 4), (-2, 2), (0, 0), (2, 2), (4, 4), (6, 6)\}$ where U is the set of integers

20. $\{(a, b) \mid a \leqq b\}$ where $U = \{1, 2, 4\}$

21. $\{(a, b) \mid a + 2 = b\}$ where $U = \{1, 2, 3, 4, 6\}$

22. $\{(a, b) \mid a^2 = b\}$ where $U = \{1, 2, 3, 4, 9, 16, 81\}$

23. $\{(a, b) \mid a = |b|\}$ where $U = \{-3, -2, -1, 0, 1, 2, 3\}$

24. $\{(a, b) \mid a = |b|\}$ where $U = \{0, 1, 2, 3\}$

State a rule for each of the relations illustrated in each of the following tables. (State each rule as an equation in x and y.)

25.

x	-1	0	1	2
y	-3	0	3	6

26.

x	-5	-3	0	1	2
y	-3	-1	2	3	4

27.

x	-1	0	1	2	3
y	$-\frac{1}{2}$	0	$\frac{1}{2}$	1	$\frac{3}{2}$

28.

x	-1	1	2	$\frac{3}{2}$	3
y	-1	1	$\frac{1}{2}$	$\frac{2}{3}$	$\frac{1}{3}$

The following ordered pairs are the elements of function f:

$$\{(3, 1), (2, 6), (1, 4), (0, 0), (-1, 4), (-2, 6)\}.$$

29. What is $f(-2)$?

30. If $f(x) = 1$, what is x?

31. What are the domain and range of f?

If the function h maps each number x into $4x - 1$, find the corresponding element in the range for each of the following.

32. $h(0)$ **33.** $h(1)$ **34.** $h(-1)$

35. $h(10)$ **36.** $h(-2)$ **37.** $h(5)$

38. $h(m)$ **39.** $h(2r)$ **40.** $h(3 + k)$

If the function f maps each number x into $2x + 1$, find the number x such that:

41. $f(x) = 3.$ **42.** $f(x) = 21.$ **43.** $f(x) = 2m + 1.$

44. $f(x) = -1.$ **45.** $f(x) = 1.$

Find the range of each of the following functions with the domain as stated.

46. $f(x) = -2x$, $\{-1, 0, 1, 2\}$

47. $f(x) = 3x + 4$, $\{-2, -1, 0, 1\}$

48. $f(x) = x^2 + 3$, $\{-2, -1, 0, 1, 2\}$

49. $f(x) = 2x^2 + 1$, $\{-2, -1, 0, 1, 2\}$

50. $f(x) = x^2 + 3x + 2$, $\{-2, -1, 0, 1, 2\}$

Answers to Review for Sections 6–1 through 6–3

1. No **2.** Yes **3.** No

4. No **5.** Yes **6.** No

7. $M \times R = \{$(Dog, Fight), (Dog, Food), (Cat, Fight), (Cat, Food)$\}$

8. $M \times R = \{(1, 2), (1, 4), (1, 6), (3, 2), (3, 4), (3, 6), (5, 2), (5, 4), (5, 6)\}$

9. $M \times R = \{(4, 7)\}$

10. $M \times R = \{(3, 0), (4, 0)\}$

11. $K \times K = \{(2, 2), (2, 4), (2, 8), (4, 2), (4, 4), (4, 8), (8, 2), (8, 4), (8, 8)\}$

12. $K \times K = \{(0, 0), (0, 1), (1, 0), (1, 1)\}$

13. $K \times K =$ an infinite set. For example: $(\frac{1}{2}, \frac{3}{4})$, $(\frac{2}{3}, \frac{1}{8})$, $(\frac{3}{5}, \frac{4}{7})$, and $(\frac{1}{6}, \frac{2}{5})$

14. $K \times K =$ an infinite set. For example: (3, 9), (7, 126), (53, 1), and (5971, 264)

15. $P = \{(1, 3), (1, 5), (3, 5)\}$

16. $P = \{(1, 1), (1, 3), (1, 5), (3, 3), (3, 5), (5, 5)\}$

17. $P = \{(4, 5), (7, 8), (8, 9)\}$

18. Domain $= \{0, 1, 2, 3, 4\}$, Range $= \{0, 1, 4, 9, 16\}$; Function

19. Domain $= \{-6, -4, -2, 0, 2, 4, 6\}$, Range $= \{0, 2, 4, 6\}$; Function

20. Domain $= \{1, 2, 4\}$, Range $= \{1, 2, 4\}$; Not a function

21. Domain $= \{1, 2, 4\}$, Range $= \{3, 4, 6\}$; Function

22. Domain $= \{1, 2, 3, 4, 9\}$, Range $= \{1, 4, 9, 16, 81\}$; Function

23. Domain $= \{0, 1, 2, 3\}$, Range $= \{-3, -2, -1, 0, 1, 2, 3\}$; Not a function

24. Domain $= \{0, 1, 2, 3\}$, Range $= \{0, 1, 2, 3\}$; Function

25. $y = 3x$ **26.** $y = x + 2$

27. $2y = x$ (or $y = \frac{1}{2}x$) **28.** $y = \dfrac{1}{x}$

29. $f(-2) = 6$ **30.** $x = 3$

31. Domain $= \{3, 2, 1, 0, -1, -2\}$, Range $= \{1, 6, 4, 0\}$

32. $h(0) = -1$ **33.** $h(1) = 3$

34. $h(-1) = -5$ **35.** $h(10) = 39$

36. $h(-2) = -9$ **37.** $h(5) = 19$

38. $h(m) = 4m - 1$ **39.** $h(2r) = 8r - 1$

40. $h(3 + k) = 4k + 11$ **41.** $x = 1$

42. $x = 10$ **43.** $x = m$

44. $x = -1$ **45.** $x = 0$

46. Range $= \{2, 0, -2, -4\}$ **47.** Range $= \{-2, 1, 4, 7\}$

48. Range $= \{7, 4, 3\}$ **49.** Range $= \{9, 3, 1\}$

50. Range $= \{0, 2, 6, 12\}$

6–4 CARTESIAN COORDINATE SYSTEM

Just as each point on a number line is assigned a number, each point in a plane is assigned a pair of numbers relative to two intersecting number lines. The point of intersection of the two number lines is called the *origin;* it is assumed to have coordinate 0 on both number lines. The number lines are called *coordinate axes* and are perpendicular to each other. One coordinate axis is usually called the *x-axis,* the other the *y-axis.*

Each point P in the plane is located by a unique ordered pair of numbers, as illustrated in Fig. 6–3. If points A on the x-axis and B on the y-axis are chosen so that $OBPA$ is a rectangle, then P has coordinates (a, b) where a is the coordinate of A on the x-axis and b the coordinate of B on the y-axis. The first number a is called the *x-coordinate,* or *abscissa,* of point P. The second number b is called the *y-coordinate,* or *ordinate,* of P. Some examples of points in a plane and their coordinates are shown in Fig. 6–4.

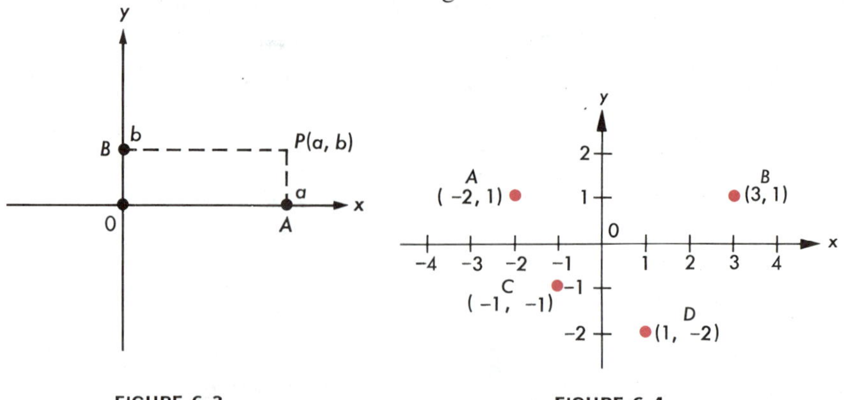

FIGURE 6–3

FIGURE 6–4

The system described above of assigning ordered number pairs to points in a plane is called a *cartesian coordinate system.* It is named in honor of the seventeenth-century French mathematician and philosopher René Descartes, who devised the system.

Arrowheads are placed on the axes of a cartesian coordinate system to indicate the direction of increasing numbers. The axes divide the plane into four parts, called *quadrants,* which are numbered as shown in Fig. 6–5. For example, Quadrant I is made up of all points having both coordinates positive, Quadrant II of all points having negative x and positive y coordinates.

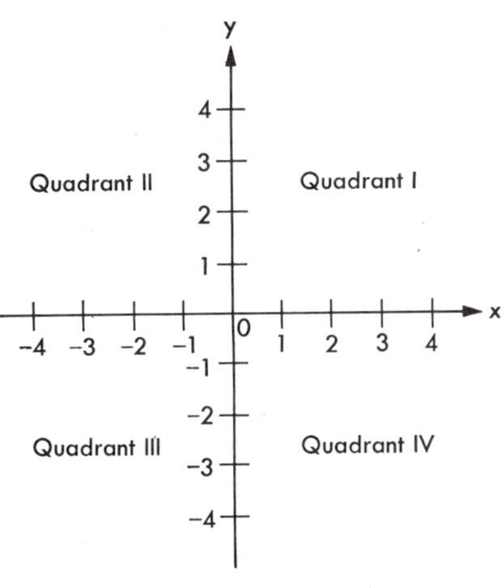

FIGURE 6–5

The point in the plane which corresponds to a number pair is called the *graph of the number pair.* Marking the point corresponding to a number pair, is graphing the pair, or *plotting* the point, as shown in Fig. 6–4.

Exercises

1. List the coordinates of each point designated by a letter in the figure.

(a)

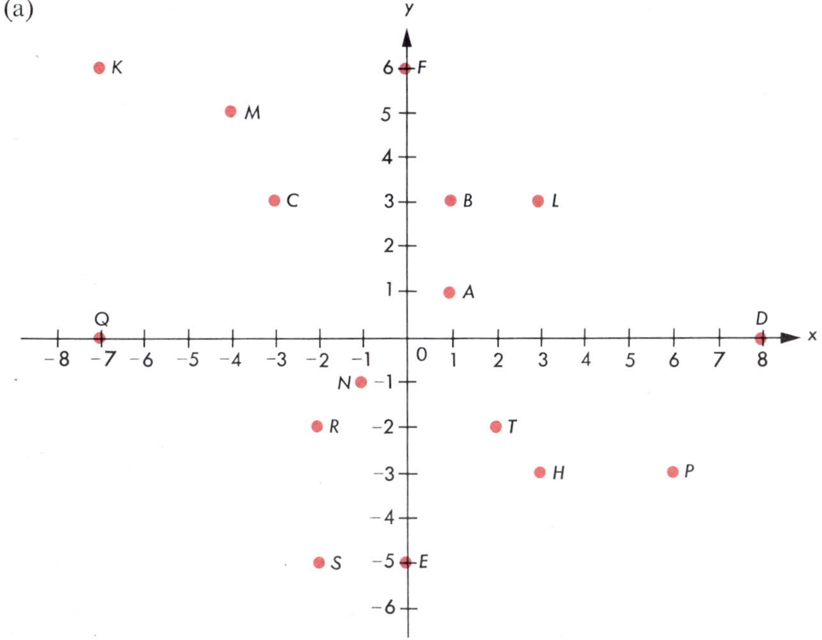

(b)

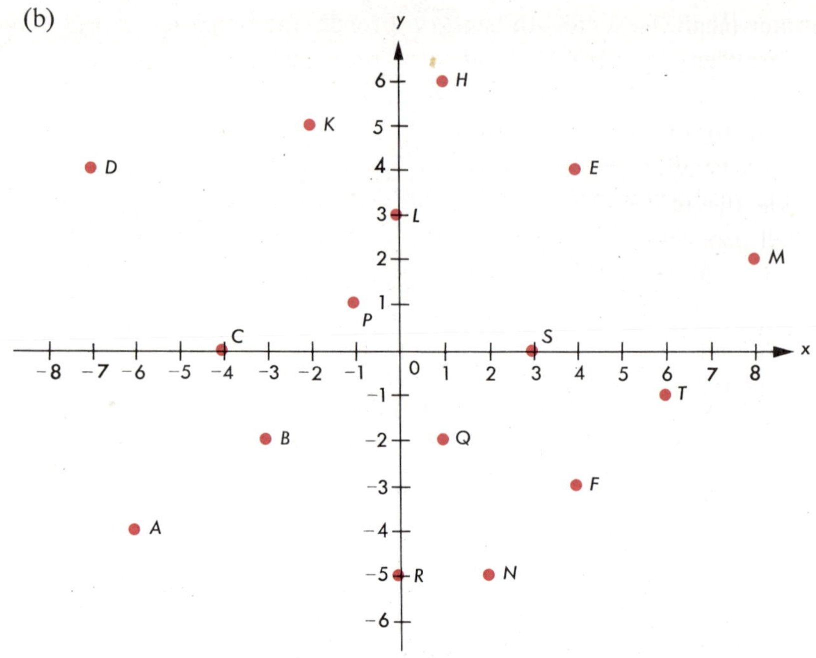

2. (a) Plot the points $A(-2, -5)$ and $B(-2, 3)$, and draw a straight line through them. Use the following information to plot the points C, D, E, and F on this line and find their coordinates.
   (i) Point B is midway between A and C.
   (ii) Point B is one-third the distance from A to D.
   (iii) Point A is midway between B and E.
   (iv) Point A is four-fifths the distance from B to F.

   (b) Plot the points $A(-5, -2)$ and $B(3, -2)$, and draw a straight line through them. Use the information in part (a) to plot the points C, D, E, and F on this line and find their coordinates.

3. (a) Plot each of the points:

$$A(2, 0), \quad B(2, 2), \quad C(4, 2), \quad D(4, 0).$$

   Join the points A, B, C, D, A, in order, and find the area of the figure.

   (b) Plot the points $A(2, 4)$, $B(2, -2)$, $C(-4, -2)$, $D(-4, 4)$. Then join A, B, C, D, and A, in order, and find the area of the figure.

4. (a) Plot each of the points:

$$A(-1, 2), \quad B(-6, -3), \quad C(4, -3).$$

   Find the coordinates of the point D such that $ABCD$ is a parallelogram. Is there more than one possible choice?

(b) Plot each of the points:

$$A(-3, 4), \quad B(-3, -2), \quad C(4, 1).$$

Find the coordinates of the point D such that $ABCD$ is a parallelogram. Is there more than one possible choice?

5. (a) Plot each of the points:

$$P(-4, 3), \quad Q(6, 3), \quad R(6, -4), \quad S(-4, -4).$$

Join the points P, Q, R, S, P in order. What is the name of the resulting figure?

(b) Plot each of the points:

$$P(3, -2), \quad Q(-2, -2), \quad R(-2, 1), \quad S(3, 1).$$

Join the points P, Q, R, S, P in order. What is the name of the resulting figure?

Describe the coordinates of a point that

6. (a) lies in Quadrant I.
(b) lies in Quadrant II.

7. (a) lies in Quadrant III.
(b) lies in Quadrant IV.

8. (a) lies on the x-axis between Quadrants I and IV.
(b) lies on the y-axis between Quadrants III and IV.

In what quadrant or quadrants may a point be located if

9. (a) its abscissa is positive?
(b) its ordinate is negative?

10. (a) its abscissa is negative and its ordinate is positive?
(b) its abscissa is positive and its ordinate is negative?

11. (a) its abscissa equals its ordinate?
(b) its abscissa equals the negative of its ordinate?

12. (a) its abscissa is -5, and its ordinate is positive?
(b) its abscissa is positive, and its ordinate is -5?

13. (a) the absolute value of the ordinate is 4?
(b) the absolute value of the abscissa is 4?

---

14. Plot each of the points:

$$A(-2, 0), \quad B(-2, 2), \quad C(2, 2).$$

(a) Find coordinates of a point D such that $ABCD$ is a trapezoid.
(b) How many possible points D are there such that $ABCD$ is a trapezoid?

15. Do you think that these three points lie on a straight line: $A(1, 1)$, $B(0, 0)$, and $C(-2, -2)$? If so, support your answer by finding the coordinates of three other points on this line.

16. Do you think that these three points lie on a straight line: $A(0, 1)$, $B(1, 0)$, and $C(3, -2)$? If so, support your answer by finding the coordinates of three other points on this line.

17. (a) A line segment is drawn from the point (30, 50) perpendicular to the x-axis. How long is this segment?
(b) A line segment is drawn from the point $(-18, 23)$ perpendicular to the y-axis. How long is this segment?

Plot three points in at least two quadrants whose coordinates are integers satisfying the given requirement.

18. The abscissa is three more than the ordinate.

19. The abscissa is four times the ordinate.

20. The ordinate is one less than twice the abscissa.

21. The ordinate is the absolute value of the abscissa.

6–5 GRAPH OF A FUNCTION

If each ordered pair of a function is graphed in a cartesian coordinate system, the resulting set of points is called the *graph of the function.*

Problem 1. Graph the function f with domain $\{1, 2, 3, 4, 5\}$ and range $\{3, 5, 7\}$, if $f(1) = 3$, $f(2) = 5$, $f(3) = 7$, $f(4) = 5$, $f(5) = 3$.

Solution. The function f consists of the ordered pairs (1, 3), (2, 5), (3, 7), (4, 5), (5, 3). Its graph is made up of the five points plotted in Fig. 6–6.

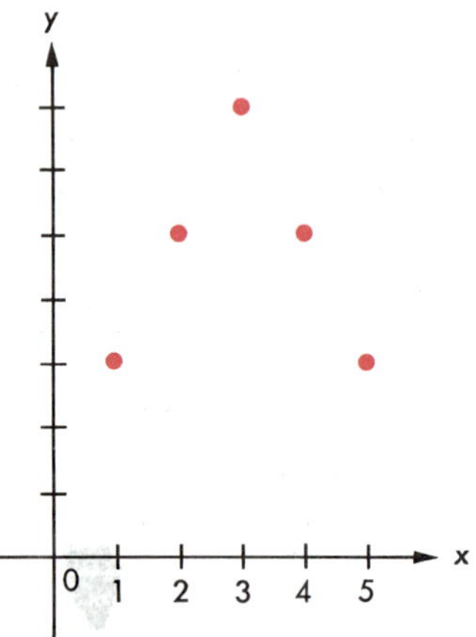

FIGURE 6–6

Problem 2. Graph the temperature function defined by the equation

$$F = \tfrac{9}{5}C + 32.$$

Solution. You are asked to graph the set

$$\{(C, F) \mid F = \tfrac{9}{5}C + 32, \quad C \geqq -273\}.$$

Clearly you cannot graph every number pair in this infinite set, so you pick out a few representative pairs to graph. You already know that $(-273, -459.4)$ is the "lowest" pair in the set. To obtain other pairs, you can give values to C ($C > -273$) and solve the resulting equation for F.

$$C = -40, \qquad F = \tfrac{9}{5} \cdot (-40) + 32$$
$$= -72 + 32$$
$$= -40.$$

Thus, $(-40, -40)$ is a number pair of the function.

$$C = 10, \qquad F = \tfrac{9}{5} \cdot 10 + 32$$
$$= 18 + 32$$
$$= 50.$$

Thus, $(10, 50)$ is a number pair of the function. These and other pairs of the temperature function are given in the table of values below.

C	-273	-40	-10	0	10	100
F	-459.4	-40	14	32	50	212

The graph of the temperature function is obtained by plotting these pairs and observing that the resulting points seemingly lie on a line. You note that, for convenience, different scales (that is, unit lengths) have been chosen on the C-axis and F-axis. Actually, the graph is an infinite ray with endpoint $(-273, -459.4)$, part of which is sketched in Fig. 6–7.

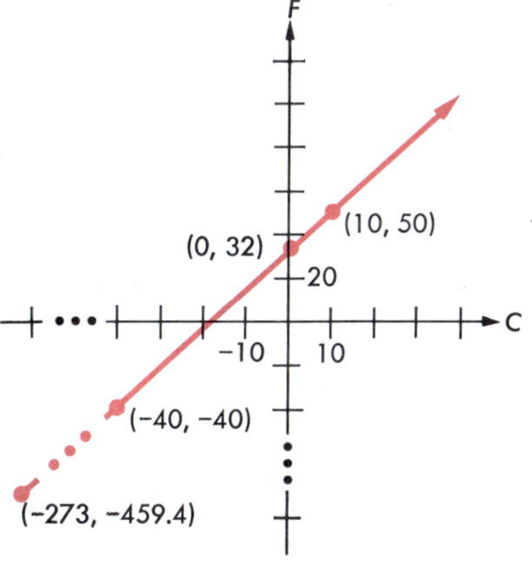

FIGURE 6–7

Problem 3. Graph the area-of-a-square function defined by the equation

$$A = x^2 \qquad (x > 0).$$

Solution. To aid in sketching the graph, you can make a table of values such as the one below.

x	$\frac{1}{4}$	$\frac{1}{2}$	1	$\frac{3}{2}$	2	3	4
A	$\frac{1}{16}$	$\frac{1}{4}$	1	$\frac{9}{4}$	4	9	16

Thus, $(\frac{1}{4}, \frac{1}{16})$, $(\frac{1}{2}, \frac{1}{4})$, . . . , (4, 16) are ordered number pairs of the function. These are plotted in Fig. 6–8, and then connected by a smooth curve to give an idea of the total graph. Once again, different scales have been chosen on the two axes. Although the graph "approaches" the origin (with coordinates (0, 0)), the origin is not on the graph since the domain and the range of the function are the set of positive real numbers. This is indicated by the hollow dot at the origin.

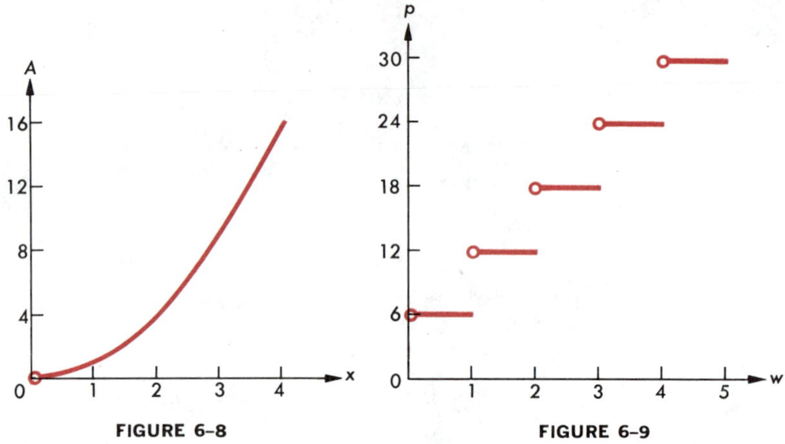

FIGURE 6–8

FIGURE 6–9

Problem 4. Graph the first-class postage function.

Solution. (See p. 247.) Since $p = 6$ for every w between 0 and 1, $0 < w \leqq 1$, the graph is a horizontal line segment between $w = 0$ and $w = 1$, including the point at $w = 1$. Similarly, $p = 12$ if $1 < w \leqq 2$, and the graph is again a horizontal line segment between $w = 1$ and $w = 2$, including the point at $w = 2$, and so on. The graph is sketched in Fig. 6–9. The hollow dot at the left end of each segment indicates that that point is not on the graph. Thus, the point (1, 6) is on the graph while (1, 12) is not; (2, 12) is on the graph while (2, 18) is not; and so on. The graph consists of 13 "steps."

Exercises

Write a formula and graph the function defined by each rule. In each case, make a table of values and give the domain and range of each function.

1. (a) The number of inches, i, is 12 times the number of feet, f.
(b) The number of ounces, n, is 16 times the number of pounds, p.

2. (a) The number of pennies, p, is 5 times the number of nickels, n.
(b) The number of dimes, d, is 10 times the number of dollars, n.

3. (a) The number of days, d, is 7 times the number of weeks, w.
   (b) The number of months, m, is 12 times the number of years, y.
4. (a) The number of gallons, g, is 3.8 times the number of liters, l.
   (b) The number of pounds, p, is .4 times the number of kilograms, k.
5. (a) The number of kilometers, k, is 1.6 times the number of miles, m.
   (b) The number of centimeters, c, is 30.5 times the number of feet, f.
6. (a) The cost of renting a car is, c, \$10 per day plus, m, 10¢ per mile driven.
   (b) The cost, c, to make photostatic copies is 15¢ plus 5¢ for each page, p.

Graph the function defined by each of the equations in Exercises 7–11; first use the domain of x stated in (i), and then use the domain of x stated in (ii). Make the graph on two separate coordinate systems. In each case, make a table of values and give the range of the function.

7. (a) $y = 3x + 1$
   (i) $\{x \mid -3 \leqq x \leqq 3, x \text{ an integer}\}$
   (ii) $\{x \mid -3 \leqq x \leqq 3, x \text{ a real number}\}$

   (b) $y = 2x - 5$
   (i) $\{x \mid 0 \leqq x \leqq 4, x \text{ an integer}\}$
   (ii) $\{x \mid 0 \leqq x \leqq 4, x \text{ a real number}\}$

8. (a) $y = 2x^2$
   (i) $\{x \mid -2 \leqq x \leqq 2, x \text{ an integer}\}$
   (ii) $\{x \mid x > 0\}$

   (b) $y = 2x^2 + 1$
   (i) $\{x \mid -2 \leqq x \leqq 2, x \text{ an integer}\}$
   (ii) $\{x \mid x < 0\}$

9. (a) $y = |x|$
   (i) $\{x \mid -4 \leqq x \leqq 4, x \text{ an integer}\}$
   (ii) $\{x \mid -4 \leqq x \leqq 4, x \text{ a real number}\}$

   (b) $y = |x| + 1$
   (i) $\{x \mid -4 \leqq x \leqq 4, x \text{ an integer}\}$
   (ii) $\{x \mid -4 \leqq x \leqq 4, x \text{ a real number}\}$

10. (a) $y = 25 - x^2$
   (i) $\{x \mid -5 \leqq x \leqq 5, x \text{ an integer}\}$
   (ii) $\{x \mid |x| \leqq 5\}$

   (b) $y = 1 - x^2$
   (i) $\{x \mid -4 \leqq x \leqq 4, x \text{ an integer}\}$
   (ii) $\{x \mid |x| \leqq 4\}$

11. (a) $y = |2x| - 3$

(i) $\{x \mid |x| < 5, x \text{ an integer}\}$

(ii) $\{x \mid x \text{ any real number}\}$

(b) $y = |3x| - 1$

(i) $\{x \mid |x| < 3, x \text{ an integer}\}$

(ii) $\{x \mid x \text{ any real number}\}$

Which of the graphs in Exercises 12–19 are graphs of functions of x? Give the domain and the range of each function that you find.

12. (a)

(b)

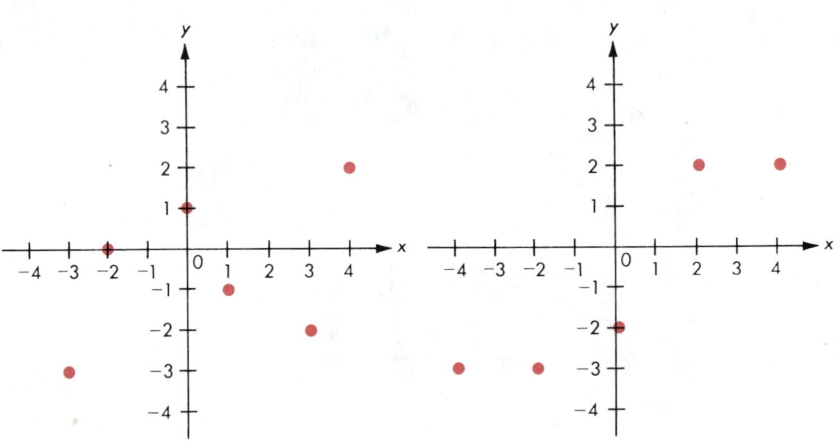

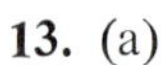
13. (a)

(b)

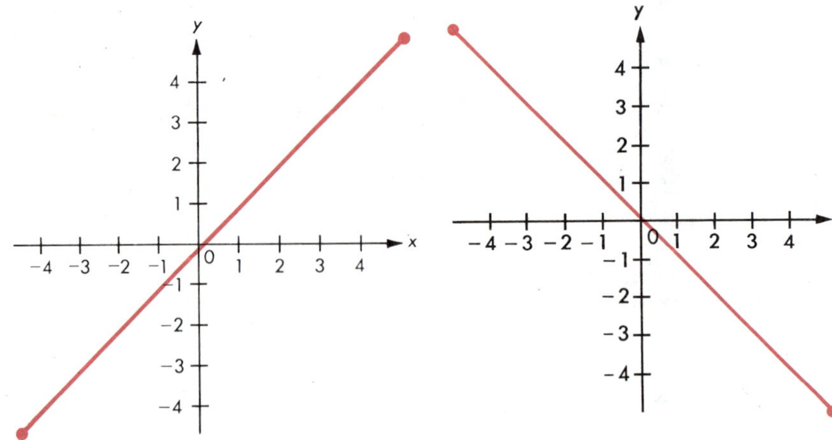

14. (a)

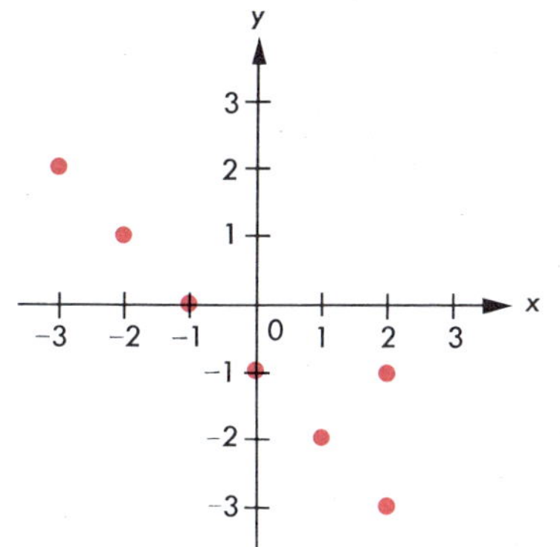

(b)

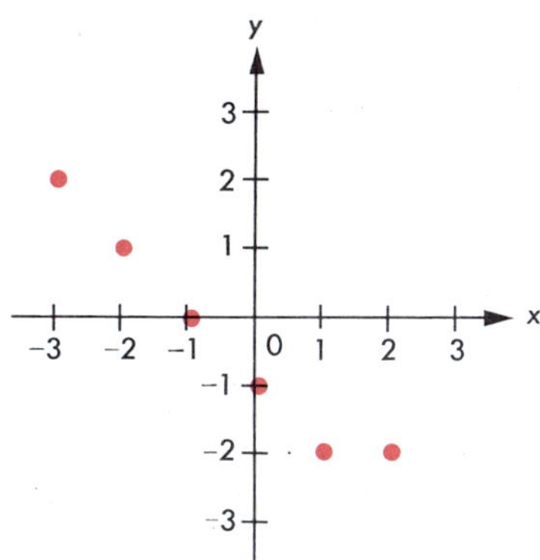

15. (a)

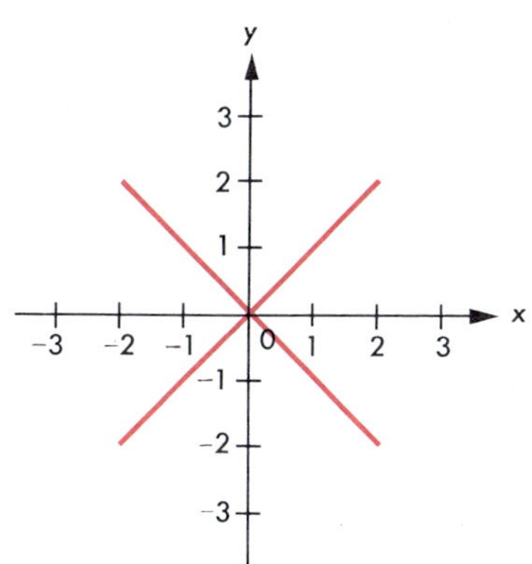

(b)

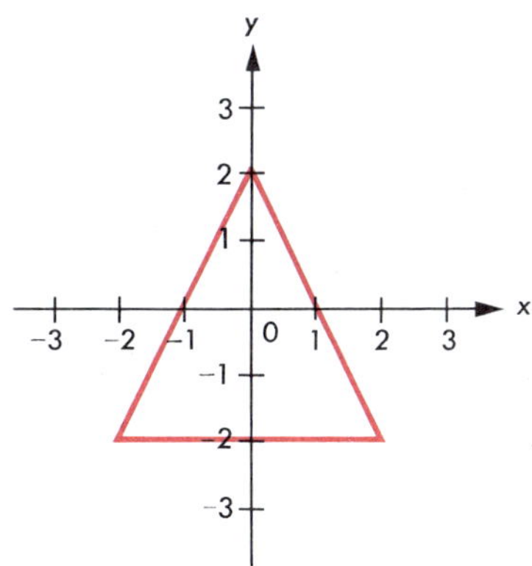

16. (a)

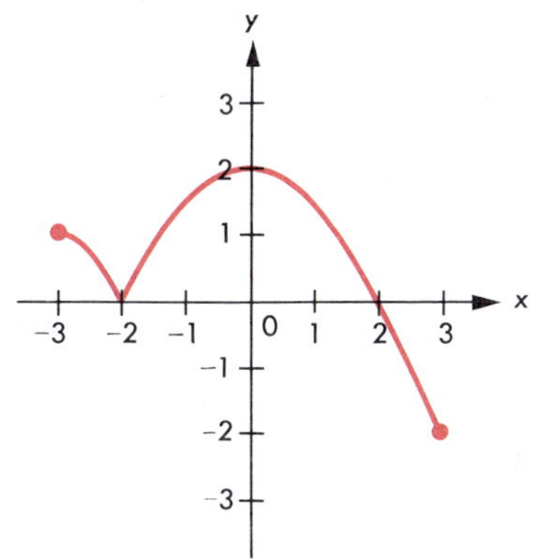

(b)

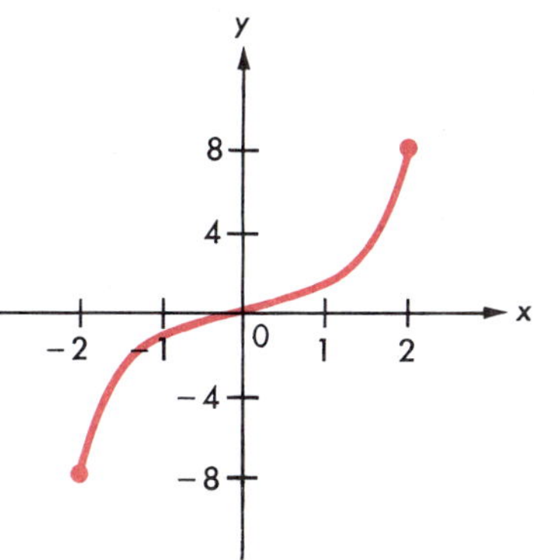

17. (a)

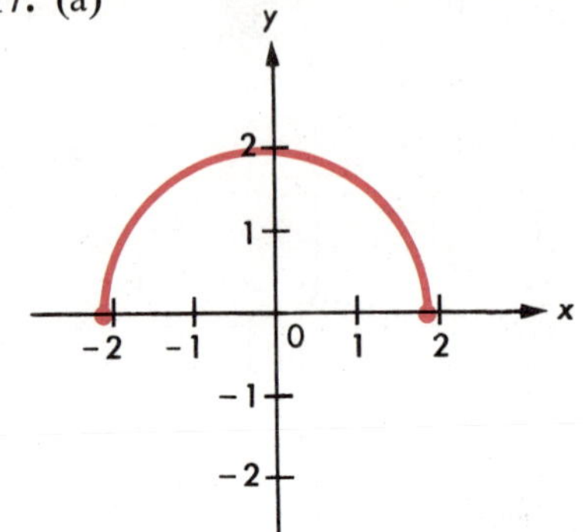

(b)

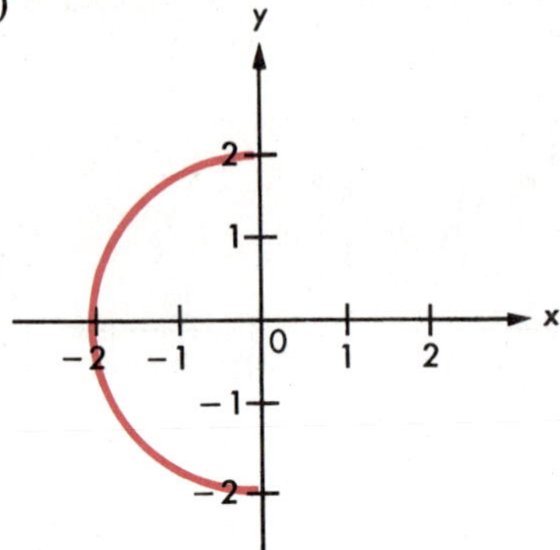

18. (a)

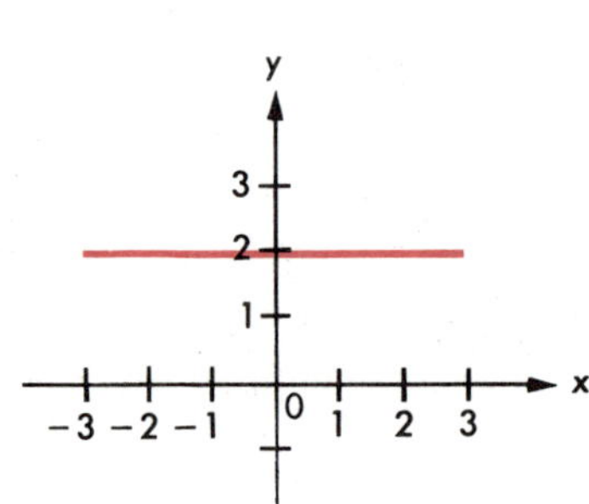

(b)

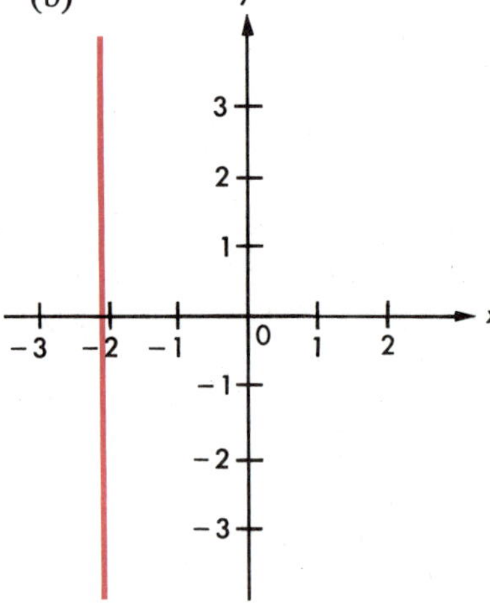

19. (a)

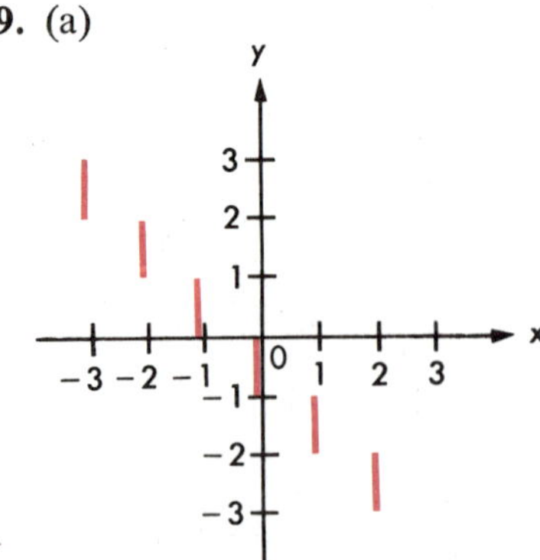

(b)

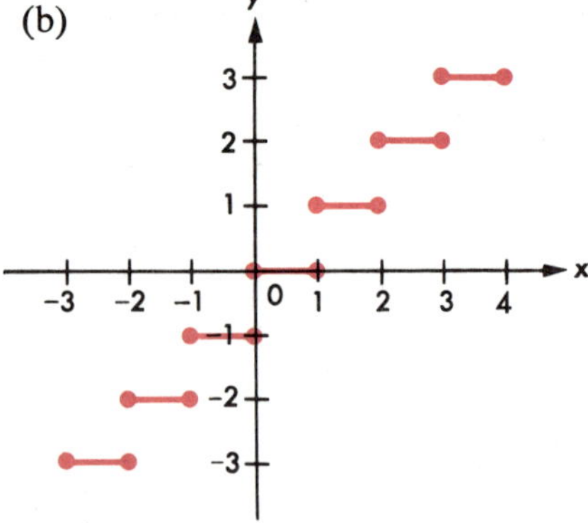

20. Graph the function which assigns the number 2 to each positive number, and the number -1 to each nonpositive number.

21. Graph the function which assigns to each number twice its absolute value.

EXTRA!

Every real number is either an integer or between two consecutive integers. For example,

$$\begin{aligned} 0 &< \tfrac{3}{4} < 1, \\ 1 &< \sqrt{2} < 2, \\ 3 &< \pi < 4, \\ -3 &< -2.1 < -2. \end{aligned}$$

In either case, for each real number x there exists a unique integer n such that

$$n \leqq x < n + 1.$$

You call n the *greatest integer in x*. For example, 3 is the greatest integer in π, -3 is the greatest integer in -2.1, and 7 is the greatest integer in 7.

The *greatest integer function* is usually denoted by heavy brackets, **[ ]**. By definition,

$$[x] \text{ is the greatest integer in } x.$$

Thus,

$$[\sqrt{2}] = 1, \quad [\tfrac{3}{4}] = 0, \quad [101] = 101.$$

The graph of

$$\{[x] \mid -3 \leqq x < 3\}$$

is shown in Fig. 6–10.

Do you see why the greatest integer function is often called a *step function?*

1. Find:
   (a) $[\frac{21}{8}]$.
   (b) $[-\frac{37}{4}]$.
   (c) $[7\sqrt{2} - 2\pi]$.
   (d) $[\pi^2 - 10]$.

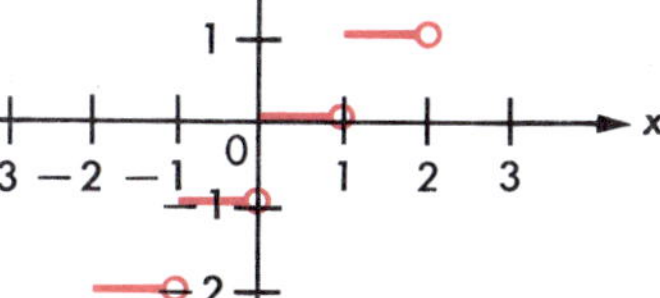

FIGURE 6–10

2. The function f defined by

$$f(x) = x - [x]$$

is called the *fractional part function.* Find:

(a) $f(\frac{21}{8})$. (b) $f(-\frac{37}{4})$.
(c) $f(\sqrt{2})$. (d) $f(-\sqrt{19})$.

3. Sketch the graph of the fractional part function in the interval $\{x \mid -3 \leqq x < 3\}$.

4. Graph the equation $[x] + [y] = 1$ in the interval $\{x \mid 0 \leqq x < 2\}$.

6–6 POLYNOMIAL FUNCTIONS

A polynomial in one variable defines a function in a natural way. For example, the polynomial $2x - 3$ defines the function f, where

$$f(x) = 2x - 3.$$

The polynomial $x^2 - 3x - 4$ defines the function g, where

$$g(x) = x^2 - 3x - 4.$$

Such a function, defined by a polynomial in one variable, is called a *polynomial function.*

The simplest polynomial is a constant. The corresponding function is called a *constant function.* For example, the function f defined by

$$f(x) = 2$$

is a constant function. It consists of all number pairs of the form $(x, 2)$, where x is any number. Its graph is a line parallel to the x-axis, as shown in Fig. 6–11.

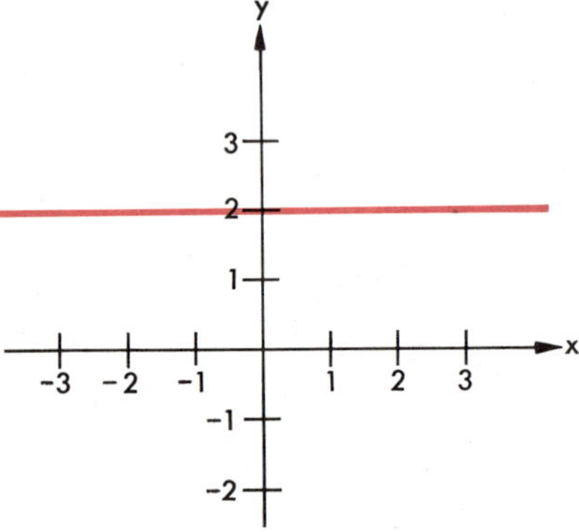

FIGURE 6–11

Two variables x and y are said to *vary directly* if they are related by an equation of the form

$$y = kx$$

for some nonzero number k, called the *constant of variation.* If x and y vary directly, they are also said to be *directly proportional* to each other. This equation clearly defines a polynomial function if we let $y = f(x)$. Then it consists of all number pairs (x, kx), x any number.

For example, if a car is traveling at a constant rate of 50 miles per hour, then the distance d that the car travels varies directly with the time t elapsed:

$$d = 50t.$$

In this case, the constant of variation is 50. The corresponding function $\{(t, d) \mid d = 50t\}$ has the graph shown in Fig. 6–12.

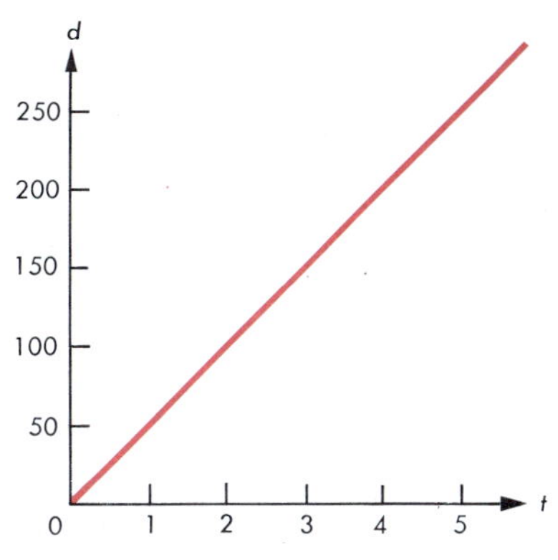

FIGURE 6–12

Exercises

Graph the function defined by each of the following polynomials, where x is any real number. In each case, make a table of values and give the range of the function.

1. (a) $3x$ (b) $2x$

2. (a) $2x - 5$ (b) $3x - 4$

3. (a) -4 (b) -3

4. (a) $2x^2 - 3$ (b) $3x^2 - 2$

5. (a) $x^2 + x - 6$ (b) $x^2 - x - 6$

Graph the function defined by each of the following polynomials.

6. (a) x^3, $\{x \mid -2 \leqq x \leqq 2, x \text{ an integer}\}$
(b) $x^3 - 4x$, $\{x \mid -2 \leqq x \leqq 2, x \text{ an integer}\}$

7. (a) x^2, $\{-3 \leqq x \leqq 2, x \text{ a real number}\}$
(b) $x^2 + 2x$, $\{-3 \leqq x \leqq 2, x \text{ a real number}\}$

8. (a) From the equation $d = 50t$, write an equation giving t as a function of d.
(i) What would the graph of this function be?
(ii) Is t directly proportional to d? What is the constant of variation now?
(iii) If t is tripled, what happens to d?

(b) (i) Is the number, n, of 5¢ candy bars a function of the total cost T? Write an appropriate equation.
(ii) Does n vary directly with T? What is the constant of variation?
(iii) If n is doubled, what happens to T?

9. Examine the following tables. If there is an example of direct variation, write a linear equation expressing the first variable as a function of the second. State the constant of variation.

(a)

p	I
50	2
75	3
87.50	3.50
100	4
150	6

(b)

d	D
7	22
14	44
21	66
28	88
56	176

10. (a) (i) If chocolates cost 75¢ a pound, write a formula expressing the cost, c, as a function of the number of pounds, p.

(ii) Make a graph of your function.

(iii) Use your graph to find how many pounds of chocolates you could buy for $3.75.

(iv) Use your graph to find the cost of $2\frac{1}{2}$ pounds of chocolates.

(b) (i) If notebooks cost 30¢ each, write a formula expressing the cost, c, as a function of the number of notebooks, n.

(ii) Make a graph of your function.

(iii) Use your graph to find how many notebooks you could buy for $2.00.

11. Is there any difference between the ranges of the variables in Exercise 10(a) and those in Exercise 10(b)?

12. What is the difference between the graphs in Exercises 10(a) and 10(b)?

13. (a) Use the graph below to complete the table.

n	0	1	2	3	4	5
w	?	?	?	?	?	?

If $n = 2\frac{1}{2}$, find w from the graph.
If $w = 5$, find n from the graph.
Write a formula expressing w as a function of n.
If w is a symbol for wages in dollars and n is a symbol for hours, interpret the formula.

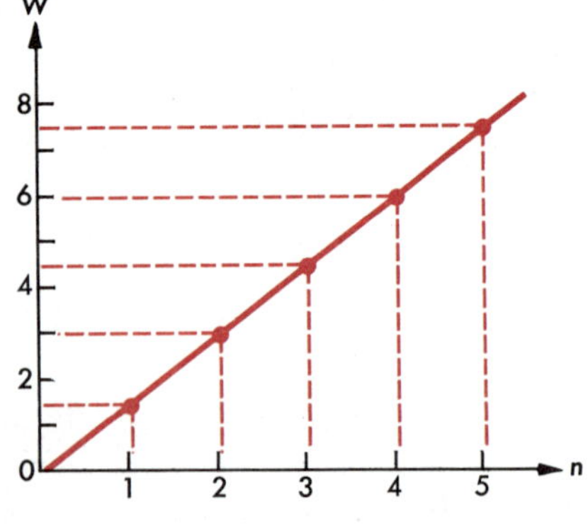

(b) Use the graph below to complete the table.

h	0	1	2	3	4	5
l	?	?	?	?	?	?

If $h = \frac{1}{2}$, find l from the graph.
If $l = 4\frac{1}{3}$, find h from the graph.
Write a formula expressing l as a function of h.
If l is a symbol for the length of the shadow of an object and h is a symbol for the object's height, interpret the formula.

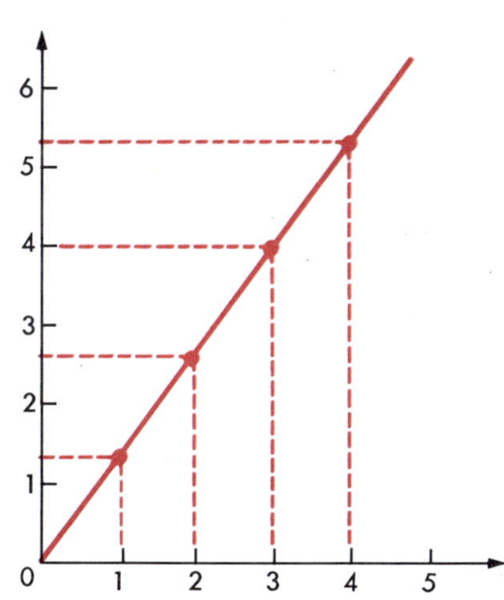

14. Graph the ordered number pairs below and find the formula that expresses one of the two variables as a function of the other. (The symbol W denotes the weight that is hung on a spiral spring, and E denotes the extension in the spring's length that is caused by the weight.)

(a)

W (grams)	E (centimeters)
0	0
5	2
10	4
20	8
30	12
35	14

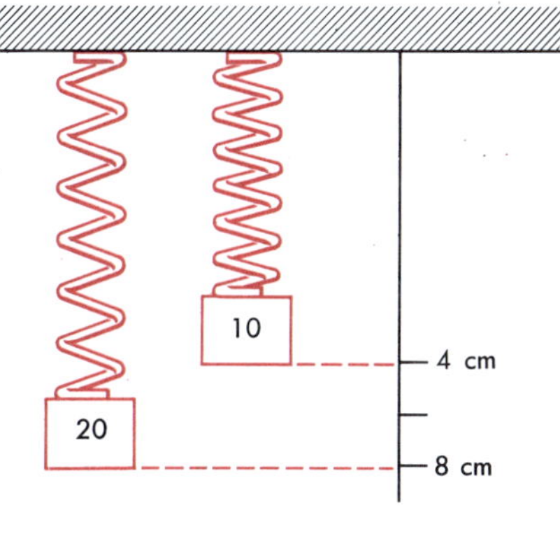

(b)

W (grams)	E (centimeters)
0	0
2	3
3	4.5
6	9
8	12
10	15

(c) How do you think the spiral spring in part (a) compares with the spiral spring in part (b)? What is the constant of variation for each?

15. (a) The constant which is multiplied by E to yield W in Exercise 14 is called the *spring constant.* With a weight and a ruler you can find the spring constant for any spiral spring. You can compute k with even less effort if you have a pair of values for E and W. For instance, knowing that the elongation, or extension, of the spring increases as the weight increases, you may write the general pattern for direct variation as

$$W = (\text{a number}) \cdot E.$$

Using k to replace "a number," the formula becomes

$$W = kE.$$

Suppose that you are told that a weight of 25 grams will stretch the spring 10 centimeters. Then you know that when $W = 25$ grams, $E = 10$ centimeters. If you insert these numbers into their respective places in the formula, you obtain

$$\begin{aligned} 25 &= k \cdot 10, \\ k &= \tfrac{25}{10}, \\ k &= \tfrac{5}{2}. \end{aligned}$$

Write an equation for W in terms of E using the above value for k.

(b) Find the spring constant if a weight of 20 grams causes an elongation of .4 centimeter on a certain spiral spring.

16. The weight of an object varies directly with its volume.

(a) Translate this sentence into a formula with W for weight, V for volume, and k for the constant of variation.

(b) If an object with volume 10 cubic centimeters weighs 14 grams, find the value of k.

(c) Now write your formula, inserting the value of k you just found.

(d) Use this formula to find the weight of an object of the same substance with a volume of 18 cubic centimeters.

(e) Find the volume of an object of the same substance having a weight of 35 grams.

17. (a) If x varies directly with y, write a formula that expresses x as a function of y. Use k as the constant of variation.

(b) If $x = 5$ when $y = 3$, what value must k have?

(c) Find x when $y = 135$.

(d) Find y when $x = 55$.

---

18. Write an equation that expresses the first variable described in each of the following statements as a function of the second.

(a) The length, l, of the shadow of an object at a given time varies directly with the height, h, of the object.

(b) The pressure, p, of a liquid in pounds per square foot is directly proportional to the depth, d, of the liquid in feet.
(c) The force, f, required to push an object along a flat surface varies directly with the weight, w, of the object.
(d) The distance, d, a person travels varies directly with the time, t, spent in traveling.

19. Tell if each of the variables in the following pairs are directly proportional to each other.
(a) The diameter and circumference of a circle
(b) The number of men working on a job and the time required to do it
(c) The side and area of a square
(d) The length and width of a rectangle having an area of 100 square inches

20. (a) The distance an object will fall in a given time varies directly with the square of the time in seconds. Suppose that an object will fall 16 feet in 1 second; write a formula which expresses this relationship.
(b) How far will the object fall in 3 seconds?

21. The tables below express a relationship similar to the relationship you found in Exercise 20. Write an equation which expresses the relationship given by each of the tables.

(a)

x	-2	-1	0	1	2
y	12	3	0	3	12

(b)

x	-2	-1	0	1	2
y	7	4	3	4	7

KEY IDEAS AND KEY WORDS

Any pair of elements (x, y) having a first element x and a second element y is called an **ordered pair.** Two ordered pairs are equal if, and only if, they have the same first elements and the same second elements.

The set of all ordered pairs (x, y) where x is an element of A and y is an element of B is called the **cartesian product** of A and B, written $A \times B$.

If S is a set, a subset of $S \times S$ is called a **relation** in S.

A relation in a set S is called a **function** if whenever two ordered pairs in the relation have the same first element, then the second elements are also the same. For a function in a set S, the set of all numbers x such that some (x, y) is in the function is called the **domain** of the function. The set of all numbers y such that some ordered pair (x, y) is in the function is called the **range** of the function. A function will **map** each element of the domain into a unique element of the range.

A **cartesian coordinate system** is a method of assigning ordered number pairs to points in a plane relative to two intersecting number lines. The point of intersection of the two number lines is called the **origin.** The point in the plane that corresponds to a number pair is called the **graph of the number pair.**

A **polynomial function** is a function defined by a polynomial in one variable.

Two variables, x and y, **vary directly** if they are related by an equation of the form $y = kx$, $k \neq 0$, and k is called the **constant of variation.** Two variables are **directly proportional** to each other if they vary directly.

CHAPTER REVIEW

1. Find the cartesian product of A and B if $A = \{4, 6, 9, 12\}$ and $B = \{-4, -6\}$.

2. Find the subset of $U \times U$ which is the given relation in U.

(a) $U = \{-3, -2, -1, 0, 10\}$, "less than or equal to"
(b) $U = \{-5, 0, 10, 15, 18, 36\}$, "is a factor of"

Which of the following relations are functions? State the domain and range of each function.

3. $\{(-10, 10), (-5, 10), (0, 10), (5, 0), (10, 0)\}$

4. $\{(-9, 8), (-8, 9), (-7, 6), (-6, 7), (7, -6), (-7, -6)\}$

5. $\{(4x, 2x^2 - 3) \mid x \text{ an integer}\}$

6. If the function f maps each number x into $3x^2 - 1$, find $f(0), f(1), f(-1)$, and $f(\frac{1}{3})$.

7. If $f(x) = 2x - 7$ defines a function, find the element in the domain of f associated with the given element in the range.

(a) $f(x) = 7$
(b) $f(x) = 0$
(c) $f(x) = -7$
(d) $f(x) = 1$

8. In what quadrant or quadrants does a point lie, if

(a) the abscissa is the negative of the ordinate.
(b) the abscissa is three times the ordinate.
(c) the abscissa is negative.
(d) the abscissa is negative and the ordinate is 10.
(e) the ordinate is five more than twice the abscissa.
(f) the ordinate is negative six decreased by twice the abscissa.

9. (a) If $f(x) = |x| - 4$, find $f(-5), f(5), f(-1), f(1), f(0), f(-\frac{1}{2}), f(\frac{1}{2}), f(-.01), f(.01)$.

(b) What is the range of the function described in (a)?

10. Given that $g(x) = x + \dfrac{1}{x}$ for every real number x except zero,

(a) find $g(1)$, $g(3)$, $g(10)$, and $g(10^2)$.
(b) find $g(\frac{1}{4})$, $g(\frac{1}{8})$, $g(\frac{1}{10})$, and $g(\frac{1}{100})$.
(c) is $g(x)$ ever negative? Illustrate your answer by giving examples.

(d) is there an x for which $g(x) = 0$?
(e) what is the domain of the function?

11. (a) Describe algebraically the function that maps each real number x into three less than its square.
(b) Into what range-value will this function map each of the following numbers?

$$-2, 2, -1, 1, 0, -5, 5, -\tfrac{1}{2}, \tfrac{1}{2}, -\tfrac{1}{10}, \tfrac{1}{10}$$

(c) What is the range of this function?

12. (a) Graph the function $y = 3 + |x|$ by first finding the y-numbers that correspond to six or more different x-numbers between -6 and 6.
(b) Describe the domain and the range of the function in part (a).

13. (a) Graph the function defined by $f(x) = 2 - x^2$ if the domain is the set of integers between -6 and 6.
(b) What is the range of the function in part (a)?

14. Describe algebraically the function determined by each of the following tables.

(a)

x	$f(x)$
2	6
4	12
5	15

(b)

x	$f(x)$
$\frac{1}{12}$	24
2	1
3	$\frac{2}{3}$

15. What kind of relation exists between $f(x)$ and x in Exercise 14(a)?

16. If y varies directly with x, and if $y = 10$ when $x = 8$,
(a) find the constant of variation.
(b) find y when $x = 12$.

17. A delivery company has a special rate for in-city deliveries. The charge is 12¢ for the first pound and 6¢ for each additional pound or part of a pound, and the total weight is not to exceed 70 pounds.
(a) Make a graph showing the price charged for weights up to and including 10 pounds.
(b) What is the charge on packages weighing 8 pounds and 1 ounce?
(c) What is the maximum weight which can be shipped for 39¢?
(d) Is there a minimum weight which requires 28¢?
(e) What weight can be shipped for exactly 42¢?

18. A mathematics club has 15 members.
(a) Write a formula for finding the number of girls, g, if you know the number of boys, b.
(b) Is the domain of g restricted in any way?
(c) Graph the function in part (a).
(d) How many points are on the graph?

19. Mia is paid to deliver old papers to the dump. She gets 2¢ per pound or part of a pound. The minimum fee is 6¢ and the maximum weight must be less than 16 pounds.

(a) Graph this function. (b) Describe the domain and the range.

20. Which of the following are graphs of functions?

(a)

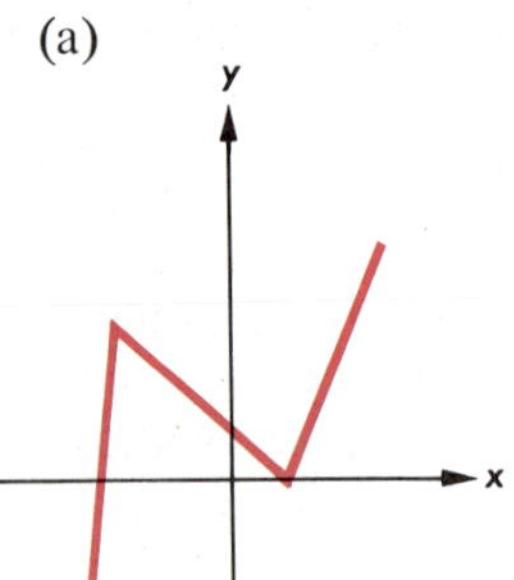

(b)

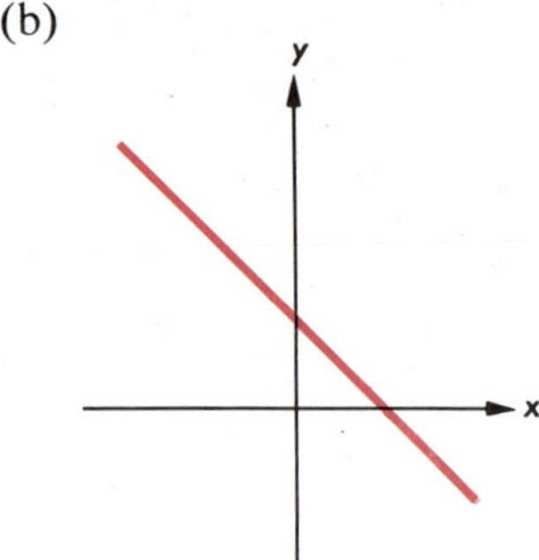

(c)

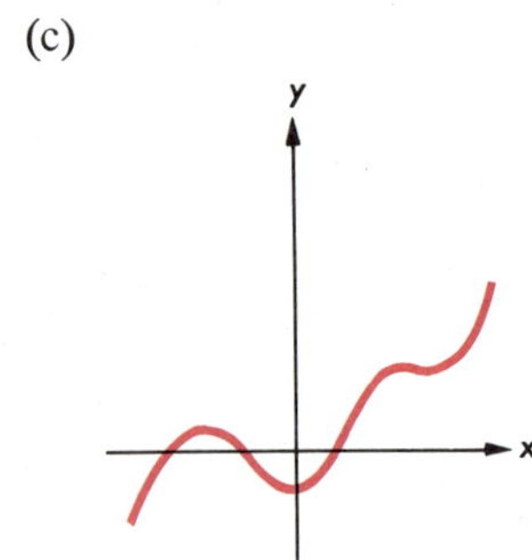

(d)

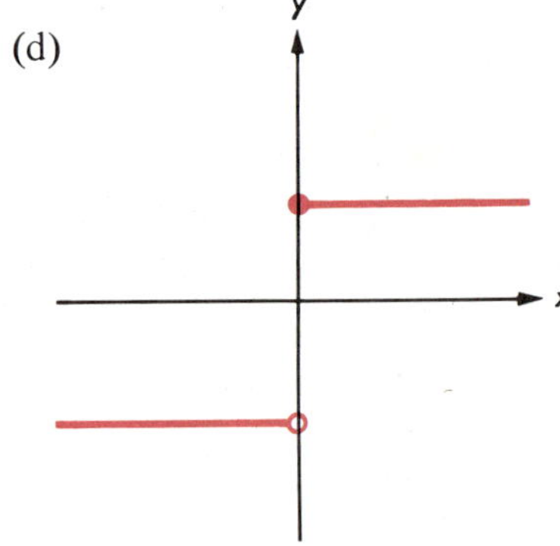

(e)

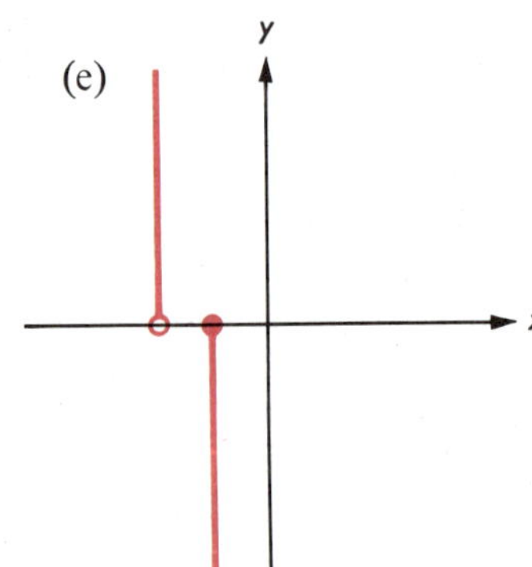

(f)

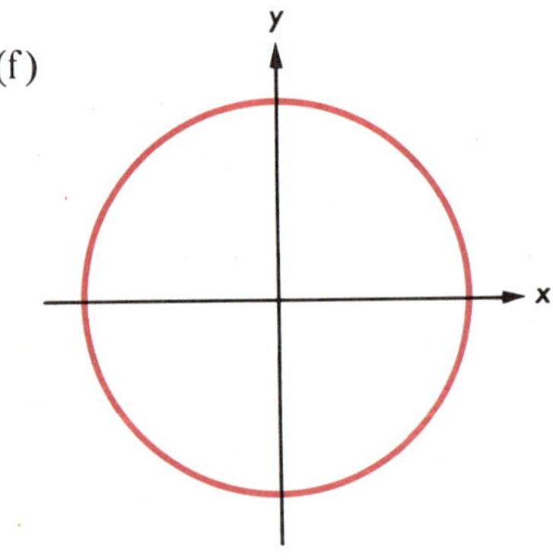

21. A certain function has as its domain the set of positive integers. The function maps the integer 1 into the first prime number 2; the integer 2 into the second prime number 3; the integer 3 into the third prime number, 5; and continues in this fashion. List the first ten pairs of this function and represent them graphically.

CHAPTER TEST

1. Set S and set U are such that S is a relation in U. List the members of the relation.

$$S = \{(x, y) \mid x + y = 1\}, \text{ where } U = \{-2, -1, 0, 1, 2, 3\}$$

2. Which of the following relations in the set Z of integers are functions? State the domain and range of each function.
   (a) $\{(-10^6, 10^6), (-200, 2), (10^6, -10^6), (-2, 200)\}$
   (b) $\{(x, |x|) \mid x \text{ in } Z\}$
   (c) $\{(|x|, x) \mid x \text{ in } Z\}$
   (d) $\{(x, x^2) \mid x \text{ in } Z\}$
3. If the function f maps each number x into $3x^2$, find $f(0)$, $f(-1)$, and $f(2)$. Find the element(s) in the domain of f associated with $f(x) = 3$.
4. A function is defined for every integer x by the equation $y = 2 - x^2$. The domain of x is $\{x \mid -4 \leqq x \leqq 4\}$. Make a list of the ordered number pairs related to this function.
5. Graph each function and describe the range of the function.
   (a) $f(x) = |x + 3|$
   (b) $g(x) = 2x^2 + 3$, x an integer, and $|x| < 5$
6. A function maps each positive integer greater than 1 into the square of the integer divided by 2.
   (a) Make a table of the first ten of these ordered number pairs and graph them.
   (b) Which integer is mapped into itself?
7. A body moves a distance, d, directly proportional to the time, t, it is in motion.
   (a) Express d as a function of t.
   (b) If a body moves 288 feet in 3 seconds, find the constant of variation.
   (c) How far does the body move in 6 seconds?
8. (a) If x and y are directly proportional to each other, write an equation expressing y as a function of x.
   (b) If $y = 10$ when $x = .3$, what is the constant of variation?
   (c) Find y when $x = 22$.

CHAPTER 7

Graphs

Objectives . . .

- To find solutions of linear equations in more than one variable.
- To find the graph of a linear equation by plotting solutions.
- To find the half-plane on either side of a given line.
- To find the graph of a linear equation by using slope-intercept techniques.

7–1 LINEAR EQUATIONS

Graphs are often used to demonstrate relationships or trends. Graphs of linear equations are the simplest types of graphs you will encounter. Many graphs follow a mathematical pattern; therefore, you can frequently visualize a complete graph even though you have only enough space to draw part of it.

The algebraic expression

$$3x - 2y + 7$$

is called a *linear form* in two variables, x and y. Each linear form in x and y is an expression of the type:

$$(\text{a number}) \cdot x + (\text{a number}) \cdot y + (\text{a number}).$$

Similarly,

$$-5a + 3b + \tfrac{2}{3}$$

is a linear form in the variables a and b.

If you give values to the variables in a linear form, you obtain a *value* of the linear form. For example, the value of the linear form $3x - 2y + 7$ when $x = 2$ and $y = 5$ is

$$3 \cdot 2 - 2 \cdot 5 + 7, \quad \text{or } 3.$$

Its value when $x = -3$ and $y = 0$ is

$$3 \cdot (-3) - 2 \cdot 0 + 7, \quad \text{or } -2.$$

The algebraic expression $-5x$ is a linear form in x and y, namely

$$-5x + 0y + 0.$$

Similarly, -4 may be thought of as

$$0x + 0y - 4.$$

If two linear forms in x and y are connected by an equals sign, the resulting expression is called a *linear equation in x and y.* For example,

$$3x - 2y - 7 = x - y - 2 \tag{1}$$

and

$$2x + y + 3 = 4 \tag{2}$$

are linear equations in the variables x and y.

If you give values to the variables in a linear equation, the resulting equation is a statement of the equality of two numbers and, hence, is either true or false.

For example, let $x = 3$ and $y = 2$ in Equation (1) above. Then you obtain the statement

$$3 \cdot 3 - 2 \cdot 2 - 7 = 3 - 2 - 2,$$

or

$$-2 = -1,$$

which is false.

If you let $x = 7$ and $y = 9$ in Equation (1), the resulting statement is

$$3 \cdot 7 - 2 \cdot 9 - 7 = 7 - 9 - 2,$$

or

$$-4 = -4,$$

which is true. According to the definition below, the ordered pair (7, 9) is a *solution* of Equation (1) above.

Definition of a solution of a linear equation in two variables

An ordered pair of numbers, (a, b), is called a solution of an equation in x and y if the equation obtained by letting x = a and y = b is true.

You can find solutions of a linear equation by giving a value to one variable and solving the resulting equation for the other variable. For example, to find a solution of (2) above,

$$2x + y + 3 = 4,$$

choose any value for x, say $x = 5$, and solve the resulting equation

$$2 \cdot 5 + y + 3 = 4$$

for y as follows:

$$\begin{aligned} y + 13 &= 4, \\ y &= -9. \end{aligned}$$

Thus, $(5, -9)$ is a solution. If you let $x = 0$, then

$$\begin{aligned} 2 \cdot 0 + y + 3 &= 4, \\ y + 3 &= 4, \\ y &= 1, \end{aligned}$$

and $(0, 1)$ is a solution. You can give values to y as well as to x. Letting $y = 3$, you would obtain

$$\begin{aligned} 2x + 3 + 3 &= 4, \\ 2x &= -2, \\ x &= -1, \end{aligned}$$

and $(-1, 3)$ is a solution. Since for every value of x or y you would obtain a solution, the linear equation $2x + y + 3 = 4$ has an infinite number of solutions.

Exercises

1. Tell which algebraic expressions represent linear forms in two variables and which do not. Explain your answers.

(a) $x + \frac{2}{3}y - 4$; $3\sqrt{y}$; $2b + 3p - 5$; $3 + \dfrac{2}{w}$; $-\frac{5}{9}$; $\dfrac{3}{x + y - 5}$

(b) $.02x - .99y - \sqrt{2}$; $-\sqrt{5}x$; $x + y^3 + 2$; $143x + 177$; 0; $\dfrac{3}{x} + 2y - 7$

Find the value of each of the following linear forms for each of the ordered pairs given in Exercises 2–5.

2. (a) $7x - 2y - 5$; $(1, 1), (0, 0), (-1, -1)$
(b) $-x + 3y + 1$; $(1, 1), (0, 0), (-1, -1)$

3. (a) $2x + y - 2$; $(3, 2), (0, 2)$
(b) $7x - 5y + 11$; $(-3, -2), (0, 0)$

4. (a) $6 - 3y - 4x$; $(-\frac{1}{3}, -\frac{1}{2}), (-\frac{1}{4}, -\frac{2}{3})$
(b) $5 - 2y - 3x$; $(.1, .03), (3, -2)$

5. (a) $\frac{x}{2} - 8y - 7$; $(8, -\frac{1}{4})$, $(14, 0)$

(b) $3x - \frac{y}{4} - 3$; $(1, 8)$, $(\frac{1}{3}, -4)$

6. In each of the following, tell which of the ordered pairs are solutions of the given linear equation.

(a) $2x + y = x - y + 5$

$$(5, 0), (0, 0), (-2, -3), (1, 4), (7, -1)$$

(b) $3x - 2y = x - y + 2$

$$(0, 0), (0, -2), (1, 2), (-2, -6), (-1, -4)$$

7. Give x, in turn, each of the following values to find four solutions of the linear equation $2x - y - 3 = 0$.

(a) $x = -3, -1, 0, \frac{11}{2}$ (b) $x = -2, 1, -\frac{1}{2}, \frac{3}{2}$

Complete the tables of ordered pairs so that they are solutions of the given linear equations.

8. (a) $2x + y - 1 = 0$

x	y
0	?
?	3
-2	?
?	-5
?	0

(b) $x - 6y = 18$

x	y
?	2
0	?
6	?
?	-4
?	$\frac{1}{3}$

9. (a) $r - s = 0$

r	s
0	?
?	2
-3	?
?	-4
$-\frac{1}{2}$	?

(b) $r + s = 0$

r	s
0	?
?	2
-3	?
?	-4
$-\frac{1}{2}$	?

10. (a) $2c - 5d = 26$

c	d
?	4
$\frac{1}{2}$	?
?	-4
13	?
?	-3

(b) $\frac{3}{4}x + 8y = 24$

x	y
?	0
-32	?
?	$\frac{3}{2}$
-4	?
?	-3

11. (a) Which of the following ordered pairs are solutions of the equation $3x + 0y = 9$?

$$(5, 7), (3, 10), (3, 2), (2, 3), (9, 10), (3, 1)$$

(b) Which of the following ordered pairs are solutions of the equation $0x + 2y = 8$?

$$(3, 4), (-1, 1), (1, 7), (-5, 4), (4, 4), (8, 0)$$

Find five solutions for each of the following linear equations in two variables. Do this by selecting a value for one variable and solving the resulting equation for the other variable.

12. (a) $3x - 4y = 12$
(b) $5x + 4y = 40$

13. (a) $2x + 3y = 6$
(b) $7x - 3y = 21$

14. (a) $(0 \cdot x) + y - 5 = 0$
(b) $x = y$

---

15. (a) What values of x correspond to the following values of y in the linear equation $x - 3 + y = 5 + y$?

$$y = 10, \quad y = -4, \quad y = 0, \quad y = 1, \quad y = 1.79$$

(b) Is there a pattern, or similarity, in the ordered pairs of 15(a)?
(c) What happens if you give x the following values in the linear equation of 15(a)?

$$x = 7, \quad x = -4, \quad x = 0, \quad x = 8$$

(d) What conclusion can you draw about the set of all solutions of the linear equation $x - 3 + y = 5 + y$?

16. (a) What values of y correspond to the following values of x in the linear equation $(0 \cdot x) + y + 3 = 0$?

$$x = 8, \quad x = -8, \quad x = 0, \quad x = \tfrac{61}{99}, \quad x = .01234$$

(b) What happens if you give y the following values in the equation $(0 \cdot x) + y + 3 = 0$?

$$y = 0, \quad y = -10, \quad y = 20, \quad y = -3$$

(c) What conclusion can you draw about the set of all solutions of the linear equation $(0 \cdot x) + y + 3 = 0$?

17. Combine the following pairs of linear forms by addition or subtraction, as indicated. Is the sum of two linear forms in x and y a linear form? Is the difference of two linear forms in c and d a linear form?

(a) $(2x + 5y - 6) + (5x + y + 2)$
(b) $(3c + d - 2) - (7c - 3d + 3)$

18. Use the distributive axiom to simplify each of the following expressions. Is the product of a linear form and a constant a linear form?

(a) $\frac{2}{3}(12x - 15y + 9)$
(b) $-3(x - 5y - 7)$
(c) $0(45x + 90y + 11^5)$
(d) $\frac{1}{2}(6x - 4y - 10) - 3(-2x - 7y - 11)$

Simplify the following expressions.

19. $6(3x + y + 2) + (x - 2y + 10)$

20. $(4x - 2y + 6) - 2(2x - y + 3)$

21. $3(2x - 5y - 1) + 7(x - 2y - 5)$

22. $\frac{1}{7}(105x - 63y - 84) - \frac{8}{7}(105x - 63y - 84)$

---

23. Let N and M denote real numbers. Write an equation which states that three times the first number, decreased by one-third the second, is 18. Find four solutions of this equation.

24. If M stands for the number of men and W the number of women on a college teaching staff, write an equation which states that there are 160 staff members. Find four solutions of this equation. What is the domain of M? of W?

25. Let J represent John's age in years and P represent Pat's age in years. Write an equation which states that John's age is three years less than twice Pat's age. Find four solutions of this equation.

7–2 GRAPH OF A LINEAR EQUATION

The set of all solutions of a linear equation in two variables is called the *solution set* of the equation. For any particular equation, say $x + y = 1$, you may use the notation

$$\{(x, y) \mid x + y = 1\}$$

to indicate its solution set. Since the solution set consists of ordered pairs, it has a graph in a cartesian coordinate system.

Definition of the graph of an equation

The graph of an equation in two variables is the graph of its solution set.

The solution set of a linear equation in x and y is ordinarily an infinite set; therefore, its graph contains an infinite number of points. Some simple equations are graphed below.

Problem 1. Graph the linear equation $x = 3$.

Solution. Although the variable y does not appear, it is understood to be there with zero coefficient;

$$x + 0y = 3.$$

This equation is true for $x = 3$ and $y =$ any number. Therefore, its solution set is

$$\{(x, y) \mid x = 3\}.$$

Another way of writing this might be

$$\{(3, y) \mid y \text{ a real number}\}.$$

The graph of this set consists of all points 3 units to the right of the y-axis. Thus, the graph of the equation $x = 3$ is a line parallel to the y-axis. It is shown in color in Fig. 7–1.

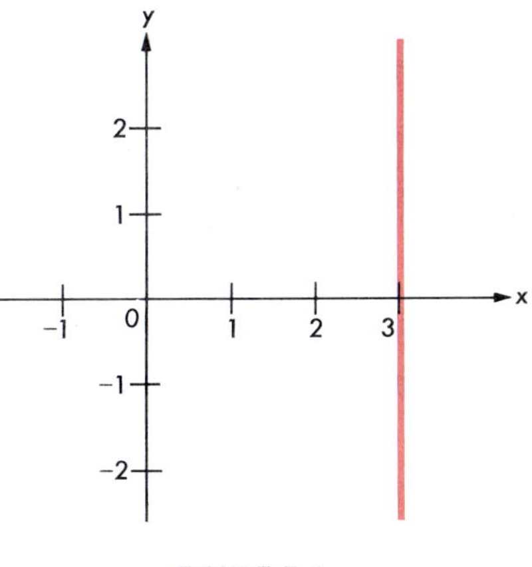

FIGURE 7–1

Problem 2. Graph the linear equation $y + 2 = 0$.

Solution. Again, this equation is considered to have the form

$$0x + y + 2 = 0.$$

This equation is true for any number x and $y = -2$. Thus,

$$\{(x, y) \mid y = -2\}, \quad \text{or}$$
$$\{(x, -2) \mid x \text{ a real number}\}$$

is its solution set. The graph of each ordered pair $(x, -2)$ is a point on the line parallel to the x-axis and 2 units below it. Thus, the graph of the equation $y + 2 = 0$ is the line shown in color in Fig. 7–2.

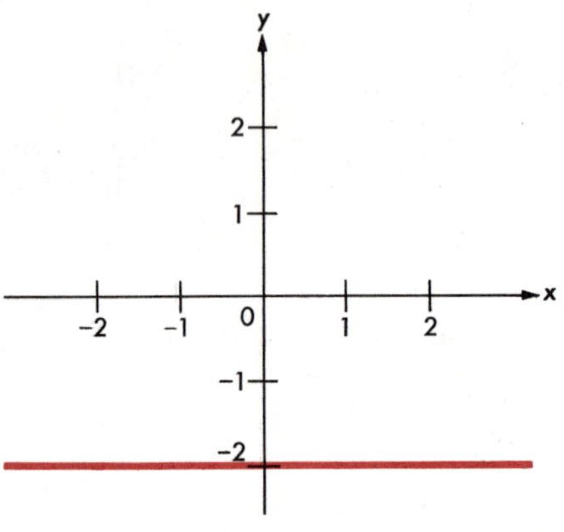

FIGURE 7–2

Problem 3. Graph the linear equation $y = x$.

Solution. Some solutions of this equation are $(0, 0)$, $(-3, -3)$, $(2, 2)$, and $(5, 5)$. The solution set may be described as

$$\{(x, y) \mid x = y\}, \quad \text{or}$$
$$\{(x, x) \mid x \text{ a real number}\}.$$

If you plot all such points in a cartesian coordinate system having the same scale on both axes, then the resulting set of points make up a line bisecting the first and third quadrants. It is sketched in Fig. 7–3.

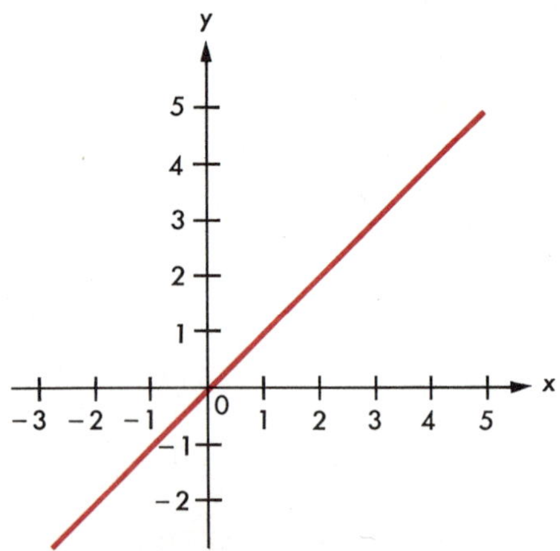

FIGURE 7–3

Exercises

1. (a) Which of the following ordered pairs are solutions of the equation $y - 1 = 0$?

$$(0, 1), (1, 1), (0, 0), (1, 0), (3, 2), (8, 1), (10^6, 1), (-10^6, 1).$$

Graph the linear equation $y - 1 = 0$.

(b) Which of the following ordered pairs are solutions of the equation $x + 1 = 0$?

$$(-1, 1), (0, 1), (-1, 0), (-1, 5), (1, -1), (-1, -80), (2, 3), (-1, 80).$$

Graph the linear equation $x + 1 = 0$.

Graph each of the following linear equations. In each case, describe the solution set, and give three of its elements.

2. (a) $x + 2 = 0$ (b) $x - 6 = 0$
3. (a) $y = 0$ (b) $x = 0$
4. (a) $y - 7 = 0$ (b) $y + 5 = 0$

Graph each of the following sets.

5. (a) $A = \{(4, y) \mid y \text{ a real number}\}$ (b) $B = \{(x, 5) \mid x \text{ a real number}\}$
6. (a) $C = \{(x, y) \mid y = -2\}$ (b) $D = \{(x, y) \mid x = -3\}$
7. (a) $E = \{(x, -2) \mid x \text{ an integer}\}$ (b) $F = \{(-3, y) \mid y \text{ an integer}\}$

Find eight elements of the solution set of each of the following linear equations. Then graph each of the equations on a separate coordinate system.

8. (a) $x + y = 0$ (b) $2x + y = 0$
9. (a) $x + y = 3$ (b) $x + y = 4$
10. (a) $x + y = -1$ (b) $x + y = -2$
11. (a) $y = x + 5$ (b) $y = x - 5$
12. (a) $y = -3x$ (b) $y = -4x$
13. (a) $y = 3 + x$ (b) $y = -3 + x$
14. (a) (i) Tell if each of the following points is above, on, or below the graph of the solution set of $y + 2 = 0$. (See Fig. 7-2)

$$(1, -2), (1, 5), (1, -6), (-3, 7), (-31, -10), (-3, -2),$$
$$(0, 0), (0, -2), (0, -4), (-1, -1), (-1, -2), (-1, -\tfrac{5}{2}).$$

(ii) Study the coordinates of all the points that were above the graph of $y + 2 = 0$. How does the y-coordinate for each of these compare with -2?

(iii) Study the coordinates of all the points that were below the graph of $y + 2 = 0$. How does the y-coordinate for each of these compare with -2?

(b) (i) Tell if each of the following points lies to the left of, on, or to the right of the graph of $x = 3$. (See Fig. 7–1.)

$$(-5, 1), (3, 1), (6, 1), (2, -4), (3, -4), (4, -4),$$
$$(\tfrac{7}{2}, 0), (\tfrac{5}{2}, 0), (3, 0), (2.99, 10), (3.01, 10), (3, 10).$$

(ii) Study the coordinates of all the points to the left of the graph of $x = 3$. How does the x-coordinate of each of these compare with 3?

(iii) Study the coordinates of all the points to the right of the graph of $x = 3$. How does the x-coordinate of each of these compare with 3?

---

15. Graph the linear equation

$$x - y = 4.$$

Refer to your graph to answer the following questions.

(a) Is $(-3, 4)$, on the graph of $x - y = 4$? Is $(-7, 8)$? Is $(-2, 0)$? Give the algebraic reason that explains why a point in Quadrant II cannot lie on the graph of $x - y = 4$.

(b) Are the ordered pairs $(4, 0)$ and $(5, 1)$ in the solution set of $x - y = 4$?

(c) If the point $(4, 0)$ were joined to the point $(5, 1)$ by a straight line segment, would every point on this segment have a number pair in the solution set of $x - y = 4$?

16. Graph the equation

$$x + y = 1.$$

Why can no point in Quadrant III lie on the graph of $x + y = 1$?

Graph each of the following linear equations.

17. $2(x + 5) + 7(x - y) = 4x - 8y$

18. $20 = 3(-x - y) - 5(-3y + 4x)$

7–3 STRAIGHT-LINE GRAPHS

From the preceding section, you might suspect that the graph of a linear equation in two variables is a straight line. This is true, although it will not be proved at this time. From now on, it will be assumed that if the solution set of a linear equation is not the empty set, then the following statement is true.

In a cartesian coordinate system, the graph of a linear equation in two variables consists of those, and only those, points that lie on some straight line.

No other type of equation in two variables has a straight line for its graph. You should understand now why the adjective *linear* is used to describe the equations you are studying.

Since two points are sufficient to determine a straight line, you need to plot only two points in order to describe the straight-line graph of a given linear equation.

Problem 1. Sketch the graph of the linear equation

$$y = x + 2.$$

Solution. You know that the graph of every linear equation is a straight line. Therefore, the graph of a particular linear equation, such as $y = x + 2$, is some straight line. If you give x the values -4, -2, 0, and 2, in turn, and solve the resulting equations for y, you find the following four elements in the solution set of the equation $y = x + 2$.

$$(-4, -2), (-2, 0), (0, 2), (2, 4)$$

The straight line drawn through these four points is the graph of the given linear equation. It is sketched in Fig. 7–4. Although you needed to plot only two points, four points were plotted to make sure that the work was correct.

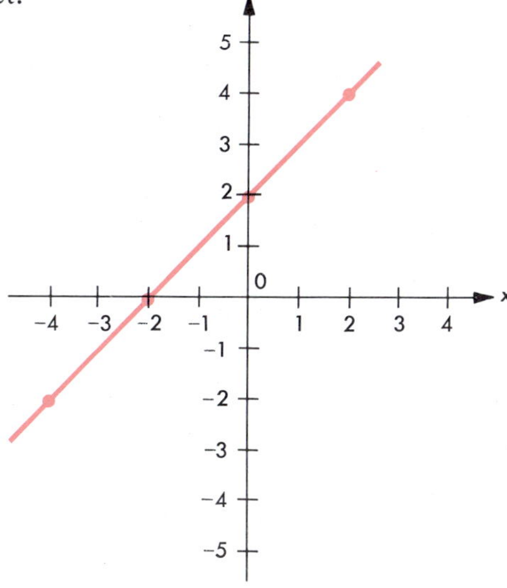

FIGURE 7–4

Problem 2. Sketch the graph of the linear equation

$$x + 2y - 1 = 0.$$

Solution. If you give y the values -2, 0, 1, and 3, you obtain four ordered pairs of the solution set:

$$(5, -2), (1, 0), (-1, 1), (-5, 3).$$

These points and the straight line joining them are shown in Fig. 7–5. The set of all points on the line is the graph of the given linear equation.

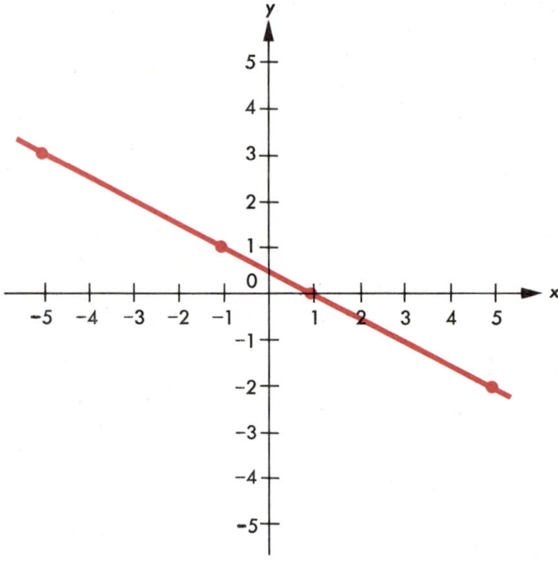

FIGURE 7–5

Exercises

1. Which of the following equations will have straight lines for graphs? Explain your answer.

   (a) $\dfrac{x}{3} - 5y = 1$; $\dfrac{3}{x} - 5y = 1$; $xy = 2$; $y = 3x^2$; $3x = 3y$

   (b) $\dfrac{2x}{5} - 3y = 2$; $4x - \dfrac{6}{y} = 21$; $x^2 = 2y + 1$; $x^2 + y^2 = 1$; $y - x^2 = 1$

2. (a) Give x the values -6, -3, 0, and 12 to obtain four solutions of the equation $y = \frac{1}{3}x - 2$. Then graph the equation.

   (b) Give y the values -2, 0, 2, and 3 to obtain four solutions of the equation $x = 3 - 3y$. Then graph the equation.

3. (a) Give four elements of the solution set of the linear equation $4x - 5 = 0$, and graph the solution set.
(b) Give four elements of the solution set of the linear equation $5y - 15 = 0$, and graph the solution set.

In Exercises 4–12, graph the linear equations.

4. (a) $x + y - 3 = 0$ (b) $x - y + 6 = 0$

5. (a) $y = 2x - 3$ (b) $x = 2y + 1$

6. (a) $x + 2y + 4 = 0$ (b) $3x - 2y - 6 = 0$

7. (a) $2y + 5 = 0$ (b) $2x + 3 = 0$

8. (a) $x + .5y = 4$ (b) $x = \frac{1}{2}y - 1$

9. (a) $y + 3x = 0$ (b) $x + \frac{1}{2}y = 0$

10. (a) $3x = 0$ (b) $\frac{1}{2}y = 0$

11. (a) $2x - 3y = 0$ (b) $3x - 2y = 0$

12. (a) $\frac{1}{4}x - \frac{1}{2}y = 3$ (b) $\frac{1}{2}x - \frac{1}{4}y = 3$

13. (a) Use one set of axes and draw the graphs of
(i) $y = x + 2$. (ii) $y = x + 5$. (iii) $y = x - 3$.
In what way are the graphs alike? In what way are they different?
(b) Use one set of axes and draw the graphs of
(i) $y = 2x$. (ii) $y = -x$. (iii) $y = 4x$.
In what way are the graphs alike? In what way are they different?

14. (a) Write an equation describing the set of all points, each of which has an ordinate three times its abscissa. List six members of the solution set of this equation and graph the solution set.
(b) Write an equation describing the set of all points, each of which has an abscissa four times its ordinate. List six members of the solution set of this equation and graph the solution set.

---

15. (a) Evaluate $x + 2y - 1$ for each of the following ordered pairs.

$$(2, 0), (1, 3), (0, 2), (-1, 2), (-5, 4)$$

Where do each of these points lie with respect to the graph of $x + 2y - 1 = 0$? (See Fig. 7–5.)
(b) Evaluate $x + 2y - 1$ for each of the following ordered pairs.

$$(-5, 2), (1, -1), (0, 0), (-2, -2), (5, -10)$$

Where do each of these points lie with respect to the graph of $x + 2y - 1 = 0$? (See Fig. 7–5.)

(c) Evaluate $x + 2y - 1$ for each of the following ordered pairs.

$$(-3, 2), (2, -\tfrac{1}{2}), (3, -1), (0, \tfrac{1}{2}), (-2, \tfrac{3}{2})$$

Where do each of these points lie with respect to the graph of $x + 2y - 1 = 0$? (See Fig. 7–5.)

---

16. According to the definition of the absolute value of a real number x,

$$|x| = x \text{ if } x \text{ is a positive number,}$$
$$|x| = -x \text{ if } x \text{ is a negative number,}$$

and

$$|0| = 0.$$

Consider the graphs of $y = x$ and $y = -x$ to graph the set

$$\{(x, y) \mid y = |x|\}.$$

7–4 SIMPLIFYING LINEAR EQUATIONS

An equation in two variables of the form

$$y = 3x - 1$$

is said to be *solved for y in terms of x.* You can easily find solutions of such an equation by giving values to x. Thus,

$$\text{if } x = 4, \quad y = 3 \cdot 4 - 1, \quad \text{or } y = 11,$$
$$\text{if } x = -10, \quad y = 3 \cdot (-10) - 1, \quad \text{or } y = -31,$$

and so on.

An equation in two variables of the form

$$x = 4 - 9y$$

is said to be *solved for x in terms of y.* Solutions of this equation can be easily found by giving values to y. Thus,

$$\text{if } y = 3, \quad x = 4 - 9 \cdot 3, \quad \text{or } x = -23,$$
$$\text{if } y = 0, \quad x = 4 - 9 \cdot 0, \quad \text{or } x = 4,$$

and $(-23, 3)$ and $(4, 0)$ are solutions of the equation.

Every linear equation in two variables (in which at least one variable has different coefficients on the two sides of the equation) is equivalent to an equation which is solved for one variable in terms of the other. You can find such an equivalent equation by using the properties of addition and multiplication.

Problem 1. Solve the equation $2x - y = 3y - x + 6$ for y in terms of x and list three members of its solution set.

Solution. You might proceed as follows:

$$\begin{aligned}
-2x + 2x - y &= -2x + 3y - x + 6, \\
(-2x + 2x) - y &= 3y + (-2x - x) + 6, \\
-y &= 3y - 3x + 6, \\
-3y - y &= -3y + 3y - 3x + 6, \\
-4y &= -3x + 6, \\
(-\tfrac{1}{4}) \cdot (-4y) &= (-\tfrac{1}{4}) \cdot (-3x + 6), \\
y &= (-\tfrac{1}{4}) \cdot (-3x) + (-\tfrac{1}{4}) \cdot 6, \\
y &= \tfrac{3}{4}x - \tfrac{3}{2}.
\end{aligned}$$

Thus, the solution set, S, of the given equation can be described as

$$S = \{(x, y) \mid y = \tfrac{3}{4}x - \tfrac{3}{2}\}.$$

Giving x the values -2, 0, and 2, you obtain the following three elements of S: $(-2, -3)$, $(0, -\frac{3}{2})$, $(2, 0)$.

Problem 2. Solve the equation $2x - 15y + 2 = 6y - x + 11$ for x in terms of y and list three members of its solution set.

Solution. One way to proceed is given below.

$$\begin{aligned}
x + 2x - 15y + 2 &= x + 6y - x + 11 \\
3x - 15y + 2 &= 6y + 11 \\
15y + 3x - 15y + 2 &= 15y + 6y + 11 \\
3x + 2 &= 21y + 11 \\
3x + 2 + (-2) &= 21y + 11 + (-2) \\
3x &= 21y + 9 \\
\tfrac{1}{3} \cdot 3x &= \tfrac{1}{3} \cdot 21y + \tfrac{1}{3} \cdot 9 \\
x &= 7y + 3
\end{aligned}$$

Thus,

$$S = \{(x, y) \mid x = 7y + 3\}$$

is the solution set of the given equation. By giving y the values -2, 0, and 3, you see that $(-11, -2)$, $(3, 0)$, and $(24, 3)$ are in S.

Exercises

1. (a) Solve the following equation for y in terms of x.

$$7 - 7x + 11y = 4x + 10y + 29$$

(b) Solve the following equation for x in terms of y.

$$2 - 8y + 2x = -2x - 7y - 14$$

2. (a) (i) Verify that the following ordered pairs are solutions of the equation $y = 11x + 22$.

$$(0, 22), (-1, 11), (-2, 0), (\tfrac{1}{11}, 23)$$

(ii) Test the four ordered pairs of part (a) in the equation

$$7 - 7x + 11y = 4x + 10y + 29.$$

(iii) Are the following two equations equivalent?

$$y = 11x + 22$$
$$7 - 7x + 11y = 4x + 10y + 29$$

(b) (i) Use the equation $x = \frac{1}{4}y - 4$ to find the value of x for each of the following values of y.

$$-4, 0, 4, 16, 20$$

(ii) Test the solutions you found in part (i) in this equation:

$$2 - 8y + 2x = -2x - 7y - 14.$$

(iii) Are the equations of parts (i) and (ii) equivalent?

Solve each of the following equations for y in terms of x. Using the simplified form of the equation, find a solution and check it in the original equation.

3. (a) $3y + 2x - 1 = 2y + 3x + 4$
(b) $5y - 3x + 4 = 4y + x - 2$

4. (a) $2x - 2y + 5 = 3x - 3y + 8$
(b) $7y - 7 + 5x = 2x - 1 + 6y$

5. (a) $6x - 3y + 1 = 5x - y$
(b) $-3x + 4y - 2 = 3x + y$

6. (a) $14x - 4y + 10 = 13x - 6y + 2$
(b) $3y - 8x + 2 = 4 + 2y - 9x$

7. (a) $2y - 8x + 1 = -9x - y - 1$
(b) $-y + 3x + 10 = -8x + 4y + 1$

8. (a) $3x - 21 = 5x - 3y - 12$
(b) $22x = 27x - 1 - 2y$

9. (a) $3y - 7 = 7y - 4 - x$
(b) $5y + 11 = 11y + 5 - 2x$

10. (a) $70x - 25y - 25 = -27y + 65x$
(b) $68x + 30y - 31 = 28y + 70x - 40$

Solve each of the following equations for x in terms of y. Using the simplified form of the equation, find a solution and check it in the original equation.

11. (a) $12x - 10y + 7 = 11x + 8y - 2$
(b) $5x + 8y - 3 = 4x - 2y + 5$

12. (a) $10x + y + 3 = 5 - 3y + 13x$
(b) $14x - 13y + 17 = 14 - 11y + 11x$

13. (a) $8x - 3y = 7x$
(b) $12x - 2y + 1 = 10x$

14. (a) $12x + 75y + 17 = 14x + 75y + 33$
(b) $8x - 2y + 3x = 5 + 12x - 2y$

15. (a) $7x + 1 = \frac{1}{3}(4y - 1 + 18x)$
(b) $\frac{1}{4}(16x - 12y + 3) = 3x - 1$

16. (a) $.2x + 1.3y + .5 = .25 + .3y - .3x$
(b) $.3x + 2.5y - .1 = .9 + .5y - .7x$

17. (a) $3(2x - 7) - 7 = 23 - 4(2y - 3x)$
(b) $7x - \frac{1}{2}(5x - 7) = 10 - \frac{1}{4}(2x - 3y)$

18. (a) $\frac{x}{9} - \frac{y}{3} - \frac{2}{3} = \frac{x}{3} - \frac{y}{9} - \frac{2}{9}$

(b) $\frac{x}{6} - \frac{y}{2} + \frac{1}{3} = \frac{x}{2} - \frac{y}{3} + \frac{1}{6}$

Review for Sections 7–1 through 7–4

Which of the following are linear forms?

1. $x + 3y - 4$

2. $3x - \sqrt{y} + 6$

3. $x - 5$

4. $7x - y + \sqrt{3}$

5. $xy + 7$

6. $y + 2$

What is the value of each of the following linear forms for each of the given ordered pairs, (x, y)?

7. $4x + 2y - 1$; $(0, 0), (0, 1), (1, 2)$

8. $3y - 7x + 4$; $(2, 2), (1, 1), (-3, -5)$

9. $y + 4x - 2$; $(0, 5), (2, -6), (-7, \frac{1}{2})$

10. $2x + 4y - 3$; $(\frac{1}{2}, \frac{1}{4}), (\frac{3}{10}, \frac{1}{5}), (\frac{2}{3}, \frac{5}{6})$

Complete the ordered pairs, (x, y), so that they are solutions of the given linear equation.

11. $x + 4y - 5 = 3$; (0, _?_), (_?_, 1), (−10, _?_)

12. $2x + 5 = y - 3$; (0, _?_), (_?_, 2), (_?_, 0)

13. $x - 2y + 6 = 4 + y$; (0, _?_), (_?_, 4), (_?_, $\frac{1}{3}$)

14. $2x + 3y - 5 = x + 2y + 5$; (0, _?_), (_?_, 5), (15, _?_)

15. $x + 7 = 3y + 2x - 5$; (0, _?_), (_?_, 3), (12, _?_)

Find four elements of the solution set of each of the following linear equations. Graph each of the equations on a separate coordinate system.

16. $x + y = 4$

17. $x - y = 7$

18. $x + y + 10 = 13$

19. $2x + y - 4 = 0$

20. $3x - 2y + 5 = 7$

21. $x = 6$

22. $y + 5 = 7$

23. $4x + 2y - 3 = 13$

Solve the following equations for y in terms of x. Using the simplified form of the equation, find a solution and check it in the original equation.

24. $3y + x - 4 = 8 - y + 7x$

25. $3x - 2y + 6 = 8x - 6y + 14$

26. $2x + 4y + 9 = 3x + y - 6$

27. $3y - x + 2 = y + 7 - 4x$

28. $18 - y - x = 3x - 5y + 20$

29. $7y + x - 9 = y + 5x + 15$

30. $8x - 2y - 4 = 4y + 2x - 8$

Answers to Review for Sections 7–1 through 7–4

1. Linear form

2. Not a linear form

3. Linear form

4. Linear form

5. Not a linear form

6. Linear form

7. $-1, 1, 7$

8. $-4, 0, 10$

9. $3, 0, -29\frac{1}{2}$ (or $-\frac{59}{2}$)

10. $-1, -\frac{8}{5}, \frac{5}{3}$

11. $2, 4, \frac{9}{2}$ (or $4\frac{1}{2}$)

12. $8, -3, -4$

13. $\frac{2}{3}, 10, -1$

14. $10, 5, -5$

15. $4, 3, 0$

16. For example: (0, 4), (4, 0), (3, 1) and (6, −2)

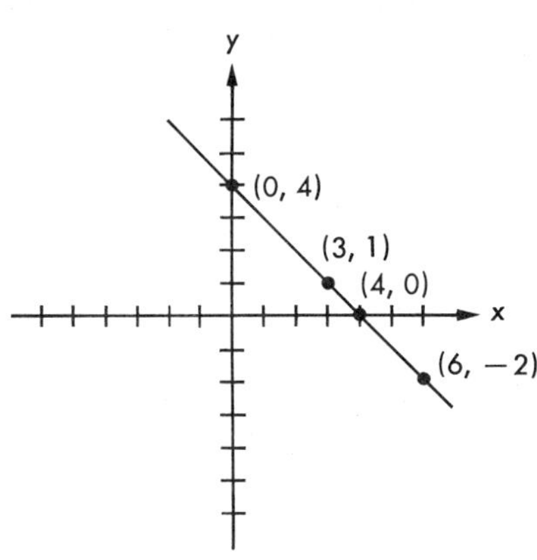

17. For example: (0, −7), (7, 0), (2, −5), (3, −4)

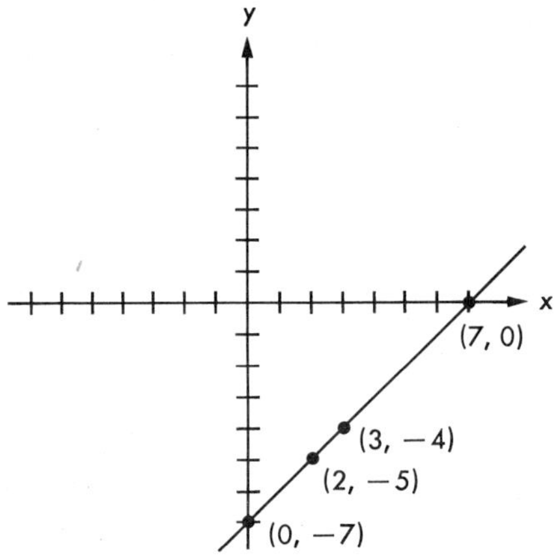

18. For example: (0, 3), (3, 0), (−2, 5), (4, −1)

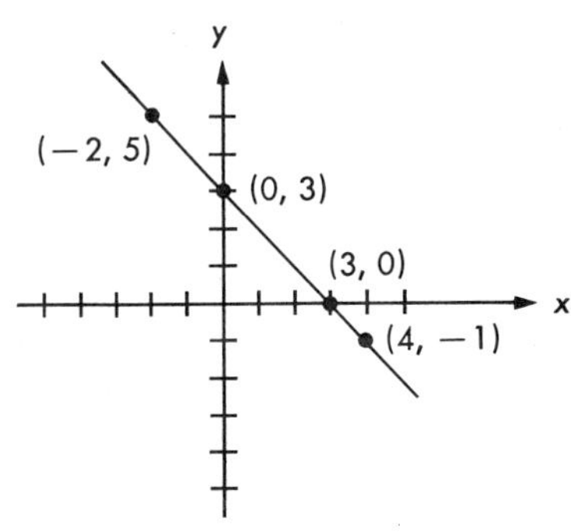

19. For example: (0, 4), (2, 0), (1, 2), (−1, 6)

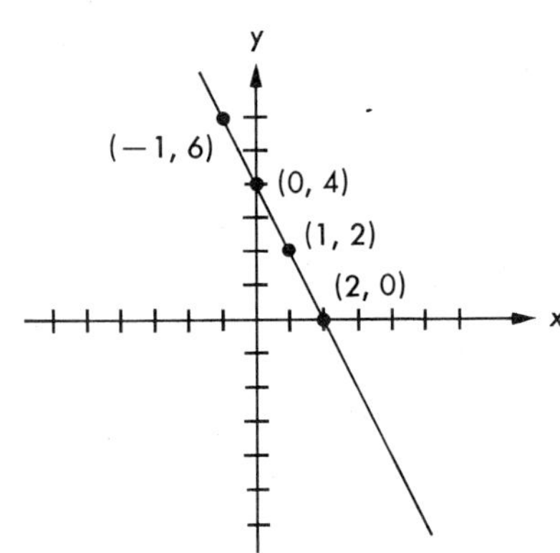

20. For example: (0, −1), ($\frac{2}{3}$, 0), (2, 2), (4, 5)

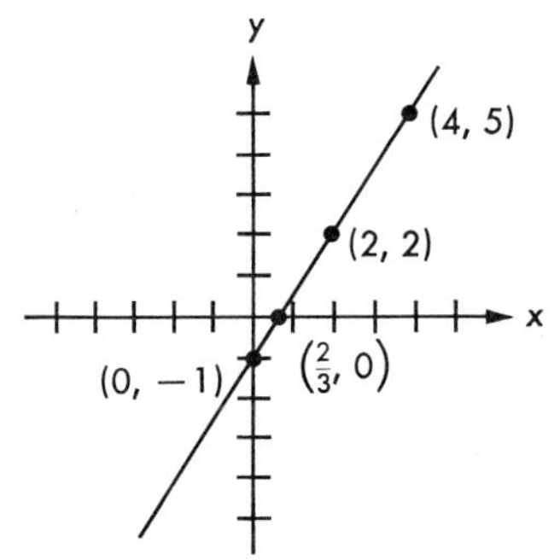

21. For example: (6, 0), (6, 2), (6, −2), (6, 4)

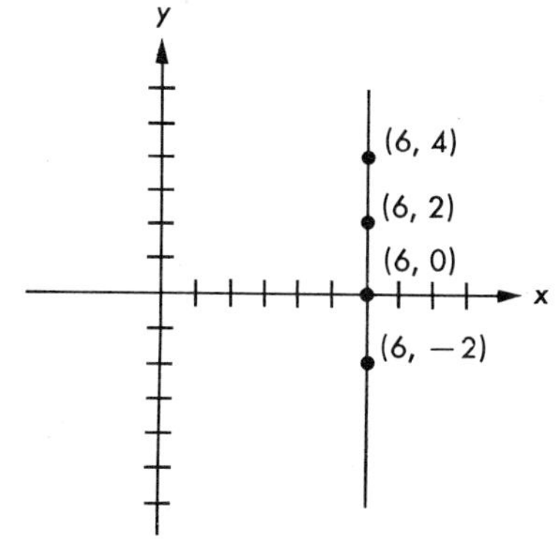

22. For example: (0, 2), (2, 2), (−2, 2), (5, 2)

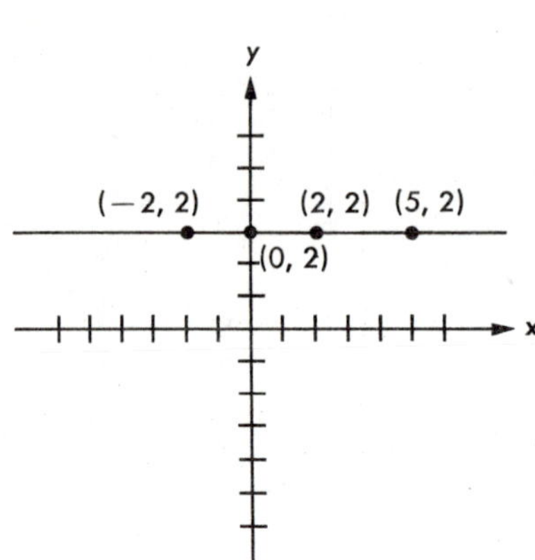

23. For example: (0, 8), (4, 0), (1, 6), (−2, 12)

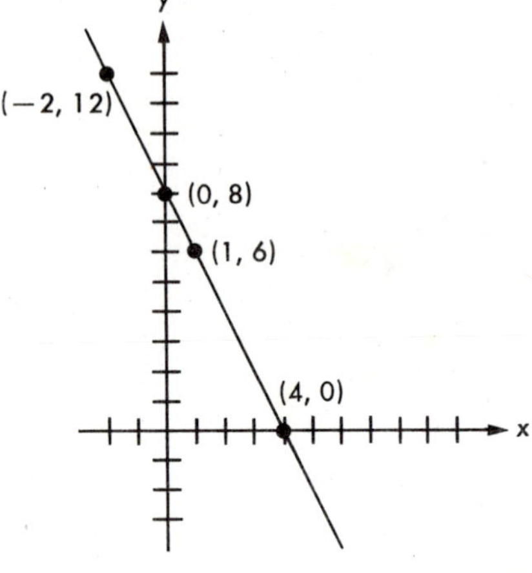

24. $y = \frac{3}{2}x + 3$
25. $y = \frac{5}{4}x + 2$
26. $y = \frac{1}{3}x - 5$
27. $y = \frac{5}{2} - \frac{3}{2}x$
28. $y = x + \frac{1}{2}$
29. $y = \frac{2}{3}x + 4$
30. $y = x + \frac{2}{3}$

7–5 HALF-PLANES

Just as each cut across a piece of paper divides the paper into two pieces, each line in a plane divides the plane into two pieces, called *half-planes.* For example, the graph of the equation $y = 1$ is shown by the line in color in Fig. 7–6. It divides the plane into two half-planes, one *above the line* and one *below the line.*

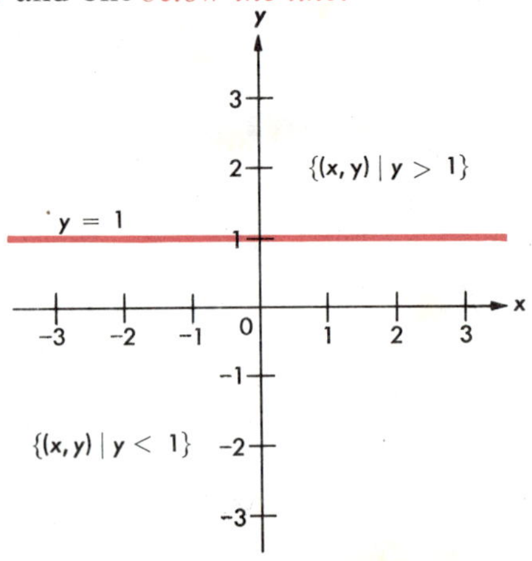

FIGURE 7–6

You can describe algebraically the two half-planes of Fig. 7–6. Notice that the half-plane above the line $y = 1$ is made up of points having y-coordinates greater than 1. In other words, the half-plane above $y = 1$ is the graph of the set

$$\{(x, y) \mid y > 1\}.$$

Similarly, the half-plane below the line $y = 1$ is the graph of the set

$$\{(x, y) \mid y < 1\}.$$

If you wish to discuss both the line $y = 1$ and the half-plane, you use a special term: *the half-plane and its edge.* The graph of the set

$$\{(x, y) \mid y \leqq 1\}$$

is the line $y = 1$ and the half-plane below.

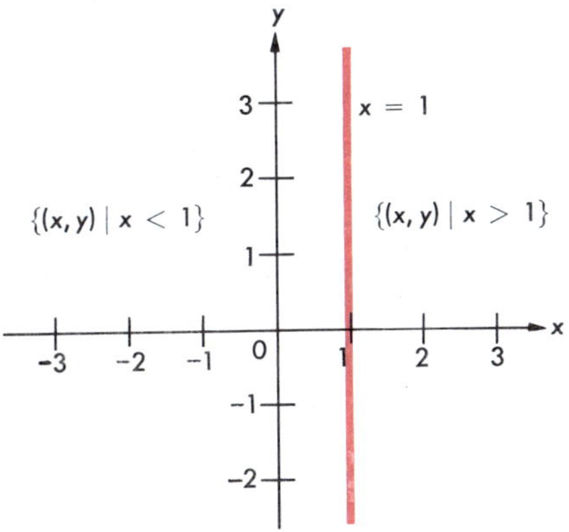

FIGURE 7–7

The graph of the equation $x = 1$ is shown in Fig. 7–7. This line also divides the plane into two half-planes. The half-plane to the right of $x = 1$ is the graph of the set

$$\{(x, y) \mid x > 1\}.$$

The graph of the set

$$\{(x, y) \mid x \geqq 1\}$$

is the half-plane and its edge. The half-plane to the left of $x = 1$ is the graph of the set

$$\{(x, y) \mid x < 1\}.$$

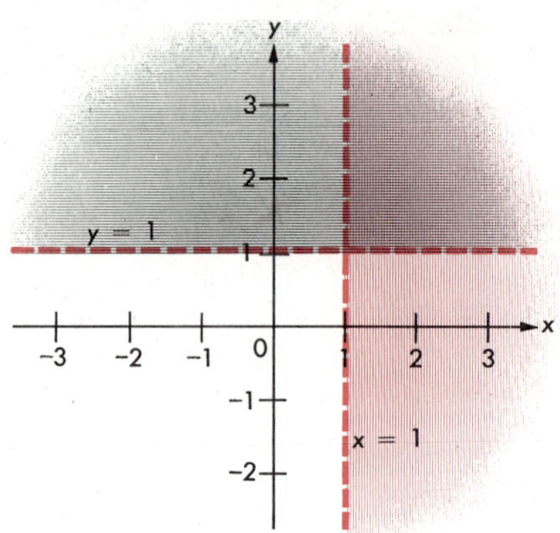

FIGURE 7-8

An example of the intersection of two half-planes is $A \cap B$, where

$$A = \{(x, y) \mid x > 1\} \quad \text{and} \quad B = \{(x, y) \mid y > 1\}.$$

By definition, $A \cap B$ is the set of points in both A and B. Thus, the graph of $A \cap B$ is a *quarter-plane*, as shown in Fig. 7-8. The lines $x = 1$ and $y = 1$ are broken to indicate that they are not part of the graph.

As another example, consider the intersection of sets C and D.

$$C = \{(x, y) \mid x \geqq -2\} \quad \text{and} \quad D = \{(x, y) \mid x \leqq 0\}$$

Then

$$C \cap D = \{(x, y) \mid -2 \leqq x \leqq 0\}.$$

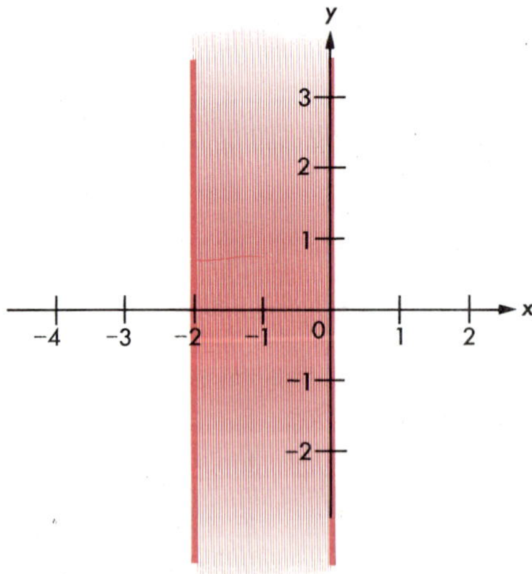

FIGURE 7-9

The graph of $C \cap D$ is a vertical strip of the plane, as shown in Fig. 7-9. The lines $x = -2$ and $x = 0$ are solid to indicate that they are part of the graph.

Exercises

1. (a) Graph the equation $y + 5 = 0$. Also plot the following ordered pairs.

$$(0, 0), (-1, -4), (4, -6), (5, 5), (3, -7), (10, -\tfrac{11}{2}), (1, -\tfrac{9}{2})$$

List the points in your graph that lie above the line $y + 5 = 0$. Describe the ordinates (y-coordinates) of these points.

List the points in your graph that lie below the line $y + 5 = 0$. Describe the ordinates of these points.

(b) Graph the equation $x + 2 = 0$. Also plot the following ordered pairs.

$$(-5, 0), (-1, 2), (-3, 1), (0, 5), (2, -3), (-2.1, -4), (-1.9, -1)$$

List the points in your graph that lie to the right of the line $x + 2 = 0$. Describe the abscissas (x-coordinates) of these points.

List the points in your graph that lie to the left of the line $x + 2 = 0$. Describe the abscissas of these points.

In Exercises 2–5, name three elements in each of the sets, and sketch the graph of each set.

2. (a) $A = \{(x, y) \mid y \geqq -2\}$ (b) $K = \{(x, y) \mid y \leqq 2\}$

3. (a) $B = \{(x, y) \mid x \leqq 0\}$ (b) $F = \{(x, y) \mid x \geqq -2\}$

4. (a) $C = \{(x, y) \mid y \leqq 0\}$ (b) $R = \{(x, y) \mid y \geqq 0\}$

5. (a) $D = \{(x, y) \mid x \geqq -1\}$ (b) $T = \{(x, y) \mid x \leqq 3\}$

Using sets A, B, C, D, K, F, R, T of Exercises 2–5, sketch the graph of each set in Exercises 6–11.

6. (a) $A \cap B$ (b) $K \cap F$

7. (a) $B \cup C$ (b) $F \cup R$

8. (a) $B \cap D$ (b) $F \cap T$

9. (a) $A \cup D$ (b) $K \cup T$

10. (a) $A \cap C$ (b) $K \cap R$

11. (a) $C \cap D$ (b) $R \cap T$

Name three elements in each of the following sets, and sketch the graph of each set.

12. (a) $A = \{(x, y) \mid -2 < x < 3\}$ (b) $B = \{(x, y) \mid -5 < y < 7\}$

13. (a) $C = \{(x, y) \mid y > 0\}$ (b) $D = \{(x, y) \mid x > 0\}$

14. Name three elements in each of the following sets and sketch the graph of each set. (The letters refer to the sets described in Exercises 12 and 13.)

(a) $C \cap D$ (b) $C \cup D$ (c) $A \cap B$ (d) $B \cup C$

15. Graph each of the following sets and describe the graph.

(a) $A = \{(x, y) \mid x > 0\}$ (b) $B = \{(x, y) \mid x \geqq -5\}$
(c) $C = \{(x, y) \mid y \leqq 1\}$ (d) $D = \{(x, y) \mid y > -3\}$

16. Graph the following sets and describe their graphs. (The letters refer to the sets defined in Exercise 15.)

(a) $A \cap D$ (b) $B \cap C$ (c) $C \cap D$ (d) $A \cup D$

17. Given that $A = \{(x, y) \mid x > 0\}$ and $B = \{(x, y) \mid x \geqq -5\}$, answer the following questions.

(a) What elements of A are also elements of B?
What elements of B are also elements of A?
(b) Describe $A \cap B$.
(c) Describe $A \cup B$.
(d) If A and B are any sets such that $A \cap B = A$ and $A \cup B = B$, then which is true, $A \subset B$ or $B \subset A$?

18. Describe each of the following sets in set notation.

(a) All the points above the x-axis
(b) All the points in Quadrants I or IV or on the positive half of the x-axis
(c) All the points in Quadrants III or IV or on the negative half of the y-axis
(d) All the points inside Quadrant II
(e) All the points to the left of the y-axis

---

19. (a) If $P = \{(x, y) \mid x > 3\}$ and $Q = \{(x, y) \mid x < -3\}$, describe $P \cap Q$.
(b) Describe $P \cup Q$ and sketch the graph.
(c) Describe the graph of S if $S = \{(x, y) \mid |x| > 3\}$.

20. (a) If $A = \{(x, y) \mid x > -4\}$ and $B = \{(x, y) \mid x < 4\}$, graph $A \cap B$.
(b) Describe the graph of C if $C = \{(x, y) \mid |x| < 4\}$.
(c) Describe the graph of $A \cup B$.

21. (a) If $R = \{(x, y) \mid y \geqq \frac{3}{2}\}$ and $S = \{(x, y) \mid y \leqq -\frac{3}{2}\}$, describe $R \cap S$.
(b) Describe $R \cup S$ and sketch the graph.
(c) Describe the graph of T if $T = \{(x, y) \mid |y| \geqq \frac{3}{2}\}$.

22. (a) If $U = \{(x, y) \mid y \leqq \frac{5}{2}\}$ and $W = \{(x, y) \mid y \geqq -\frac{5}{2}\}$, graph $U \cap W$.
(b) Describe the graph of Z if $Z = \{(x, y) \mid |y| \leqq \frac{5}{2}\}$.
(c) How would you describe the graph of $U \cup W$?

23. Given

$$A = \{(x, y) \mid x > 1\},$$
$$B = \{(x, y) \mid y > 1\},$$
$$C = \{(x, y) \mid y < -1\}.$$

Sketch the graphs of each of the following sets.

(a) $A \cap (B \cup C)$ (b) $(A \cap B) \cup (A \cap C)$

(c) What property of the real number system do (a) and (b) remind you of?

7-6 LINES AND HALF-PLANES

If a line is not parallel to the x-axis or y-axis, the half-plane formed by the line can be described as shown below.

Problem 1. Describe algebraically the half-planes on either side of line L with equation

$$y = 2x + 1.$$

Solution. In Fig. 7-10, line L divides the plane into two half-planes; one is *above* L and the other *below* L. Let H denote the half-plane above L. A point P is in H if, and only if, P is directly above some point on L. For example, the point $(-1, 1)$ is in H since it is directly above the point $(-1, -1)$ on L.

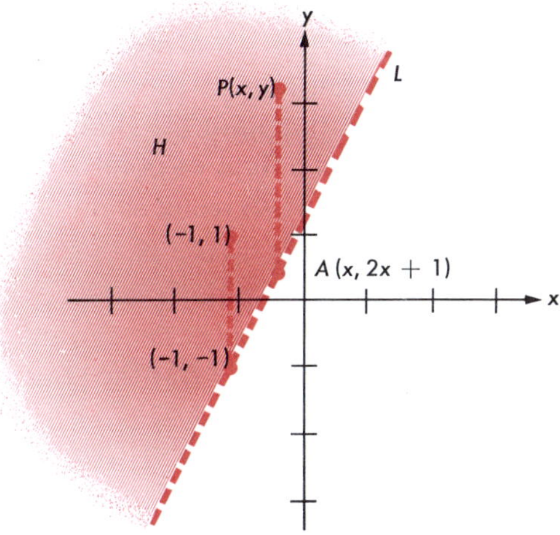

FIGURE 7-10

For every real number x, the point A $(x, 2x + 1)$, is on line L of Fig. 7-10. Under what conditions is a point P (x, y), directly above A?

Since P and A have the same x-coordinate, P is above A if, and only if, the y-coordinate of P is greater than the y-coordinate of A. In other words,

$$y > 2x + 1.$$

Thus, the graph of the set

$$\{(x, y) \mid y > 2x + 1\}$$

is H, or the half-plane above line L. (See the shaded region in Fig. 7–10.) Similarly, the graph of the set

$$\{(x, y) \mid y < 2x + 1\}$$

is the half-plane below line L (the unshaded region in Fig. 7–10).

Problem 2. Describe algebraically the half-planes on either side of line L with equation

$$x + 2y = 2.$$

Solution. You can solve this equation for x in terms of y, obtaining

$$x = -2y + 2.$$

The graph, L, of $x = -2y + 2$ in Fig. 7–11 divides the plane into two half-planes, one to the *right* of L and the other to the *left* of L. Let H be the shaded half-plane to the right of L. A point P is in H if, and only if, it is directly to the right of a point on L. For example, $(4, \frac{1}{2})$ is in H since it is directly to the right of the point $(1, \frac{1}{2})$ on L.

A point P (x, y), is directly to the right of point A $(2 - 2y, y)$, on L if, and only if, the x-coordinate of P is greater than the x-coordinate of A. In other words,

$$x > 2 - 2y.$$

Thus, the half-plane H to the right of line L is the graph of the set

$$\{(x, y) \mid x > 2 - 2y\}.$$

Similarly, the half-plane to the left of L is the graph of the set

$$\{(x, y) \mid x < 2 - 2y\}.$$

How would you describe the graph of the set

$$\{(x, y) \mid x \geqq 2 - 2y\}\ ?$$

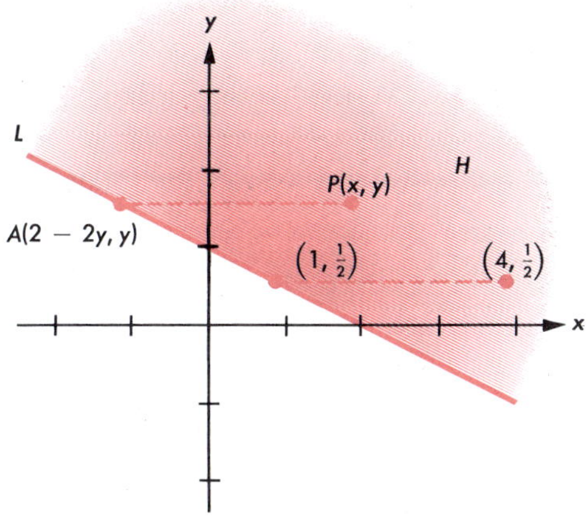

FIGURE 7–11

The set contains all points *on and to the right of line L* in Fig. 7–11. In other words, the graph consists of the half-plane to the right of L and its edge. The graph is the shaded region in Fig. 7–11. The solid red line indicates that line L is included in the graph.

Exercises

1. (a) Graph the line $y = 3x + 3$ and the following points on the same coordinate axis.

$$(1, -1), (2, 4), (-3, -8), (-1, -7), (2, 10), (-4, -8), (0, 2)$$

List the coordinates of the points in your graph that lie above the line $y = 3x + 3$. Verify that they are solutions of the inequality $y > 3x + 3$.

List the coordinates of the points in your graph that lie below the line $y = 3x + 3$. Verify that they are solutions of the inequality $y < 3x + 3$.

Without plotting them, tell which of the following points lie above the line $y = 3x + 3$.

$$(50, 148), (-20, -65), (20, 64), (-10, -26), (-10, -28)$$

(b) Graph the line $x = 2y - 3$. On the same axes plot the following points.

$$(0, 2), (2, 2), (\tfrac{5}{2}, 3), (\tfrac{7}{2}, 3), (-2, 1), (-\tfrac{1}{2}, 1), (6, 4), (4, 4)$$

List the coordinates of the points in your graph that lie to the right of the line $x = 2y - 3$. Verify that they are solutions of the inequality $x > 2y - 3$.

List the coordinates of the points in your graph that lie to the left of the line $x = 2y - 3$. Verify that they are solutions of the inequality $x < 2y - 3$.

Without plotting them, tell which of the following points lie to the right of the line $x = 2y - 3$.

$$(60, 70), (-110, 40), (-10, -35), (-17, -21), (18, 12)$$

2. (a) Which of the following ordered pairs make a true statement of the inequality $y > -4(x - 1)$?

$$(-2, 3), (1, 2), (0, 5), (50, -200), (-50, 205), (0, 3)$$

Without plotting them, tell which of these points lie above the line $y = -4(x - 1)$.

(b) Tell which of the following ordered pairs make the statement $x > 4y + 6$ true.

$$(7, 0), (5, 0), (9, 1), (11, 1), (-7, -3), (-5, -3), (2, -\tfrac{3}{2})$$

Which of these points lie to the right of the line $x = 4y + 6$?

3. (a) Test the following ordered pairs in the statements $x = -2y + 1$, $x > -2y + 1$, and $x < -2y + 1$.

$$(3, -1), (-3, \tfrac{1}{2}), (2, -1), (-14, 7), (1, \tfrac{1}{2}), (4, -1), (0, \tfrac{1}{2}), (12, -5)$$

Determine which points are on, which are to the right of, and which are to the left of the line $x = -2y + 1$.

(b) Test each of the following ordered pairs in the statements $y = 4 - 4x$, $y > 4 - 4x$, and $y < 4 - 4x$.

$$(1, 2), (0, 4), (-2, 13), (2, -5), (0, 5), (-2, 11), (1, 0), (2, -3)$$

Determine which points corresponding to the ordered pairs are on, which are above, and which are below the line $y = 4 - 4x$.

4. Describe algebraically the half-planes above and below the line which is the graph of the equation

(a) $y = 4x - 3$. (b) $2x + 3y - 18 = 0$.

5. Describe algebraically the half-planes to the right and to the left of the line which is the graph of the equation

(a) $x = \frac{3}{4}y - 2$. (b) $3x - 2y = 12$.

Sketch and describe the graph of each of the following sets.

6. (a) $\{(x, y) \mid x \geqq \frac{3}{4}y - 2\}$
(b) $\{(x, y) \mid x < \frac{2}{3}y + 4\}$

7. (a) $\{(x, y) \mid y > 4x - 3\}$
(b) $\{(x, y) \mid y \leqq -\frac{2}{3}x + 6\}$

8. (a) Identify each of the following points as to the right of, on, or to the left of the line $x = \frac{1}{2}y + 3$.

$$(2, -8), (5, 4), (2, -6), (6, 4), (-3, -6),$$
$$(-1, -8), (0, -6), (1, -8), (4, 4)$$

(b) Identify each of the following points as above, on or below the line $y = \frac{1}{3}x - 2$.

$$(3, -1), (3, 0), (3, -2), (6, -1), (6, 1),$$
$$(0, -2), (-3, -4), (-3, -2), (-3, -3)$$

Graph each of the following sets.

9. (a) $A = \{(x, y) \mid y \geqq \frac{1}{2}x + 3\}$
(b) $B = \{(x, y) \mid y \leqq \frac{1}{5}x - 2\}$

10. (a) $C = \{(x, y) \mid y > 2x - 5\}$
(b) $D = \{(x, y) \mid y < 2x - 5\}$

11. (a) $E = \{(x, y) \mid 2y \leqq 5x\}$
(b) $F = \{(x, y) \mid 3y \geqq 4x\}$

12. (a) $G = \{(x, y) \mid x < 3y + 2\}$
(b) $H = \{(x, y) \mid x \geqq 3y + 2\}$

13. (a) $I = \{(x, y) \mid x \leqq -2y + 5\}$
(b) $J = \{(x, y) \mid x \geqq -2y + 5\}$

---

14. Refer to Exercises 10–13 and describe the following sets.

(a) $C \cup D$ (b) $C \cap D$ (c) $E \cap F$ (d) $I \cap J$
(e) $G \cup J$ (f) $I \cup J$ (g) $G \cap H$

Describe algebraically each of the following sets of points in the plane.

15. Points below the x-axis

16. Points to the right of the y-axis

17. Points on the x-axis

18. Points on the y-axis

19. Points on or above the x-axis

20. Points in the fourth quadrant

21. Points in the second or in the fourth quadrant

22. Points with positive ordinates and zero abscissas

23. Points with negative abscissas and zero ordinates

24. Points to the left of the y-axis or on the y-axis

25. Points with abscissas greater than ordinates

26. Describe algebraically each of the following sets of points. Sketch the graph of the set.

(a) Points above the graph of the linear equation $2x + 5y - 10 = 0$
(b) Points below the graph of the linear equation $3x - 7y + 21 = 0$
(c) Points on or to the right of the graph of $7x + 2y = 14$
(d) Points on or to the left of the graph of $4x - 6y = 12$

Preparation for Section 7–7

1. Solve each of the following equations for y in terms of x.
(a) $3x - 4y + 12 = 0$ (b) $-2x + 3y - 6 = 0$

2. Solve each of the equations in Exercise 1 for x in terms of y.

3. Graph the equations $3x - 4y + 12 = 0$ and $-2x + 3y - 6 = 0$ on separate sets of coordinate axes.

7–7 FAMILIES OF LINES

Consider the three linear equations listed below.

$$y = 2x - 3 \qquad (1)$$

x	0	?
y	?	0

$$3x + 2y + 6 = 0 \qquad (2)$$

x	0	?
y	?	0

$$3x - 5y - 15 = 0 \qquad (3)$$

x	0	?
y	?	0

Complete the tables above, and graph each equation on the same set of axes. What property do these graphs have in common? Check the graphs you drew to see if they agree with Fig. 7–12.

If you solve (1), (2), and (3) for y, you obtain the following equations

$$y = 2x - 3 \qquad (4)$$
$$y = -\tfrac{3}{2}x - 3 \qquad (5)$$
$$y = \tfrac{3}{5}x - 3 \qquad (6)$$

Thus (4) is equivalent to (1), and, therefore, has the same graph; (5) is equivalent to (2); (6) is equivalent to (3).

Write an equation of your own whose graph has the same property as the three lines in Fig. 7–12.

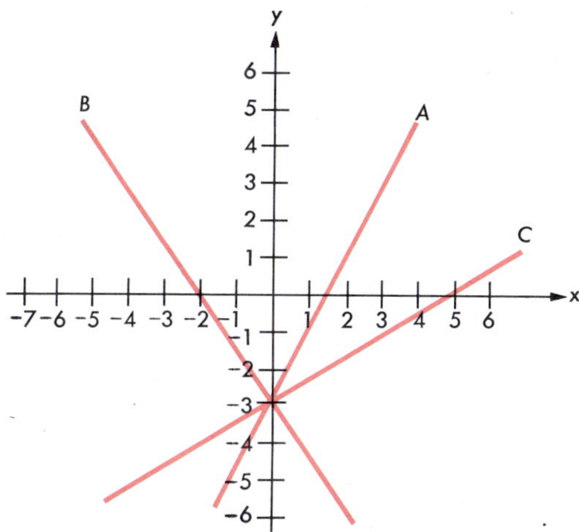

FIGURE 7-12

Each of the equations above has the ordered pair $(0, -3)$ in its solution set. The number -3 is called the *y-intercept* of each graph since that is the point at which the graph *intercepts* the y-axis. Can you define the *x-intercept?* Line A in Fig. 7-12 has x-intercept $\frac{3}{2}$, and line B has x-intercept -2. What is the x-intercept of line C?

The lines of Fig. 7-12 are contained in the set or *family of lines* which includes all lines having the same y-intercept, -3. Other lines which have y-intercept -3 are given below.

$$y = -3,$$
$$y = 17x - 3,$$
$$y = -4x - 3.$$

Exercises

1. (a) What is the x-coordinate of every point on the y-axis?
(b) What is the y-coordinate of every point on the x-axis?

Let $x = 0$. Find the y-intercept of each of the following graphs.

2. (a) $2x - 3y + 8 = 0$ (b) $5x - 3y + 8 = 0$

3. (a) $3x + 2y + 7 = 0$ (b) $9x + 2y + 7 = 0$

Let $y = 0$. Find the x-intercept of each of the following graphs.

4. (a) $4x - 5y - 12 = 0$ (b) $4x + 90y - 12 = 0$

5. (a) $7x + 8y + 35 = 0$ (b) $7x - 20y + 35 = 0$

In each exercise below, graph the three equations on the same coordinate system. Describe the common property of the graphs, and write an equation of a fourth member of each family of lines.

6. (a) $y = x$
$x + 2y = 0$
$3x - y = 0$

(b) $y + x = 0$
$x - 2y = 0$
$3x + y = 0$

7. (a) $x + y = 1$
$x + 2y = 2$
$y - 1 = 0$

(b) $x + y = 2$
$x + 2y = 4$
$x - 3y + 6 = 0$

8. (a) $x = 0$
$3x + 5 = 0$
$x = 1$

(b) $y = 0$
$4y - 5 = 0$
$2y + 3 = 0$

9. (a) $x = 2$
$2x + y - 4 = 0$
$y = 2 - x$

(b) $3x + y + 3 = 0$
$x + 1 = 0$
$x = 2y - 1$

10. (a) $x = 1$
$x + y = 2$
$x = y$

(b) $y - x = 3$
$x + y + 1 = 0$
$y = 1$

11. (a) $x + y = 0$
$x + y = 1$
$x + y + 1 = 0$

(b) $x + 2y = 0$
$x = 2 - 2y$
$4y = 3 - 2x$

---

12. Graph the equations on the same coordinate system:

$$y = 2x, \quad y = -2x, \quad y = \tfrac{1}{2}x, \quad \text{and} \quad y = -(\tfrac{1}{2}x).$$

13. Describe the common characteristic of the family of lines having an equation of the form $y =$ (a number) $\cdot\, x$.

14. If m denotes a positive real number, through what quadrants does the line $y = mx$ pass?

15. If m denotes a negative real number, through what quadrants does the line $y = mx$ pass?

---

16. Graph each of the following lines on the same coordinate system.
(a) $y = 2x$
(b) $2x - y - 2 = 0$
(c) $2x + 4 = y$

17. (a) Solve each of the equations in Exercise 16 for y in terms of x.
(b) What do the three equations now have in common?

(c) What do the three graphs have in common?
(d) Write your own equation of a fourth member of this family of lines, and check it by graphing.

7-8 SLOPE OF A LINE

The linear equation

$$y = 2x + 1$$

is graphed in Fig. 7-13. If you start at the point (0, 1) on this line, go 1 unit to the right and then 2 units up, you arrive at the point (1, 3) on the line. Starting from (1, 3), again go 1 unit to the right and then 2 units up. The point (2, 5) at which you arrive is also on the line. If you start with *any* point on the line, go 1 unit to the right and then 2 units up, you will *always* arrive at another point on the line.

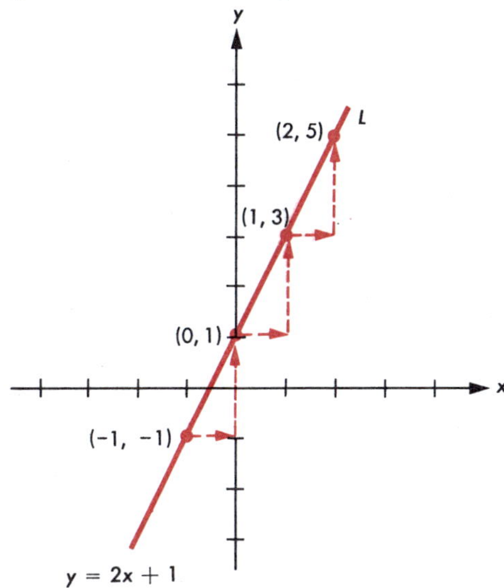

FIGURE 7-13

In the directions given above for getting from one point to another on the line,

right 1 unit, up 2 units,

the number 1 is called the *run* and 2 the *rise* of the line. The ratio of rise to run is called the *slope* of the line,

$$\text{slope} = \frac{\text{rise}}{\text{run}}.$$

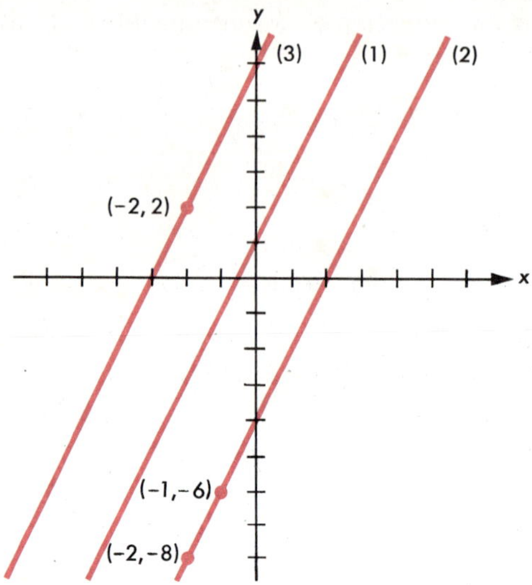

FIGURE 7–14

In the example above,

$$\text{slope} = \tfrac{2}{1}, \quad \text{or } 2.$$

The linear equation

$$y = 2x - 4$$

is graphed in Fig. 7–14. Starting from the point (2, 0) on this line, you arrive at another point (3, 2) on the line if you use the same direction as above,

right 1 unit, up 2 units.

Use these directions starting from the point $(-3, -10)$; from the point $(0, -4)$. Did you arrive at other points on the line? For this line, again

$$\text{slope} = \tfrac{2}{1}, \quad \text{or } 2.$$

The graphs above have the same slope, 2. What does this mean geometrically? It appears that the lines are parallel. What do the two equations have in common? Is 2 the coefficient of x in each equation?

The linear equation

$$y = -\tfrac{1}{3}x + 6$$

is graphed in Fig. 7–15. Two points on this line are (0, 6) and (6, 4). In going from (0, 6) to (6, 4), the run is 6 and the rise -2. (Going up -2 is the same as going down 2.) For this line,

$$\text{slope} = \frac{-2}{6}, \quad \text{or } -\tfrac{1}{3}.$$

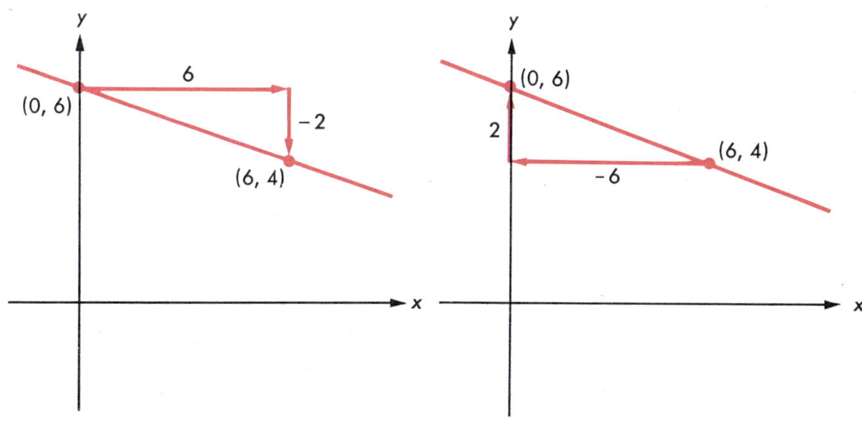

FIGURE 7–15

FIGURE 7–16

You could go from the point (6, 4) to the point (0, 6) with a run of -6 and a rise of 2 (Fig. 7–16). (Going to the right -6 units is the same as going to the left 6 units.) Again,

$$\text{slope} = \frac{2}{-6}, \quad \text{or } -\tfrac{1}{3}.$$

Once again, the slope of this line is the coefficient of x in its equation.

The observations above can be summarized as follows. If a linear equation is solved for y in terms of x, it will have the form

$$y = mx + k$$

for some numbers m and k. Then the number m is the *slope* of the graph of this equation. The set of all lines with the same slope m is a *family of parallel lines.* Incidentally, the number k is the *y-intercept* of the line.

A linear equation such as

$$x = 2$$

cannot be solved for y in terms of x. In going from the point (2, 0) to

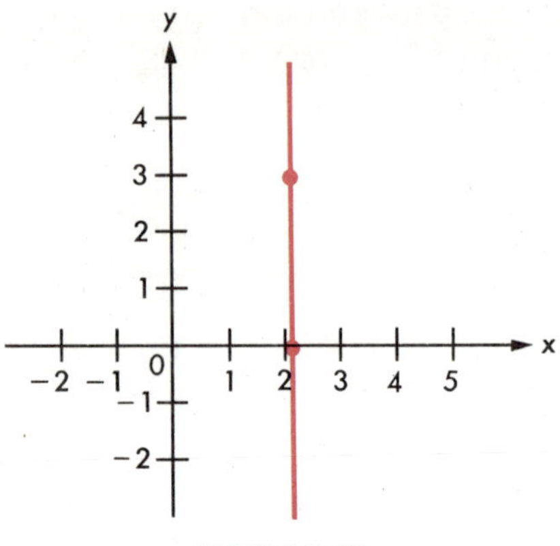

FIGURE 7-17

the point (2, 3) on this line (Fig. 7–17) the run is 0 units and the rise is 3 units. By definition, slope $= \frac{3}{0}$. However, division by zero is not defined. Therefore, this line has no slope. Generally, the lines without slope are those parallel to the y-axis.

Exercises

Complete the following instructions so that they will lead from one point to another point on the line $y = 2x + 1$. (See Fig. 7–13.)

1. (a) Right 3 units, then __?__ __?__ units
(b) Left $\frac{1}{2}$ unit, then __?__ __?__ units

2. (a) __?__ __?__ units, then down 3 units
(b) __?__ __?__ units, then up 10 units

A line is drawn through each of the following pairs of points. State the rise (vertical distance from first to second point), the run (horizontal distance from first to second point), and the slope.

3. (a) (1, 2) and (5, 8) (b) (1, 2) and $(-3, -4)$

4. (a) $(-2, -3)$ and (1, 4) (b) $(-2, -3)$ and $(1, -5)$

5. (a) (3, 4) and $(6, -5)$ (b) (2, 1) and (5, 7)

6. (a) $(-1, 0)$ and $(-4, 3)$ (b) $(-1, 0)$ and $(-2, -4)$

Find two points on each of the following lines, and compute the slope of the lines from the rise and run between these points. Sketch the graphs of these equations on the same set of axes.

7. (a) $y = 4x + 3$ (b) $y = -4x + 3$

8. (a) $y = -x$ (b) $y = x$

9. How does the slope of each line in Exercises 7 and 8 compare with the coefficient of x in the given equation?

10. How does the graph of a line with positive slope differ in appearance from the graph of a line with negative slope?

Find the slope of each line.

11. (a) $9x - 3y + 5 = 0$ (b) $2x - 7y - 15 = 0$

12. (a) $3x + 6y = 20$ (b) $4y + 12x = 9$

13. (a) $x = 3$ (b) $x + 15 = 0$

14. (a) $y = 4$ (b) $y - 15 = 0$

Write each of the following equations in the form $y = mx + k$, and find the slope and y-intercept of each. Use this information to draw their graphs.

15. (a) $2x + y = 5$ (b) $3x + y = -4$

16. (a) $2x - 4y + 8 = 0$ (b) $3x - 9y + 6 = 0$

17. (a) $x - 3y + 9 = 0$ (b) $x - 4y + 2 = 0$

Tell which of the following pairs of equations represent parallel lines by finding the slopes of each.

18. (a) $\begin{cases} 2x - 3y + 5 = 0 \\ 2x - 3y - 6 = 0 \end{cases}$ (b) $\begin{cases} 3x - 2y + 1 = 0 \\ 2x - 3y + 1 = 0 \end{cases}$

19. (a) $\begin{cases} 6x + 10y = 7 \\ 3x = 2 - 5y \end{cases}$ (b) $\begin{cases} 4x = 3y - 1 \\ 8x - 7 = 6y \end{cases}$

20. (a) $\begin{cases} x - 5y = 6 \\ 2x - 10y = 4 \end{cases}$ (b) $\begin{cases} 2x + 5y = 3 \\ 10x + 25y = 20 \end{cases}$

---

21. (a) What is the rise, the run, and the slope of the line through the points (1, 2) and (5, 2)?
(b) Draw the line through the points $(-4, -1)$ and $(3, -1)$. Give the rise, run, and slope of the line.
(c) Describe the family of lines with slope zero.

22. (a) Draw the line through (2, 1) and (2, 5). Give the rise and run. Can you compute the slope?
(b) Draw the line through $(-3, -4)$ and $(-3, 2)$. Give the rise and run. Can you compute the slope?
(c) Describe the family of lines for which there is no slope, and explain why no number can be assigned as the slope.

---

23. (a) Write an equation of the family of lines with slope $\frac{3}{5}$.
(b) Write an equation of the family of lines with y-intercept 2.
(c) Write an equation of the line with slope $\frac{3}{5}$ and y-intercept 2.

The following set of equations is referred to in Exercises 24–27.

(a) $3x - 5y - 6 = 0$ (b) $2y + 8x = 3$
(c) $2x + y + 4 = 0$ (d) $6x - 10y + 30 = 0$
(e) $6x + 10y = 15$ (f) $7x + 9y = 3$
(g) $9x - 15y = 45$ (h) $2x - 2y + 3 = 0$
(i) $2x - 5y + 4 = 0$ (j) $3y + 2x = 0$
(k) $5x - 6y = 27$ (l) $y = 2x + 4$

24. From the above set, select three equations representing lines with the same slope. Graph them on one coordinate system.

25. From the above set, select three equations representing lines with the same y-intercept. Graph them on one coordinate system.

26. From the above set, select three equations representing lines with the same x-intercept. Graph them on one coordinate system.

27. Select three equations having the number pair $(3, -2)$ in their solution sets. Graph them on the same coordinate system.

7–9 EQUATION OF A LINE

You know that the graph of a linear equation in x and y is a straight line in a cartesian coordinate system. Now, proceed in the opposite direction; that is, start with a straight line and find a linear equation whose graph is that line. In preceding sections, you found graphs of equations. Now you will find *an equation of the graph.*

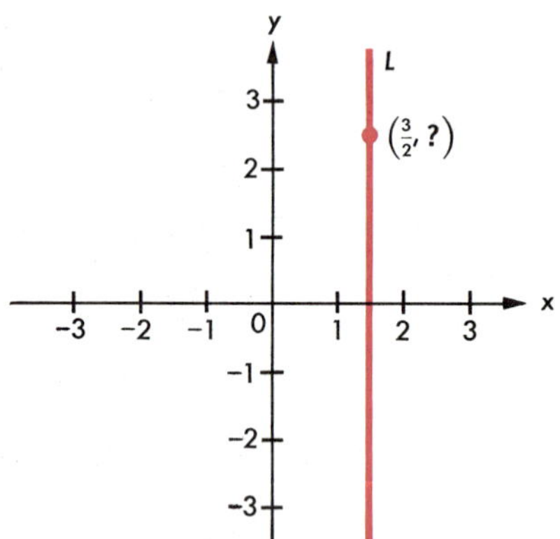

FIGURE 7–18

Problem 1. In Fig. 7–18, find an equation of the line L which has x-intercept $\frac{3}{2}$ and is parallel to the y-axis.

Solution. Since each point on this line has $\frac{3}{2}$ as the x-coordinate for every value of y, the line is the graph of the set

$$S = \{(x, y) \mid x = \tfrac{3}{2}\}.$$

Therefore, an equation of this line is

$$x = \tfrac{3}{2}.$$

Problem 2. Find an equation of the line K having x-intercept -2 and y-intercept 1. (See Fig. 7–19.)

Solution. A run of 2 and a rise of 1 will carry you from one point to another point on the line. Thus, line K has slope $\frac{1}{2}$. From the preceding section, a line with slope m and y-intercept k has equation

$$y = mx + k.$$

Since the line K has slope $\frac{1}{2}$ and y-intercept 1, an equation for K is

$$y = \tfrac{1}{2}x + 1.$$

You may easily plot points on the graph of the above equation and observe that it actually is the line K.

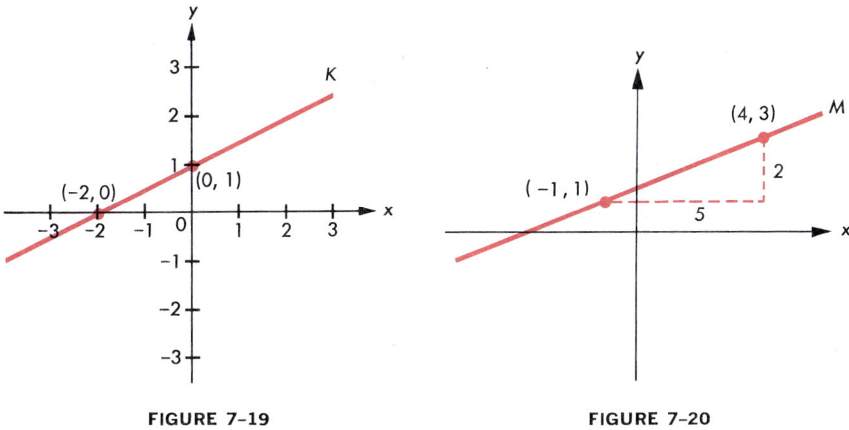

FIGURE 7–19

FIGURE 7–20

Problem 3. The line M in Fig. 7–20 has point $(-1, 1)$ and $(4, 3)$ on it. Find an equation for M.

Solution. A run of 5 and a rise of 2 takes you from point $(-1, 1)$ to $(4, 3)$. Therefore, M has slope $\frac{2}{5}$. An equation of M has the form

$$y = \tfrac{2}{5}x + k$$

where k is the y-intercept of M.

The following is a method for finding k. Since the number pair $(-1, 1)$ is in the solution set of the equation

$$y = \tfrac{2}{5}x + k,$$

the equation that results when $x = -1$ and $y = 1$ must be true. Thus,

$$1 = \tfrac{2}{5}(-1) + k.$$

This equation is solved as follows:

$$\begin{aligned} 1 &= -\tfrac{2}{5} + k, \\ 1 + \tfrac{2}{5} &= k, \\ \tfrac{7}{5} &= k. \end{aligned}$$

Thus, k is $\frac{7}{5}$ and an equation of line M is

$$y = \tfrac{2}{5}x + \tfrac{7}{5}.$$

Exercises

1. (a) (i) Draw a line which has a slope of $\frac{2}{3}$ and passes through the point $(1, -2)$. Extend it across the y-axis.
(ii) Find a value for k by using $x = 1$ and $y = -2$ in the equation $y = \frac{2}{3}x + k$.
(iii) Is the computed value for k the same as the y-intercept on the graph of the line you draw in part (i)?
(b) A line has slope $\frac{4}{5}$ and passes through the point $(-1, -2)$. Find an equation of the line and sketch its graph.

For Exercises 2–12, find an equation of a line which fulfills the requirements given.

2. (a) Slope $-\frac{3}{4}$ and y-intercept 2
(b) Slope 10 and y-intercept -60

3. (a) Slope 2 and passes through point $(-1, 3)$
(b) Slope $\frac{5}{6}$ and passes through point $(2, 1)$

4. (a) Slope -10, y-intercept $\frac{1}{2}$
(b) Slope -1, y-intercept 0

5. (a) Slope 0 and passes through point $(4, 3)$
(b) Slope 0 and passes through point $(-2, 1)$

6. (a) Slope 0, y-intercept 3
(b) Slope 0, y-intercept -2

7. (a) No slope and passes through point $(2, 3)$
(b) No slope and passes through point $(-3, -2)$

8. (a) Passes through points $(-1, -2$ and $(-3, -5)$
 (b) Passes through points $(2, -1)$ and $(-3, 4)$
9. (a) Passes through points $(3, 3)$ and $(7, 7)$
 (b) Passes through points $(-3, 3)$ and $(-7, 7)$
10. (a) Slope $-\frac{3}{4}$ and passes through point $(4, -6)$
 (b) Slope $-\frac{4}{3}$ and passes through point $(-6, 4)$
11. (a) Passes through points $(-2, -5)$ and $(3, 6)$
 (b) Passes through points $(3, -1)$ and $(-2, 6)$
12. (a) Parallel to the line $x + y = 2$ and passes through $(2, 3)$
 (b) Parallel to the line $x - y = 1$ and passes through $(-1, 2)$

Without graphing, find the slope and y-intercept of the graph of each of the following linear equations.

13. (a) $7(x - y) = y - x$
 (b) $15(x - y) = 10 - 15y$
14. (a) $2(x + y) + 3(y - x) = 4$
 (b) $7 + 4(x - y) = 0$

Find an equation for each of the lines shown in Exercises 15–20.

15. (a)

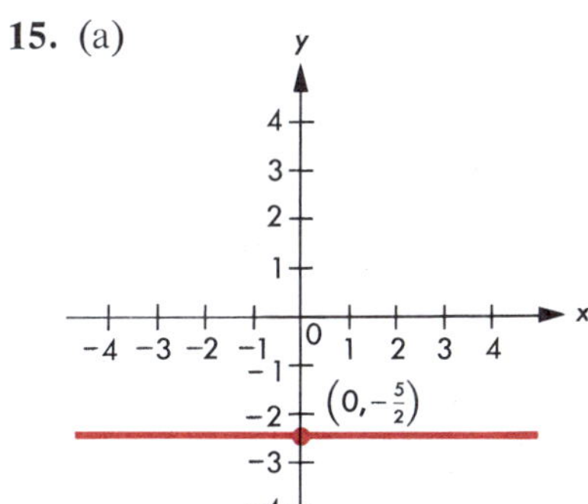

(b)

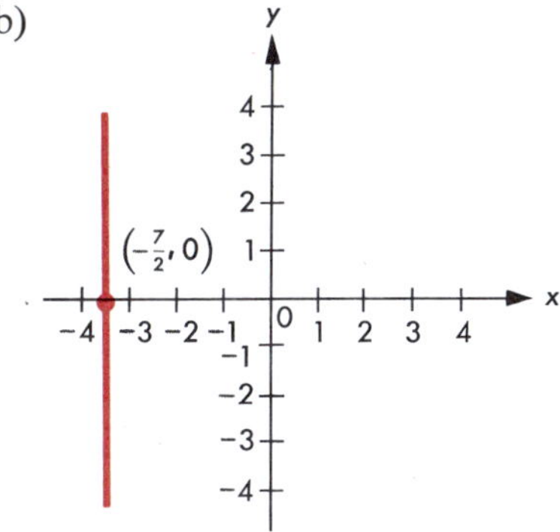

16. (a)

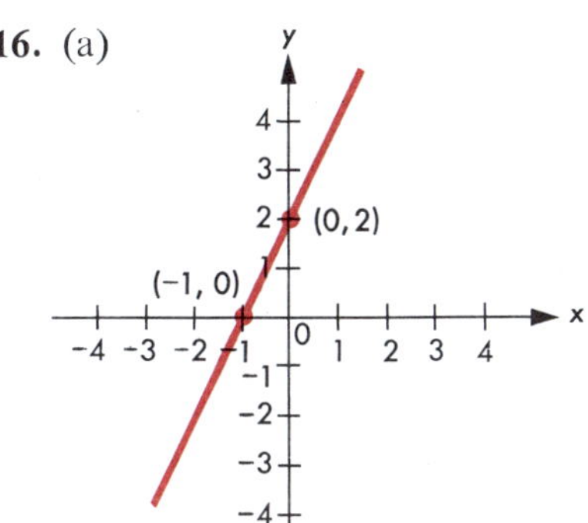

(b)

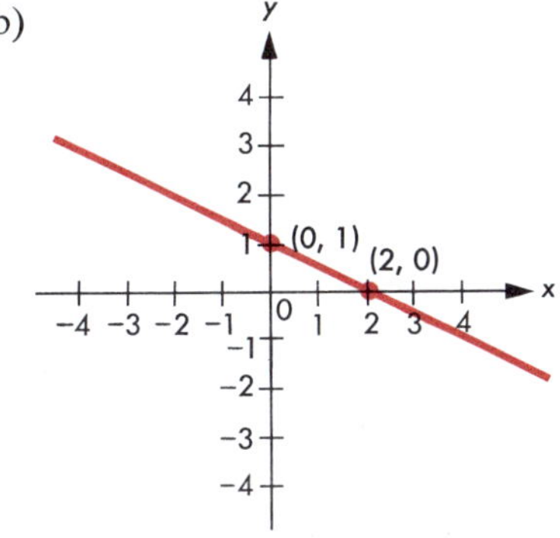

17. (a)

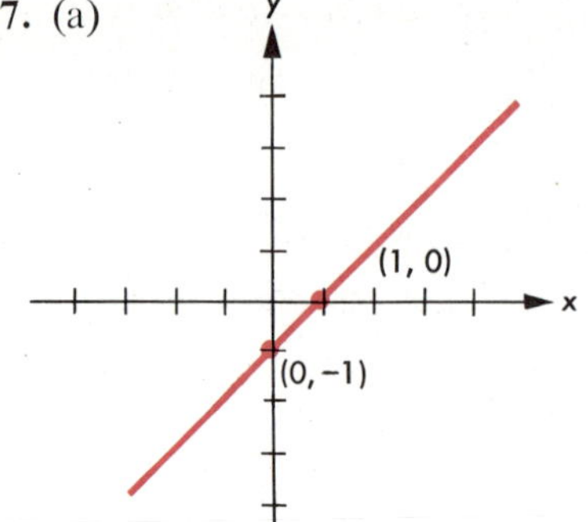

(b)

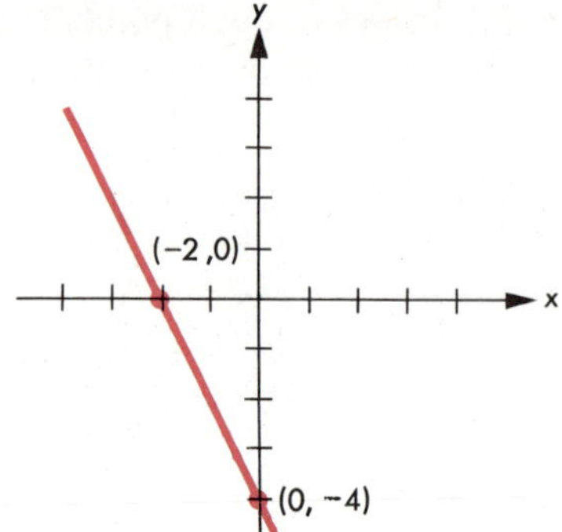

18. (a)

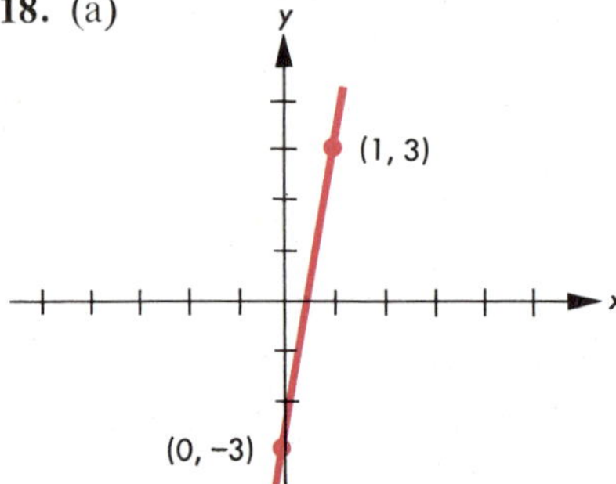

(b)

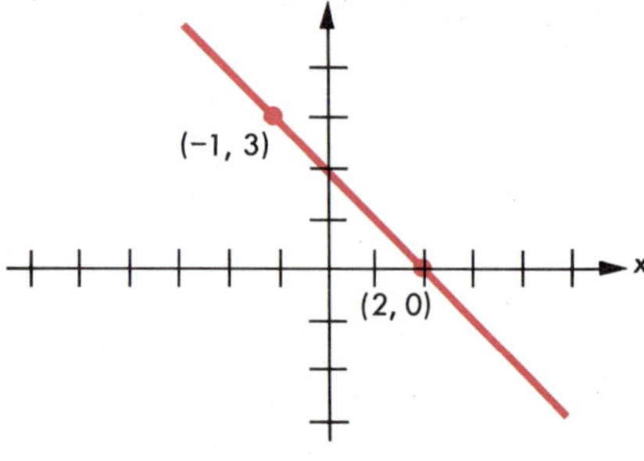

19. (a)

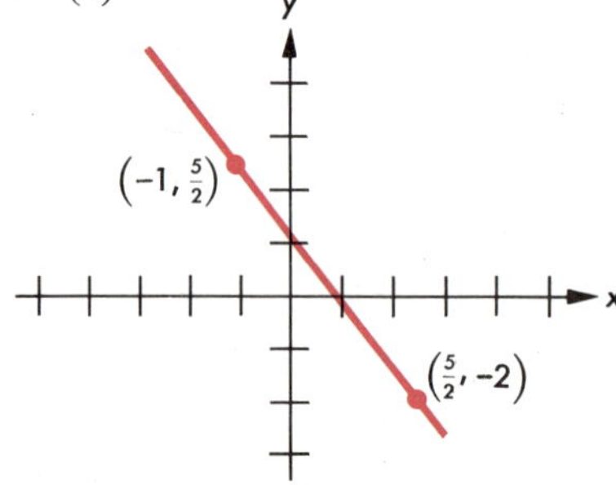

(b)

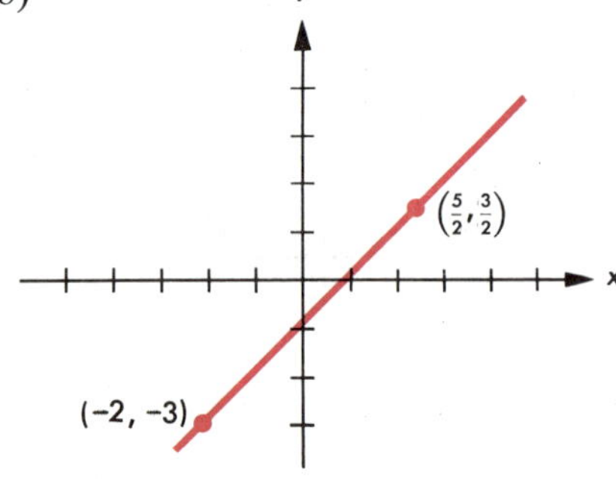

20. (a)

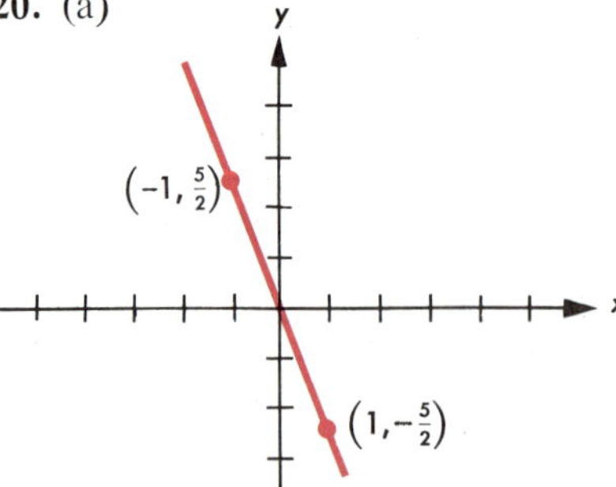

(b)

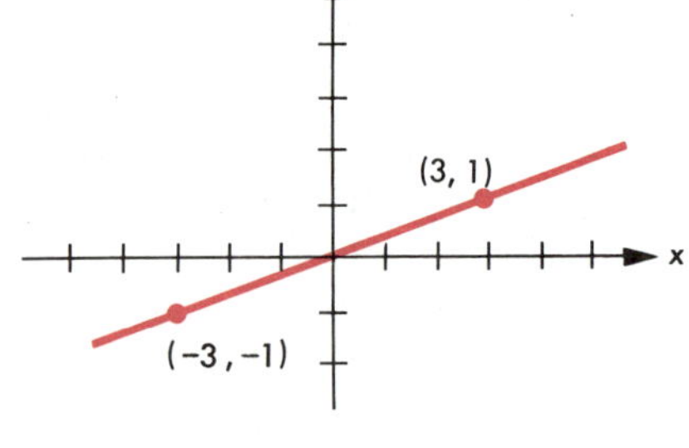

Find an inequality whose graph is the region indicated in each of the Exercises 21–26. You may assume that each graph is incomplete as it is presented here and represents a region infinite in extent.

21.

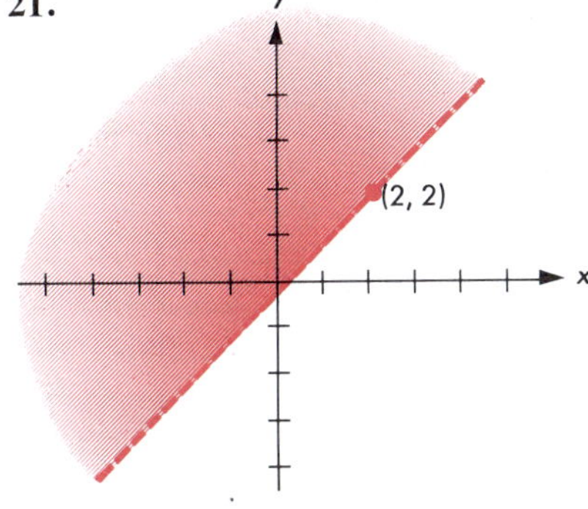

22.

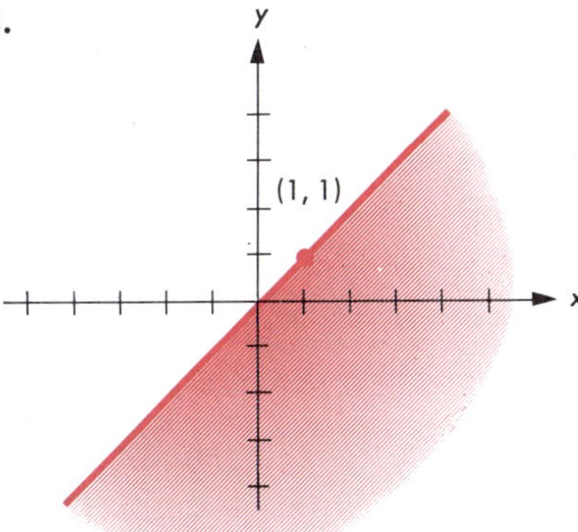

23.

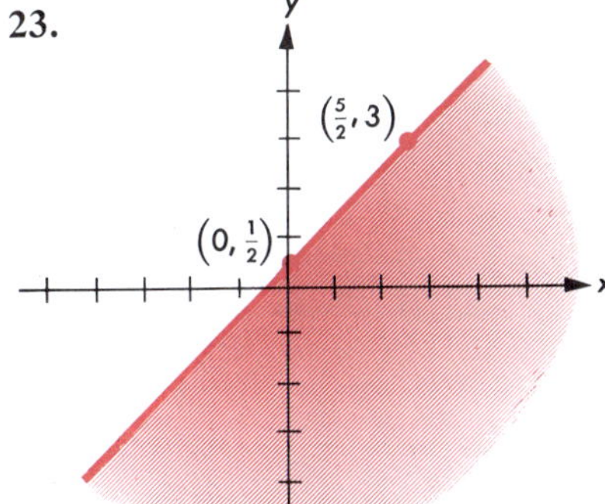

24.

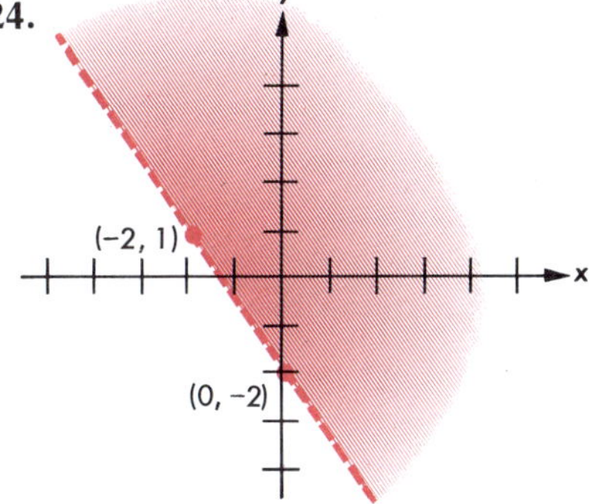

25.

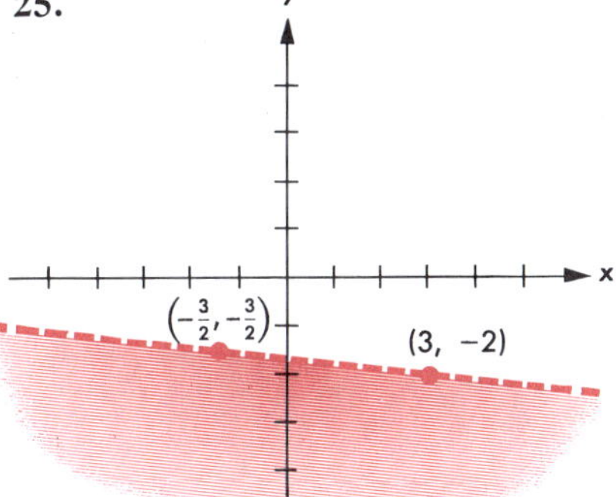

26.

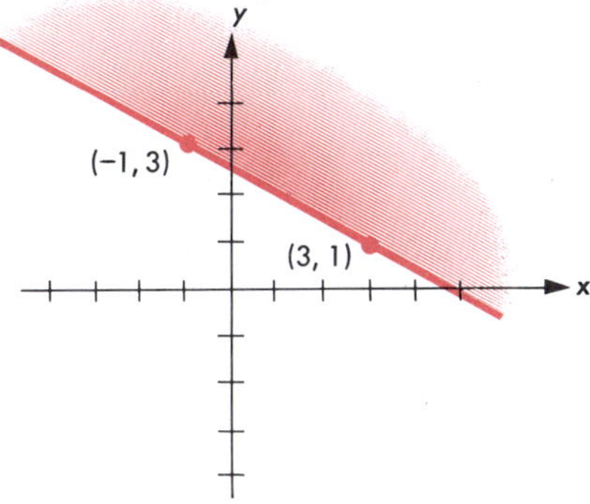

*7–10 PROOF IN ALGEBRA III

The main result of Section 7–8 may be stated and proved as follows.

Theorem

The graph of the linear equation

$$y = mx + k$$

has slope m and y-intercept k.

Proof. Let $P_1(a_1, b_1)$ and $P_2(a_2, b_2)$ be any two points on this line (Fig. 7–21). The run in going from P_1 to P_2 is simply the x-coordinate of P_2 minus the x-coordinate of P_1. Similarly, the rise in going from P_1 to P_2 is the y-coordinate of P_2 minus the y-coordinate of P_1. Therefore, the slope of the line is given by

$$\text{slope} = \frac{b_2 - b_1}{a_2 - a_1}.$$

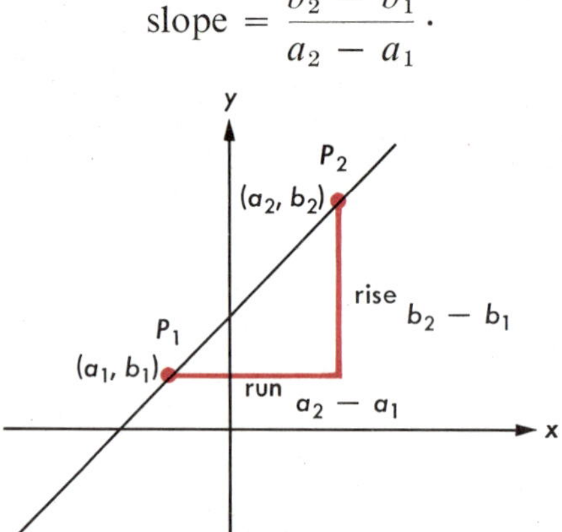

FIGURE 7–21

You next must use the fact that the points P_1 and P_2 are *on* the line; that is, that their coordinates are *solutions* of the equation $y = mx + k$. Thus, (a_1, b_1) is a solution,

$$b_1 = ma_1 + k,$$

and (a_2, b_2) is a solution,

$$b_2 = ma_2 + k.$$

Subtracting,

$$\begin{aligned} b_2 - b_1 &= (ma_2 + k) - (ma_1 + k) \\ b_2 - b_1 &= ma_2 - ma_1 \\ b_2 - b_1 &= m(a_2 - a_1) \\ \frac{b_2 - b_1}{a_2 - a_1} &= m. \end{aligned}$$

This proves that m is the slope of the graph of $y = mx + k$.

If $x = 0$, then $y = k$. Thus, $(0, k)$ is a solution of the equation $y = mx + k$. This proves that k is the y-intercept of the graph.

Exercises

1. Show that the slope formula,

$$\text{slope} = \frac{b_2 - b_1}{a_2 - a_1},$$

does not depend on which point is considered the first one. That is, show that

$$\frac{b_2 - b_1}{a_2 - a_1} = \frac{b_1 - b_2}{a_1 - a_2}.$$

2. Prove that a line with an equation $y = mx + k$ has slope m and y-intercept k by using the fact that the points $(0, k)$ and $(1, m + k)$ are on the line.

3. Prove that the lines whose equations are $x + 2y = 8$ and $x + 2y = 4$ are parallel. Find an equation of the line parallel to each and midway between them.

4. Prove that $x + 4y = 3$ and $2x + 8y = 6$ have the same graphs.

5. Prove that $ax + by = c$ and $kax + kby = kc$ have the same graphs, if $k \neq 0$.

6. Show that if a and b are nonzero numbers denoting the x- and y-intercepts of a straight line, then the equation for the line is equivalent to the equation

$$\frac{x}{a} + \frac{y}{b} = 1.$$

7. (a) Use the equation in Exercise 6 to find an equation for the line with x-intercept 4 and y-intercept -3.

(b) Use the equation in Exercise 6 to find an equation for the line through the points $(-5, 0)$ and $(0, 2)$.

8. Prove that the equation of a line with slope m and passing through (a, b) is given by

$$y - b = m(x - a).$$

9. Show that, for $a_2 \neq a_1$,

$$y - b_1 = \left(\frac{b_2 - b_1}{a_2 - a_1}\right)(x - a_1)$$

is an equation of the line joining the points (a_1, b_1) and (a_2, b_2).

10. (a) Use the equation in Exercise 8 to find an equation for the line with slope $\frac{1}{2}$ and passing through $(-3, 4)$.
(b) Use the equation in Exercise 9 to find an equation for the line joining the points $(-2, -4)$, and $(5, 2)$.

EXTRA!

The integers, rational numbers, and real numbers are examples of number systems having an infinite number of elements. There are finite number systems; in fact, for every positive integer n there is a number system, called the *integers modulo n,* having n elements. The notation $\mathbb{Z}_n$ is often used for the set of integers modulo n.

Consider, for example, the set $\mathbb{Z}_{10}$ of integers modulo 10. The elements of $\mathbb{Z}_{10}$ are denoted by

$$0, 1, 2, 3, 4, 5, 6, 7, 8, 9.$$

Addition and multiplication are defined in $\mathbb{Z}_{10}$ in a natural way. Thus,

$$2 + 3 = 5, \quad 4 + 2 = 6,$$
$$2 \cdot 3 = 6, \quad 4 \cdot 2 = 8.$$

Since $\mathbb{Z}_{10}$ is not a system dependent on place value, you cannot similarly define $7 + 5 = 12$ and $7 \cdot 5 = 35$, because 12 and 35 are not elements of $\mathbb{Z}_{10}$. What you do is drop the "1" in "12" and the "3" in "35," defining

$$7 + 5 = 2, \quad 7 \cdot 5 = 5.$$

Now $7 + 5$ and $7 \cdot 5$ are elements of $\mathbb{Z}_{10}$. In like manner,

$$8 + 7 = 5, \quad 9 + 5 = 4,$$
$$8 \cdot 7 = 6, \quad 9 \cdot 5 = 5.$$

The commutative, associative, and distributive axioms are valid for the system of integers modulo n, and there exist additive and multiplicative identity elements 0 and 1. Thus, in $\mathbb{Z}_{10}$,

$$0 + 7 = 7, \quad 3 + 0 = 3,$$
$$1 \cdot 7 = 7, \quad 3 \cdot 1 = 3,$$

and so on. Each integer modulo n has a "negative." In $\mathbb{Z}_{10}$, the negative of 2 is 8 because $2 + 8 = 0$; and the negative of 4 is 6 because $4 + 6 = 0$.

An unusual feature of $\mathbb{Z}_{10}$ is that a product of two nonzero numbers might be zero. For example,

$$2 \cdot 5 = 0, \quad 4 \cdot 5 = 0, \quad 5 \cdot 8 = 0.$$

For any positive integer n, let the elements of $\mathbb{Z}_n$ be denoted by

$$0, 1, 2, \ldots, n - 2, n - 1.$$

Addition and multiplication may be defined in $\mathbb{Z}_n$, just as they were in $\mathbb{Z}_{10}$, by expressing each number to the base n and then keeping only the units' digit in sums and products.

For example, $\mathbb{Z}_5$ consists of the elements

$$0, 1, 2, 3, 4.$$

As above,

$$2 + 2 = 4, \quad 1 + 2 = 3,$$
$$2 \cdot 2 = 4, \quad 1 \cdot 2 = 2.$$

To find other sums and products, you make use of the fact that the units' digit of any integer x expressed in the base 5 is simply the remainder on dividing x by 5.

For example, $3 + 4 = 7$ in your usual arithmetic. Therefore,

$$3 + 4 = 2$$

in the system of integers modulo 5, because the remainder on dividing 7 by 5 is 2. As another example,

$$4 \times 4 = 1,$$

because the remainder on dividing 16 by 5 is 1.

You can make complete addition and multiplication tables for any modular system. The multiplication table for the integers modulo 5 is shown on page 322. Since the row for 2 and the column for 3 intersect at 1, you have $2 \times 3 = 1$. As another example, the row for 3 and the column for 4 intersect at 2; therefore, $3 \times 4 = 2$. You construct the addition table.

×	0	1	2	3	4
0	0	0	0	0	0
1	0	1	2	3	4
2	0	2	4	1	3
3	0	3	1	4	2
4	0	4	3	2	1

As another example, $\mathbb{Z}_4$ has the elements

$$0, 1, 2, 3.$$

Since $3 + 2 = 5$, in $\mathbb{Z}_4$, $3 + 2 = 1$, the remainder on dividing 5 by 4. Similarly, $3 \cdot 2 = 2$. The addition table for $\mathbb{Z}_4$ follows.

+	0	1	2	3
0	0	1	2	3
1	1	2	3	0
2	2	3	0	1
3	3	0	1	2

1. Construct the addition table for $\mathbb{Z}_5$.

2. Construct the multiplication table for $\mathbb{Z}_4$.

Construct addition and multiplication tables for each of the following.

3. $\mathbb{Z}_2$ **4.** $\mathbb{Z}_3$ **5.** $\mathbb{Z}_6$ **6.** $\mathbb{Z}_7$

Compute the following.

7. $4 + 9$ in $\mathbb{Z}_{12}$ **8.** $4 \cdot 9$ in $\mathbb{Z}_{12}$

9. $8 + 10$ in $\mathbb{Z}_{11}$ **10.** $8 \cdot 10$ in $\mathbb{Z}_{11}$

11. $7 + 12$ in $\mathbb{Z}_{14}$ **12.** $7 \cdot 12$ in $\mathbb{Z}_{14}$

Solve each of the following linear equations.

13. $2 + 3x = 4$ in $\mathbb{Z}_7$ **14.** $3 + 5x = 7$ in $\mathbb{Z}_{12}$

15. $17 + 11x = 5$ in $\mathbb{Z}_{19}$ **16.** $3x = 0$ in $\mathbb{Z}_{18}$

17. List the elements of $\mathbb{Z}_{10}$ which have multiplicative inverses.

18. List the elements of $\mathbb{Z}_{12}$ which have multiplicative inverses.

19. If the positive integer n is not prime, prove that some nonzero element of $\mathbb{Z}_n$ does not have a multiplicative inverse.

KEY IDEAS AND KEY WORDS

A **linear form in *x* and *y*** is an expression of the type

$$(\text{a number}) \cdot x + (\text{a number}) \cdot y + (\text{a number}).$$

A **linear equation in *x* and *y*** consists of two linear forms in x and y connected by an equals sign.

A **solution of an equation in *x* and *y*** is an ordered pair of numbers (a, b) which will make the equation true if $x = a$ and $y = b$. The set of all solutions of a linear equation in two variables is called the **solution set** of the equation.

The **graph of an equation in two variables** is the graph of its solution set. In a cartesian coordinate system, the graph of a linear equation in two variables consists of those, and only those, points that lie on some straight line unless the solution set is the empty set.

Each line in a plane divides the plane into two **half-planes.**

The **slope** of a line is the ratio of rise to run. That is,

$$\text{slope} = \frac{\text{rise}}{\text{run}}.$$

The set of all lines with the same slope is a family of parallel lines.

If a linear equation is solved for y in terms of x, $y = mx + k$, then m is the slope of the graph of the equation and k is the y-intercept of the line.

CHAPTER REVIEW

1. Which of the following statements are true and which are false?
   (a) The graph of the set $\{(x, y) \mid x = k, k \text{ any real number}\}$ is a line parallel to the x-axis.
   (b) The line $x = k$, k any real number, has slope zero.

(c) The graph of the following set is in the first and fourth quadrants.

$$\{(x, y) \mid x \geqq 2\} \cup \{(x, y) \mid y \leqq 2\}$$

(d) The following inequalities are equivalent.

$$2x - 3y > -8$$
$$3y - 2x > -8$$

(e) A line with slope $-(\frac{2}{3})$ containing the point $(4, -5)$ also contains the point $(7, -3)$.

(f) The intersection of the set of points called Quadrant I with the set of points called x-axis is $\emptyset$.

2. Correct the statements you marked false in Exercise 1.

3. For each equation, simplify and give two equivalent equations: one solved for y in terms of x and one solved for x in terms of y.

(a) $11y - 5x + 10 = 7y - 2x + 30$

(b) $\frac{y}{5} + \frac{2x}{5} + \frac{7}{10} = \frac{x}{5} - \frac{y}{10} - \frac{1}{2}$

4. Find the x- and y-intercepts for the lines in Exercise 3. For each line, check these intercepts in the original and in the simplified equations in order to detect and eliminate errors.

5. Find the slope of the line through the points $(6, 1)$ and $(-2, 3)$.

6. What is the slope of the line $y = \frac{3}{4}x - 1$?

7. Find the slope of the line $5x + 10y = 4$.

8. What do the following lines have in common?

$$y = 3x + 1, \quad 2y - 6x = 7, \quad \text{and } 9x = 3y + 8$$

9. What do these lines have in common?

$$y = -4x - 5, \quad 8x + 3y + 15 = 0, \quad \text{and } 2y - x + 10 = 0$$

10. What do these lines have in common?

$$y = x, \quad y = -3x, \quad \text{and } y = \tfrac{1}{2}x$$

11. Find an equation of the line with slope -3 and y-intercept 2.

12. Find an equation of the line through the point $(-1, 5)$ with slope $\frac{1}{6}$.

13. Find an equation of the line through the point $(2, 3)$ with zero slope.

14. Find an equation of the line through the point $(3, 2)$ with no slope.

15. Find an equation of the line through the points $(5, -2)$ and $(0, 6)$.

In Exercises 16–23, if possible, name three elements of each of the sets. Then graph the sets.

16. $A = \{(x, y) \mid x > 1\}$

17. $B = \{(x, y) \mid x < -3\}$

18. $C = \{(x, y) \mid y > -1\}$

19. $D = \{(x, y) \mid y < 2\}$

20. $A \cup C$

21. $A \cap D$

22. $A \cap B$

23. $B \cap C$

24. Classify the following points as above, below, or on the graph of $y = x - 2$.

$$(2, 0),\quad (2, 2),\quad (3, 1),\quad (4, 5),\quad (-1, -4),\quad (3, 4),\quad (4, 3)$$

25. If

$$\begin{aligned} A &= \{(x, y) \mid x > 0\}, \\ B &= \{(x, y) \mid y > 0\}, \\ C &= \{(x, y) \mid x < 0\}, \\ D &= \{(x, y) \mid y < 0\}, \end{aligned}$$

describe the set of points in each of the Quadrants I, II, III, and IV in terms of A, B, C, and D.

26. Use set notation to describe each of the following half-planes.

(a)

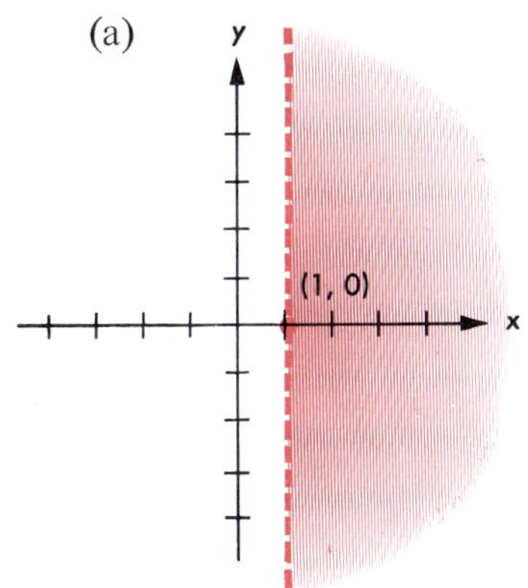

(b)

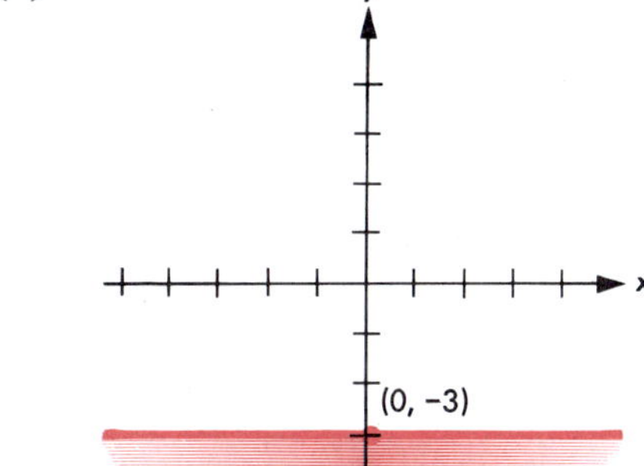

In each of the following exercises, graph the three equations on the same coordinate system. Describe the common property of the graphs, and write an equation of a fourth member of each family of lines.

27.	$y = 3x - 4$	$2y - 6x - 1 = 0$	$3y + 6 = 9x$
28.	$2y - 6 = 4x$	$y = -2x + 3$	$y - 3 = x$
29.	$y + x = 0$	$y - 2x = 0$	$y = 0$
30.	$x + 5 = 0$	$3x - 2 = 0$	$4x + 5 = 3x + 1$

CHAPTER TEST

1. In each of the equations

$$3x + y - 24 = x + 5y,$$
$$2y + 7 - x = -y + 4x + 22,$$

(a) simplify.
(b) solve for x in terms of y.
(c) solve for y in terms of x.
(d) find the x- and y-intercepts.
(e) find the slope of the graph of the equation.

2. Find the slope of the line through the points (2, 3) and (5, 6).

3. Write an equation of the line that has a slope of $\frac{1}{2}$ and a y-intercept of 3.

4. Write an equation of the line that passes through the point (2, 3) and has a slope of 4.

5. Write an equation of the line that passes through the points (2, 4) and (3, 1).

6. Graph the following sets.

(a) $A = \{(x, y) \mid x > 2\}$ (b) $B = \{(x, y) \mid y \leqq 3\}$

7. Classify each of the following points. Is it above, below, or on the graph of $y = 4 - x$?

$$(4, 5), \quad (3, 3), \quad (1, 1), \quad (1, 5), \quad (2, 3)$$

8. Use set notation to describe the following half-plane.

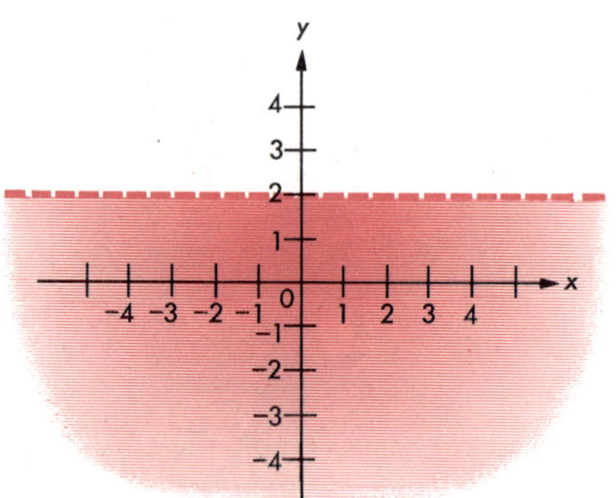

CUMULATIVE REVIEW II

Perform the indicated operations.

1. $(3x^2y^2)^2$ **2.** $(4xy)(x^2y^3)$

3. $(x + 3)^2$ **4.** $(x - 2)(x + 2)$

5. $(3xy^4 + 9x^2y + 12xy^2) \div 3xy$
6. $3(x^2 + 2x - y) - 2(y - x + 3x^2)$
7. $(x^2 - 2x - 35) \div (x + 5)$
8. $(x^4 - x^3 + x^2 - x) \div (x^2 + 1)$

Factor each of the following.

9. $6a^2 - 54$
10. $6c^2 + c - 1$
11. $6 + 9x + 3x^2$
12. $a(x - 2) + 3(x - 2)$
13. $x^2 - 4x + 4$
14. $4x^2 - 13x + 9$
15. $x^2 - 3x - 28$
16. $x^8 - y^8$

Find the solution set of each of the following equations.

17. $y^2 + 5y + 6 = 0$
18. $x^2 + 3x + 2 = 0$
19. $y^2 + 6y + 8 = 0$
20. $x^2 - 4x + 3 = 0$
21. $x^2 + x - 6 = 0$
22. $m^2 - 8m - 20 = 0$
23. In which of the following equations is y a function of x? In which is x a function of y?
    (a) $y = x + 2$
    (b) $y = x^2 + x$
    (c) $y^2 = x^2 + 5$
    (d) $y^2 = 2x + 3$

In Exercises 24–27, graph the equations by finding three solutions.

24. $x - y = 6$
25. $x + y = -2$
26. $2x + y = 4$
27. $x - 2y = 6$
28. Find the slope, y-intercept, and x-intercept of the equations in Exercises 24–27.
29. Find the slope of the line through the points (5, 6) and (−7, 9).
30. Find an equation of the line that passes through the point (2, 3) and has a slope of $\frac{4}{5}$.
31. Graph the following three lines. What do the graphs have in common?

$$2x - 3y + 15 = 0$$
$$2x - 3y - 6 = 0$$
$$3y = 2x - 3$$

32. Graph the following three lines. What do the graphs have in common?

$$2y = x + 3$$
$$4y = 2x + 4$$
$$5 - 2x = -4y$$

33. Graph each of the following sets.
    (a) $\{(x, y) \mid x + y > 1\}$
    (b) $\{x \mid x \geqq 1\} \cap \{y \mid y = 0\}$

34. Which of the following points are to the right, on, and to the left of the line $x - y = 2$?
(a) (5, 3) (b) (−2, 1) (c) (1, 3) (d) (7, 5) (e) (4, 1)

35. Graph the following equations and describe the half-planes to the right and left of the line.
(a) $x = 1$ (b) $x = -1$
(c) $-x = 2$ (d) $x + y = 0$
(e) $2x + 3y = 6$ (f) $5 - x = y$

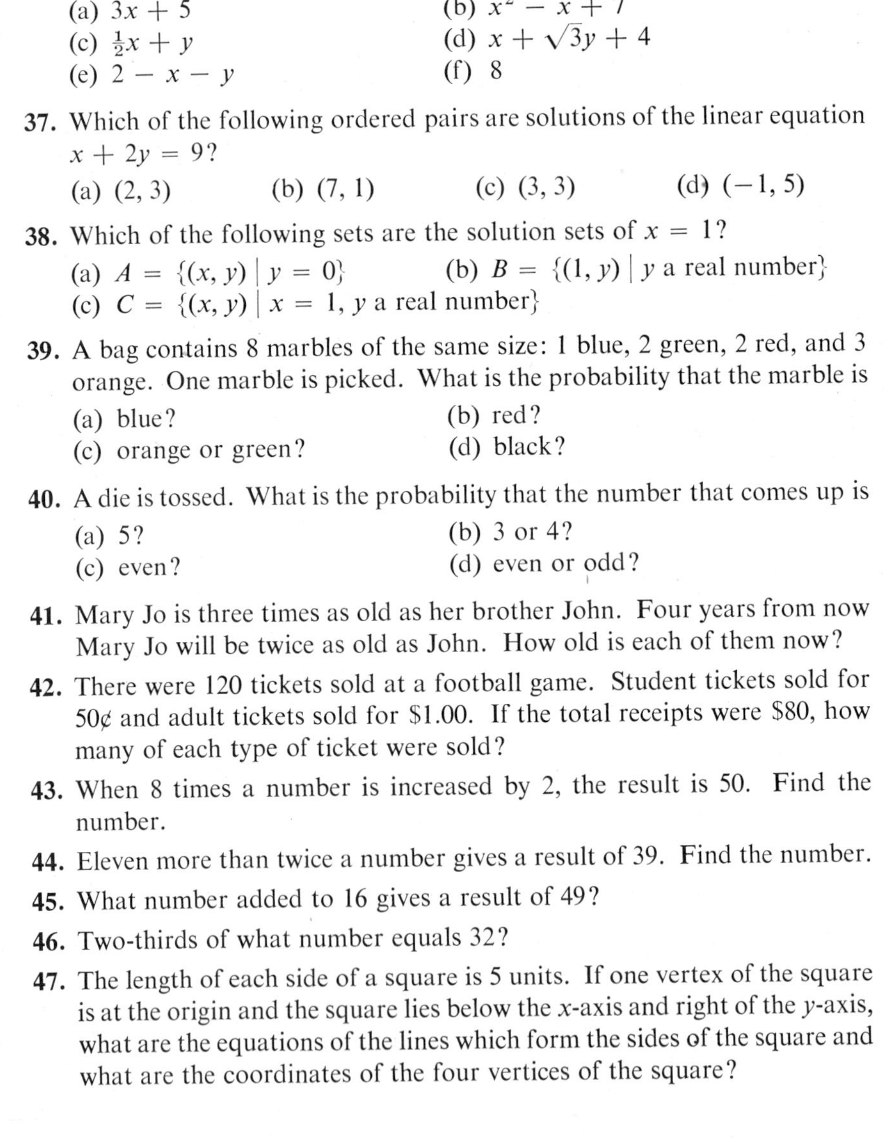

36. Which of the following algebraic expressions are linear forms?
(a) $3x + 5$ (b) $x^2 - x + 7$
(c) $\frac{1}{2}x + y$ (d) $x + \sqrt{3}y + 4$
(e) $2 - x - y$ (f) 8

37. Which of the following ordered pairs are solutions of the linear equation $x + 2y = 9$?
(a) (2, 3) (b) (7, 1) (c) (3, 3) (d) (−1, 5)

38. Which of the following sets are the solution sets of $x = 1$?
(a) $A = \{(x, y) \mid y = 0\}$ (b) $B = \{(1, y) \mid y \text{ a real number}\}$
(c) $C = \{(x, y) \mid x = 1, y \text{ a real number}\}$

39. A bag contains 8 marbles of the same size: 1 blue, 2 green, 2 red, and 3 orange. One marble is picked. What is the probability that the marble is
(a) blue? (b) red?
(c) orange or green? (d) black?

40. A die is tossed. What is the probability that the number that comes up is
(a) 5? (b) 3 or 4?
(c) even? (d) even or odd?

41. Mary Jo is three times as old as her brother John. Four years from now Mary Jo will be twice as old as John. How old is each of them now?

42. There were 120 tickets sold at a football game. Student tickets sold for 50¢ and adult tickets sold for $1.00. If the total receipts were $80, how many of each type of ticket were sold?

43. When 8 times a number is increased by 2, the result is 50. Find the number.

44. Eleven more than twice a number gives a result of 39. Find the number.

45. What number added to 16 gives a result of 49?

46. Two-thirds of what number equals 32?

47. The length of each side of a square is 5 units. If one vertex of the square is at the origin and the square lies below the x-axis and right of the y-axis, what are the equations of the lines which form the sides of the square and what are the coordinates of the four vertices of the square?

48. A square has two vertices at (3, 1) and (3, 6). The square lies entirely above the x-axis. What are the equations of the lines which form the sides of the square?

49. Write an algorithm and draw a flow chart for generating elements of the set $\{(x, y) \mid y = x - 1, 0 < x \leqq 4, x \text{ an integer}\}$.

Determine whether each statement is true or false. Correct the statements you marked false.

50. The set of all real numbers is made up of rational and irrational numbers.

51. The union of sets A and B consists of all elements in either A or B or both.

52. Set A is called a subset of B if every element of set A is also contained in set B.

53. Zero is a member of the empty set.

54. The set of positive numbers is closed under subtraction.

55. In the set of real numbers, division is a commutative operation.

56. $|x| > 0$ if $x \neq 0$

57. The factors of a product may be rearranged in any way.

58. The solution set of $x + y = 1$ is $\{(2, -1)\}$.

59. In division of polynomials, the remainder has a higher degree than the divisor.

60. The graph of a number is the point on a number line associated with it.

61. The graph of a linear equation in a cartesian coordinate system is a straight line.

62. The second law of exponents states that $(xy)^n = x^n y^n$.

CHAPTER 8

Systems of Linear Equations

Objectives . . .

- To solve a system of linear equations in two unknowns by substitution, by finding the intersection of the solution sets of each equation, and by using the method of elimination (addition method).
- To apply the techniques of this chapter to word problems that lead to systems of equations.
- To use determinants to solve a system of two or three equations.

8-1 SYSTEMS OF LINEAR EQUATIONS

Because mathematics is applied to many situations involving several variables, it is important that you learn how to solve systems of equations in two or more variables.

Suppose that two number scales are related to each other as shown in Fig. 8-1. Each unit of the b-scale contains $2\frac{1}{2}$ units of the a-scale. The units of these two scales have approximately the same relationship to each other as inches have to centimeters if the a-scale represents centimeters and the b-scale represents inches. Note that the numbers on the b-scale increase from right to left and that the numbers on the a-scale increase from left to right. Is there a point that has the same a-number and b-number? What is this number? Merely by looking at the figure, you notice that the number appears to be between 1 and 2. Perhaps you can figure it out by trial and error.

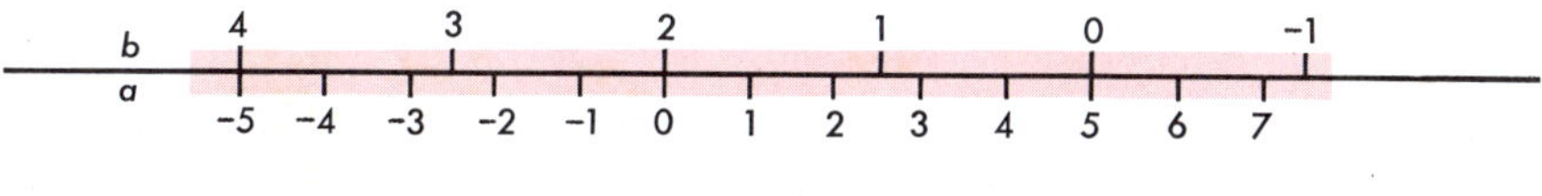

FIGURE 8-1

The relationship between the a-numbers and the b-numbers is given by the equation

$$2a + 5b = 10.$$

You may verify this equation by checking some pairs of a- and b-numbers from Fig. 8-1. Thus, $a = -5$ when $b = 4$, and

$$2 \cdot (-5) + 5 \cdot 4 = 10.$$

Also, $a = 5$ when $b = 0$, and

$$2 \cdot 5 + 5 \cdot 0 = 10.$$

To find the point with the same a-number and b-number, you must find a solution of the equation $2a + 5b = 10$ which has the additional property that $a = b$. In other words, you must find a solution of the system of equations

$$\begin{cases} 2a + 5b = 10, \\ \qquad\ \ a = b. \end{cases}$$

You may solve this system by replacing a by b in the first equation and solving the resulting equation in one variable:

$$\begin{aligned} 2b + 5b &= 10, \\ 7b &= 10, \\ b &= \tfrac{10}{7}. \end{aligned}$$

Therefore, the point on the line with $\frac{10}{7}$ as its b-number also has $\frac{10}{7}$ as its a-number.

Check.

$$\begin{aligned} 2 \cdot \tfrac{10}{7} + 5 \cdot \tfrac{10}{7} &\stackrel{?}{=} 10 \\ \tfrac{20}{7} + \tfrac{50}{7} &\stackrel{?}{=} 10 \\ \tfrac{70}{7} &\stackrel{\checkmark}{=} 10 \end{aligned}$$

A set of two or more linear equations is called a *system of linear equations.* The *solution set* of a system of linear equations in two variables consists of all ordered pairs which are solutions of every equation in the system.

Consider, for example, the system of linear equations

$$\begin{cases} 3x + 4y = 8, \\ -2x + y = 13. \end{cases}$$

If

S is the solution set of the equation $3x + 4y = 8$,

and

T is the solution set of the equation $-2x + y = 13$,

then the solution set of the system above is the *intersection* of sets S and T,

$$S \cap T.$$

This system will be solved in a later section. Some simpler systems are solved below.

Problem 1. Solve the system of linear equations in two variables,

$$\begin{cases} x = 3, \\ y = 7. \end{cases}$$

Solution. You see at a glance that (3, 7) is the only solution of both equations. That is, if S is the solution set of $x = 3$,

$$S = \{(x, y) \mid x = 3\}$$

and T is the solution set of $y = 7$,

$$T = \{(x, y) \mid y = 7\}$$

then

$$S \cap T = \{(x, y) \mid x = 3 \text{ and } y = 7\} \quad \text{or } \{(3, 7)\}.$$

Problem 2. Solve the system of linear equations in two variables,

$$\begin{cases} 3x - y = 4, \\ \phantom{3x -{}} y = 8. \end{cases}$$

Solution. The only value of y which makes the second equation true is $y = 8$. Therefore, $y = 8$ is the only value of y which can make both equations true. Giving y the value 8 in the first equation, you obtain the following equation in one variable:

$$3x - 8 = 4.$$

Its solution is $x = 4$. The solution set of the system must be

$$\{(4, 8)\}.$$

Exercises

1. (a) (i) Give three solutions of the equation $5y = x - 9$.
   (ii) How many solutions does the following equation have?

   $$x = 17$$

   (iii) Solve the system

   $$\begin{cases} 5y = x - 9, \\ x = 17 \end{cases}$$

   by replacing x in the first equation with 17.

(b) (i) Give three solutions of the equation $3x - 2y = 9$.
(ii) What y-number makes the following equation true?

$$y = -6$$

(iii) Solve the system

$$\begin{cases} 3x - 2y = 9, \\ \qquad\ \ y = -6. \end{cases}$$

2. (a) (i) Give three solutions of the equation $x + y = 2$.
(ii) What is the solution set of the equation $x = x + 2$?
(iii) Solve the system

$$\begin{cases} x + y = 2, \\ \quad\ \ x = x + 2. \end{cases}$$

(b) (i) Give three solutions of the equation $x = 7$.
(ii) What is the solution set of the equation $4x + 3 = 4x - 3$?
(iii) Solve the system

$$\begin{cases} \qquad\ x = 7, \\ 4x + 3 = 4x - 3. \end{cases}$$

3. (a) If $S = \{(x, y) \mid 5x + 4y = 9\}$ and $T = \{(x, y) \mid x = -3\}$, find $S \cap T$. Give an element of S not in T. Give an element of T not in S.

(b) If $S = \{(x, y) \mid 3x = 6\}$ and $T = \{(x, y) \mid x - 2y = 4\}$, find $S \cap T$. Give an element of S not in T. Give an element of T not in S.

Find the solution set of each of the following systems.

4. (a) $\begin{cases} x = 4 \\ y = 2 \end{cases}$ (b) $\begin{cases} x = \frac{2}{3} \\ y = \frac{3}{2} \end{cases}$

5. (a) $\begin{cases} x = 9 \\ x + y = 10 \end{cases}$ (b) $\begin{cases} y = \frac{1}{3} \\ x + y = 4 \end{cases}$

6. (a) $\begin{cases} x = -1 \\ x = 3 \end{cases}$ (b) $\begin{cases} x - y = 3 \\ x = x + 3 \end{cases}$

7. (a) $\begin{cases} y + x = 2 \\ x = 2 \end{cases}$ (b) $\begin{cases} 4x - y = 3 \\ y = 1 \end{cases}$

8. (a) $\begin{cases} x + y = 6 \\ y = -8 \end{cases}$ (b) $\begin{cases} x + y = 6 \\ x = -8 \end{cases}$

9. (a) $\begin{cases} x + 8y = 66 \\ x = 2 \end{cases}$ (b) $\begin{cases} 7y - x = 11 \\ x = 10 \end{cases}$

10. (a) $\begin{cases} 7x + 2y = 9 \\ y = 22 \end{cases}$ (b) $\begin{cases} 8x - 3y = 10 \\ y = -3 \end{cases}$

11. (a) $\begin{cases} -21x + 16y = -63 \\ x = 3 \end{cases}$ (b) $\begin{cases} 3x - 5y = 17 \\ y = 2 \end{cases}$

12. (a) $\begin{cases} \frac{x}{2} + 3y = 4 \\ x = 2 \end{cases}$ (b) $\begin{cases} 5x + \frac{y}{4} = -6 \\ y = 16 \end{cases}$

13. (a) $\begin{cases} 2x + 5y = 7 \\ x = -\frac{3}{2} \end{cases}$ (b) $\begin{cases} \frac{3}{4}x + \frac{1}{3}y = 10 \\ y = -6 \end{cases}$

14. (a) $\begin{cases} 2x + 4 = x - 3 \\ y = 5 \end{cases}$ (b) $\begin{cases} 3y + 8 = y - 4 \\ x = 2 + x \end{cases}$

Find the solution set of each of the systems.

15. $\begin{cases} \frac{2}{3}y - \frac{3}{5}x = 11 \\ x = 1 \end{cases}$

16. $\begin{cases} .09x + .1y = 72 \\ y = 99 \end{cases}$

17. $\begin{cases} .25y - .01x = .001 \\ x = .1 \end{cases}$

18. $\begin{cases} \frac{3}{4}y + 2x = 11 \\ y = 4 \end{cases}$

19. If the a-numbers and b-numbers on a scale are related such that $2a + 5b = 10$, at what point is the a-number five times the corresponding b-number? (See Fig. 8–1.)

20. Suppose that two number scales are related to each other as shown in the figure. Each unit of the lower c-scale contains 2.2 units of the upper d-scale. The units of these two scales are approximately in the same ratio as 1 kilogram is to 1 pound if kilograms are represented on the lower scale and pounds are represented on the upper scale. The relationship between the c-number and d-number of a point is given by the equation

$$5d - 11c = 55.$$

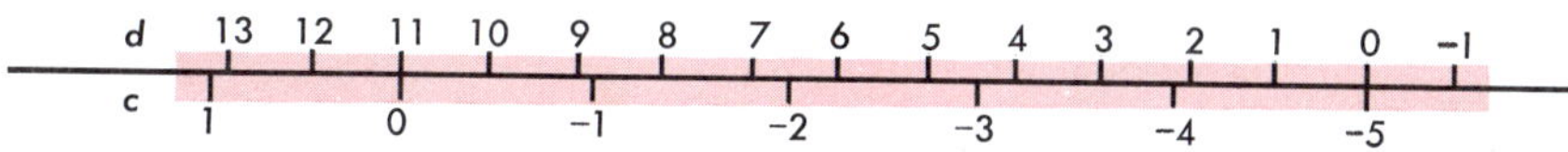

To answer each of the questions below, write an appropriate system of equations in c and d, and solve the system.

(a) What point has the same c-number and d-number?
(b) At what point is the d-number 17 units more than the c-number?
(c) At what point is the d-number twice the c-number?

Given $A = \{(x, y) \mid 9x + 2y = 3\}$,
$B = \{(x, y) \mid y = 6\}$,
$C = \{(x, y) \mid x = -1\}$.

Find:

21. $A \cap B$.

22. $A \cap C$.

23. $B \cap C$.

24. an element of A not in B or C.

25. an element of B not in A or C.

26. an element of C not in A or B.

27. a common element of A, B, and C (an element of $A \cap B \cap C$).

Preparation for Section 8–2

1. Find the x- and y-intercepts of the line $3x - 4y = 12$, and draw the graph.
2. What is the slope of the line $x - 2y = 2$? What is the y-intercept? Draw the graph of this line on the same coordinate system that you used for Exercise 1.
3. Where do the lines of Exercises 1 and 2 intersect?

8–2 GRAPHICAL SOLUTION OF A SYSTEM

If you graph each equation of a system of two linear equations in two variables, you ordinarily get two lines in the plane. The solution of the system is represented by all points which are common to both lines. If, as is usually the case, the lines intersect, then there is one solution represented by the point of intersection. In this case, the system is called *independent*. It might happen that the two lines coincide. Then every point on the common line is a solution, and the system is called *dependent*.

Problem 1. Solve graphically the system of linear equations

$$\begin{cases} 3x + 2y = 1, & (1) \\ 2x - 3y = -8. & (2) \end{cases}$$

Solution. Equations (1) and (2) are graphed by finding and then plotting two solutions of each equation, as indicated in Fig. 8–2. The point of intersection of these two lines appears to be the point $(-1, 2)$.

Check.

$$3x + 2y = 1$$
$$3 \cdot (-1) + 2 \cdot 2 \stackrel{?}{=} 1$$
$$-3 + 4 \stackrel{\checkmark}{=} 1$$

$$2x - 3y = -8$$
$$2 \cdot (-1) - 3 \cdot 2 \stackrel{?}{=} -8$$
$$-2 - 6 \stackrel{\checkmark}{=} -8$$

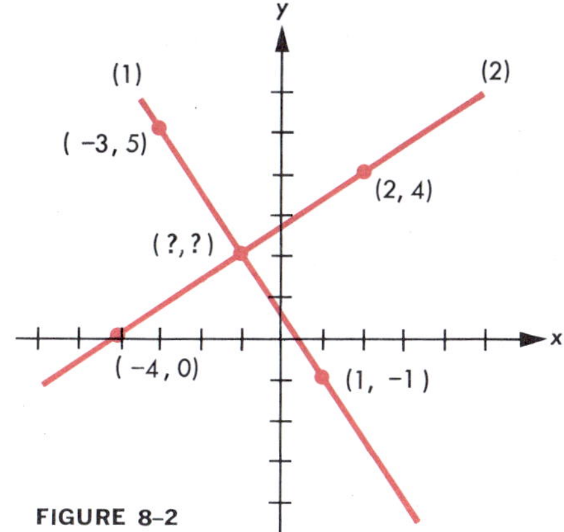

FIGURE 8–2

Problem 2. Solve graphically the system of linear equations

$$\begin{cases} 2x - y = 7, & (1) \\ 4x - 2y = 3. & (2) \end{cases}$$

Solution. Equations (1) and (2) are graphed in Fig. 8–3. The lines do not appear to meet. If you solve each equation for y in terms of x,

$$\begin{cases} y = 2x - 7, & (1) \\ y = 2x - \frac{3}{2}, & (2) \end{cases}$$

you see that they have the same slope, 2, and therefore are parallel. They do not intersect and thus, the system has no solution; that is, the solution set is the empty set, $\emptyset$.

A system of linear equations that has no solution, such as in Problem 2, is called an *inconsistent system.*

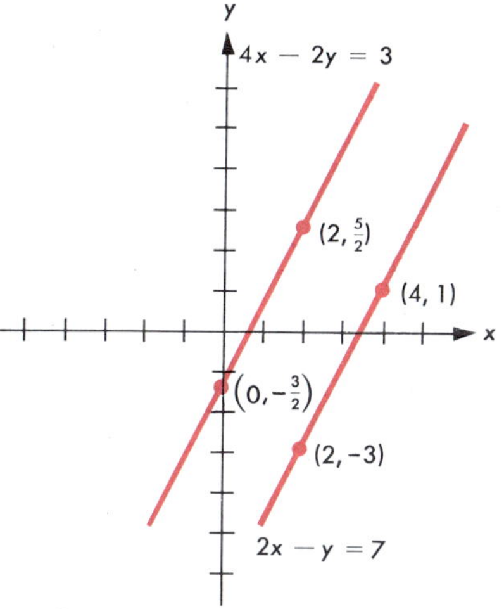

Graph of an inconsistent system.

FIGURE 8–3

Exercises

Solve graphically the systems of linear equations given in Exercises 1–10. Check each solution.

1. (a) $\begin{cases} x = 2 \\ y = -3 \end{cases}$ (b) $\begin{cases} x = -5 \\ y = 4 \end{cases}$

2. (a) $\begin{cases} x - 5 = 0 \\ y + 2 = 0 \end{cases}$ (b) $\begin{cases} x + 7 = 0 \\ y - 1 = 0 \end{cases}$

3. (a) $\begin{cases} x + 3 = 0 \\ y = -2x \end{cases}$ (b) $\begin{cases} y = -x \\ y - 3 = 0 \end{cases}$

4. (a) $\begin{cases} 2x - 3y = 6 \\ y = -2 \end{cases}$ (b) $\begin{cases} x + 2y = 7 \\ x = 3 \end{cases}$

5. (a) $\begin{cases} y = 2x - 8 \\ x = 2y + 4 \end{cases}$ (b) $\begin{cases} x - y + 1 = 0 \\ x + y = 3 \end{cases}$

6. (a) $\begin{cases} y + 2 = x \\ y - 2 = -x - 4 \end{cases}$ (b) $\begin{cases} 3y - 2x = 4 \\ x + y + 3 = 1 \end{cases}$

7. (a) $\begin{cases} 2y - x = 2 \\ x = 4y \end{cases}$ (b) $\begin{cases} 3x + y = 0 \\ y = 2x \end{cases}$

8. (a) $\begin{cases} x + y = 0 \\ y = 2x + 6 \end{cases}$ (b) $\begin{cases} x = y - 1 \\ 3y + 2x + 2 = 0 \end{cases}$

9. (a) $\begin{cases} x + 2y + 6 = 0 \\ 2y = x \end{cases}$ (b) $\begin{cases} x - y = 2 \\ 2x + y = -1 \end{cases}$

10. (a) $\begin{cases} \dfrac{x}{3} - \dfrac{y}{5} = \dfrac{1}{15} \\ x + y = 5 \end{cases}$ (b) $\begin{cases} \dfrac{x}{2} - \dfrac{y}{4} = \dfrac{1}{8} \\ x + y = 1 \end{cases}$

---

11. Where on the graph of $x - 2y = 3$ is the ordinate twice the abscissa?

12. Find the vertices of a triangle determined by the graphs of $y = x + 2$, $x - 3 = 0$, $y + x = 0$. (Hint: find the solution set for each of the following systems of equations).

$$\begin{cases} y = x + 2 \\ y + x = 0 \end{cases} \qquad \begin{cases} x - 3 = 0 \\ y = x + 2 \end{cases} \qquad \begin{cases} y + x = 0 \\ x - 3 = 0 \end{cases}$$

---

13. Solve the next two problems graphically. Make your time-axis horizontal and your distance-axis vertical. (Be careful of your units.)
    (a) A truck traveling 30 miles per hour left a town at 7 A.M. A second truck traveling in the same direction at 40 miles per hour left the same town at 9 A.M. At what time will the second truck overtake the first? How far from the town will they be at that time?
    (b) A freight train traveling at 20 miles per hour leaves a station at 2 P.M. A passenger train traveling in the same direction at the rate of 60 miles per hour leaves the same station 3 hours later. At what time will the passenger train overtake the freight train?

14. Show by a graph the number of prints for which the rate of 10¢ for printing each negative with free developing is cheaper than the rate of 6¢ for each print with a charge of 25¢ for developing.

15. Show by a graph the amount of sales for which a weekly salary of \$30 and 10% commission on sales is better than a weekly salary of \$60 and 5% commission.

16. Determine the solutions of each of the following systems graphically.

(a) $\begin{cases} x - y = 1 \\ 3x - 6 = 3y \end{cases}$ (b) $\begin{cases} 2x + 3y = 6 \\ 6y - 12 = -4x \end{cases}$

17. How do the slopes of the graphs of the equations in an inconsistent system compare?

8–3 SUBSTITUTION METHOD OF SOLVING A SYSTEM

A system of linear equations such as

$$\begin{cases} 3x - 4y = 8, \\ y = 4 \end{cases}$$

is easily solved by letting y have the value 4 in the first equation to obtain $x = 8$. Then (8, 4) is the only solution of the system. This method of solving a system, called the *substitution method,* can be used even if the variables x and y occur in both equations. You shall see this below.

To solve a system of linear equations, try to find an equivalent system whose solution set is evident, as is the case in the example above.

Definition of equivalent systems

Two systems of equations are said to be equivalent if they have the same solution set.

The substitution method is used in the problem below to obtain equivalent systems.

Problem. Solve the system of linear equations.

$$\begin{cases} 3x + 4y = 8 \\ -2x + y = 13 \end{cases} \qquad (A)$$

Solution. By solving the second equation of system (A) for y and leaving the first equation as it is, you get the equivalent system

$$\begin{cases} 3x + 4y = 8, \\ y = 2x + 13. \end{cases} \qquad (B)$$

Next, substitute the value of y from the second equation of (B) in the first equation, obtaining the system

$$\begin{cases} 3x + 4(2x + 13) = 8, \\ y = 2x + 13. \end{cases} \qquad (C)$$

System (C) is equivalent to (B)—any ordered pair (x, y) which makes both equations of (B) true makes both equations of (C) true, and vice versa.

The first equation of (C) is equivalent to each of the following equations:

$$\begin{aligned} 3x + 8x + 52 &= 8 \\ 11x &= -44 \\ x &= -4. \end{aligned}$$

Thus, system (C) is equivalent to the system

$$\begin{cases} x = -4, \\ y = 2x + 13. \end{cases} \qquad (D)$$

The solution set of system (D) is evident; let x have the value -4 in the second equation, obtaining $y = 5$. Hence,

$$\{(-4, 5)\}$$

is the solution set of system (D) and also of the equivalent system (A), with which you started.

Check.

$$\begin{aligned} 3x + 4y &= 8 \\ 3 \cdot (-4) + 4 \cdot 5 &\overset{?}{=} 8 \\ -12 + 20 &\overset{\checkmark}{=} 8 \end{aligned} \qquad \begin{aligned} -2x + y &= 13 \\ (-2) \cdot (-4) + 5 &\overset{?}{=} 13 \\ 8 + 5 &\overset{\checkmark}{=} 13 \end{aligned}$$

Exercises

1. Complete the outline, and give the solution set of the following systems of linear equations. Graph systems (A), (C), and (D) on separate coordinate axes, and compare the graphs.

(a) $\begin{cases} x = y + 5 \\ 25 = 3x + 2y \end{cases}$ (A) (b) $\begin{cases} 2y = 3x + 1 \\ y = x - 4 \end{cases}$ (A)

$\begin{cases} x = y + 5 \\ 25 = 3(\underline{\ ?\ }) + 2y \end{cases}$ (B) $\begin{cases} 2(\underline{\ ?\ }) = 3x + 1 \\ y = x - 4 \end{cases}$ (B)

$\begin{cases} x = y + 5 \\ y = \underline{\ ?\ } \end{cases}$ (C) $\begin{cases} x = \underline{\ ?\ } \\ y = x - 4 \end{cases}$ (C)

$\begin{cases} x = \underline{\ ?\ } \\ y = \underline{\ ?\ } \end{cases}$ (D) $\begin{cases} x = \underline{\ ?\ } \\ y = \underline{\ ?\ } \end{cases}$ (D)

$S = \{(\underline{\ ?\ }, \underline{\ ?\ })\}$ $S = \{(\underline{\ ?\ }, \underline{\ ?\ })\}$

Solve each of the following systems of linear equations by the substitution method. Check your answers.

2. (a) $\begin{cases} x = 4 \\ 2x + y = 8 \end{cases}$ (b) $\begin{cases} x = 5 \\ 3x - y = 2 \end{cases}$

3. (a) $\begin{cases} x = 4 + y \\ x - 3y = 4 \end{cases}$ (b) $\begin{cases} x = y + 5 \\ 25 = 3x + 2y \end{cases}$

4. (a) $\begin{cases} 2x + y = 3 \\ y = 5 - 4x \end{cases}$ (b) $\begin{cases} y = 1 - x \\ 2x + y = 4 \end{cases}$

5. (a) $\begin{cases} 3x - 2y = 0 \\ x = 8 - 2y \end{cases}$ (b) $\begin{cases} x + 2y = 4 \\ 3x + 2y = 0 \end{cases}$

6. (a) $\begin{cases} x - 2y = 4 \\ y = 3x - 2 \end{cases}$ (b) $\begin{cases} x = 5 - 2y \\ 3x + 2y = 17 \end{cases}$

7. (a) $\begin{cases} 2y + x + 10 = 0 \\ y = 4x + 13 \end{cases}$ (b) $\begin{cases} 2a - 5b + 3 = 0 \\ b = 3 + a \end{cases}$

8. (a) $\begin{cases} 2x + 5y = 20 \\ x = 1 + 2y \end{cases}$ (b) $\begin{cases} y + 4x + 1 = 0 \\ y = 4x - 7 \end{cases}$

9. (a) $\begin{cases} 4p + q = 8 \\ 2p = 3q + 11 \end{cases}$ (b) $\begin{cases} 2x + y = 9 \\ x - y = 3 \end{cases}$

10. (a) $\begin{cases} x + y = 1 \\ x - y = -5 \end{cases}$ (b) $\begin{cases} 2x - y = 2 \\ x + y = 7 \end{cases}$

11. (a) $\begin{cases} x + y = -8 \\ x - 2y = 7 \end{cases}$ (b) $\begin{cases} x + 2y = 8 \\ 2x - 2y = -8 \end{cases}$

12. (a) $\begin{cases} x + 2y = -2 \\ -2x + y = 6 \end{cases}$ (b) $\begin{cases} x + y = -8 \\ 2x - y = 2 \end{cases}$

13. (a) $\begin{cases} 2x + y = -7 \\ 5x + 6y = 0 \end{cases}$ (b) $\begin{cases} 2x + y = 8 \\ 4x - 9y = 5 \end{cases}$

14. (a) $\begin{cases} 4x + y = -6 \\ -2x + 3y = 24 \end{cases}$ (b) $\begin{cases} 2x + 3y = 9 \\ 4x + y = 18 \end{cases}$

15. (a) $\begin{cases} x - 9y = 0 \\ \dfrac{x}{3} = 2y + \dfrac{1}{3} \end{cases}$ (b) $\begin{cases} x + 5 = 2y \\ \dfrac{4y + 1}{5} = 3x - 3 \end{cases}$

Preparation for Section 8–4

Solve each of the following equations.

1. $\dfrac{x}{4} + \dfrac{1}{3} = \dfrac{1}{12}$

2. $\dfrac{x}{7} - \dfrac{1}{2} = \dfrac{x}{2} - \dfrac{1}{14}$

3. $\dfrac{3x}{5} - \dfrac{2}{3} = \dfrac{2x}{15} + \dfrac{1}{5}$

4. $\dfrac{5x}{3} - \dfrac{3x}{6} + \dfrac{1}{5} = 0$

8–4 EXTENDING THE SUBSTITUTION METHOD

The *substitution method* of solving a system of two linear equations in two variables involves the following steps. You start with a system of this form:

$$\begin{cases}\text{First equation,}\\ \text{Second equation.}\end{cases} \tag{1}$$

Solve one of the equations for x in terms of y (or y in terms of x) and obtain the following:

$$\begin{cases}\text{First equation,}\\ \text{Second equation (solved for } x \text{ in terms of } y \text{ or } y \text{ in terms of } x).\end{cases} \tag{2}$$

Then substitute the value of the variable given by the second equation into the first equation, thereby eliminating a variable from the first equation. This gives a system of this form:

$$\begin{cases}\text{First equation in one variable,}\\ \text{Second equation (solved for } x \text{ in terms of } y \text{ or } y \text{ in terms of } x).\end{cases} \tag{3}$$

System (3) can be solved by the methods of previous sections.

Problem. Solve the system of linear equations

$$\begin{cases}2x + 6y = 5,\\ \quad\;\; 4x = 3y.\end{cases} \tag{1'}$$

Solution. Follow the plan outlined above.

$$\begin{cases}2x + 6y = 5\\ \qquad\;\; x = \frac{3}{4}y\end{cases} \tag{2'}$$

$$\begin{cases}2\left(\frac{3}{4}y\right) + 6y = 5\\ \qquad\qquad\;\; x = \frac{3}{4}y\end{cases} \tag{3'}$$

If you simplify $2(\frac{3}{4}y) + 6y = 5$, you obtain

$$\begin{cases}y = \frac{2}{3},\\ x = \frac{3}{4}y.\end{cases}$$

The solution set of this system and, therefore, also of the given system is

$$\{(\tfrac{1}{2}, \tfrac{2}{3})\}.$$

Exercises

1. (a) Given the system of equations

$$\begin{cases} 3x + 4y + 3 = 0, \\ 5x - 2y + 5 = 0, \end{cases} \tag{1}$$

write an equivalent system by solving the second equation of (1) for x.

$$\begin{cases} 3x + 4y + 3 = 0 \\ x = \underline{\ ?\ } \end{cases} \tag{2}$$

Complete the system by substituting for x in the first equation of (2).

$$\begin{cases} 3(\underline{\ ?\ }) + 4y + 3 = 0 \\ x = \underline{\ ?\ } \end{cases} \tag{3}$$

Solve the first equation of (3) for y and complete the system.

$$\begin{cases} y = \underline{\ ?\ } \\ x = \underline{\ ?\ } \end{cases} \tag{4}$$

Substitute from the first equation of (4) into the second. What is the solution set of (1)?

(b) Complete the outline to find the solution set of the system of equations

$$\begin{cases} 2x - 6y + 4 = 0, \\ 3x + 2y - 5 = 0. \end{cases}$$

$$\begin{cases} 2x - 6y + 4 = 0 \\ y = \underline{\ ?\ } \end{cases}$$

$$\begin{cases} 2x - 6(\underline{\ ?\ }) + 4 = 0 \\ y = \underline{\ ?\ } \end{cases}$$

$$\begin{cases} x = \underline{\ ?\ } \\ y = \underline{\ ?\ } \end{cases}$$

Solve each of the systems of equations in Exercises 2–8 by substitution. Check your answers.

2. (a) $\begin{cases} 2x + 3y = -3 \\ 5x + 6y = 0 \end{cases}$ (b) $\begin{cases} 3x - 2y = 4 \\ -2x + 4y = 0 \end{cases}$

3. (a) $\begin{cases} 4x - 3y + 2 = 0 \\ 3y = 2x + 4 \end{cases}$ (b) $\begin{cases} 4x + 3y = 20 \\ 2x + 6y = 46 \end{cases}$

4. (a) $\begin{cases} 4x - 3y = 0 \\ 4x = 7y \end{cases}$ (b) $\begin{cases} 3x - 15y = 0 \\ 15x - 3y = 0 \end{cases}$

5. (a) $\begin{cases} 3y - 2x = 4 \\ 5x + 3y + 10 = 0 \end{cases}$ (b) $\begin{cases} 2x + 3y = 2 \\ 8x + 9y = 7 \end{cases}$

6. (a) $\begin{cases} 5y + 3x = 1 \\ 2x + 3y = 0 \end{cases}$ (b) $\begin{cases} 2x - 3y = 19 \\ 3x - 2y = 16 \end{cases}$

7. (a) $\begin{cases} 2x - \frac{1}{3}y = -1 \\ 5x - 2y = 1 \end{cases}$ (b) $\begin{cases} \frac{1}{4}x - 2y = 1 \\ 3x + 5y = 12 \end{cases}$

8. (a) $\begin{cases} 2a - 3b = \frac{1}{2} \\ 4a + 3b = \frac{11}{2} \end{cases}$ (b) $\begin{cases} 3c + 4d = 3 \\ 6c - 2d = \frac{7}{2} \end{cases}$

Simplify each equation; then solve the system by substitution. Check each solution in the given system of equations.

9. $\begin{cases} m - \dfrac{n-2}{3} = 0 \\ .2m + .3n = 5 \end{cases}$ **10.** $\begin{cases} \frac{1}{4}w + \frac{7}{4}z = 3 \\ \frac{3}{2}w + 2z = 1 \end{cases}$

11. $\begin{cases} \dfrac{2x}{3} = y \\ \dfrac{3x}{5} - \dfrac{2y}{5} = 1 \end{cases}$ **12.** $\begin{cases} 4x + 3y = -1.25 \\ 1.5x - y = -1 \end{cases}$

13. $\begin{cases} \frac{1}{5}g + \frac{1}{4}h = 12 \\ \frac{1}{3}g - \frac{1}{3}h = 4 \end{cases}$ **14.** $\begin{cases} m - n = 8 \\ \dfrac{m}{3} - \dfrac{n}{2} = 1 \end{cases}$

15. $\begin{cases} 3x + \frac{5}{4}y = 4 \\ 2x - 5y = -9 \end{cases}$ **16.** $\begin{cases} 3x - 4y - 1 = 3(2x - y + 2) \\ 2(4x + 2y + 3) = 7x + y + 9 \end{cases}$

---

17. Given $A = \{(x, y) \mid x - 2y = 0\}$,
$B = \{(x, y) \mid 2x - 6y = 3\}$.

Describe the graphs of $A \cup B$ and $A \cap B$.

8–5 THE ADDITION METHOD OF SOLVING A SYSTEM

Another method of solving a system of linear equations, similar to the substitution method, is the *addition method.* It is illustrated in the problems below.

Every linear equation in x and y is equivalent to one of the form $ax + by + c = 0$ for some numbers a, b, and c. For example, the linear equation

$$3x = 2y - 7$$

is equivalent to

$$3x - 2y + 7 = 0.$$

For convenience, all the linear equations in this section have been put in this form.

Problem 1. Solve the system of linear equations

$$\begin{cases} 2x + y - 1 = 0, \\ -2x - 3y + 11 = 0. \end{cases}$$

Solution. Any solution of both of these equations is also a solution of their *sum*, obtained by adding left side to left side and right side to right side:

$$\begin{array}{r} 2x + y - 1 = 0 \\ -2x - 3y + 11 = 0 \\ \hline -2y + 10 = 0. \end{array}$$

In turn, any solution of the new equation $-2y + 10 = 0$ and either of the given equations is a solution of the given system. Thus, the new system

$$\begin{cases} 2x + y - 1 = 0, \\ -2y + 10 = 0 \end{cases}$$

is equivalent to the original system. Now, $y = 5$ from the second equation, and therefore,

$$2x + 5 - 1 = 0, \quad \text{or } x = -2,$$

from the first equation. Thus,

$$\{(-2, 5)\}$$

is the solution set of the given system.

Check.

$$\begin{array}{rl} 2x + y - 1 &= 0 \\ 2 \cdot (-2) + 5 - 1 &\stackrel{?}{=} 0 \\ -4 + 5 - 1 &\stackrel{\checkmark}{=} 0 \end{array} \qquad \begin{array}{rl} -2x - 3y + 11 &= 0 \\ (-2) \cdot (-2) - 3 \cdot 5 + 11 &\stackrel{?}{=} 0 \\ 4 - 15 + 11 &\stackrel{\checkmark}{=} 0 \end{array}$$

In the problem above, you eliminated one of the variables, x, by adding the two equations. This led to a simplified system of equations. The method above worked because the coefficients of x in the two given equations were negatives of each other. If the coefficients of x or y are not negatives of each other, an extra step will lead to a system in which they are negatives of each other. This is illustrated below.

Problem 2. Solve the system of linear equations

$$\begin{cases} 2x + 5y + 1 = 0, \\ 3x + 4y - 9 = 0. \end{cases}$$

Solution. If you multiply both sides of the first equation by 4 and both sides of the second equation by -5, you obtain the equivalent system

$$\begin{cases} 8x + 20y + 4 = 0, \\ -15x - 20y + 45 = 0. \end{cases} \tag{1}$$

In this system, the coefficients of y are negatives of each other. On adding these equations,

$$\begin{array}{r} 8x + 20y + 4 = 0 \\ -15x - 20y + 45 = 0 \\ \hline -7x \qquad + 49 = 0, \end{array}$$

you obtain a linear equation with the y-term missing. This equation, together with either of the equations of system (1), gives you a system equivalent to the given one:

$$\begin{cases} 8x + 20y + 4 = 0, \\ \quad -7x + 49 = 0. \end{cases} \tag{2}$$

From the second equation of system (2), $x = 7$, and from the first equation,

$$8 \cdot 7 + 20y + 4 = 0, \quad \text{or } y = -3.$$

Thus,

$$\{(7, -3)\}$$

is the solution of the given system.

Check.

$$\begin{array}{rl} 2x + 5y + 1 &= 0 \\ 2 \cdot 7 + 5 \cdot (-3) + 1 &\stackrel{?}{=} 0 \\ 14 - 15 + 1 &\stackrel{\checkmark}{=} 0 \end{array} \qquad \begin{array}{rl} 3x + 4y - 9 &= 0 \\ 3 \cdot 7 + 4 \cdot (-3) - 9 &\stackrel{?}{=} 0 \\ 21 - 12 - 9 &\stackrel{\checkmark}{=} 0 \end{array}$$

Exercises

Solve each of the following systems of linear equations by the addition method. Check your answers.

1. (a) $\begin{cases} x + y - 17 = 0 \\ x - y - 7 = 0 \end{cases}$ (b) $\begin{cases} x + y - 2 = 0 \\ x - y - 0 = 0 \end{cases}$

2. (a) $\begin{cases} 4a - \mathrm{b} - 7 = 0 \\ 3a + b - 7 = 0 \end{cases}$ (b) $\begin{cases} c + 3d - 11 = 0 \\ c + d - 3 = 0 \end{cases}$

3. (a) $\begin{cases} x + 2y - 6 = 0 \\ x + 15y - 6 = 0 \end{cases}$ (b) $\begin{cases} m + 3n - 11 = 0 \\ m - n + 1 = 0 \end{cases}$

4. (a) $\begin{cases} 3w - 4z - 1 = 0 \\ 3w - 2z + 1 = 0 \end{cases}$ (b) $\begin{cases} 2x + y - 3 = 0 \\ 2x - 3y - 31 = 0 \end{cases}$

5. (a) $\begin{cases} 2m - 3n - 7 = 0 \\ 2m + 9n + 21 = 0 \end{cases}$ (b) $\begin{cases} 2x + 4y - 3 = 0 \\ 2x - 8y + 3 = 0 \end{cases}$

6. (a) $\begin{cases} 3a - b = 2 \\ 3a - 5b = 10 \end{cases}$ (b) $\begin{cases} 3r + 2s = 0 \\ 5r + 2s = 0 \end{cases}$

7. (a) $\begin{cases} 3x + 2y = 5 \\ 9x - 2y = 15 \end{cases}$ (b) $\begin{cases} 3p - q = 17 \\ 5p + q = 4 \end{cases}$

8. (a) $\begin{cases} 19x + 19y = 0 \\ 19x - 19y = 10 \end{cases}$ (b) $\begin{cases} 10x - 10y = 4 \\ 10x + 10y = 0 \end{cases}$

9. (a) $\begin{cases} -4x + 5y = 18 \\ -4x - 3y = 2 \end{cases}$ (b) $\begin{cases} 3x - 7y = -4 \\ 4x - 7y = -3 \end{cases}$

10. (a) $\begin{cases} 3x - 11y = 6 \\ 5x + 11y = 10 \end{cases}$ (b) $\begin{cases} 10x - 3y = 6 \\ 7y - 10x = -14 \end{cases}$

---

11. Simplify each equation, and then solve the system by the addition method.

(a) $\begin{cases} \frac{3}{2}p + \frac{4}{3}q = 2 \\ 8q = 3p + 4 \end{cases}$ (b) $\begin{cases} 3r + 3s - 2 = r - 2s + 12 \\ 2 - 3s + 3r = s - 6 - r \end{cases}$

(c) $\begin{cases} x + 3(1 - y) = 0 \\ 3y - 2(6 - x) = 0 \end{cases}$ (d) $\begin{cases} 7(p + 1) = \frac{1}{3}(q - 1) \\ q - 4 = 6(p + 3) \end{cases}$

Preparation for Section 8–6

1. Solve by the substitution method:

$$\begin{cases} 4x + y = 5, \\ 3x - 2y = 1. \end{cases}$$

2. Solve by the addition method:

$$\begin{cases} 4x + 2y = 5, \\ 3x - 2y = 2. \end{cases}$$

3. Solve by the graphical method:

$$\begin{cases} 4x + y = 4, \\ x - 3y = 14. \end{cases}$$

8-6 MORE ON THE ADDITION METHOD

The addition method, illustrated in the preceding section, involves the following steps.

Starting with a system of two linear equations in x and y, you can find an equivalent system having the form

$$\begin{cases} A = 0, \\ B = 0, \end{cases} \tag{1}$$

where A and B are linear forms in x and y (that is, each of A and B has the form $ax + by + c$ for some numbers a, b, c). You next multiply each equation in (1) by a nonzero number in order to obtain an equivalent system,

$$\begin{cases} C = 0, \\ D = 0, \end{cases} \tag{2}$$

in which the coefficients of either x or y in C and D are negatives of each other. Then the system

$$\begin{cases} C = 0 \qquad (\text{or } D = 0), \\ C + D = 0 \end{cases} \tag{3}$$

is equivalent to (2) and (1). System (3) is easily solved, since the second equation only involves one variable.

A remark might be in order. If, in system (2), the coefficients of either x or y are the *same*, then the system

$$\begin{cases} C = 0 \qquad (\text{or } D = 0), \\ C - D = 0 \end{cases} \tag{4}$$

is equivalent to (2) and has the property that the second equation involves only one variable.

Problem 1. Solve the system

$$\begin{cases} 4x - 3y = 2, \\ 7x + y = 6. \end{cases}$$

Solution. This system is equivalent to

$$\begin{cases} 4x - 3y - 2 = 0, \\ 7x + y - 6 = 0. \end{cases} \tag{1'}$$

Referring to (1) above, $A = 4x - 3y - 2$ and $B = 7x + y - 6$. You can retain the first equation of (1′) and multiply the second equation by 3. This yields an equivalent system:

$$\begin{cases} 4x - 3y - 2 = 0, \\ 21x + 3y - 18 = 0. \end{cases} \qquad (2')$$

Referring to (2) above, $C = 4x - 3y - 2$ and $D = 21x + 3y - 18$. Now $C + D = 25x - 20$ involves only the variable x. Thus, the system

$$\begin{cases} 4x - 3y - 2 = 0, \\ \quad 25x - 20 = 0 \end{cases} \qquad (3')$$

is equivalent to (2′) and also to (1′).

You can solve (3′) as follows. From the second equation, $5x = 4$ and $x = \frac{4}{5}$. Then from the first equation,

$$\begin{aligned} 4 \cdot \tfrac{4}{5} - 3y - 2 &= 0, \\ -3y &= 2 - \tfrac{16}{5}, \\ -3y &= -\tfrac{6}{5}, \\ y &= \tfrac{2}{5}. \end{aligned}$$

Thus,

$$\{(\tfrac{4}{5}, \tfrac{2}{5})\}$$

is the solution set of the given system.

Check.

$$\begin{aligned} 4x - 3y &= 2 \\ 4 \cdot \tfrac{4}{5} - 3 \cdot \tfrac{2}{5} &\stackrel{?}{=} 2 \\ \tfrac{16}{5} - \tfrac{6}{5} &\stackrel{\checkmark}{=} 2 \end{aligned} \qquad \begin{aligned} 7x + y &= 6 \\ 7 \cdot \tfrac{4}{5} + \tfrac{2}{5} &\stackrel{?}{=} 6 \\ \tfrac{28}{5} + \tfrac{2}{5} &\stackrel{\checkmark}{=} 6 \end{aligned}$$

Problem 2. Solve the system

$$\begin{cases} \tfrac{2}{3}a = \tfrac{3}{4}b + \tfrac{17}{6}, \\ \tfrac{1}{2}a = -6b - \tfrac{9}{4}. \end{cases}$$

Solution. This system is equivalent to

$$\begin{cases} \tfrac{2}{3}a - \tfrac{3}{4}b - \tfrac{17}{6} = 0, \\ \tfrac{1}{2}a + 6b + \tfrac{9}{4} = 0. \end{cases}$$

On multiplying the first equation by 12, the greatest common denominator of the fractions in the equation, and the second by 4 for the same reason, you obtain an equivalent system

$$\begin{cases} 8a - 9b - 34 = 0, \\ 2a + 24b + 9 = 0. \end{cases}$$

Now multiply the second equation by 4 and carry over the first equation, to obtain another equivalent system:

$$\begin{cases} 8a - 9b - 34 = 0, \\ 8a + 96b + 36 = 0. \end{cases}$$

If you subtract the second equation from the first, you obtain an equation in one variable b which can be solved:

$$\begin{array}{r} 8a - 9b - 34 = 0 \\ -\ (8a + 96b + 36 = 0) \\ \hline -105b - 70 = 0, \end{array}$$

$$b = \frac{-70}{105}, \quad \text{or} \quad b = \frac{-2}{3}.$$

Thus, the original system is equivalent to

$$\begin{cases} 8a - 9b - 34 = 0, \\ b = -\frac{2}{3}. \end{cases}$$

Letting $b = -\frac{2}{3}$ in the first equation, you obtain

$$\begin{aligned} 8a - 9 \cdot (-\tfrac{2}{3}) - 34 &= 0, \\ 8a + 6 - 34 &= 0, \\ 8a &= 28, \\ a &= \tfrac{7}{2}. \end{aligned}$$

Therefore, the solution set of the given equation is

$$\{(\tfrac{7}{2}, -\tfrac{2}{3})\}.$$

Check.

$$\begin{aligned} \tfrac{2}{3}a &= \tfrac{3}{4}b + \tfrac{17}{6} \\ \tfrac{2}{3} \cdot \tfrac{7}{2} &\overset{?}{=} \tfrac{3}{4} \cdot (-\tfrac{2}{3}) + \tfrac{17}{6} \\ \tfrac{7}{3} &\overset{?}{=} -\tfrac{1}{2} + \tfrac{17}{6} \\ \tfrac{7}{3} &\overset{?}{=} -\tfrac{3}{6} + \tfrac{17}{6} \\ \tfrac{7}{3} &\overset{\checkmark}{=} \tfrac{14}{6} \end{aligned}$$

$$\begin{aligned} \tfrac{1}{2}a &= -6b - \tfrac{9}{4} \\ \tfrac{1}{2} \cdot \tfrac{7}{2} &\overset{?}{=} -6 \cdot (-\tfrac{2}{3}) - \tfrac{9}{4} \\ \tfrac{7}{4} &\overset{?}{=} 4 - \tfrac{9}{4} \\ \tfrac{7}{4} &\overset{\checkmark}{=} \tfrac{16}{4} - \tfrac{9}{4} \end{aligned}$$

Exercises

1. Given the system of linear equations

$$\begin{cases} 6x - 5y = 15, \\ 3x + 2y = 21. \end{cases}$$

(a) Write an equivalent system in which the right side of each equation is zero, and the coefficients of x are the same. Use this system and the method of subtraction to find the solution set of the given system.

(b) Write an equivalent system in which the coefficients of y are negatives of each other. Use this system and the method of addition to find the solution set of the given system.

Solve each of the following systems of linear equations by the method of addition or subtraction. Check your answers.

2. (a) $\begin{cases} 5x + 8y = 1 \\ 2x - 7y = -20 \end{cases}$ (b) $\begin{cases} 3x - 4y = -2 \\ 4x - 3y = -5 \end{cases}$

3. (a) $\begin{cases} 5p + 2q = 15 \\ 15p - 3q = 45 \end{cases}$ (b) $\begin{cases} 3x + 4y = 16 \\ 12x - 5y = 43 \end{cases}$

4. (a) $\begin{cases} 17r + 12s = -24 \\ 13r - 3s = 6 \end{cases}$ (b) $\begin{cases} 8m + 5n = -3 \\ 3m - 15n = -18 \end{cases}$

5. (a) $\begin{cases} 7m + 3n = -8 \\ 5m - 4n = -61 \end{cases}$ (b) $\begin{cases} 2x + 3y = -7 \\ 3x - 7y = 24 \end{cases}$

6. (a) $\begin{cases} 5p - 2q = 11 \\ 3p + 5q = 19 \end{cases}$ (b) $\begin{cases} 7x - 2y = 4 \\ 3x + 5y = -10 \end{cases}$

7. (a) $\begin{cases} 3a + 2b = 5 \\ 2a + 3b = 0 \end{cases}$ (b) $\begin{cases} 4x + 7y = 10 \\ 7x + 4y = 1 \end{cases}$

8. (a) $\begin{cases} 9x + 11y = 40 \\ 5x = 7y - 4 \end{cases}$ (b) $\begin{cases} 3x + 10y = -4 \\ 4x = 5y + 13 \end{cases}$

9. (a) $\begin{cases} x + 2y - 4 = 0 \\ \frac{5}{2}x - 13y - 10 = 0 \end{cases}$ (b) $\begin{cases} 2x + y + 3 = 0 \\ 8x - \dfrac{y}{3} + 25 = 0 \end{cases}$

10. (a) $\begin{cases} 6m + 5 = 2n \\ \frac{1}{2}n = 4m \end{cases}$ (b) $\begin{cases} 7p + 18 = 4q \\ \frac{1}{4}q = p \end{cases}$

Simplify each of the following equations, and then solve the system by any of the three methods you have learned. Check your solutions.

11. $\begin{cases} y = \dfrac{25 - 5x}{2} \\ 3x + 4y = 29 \end{cases}$

12. $\begin{cases} \dfrac{7 + x}{5} - \dfrac{2x - y}{4} = \dfrac{8y + x}{8} \\ \dfrac{5y - 7}{1} + \dfrac{4x - 3}{6} = \dfrac{x + 9y + 1}{12} \end{cases}$

Simplify each of the following equations and then solve each system by any of the three methods you have learned. Check your solutions.

13. $\begin{cases} \dfrac{7a}{25} - \dfrac{5b}{16} = \dfrac{3}{20} \\ \dfrac{2a}{15} + \dfrac{7b}{12} = 3 \end{cases}$

14. $\begin{cases} \dfrac{x+y}{2} + \dfrac{x-y}{4} = 8 \\ \dfrac{2(x+y)}{3} - \dfrac{3(x-y)}{4} = 2 \end{cases}$

15. $\begin{cases} x + 3y + 15 = 3(4x + 5y + 5) \\ 5(5x + 3y - 2) = 8x + y - 10 \end{cases}$

Review for Sections 8–1 through 8–6

Find the solution set of each of the following systems. (You may use any of the methods studied in Sections 8–1 through 8–6.)

1. $\begin{cases} x = 4 \\ y = -3 \end{cases}$

2. $\begin{cases} x + y = 7 \\ y = 2 \end{cases}$

3. $\begin{cases} 2x + 3y = 10 \\ x = 2 \end{cases}$

4. $\begin{cases} x + y = 2 \\ 4y + x = 8 \end{cases}$

5. $\begin{cases} 4x + y = 3 \\ x - 3y = 4 \end{cases}$

6. $\begin{cases} x + 4y - 7 = 0 \\ y = 1 \end{cases}$

7. $\begin{cases} 2x - y = 2 \\ x + y = 8 \end{cases}$

8. $\begin{cases} 3y + 4x = 7 \\ x = y \end{cases}$

9. $\begin{cases} 3y + 4x = 17 \\ 2y - 2x = 2 \end{cases}$

10. $\begin{cases} 3x - y = 2 \\ y = 6 \end{cases}$

11. $\begin{cases} 2x + 3y = 15 \\ 3x = 6y + 18 \end{cases}$

12. $\begin{cases} 2y + 7 = 3x \\ x = y + 2 \end{cases}$

13. $\begin{cases} 8y - 3x = 2x + 1 \\ 2y = x + 4 \end{cases}$

14. $\begin{cases} x + 3y - 12 = 0 \\ x - 4y + 6 = 0 \end{cases}$

15. $\begin{cases} x = 6 \\ y = -15 \end{cases}$

16. $\begin{cases} 4x + 2y = 8 \\ y = x + 2 \end{cases}$

17. $\begin{cases} 2x + 3y - 7 = 0 \\ x - 3y + 2 = 0 \end{cases}$

18. $\begin{cases} 4x + 3y - 18 = 0 \\ 2x + 6y = 12 \end{cases}$

Answers to Review for Sections 8–1 through 8–6

1. $\{(4, -3)\}$
2. $\{(5, 2)\}$
3. $\{(2, 2)\}$
4. $\{(0, 2)\}$

5. $\{(1, -1)\}$
6. $\{(3, 1)\}$
7. $\{(\frac{10}{3}, \frac{14}{3})\}$
8. $\{(1, 1)\}$
9. $\{(2, 3)\}$
10. $\{(\frac{8}{3}, 6)\}$
11. $\{(\frac{48}{7}, \frac{3}{7})\}$
12. $\{(3, 1)\}$
13. $\{(15, \frac{19}{2})\}$
14. $\{(\frac{30}{7}, \frac{18}{7})\}$
15. $\{(6, -15)\}$
16. $\{(\frac{2}{3}, \frac{8}{3})\}$
17. $\{(\frac{5}{3}, \frac{11}{9})\}$
18. $\{(4, \frac{2}{3})\}$

8-7 WORD PROBLEMS

Some word problems can be solved conveniently by using two variables. Even some of the problems which you have solved using one variable can be set up more easily by using two. The following problems illustrate the use of two variables in solving word problems.

Problem 1. Find two numbers such that their sum is 337 and their difference is 43.

Solution. You can write this problem in the form

First number + Second number = 337,
First number − Second number = 43.

It is clear how you should go about solving it. If you let

x designate the first number

and

y designate the second number,

then, by what is given,

$$\begin{cases} x + y = 337, \\ x - y = 43. \end{cases}$$

You can solve this system by replacing it with equivalent systems as follows:

$$\begin{cases} x + y - 337 = 0, \\ x - y - 43 = 0. \end{cases}$$

Adding these equations, you get the system

$$\begin{cases} 2x - 380 = 0, \\ x - y - 43 = 0. \end{cases}$$

Thus $x = 190$, and

$$\begin{aligned} 190 - y - 43 &= 0, \\ y &= 147. \end{aligned}$$

Therefore, the first number is 190 and the second is 147.

Check.

$$\begin{array}{r} 190 \\ +\ 147 \\ \hline 337 \end{array} \qquad \begin{array}{r} 190 \\ -\ 147 \\ \hline 43 \end{array}$$

Problem 2. Richard has a bank in which he saves only pennies and nickels. One day before Richard went to school, he asked his mother to open the bank and count the amount of money in it. When Richard got home, his mother said to him, "I opened your bank today and found that you have 66 coins in it worth \$2.38." Richard asked, "How many pennies and how many nickels were there?" His mother said that she forgot to keep track, but she was sure Richard could figure it out without reopening his bank.

Solution. The gist of this conversation is that

number of pennies + number of nickels = 66.
value of pennies in cents + value of nickels in cents = 238.

If you let

p designate the number of pennies

and

n designate the number of nickels,

then

$$p + n = 66.$$

What is the value of the pennies? Since each is worth 1 cent, p pennies are worth p cents. The value of each nickel is 5 cents, so n nickels have a total value of $5n$ cents. Thus, by what is given,

$$p + 5n = 238.$$

In algebraic language, the problem is to solve the system of linear equations in the two variables p and n.

$$\begin{cases} p + n = 66 \\ p + 5n = 238 \end{cases}$$

This system is solved as follows. Take the equivalent system

$$\begin{cases} p + n - 66 = 0, \\ p + 5n - 238 = 0 \end{cases}$$

and subtract the first equation from the second to get another equivalent system:

$$\begin{cases} p + n - 66 = 0, \\ 4n - 172 = 0. \end{cases}$$

Thus,

$$4n = 172,$$
$$n = 43,$$

and

$$p + 43 - 66 = 0,$$
$$p = 23.$$

You can conclude that Richard's bank contained 23 pennies and 43 nickels.

Check. There are 23 plus 43, or 66, coins in all. The pennies are worth 23¢, and the nickels are worth $5 \cdot 43$, or 215¢, a total of 238¢, or $2.38.

Problem 3. The sum of the digits of a two-digit number is 11. The number itself is 7 more than twice the number which results from interchanging the digits. Find the number.

Solution. Let

t be the tens' digit and u be the units' digit

of the number. By what is given,

$$t + u = 11.$$

The number itself is

$$10t + u.$$

If you interchange the digits, the new number is

$$10u + t.$$

It is stated that

$$\underbrace{\text{given number}}_{10t+u}\ \underbrace{\text{is}}_{=}\ \underbrace{\text{7 more than}}_{7+}\ \underbrace{\text{twice the new number.}}_{2(10u+t)}$$

The problem is solved if you solve the system of linear equations

$$\begin{cases} t + u = 11, \\ 10t + u = 7 + 2(10u + t). \end{cases}$$

This system is equivalent to each of the following systems.

$$\begin{cases} t = -u + 11 \\ 10t + u = 7 + 20u + 2t \end{cases}$$

$$\begin{cases} t = -u + 11 \\ 10(-u + 11) + u = 7 + 20u + 2(-u + 11) \end{cases}$$

$$\begin{cases} t = -u + 11 \\ 27u = 81 \end{cases}$$

Hence, $u = 3$ and $t = 8$.

Check. The given number is 83, the new 38:

$$83 \stackrel{\checkmark}{=} 7 + 2 \cdot 38.$$

Exercises

In each of the following exercises, write a system of equations. Tell what is denoted by each variable that you use, and solve the systems.

1. (a) The sum of two integers is 36, and their difference is 4. Find the integers.
(b) The sum of two numbers is 18, and their difference is 7. Find the numbers.

2. (a) The sum of two numbers is 55, and one number is 9 less than the other. Find the numbers.
(b) Find two numbers whose sum is 196 if the larger exceeds the smaller by 8.

3. (a) The sum of the digits of a two-digit number is 12. If the units' digit is 2 more than the tens' digit, find the number.
(b) The sum of the digits of a two-digit number is 10. If the units' digit is 4 less than the tens' digit, find the number.

4. (a) The sum of two integers is 51. The larger integer is 3 more than twice the smaller integer. Find the integers.
(b) The sum of two integers is 63. The smaller integer is 11 more than one-third the larger integer. Find the integers.

5. (a) The board of directors of a corporation granted a bonus of $5000 to be divided between two employees, with the senior of the two receiving $1200 more than the junior. How much did each man receive?

(b) Two men start a business, one investing more money than the other. If the first year's profit is \$10,000, of which the bigger investor should get \$900 more than his partner, how much should each man receive?

6. (a) Eight peaches and one melon cost \$1.24. At the same price, eight melons and one peach cost \$3.70. Find the cost of one melon.
   (b) For \$1 you can buy 4 cinnamon pastries and 12 jelly-filled ones, or you can buy 5 jelly-filled pastries and 10 cinnamon ones. How much does one pastry of each kind cost?

7. (a) Tickets to a high-school play cost \$1.10 for each adult and 40¢ for each child. If 360 tickets were sold for a total of \$282.60, how many tickets of each kind were sold?
   (b) A merchant received a shipment of canned beans and peas costing \$78.50. There were 50 more cans of peas than beans. How many of each kind were there if a can of peas costs 29¢ and a can of beans costs 35¢?

8. (a) A merchant has two kinds of peanuts, one kind which sells for 60¢ a pound and another kind which sells for 50¢ a pound. How many pounds of each kind should be mixed in order to have 50 pounds of mixture to sell at 52¢ a pound?
   (b) Two vats contain milk, one sample testing 3% butterfat and the other testing 7% butterfat. How much milk should be taken from each vat to produce a 10-gallon mixture which will test 4.5% butterfat?

In each of the following exercises, write a system of equations. Tell what is denoted by each variable that you use, and solve the system.

9. A sum of money amounting to \$3.55 consists of dimes and quarters. If there are 25 coins in all, how many quarters are there?

10. A child's coin bank contains \$2.46 in pennies and nickels. If the number of pennies is 10 less than 3 times the number of nickels, how many coins of each type are there in the bank?

11. A paper boy collected \$6.55, part in nickels and part in dimes. If the number of nickels was 6 more than one-half the number of dimes, how many nickels were there?

12. A man made a deposit of \$148. If his deposit consisted of 60 bills, some one-dollar bills and the rest five-dollar bills, how many bills of each kind did he have?

For Exercises 13–17, write a system of equations. Tell what is denoted by each variable that you use, and solve the system.

13. If one-half of the smaller number is subtracted from the larger of two numbers, the result is 65. Find the numbers if they differ by 35.

14. The units' digit of a two-digit number is 1 more than twice the tens' digit. Find the number if the sum of the digits is 10.

15. The sum of the digits of a two-digit number is 11. If the order of the digits is reversed, the resulting number exceeds the original number by 45. Find the original number.

16. A number is less than 100, and its tens' digit is 2 more than its units' digit. If the number with the digits reversed is subtracted from the original number, the remainder is 3 times the sum of the digits. Find the number.

17. One train, heading west, leaves a certain station. Two hours later a second train, heading east, leaves the same station. The second train is traveling 15 miles per hour faster than the first. Six hours after the second train left, the two trains are 580 miles apart. Find the rate at which each train is traveling.

8–8 SYSTEMS OF LINEAR INEQUALITIES

Every linear equation in two variables is equivalent to an equation whose left side is a linear form and whose right side is zero. Similarly, each linear inequality in two variables is equivalent to an inequality whose left side is a linear form and whose right side is zero. Some examples of inequalities in this form are

$$2x + 5y - 2 < 0,$$
$$-x + y + 1 \geqq 0.$$

The solution set of a linear inequality in two variables was discussed in Chapter 7. You learned there that the graph of a linear inequality is a *half-plane*.

A system of two linear inequalities in two variables has as its solution set the intersection of the solution sets of the two inequalities. Thus, the graph of the solution set of the system is the intersection of two half-planes.

For example, the system

$$\begin{cases} x - 1 > 0, \\ y - 2 < 0 \end{cases}$$

has $S \cap T$ as its solution set. Here

$$S = \{(x, y) \mid x > 1\} \quad \text{and} \quad T = \{(x, y) \mid y < 2\}.$$

Thus,

$$S \cap T = \{(x, y) \mid x > 1 \quad \text{and} \quad y < 2\}.$$

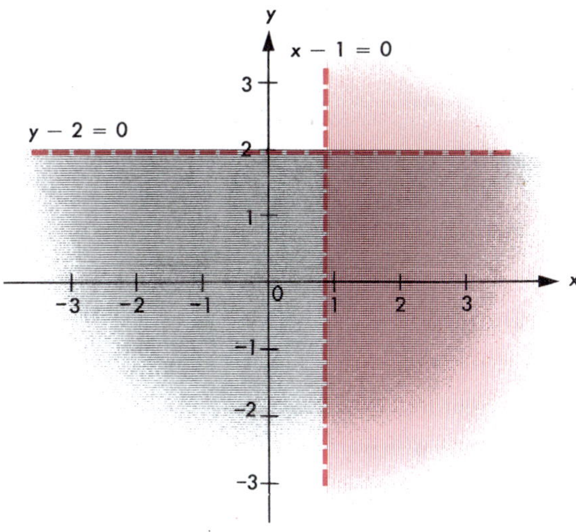

FIGURE 8-4

The graph of S is the half-plane to the right of the line $x = 1$, and the graph of T is the half-plane below the line $y = 2$. Therefore, the graph of the system is the doubly shaded quarter-plane of Fig. 8-4. The boundary lines are broken to indicate that they are not part of the graph.

Problem 1. Indicate the graph of the system of inequalities

$$\begin{cases} y - x > 0, \\ y - 1 \leqq 0. \end{cases}$$

Solution. The graph is the doubly shaded region of Fig. 8-5 above the line $y - x = 0$ and below and including the line $y - 1 = 0$.

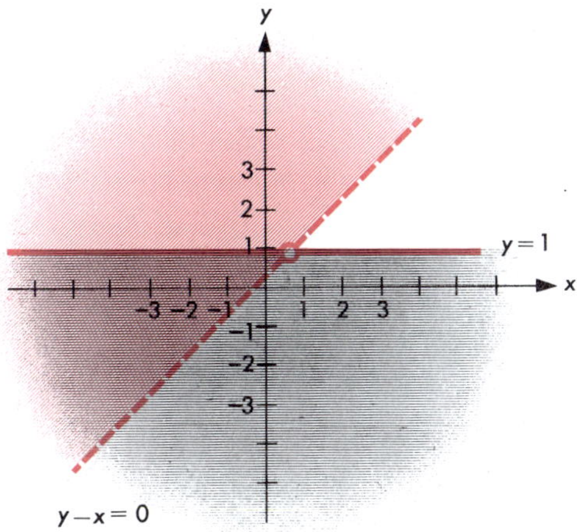

FIGURE 8-5

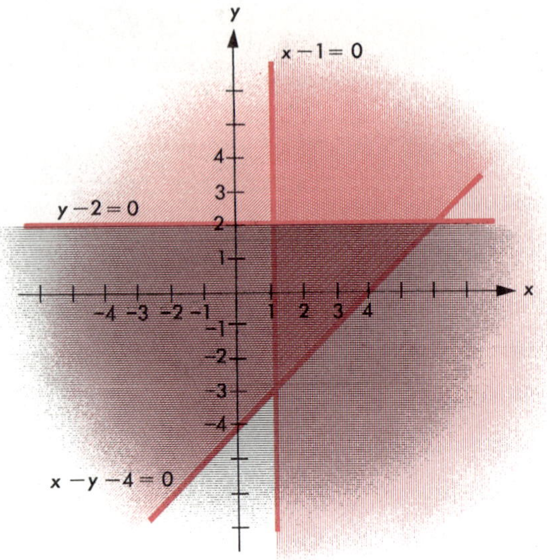

FIGURE 8–6

Problem 2. Find the graph of the system of three inequalities

$$\begin{cases} x - 1 \geqq 0, \\ y - 2 \leqq 0, \\ x - y - 4 \leqq 0. \end{cases}$$

Each inequality has a half-plane and its edge as its graph. The intersection of all three half-planes is the triply shaded triangle of Fig. 8–6. This graph is complete, unlike those in Fig. 8–4 and Fig. 8–5.

Exercises

1. (a) Given $A = \{(x, y) \mid 2y + x \geqq 0\}$ and $B = \{(x, y) \mid 3x \geqq -y\}$.
  (i) Graph A, B, and $A \cap B$.
  (ii) Give three elements of A not in B.
  (iii) Give the quadrants containing points not in $A \cup B$.

(b) Given $A = \{(x, y) \mid y - x > 0\}$ and $B = \{(x, y) \mid y - 1 \leqq 0\}$.
  (i) Graph A, B, and $A \cap B$.
  (ii) Give three elements of A not in B and three elements of B not in A.
  (iii) Give three elements not in $A \cup B$ and three elements of $A \cap B$.

Sketch the graph of each system of inequalities.

2. (a) $\begin{cases} x - 5 < 0 \\ y + 1 < 0 \end{cases}$ (b) $\begin{cases} x + 5 > 0 \\ y - 4 > 0 \end{cases}$

3. (a) $\begin{cases} x < 3 \\ y > 2 \end{cases}$ (b) $\begin{cases} x > 2 \\ y < -2 \end{cases}$

4. (a) $\begin{cases} x + y > 0 \\ x < 3 \end{cases}$ (b) $\begin{cases} x + y < 0 \\ y > -2 \end{cases}$

5. (a) $\begin{cases} y + x + 2 < 0 \\ y - 2 > x \end{cases}$ (b) $\begin{cases} y - x + 2 < 0 \\ 2y - 2 > x \end{cases}$

6. (a) $\begin{cases} x + 3 > y \\ x > -3 \\ y < -2 \end{cases}$ (b) $\begin{cases} y < x \\ x + y > 0 \\ x \leqq 2 \end{cases}$

7. (a) $\begin{cases} 2x + y \geqq 1 \\ y + x < 0 \\ y \geqq -3 \end{cases}$ (b) $\begin{cases} 4x + y \geqq 2 \\ x + 2y \leqq -1 \\ x \geqq 5 \end{cases}$

---

8. Given $A = \{(x, y) \mid x - 1 \geqq 0\}$,
$B = \{(x, y) \mid y - 2 \leqq 0\}$,
$C = \{(x, y) \mid x - y - 1 \leqq 0\}$.

(a) In what quadrant do the elements of $A \cap C$ lie?
(b) In what quadrants do the elements of $B \cap C$ lie?
(c) In what quadrants do the elements of $A \cap B$ lie?
(d) In what quadrants do the elements of $A \cap B \cap C$ lie?
(e) Give three elements of $A \cap B \cap C$.

Graph each of the following systems of inequalities. Describe each graph, giving its shape. Name three elements of each solution set.

9. (a) $\begin{cases} y \geqq x \\ y \leqq 2x \\ y \leqq 4 - x \end{cases}$ (b) $\begin{cases} y \geqq x - 5 \\ y \leqq 5 \\ x \geqq 0 \\ y + 2 \geqq 0 \end{cases}$

(c) $\begin{cases} y \geqq 0 \\ x \geqq y \\ y \leqq 3 \\ x \leqq y + 4 \end{cases}$ (d) $\begin{cases} x \geqq 2y \\ y + 2x \geqq 0 \\ x \leqq 2y + 5 \\ y \leqq 5 - 2x \end{cases}$

Describe the solution set of each of the following systems.

10. (a) $\begin{cases} y > x + 2 \\ x \geqq y - 2 \\ y + 2 \geqq 3 \end{cases}$

(b) $\begin{cases} y \leqq 2x + 4 \\ y \leqq 4 - 2x \\ y \geqq 5 \end{cases}$

8–9 DETERMINANTS

A system of two linear equations in variables x and y has the form

$$\begin{cases} a_1x + b_1y = c_1, \\ a_2x + b_2y = c_2 \end{cases}$$

for some numbers a_1, b_1, c_1, a_2, b_2, c_2. The coefficients of x and y form a square array of numbers

$$A = \begin{pmatrix} a_1 & b_1 \\ a_2 & b_2 \end{pmatrix},$$

called a 2×2 *matrix.* Two other 2×2 matrices associated with the system above are

$$B = \begin{pmatrix} c_1 & b_1 \\ c_2 & b_2 \end{pmatrix} \quad \text{and} \quad C = \begin{pmatrix} a_1 & c_1 \\ a_2 & c_2 \end{pmatrix}.$$

Every 2×2 matrix has associated with it a number called its *determinant,* defined as follows.

Definition of a determinant

If

$$A = \begin{pmatrix} a_1 & b_1 \\ a_2 & b_2 \end{pmatrix}$$

is a 2×2 matrix, then the determinant of A is denoted by $|A|$ and is defined by

$$|A| = a_1b_2 - b_1a_2.$$

For example,

$$\begin{vmatrix} 3 & -2 \\ 4 & 3 \end{vmatrix} = 3 \cdot 3 - (-2) \cdot 4, \quad \text{or } 17.$$

Thus, the determinants of the matrix $\begin{pmatrix} 3 & -2 \\ 4 & 3 \end{pmatrix}$ is the number 17.

Determinants may be used to solve systems of linear equations according to the following theorem.

THEOREM.

The system of linear equations

$$\begin{cases} a_1x + b_1y = c_1, \\ a_2x + b_2y = c_2 \end{cases}$$

has solution

$$x = \frac{\begin{vmatrix} c_1 & b_1 \\ c_2 & b_2 \end{vmatrix}}{\begin{vmatrix} a_1 & b_1 \\ a_2 & b_2 \end{vmatrix}} \quad \text{and} \quad y = \frac{\begin{vmatrix} a_1 & c_1 \\ a_2 & c_2 \end{vmatrix}}{\begin{vmatrix} a_1 & b_1 \\ a_2 & b_2 \end{vmatrix}}$$

if

$$\begin{vmatrix} a_1 & b_1 \\ a_2 & b_2 \end{vmatrix} \neq 0.$$

The proof of this theorem will not be given at this time. Suffice it to say that these are the values obtained for x and y if the system is solved by any other method.

Problem 1. Solve the system of linear equations

$$\begin{cases} 2x + 3y = 4, \\ x + 2y = -7. \end{cases}$$

Solution. By the theorem, the solution is

$$x = \frac{\begin{vmatrix} 4 & 3 \\ -7 & 2 \end{vmatrix}}{\begin{vmatrix} 2 & 3 \\ 1 & 2 \end{vmatrix}} = \frac{29}{1}, \qquad y = \frac{\begin{vmatrix} 2 & 4 \\ 1 & -7 \end{vmatrix}}{\begin{vmatrix} 2 & 3 \\ 1 & 2 \end{vmatrix}} = \frac{-18}{1},$$

or $(29, -18)$.

Check.

$$\begin{aligned} 2x + 3y &= 4 \\ 2 \cdot 29 + 3 \cdot (-18) &\stackrel{?}{=} 4 \\ 58 - 54 &\stackrel{\checkmark}{=} 4 \end{aligned} \qquad \begin{aligned} x + 2y &= -7 \\ 29 + 2 \cdot (-18) &\stackrel{?}{=} -7 \\ 29 - 36 &\stackrel{\checkmark}{=} -7 \end{aligned}$$

Problem 2. Solve the system of linear equations

$$\begin{cases} 5x - 2y = 3, \\ 4x - 3y = -2. \end{cases}$$

Solution. The solution is

$$x = \frac{\begin{vmatrix} 3 & -2 \\ -2 & -3 \end{vmatrix}}{\begin{vmatrix} 5 & -2 \\ 4 & -3 \end{vmatrix}} = \frac{-13}{-7}, \qquad y = \frac{\begin{vmatrix} 5 & 3 \\ 4 & -2 \end{vmatrix}}{\begin{vmatrix} 5 & -2 \\ 4 & -3 \end{vmatrix}} = \frac{-22}{-7},$$

or $(\frac{13}{7}, \frac{22}{7})$.

Check.

$$\begin{aligned} 5x - 2y &= 3 \\ 5 \cdot \tfrac{13}{7} - 2 \cdot \tfrac{22}{7} &\stackrel{?}{=} 3 \\ \tfrac{65}{7} - \tfrac{44}{7} &\stackrel{\checkmark}{=} 3 \end{aligned} \qquad \begin{aligned} 4x - 3y &= -2 \\ 4 \cdot \tfrac{13}{7} - 3 \cdot \tfrac{22}{7} &\stackrel{?}{=} -2 \\ \tfrac{52}{7} - \tfrac{66}{7} &\stackrel{\checkmark}{=} -2 \end{aligned}$$

For a system such as

$$\begin{cases} 3x + 4y = 1, \\ 9x + 12y = -2, \end{cases}$$

the theorem cannot be used because the determinant

$$\begin{vmatrix} 3 & 4 \\ 9 & 12 \end{vmatrix}$$

which is in the denominator of x and y is zero. This system may be shown to be inconsistent, that is, it has no solution.

Exercises

Evaluate each of the following determinants.

1. (a) $\begin{vmatrix} 3 & 4 \\ 2 & 1 \end{vmatrix}$ (b) $\begin{vmatrix} 1 & 2 \\ 3 & 3 \end{vmatrix}$

2. (a) $\begin{vmatrix} -3 & -1 \\ 2 & 0 \end{vmatrix}$ (b) $\begin{vmatrix} 0 & -3 \\ -2 & -1 \end{vmatrix}$

3. (a) $\begin{vmatrix} 5 & -5 \\ 5 & -2 \end{vmatrix}$ (b) $\begin{vmatrix} -3 & 1 \\ -4 & -2 \end{vmatrix}$

Use determinants to solve each of the following systems of linear equations.

4. (a) $\begin{cases} 2x + 5y = 7 \\ 3x - 2y = 1 \end{cases}$ (b) $\begin{cases} 4x - 5y = 3 \\ 5x + 3y = 13 \end{cases}$

5. (a) $\begin{cases} 3x - 2y = -1 \\ 2x - 3y = 1 \end{cases}$ (b) $\begin{cases} 4x + 7y = 8 \\ 7x - 4y = 14 \end{cases}$

6. (a) $\begin{cases} 5m + 7n = 2 \\ 3m - 4n = -7 \end{cases}$ (b) $\begin{cases} 3p - 7q = 1 \\ 8p + q = 42 \end{cases}$

7. (a) $\begin{cases} 8x + 7y = 3 \\ 3x - 4y = 1 \end{cases}$ (b) $\begin{cases} 6x + 4y = 0 \\ 3x - y = 1 \end{cases}$

8. (a) $\begin{cases} x + 5y = 6 \\ 5x - 4y = 4 \end{cases}$ (b) $\begin{cases} 5x - 3y = 6 \\ 3x + 7y = 1 \end{cases}$

9. (a) $\begin{cases} 8x - y = -1 \\ -3x + 5y = 0 \end{cases}$ (b) $\begin{cases} 2x - 6y = 1 \\ x + 4y = 3 \end{cases}$

10. (a) $\begin{cases} 6x - y = -2 \\ x + 4y = -3 \end{cases}$ (b) $\begin{cases} 2x - 5y = -6 \\ x + y = -3 \end{cases}$

11. (a) $\begin{cases} 3x - 5y = -10 \\ 2x + y = 11 \end{cases}$ (b) $\begin{cases} 10x - y = 0 \\ 2x + 8y = 5 \end{cases}$

12. (a) $\begin{cases} 3x - 7y = 2 \\ 6x - 14y = 5 \end{cases}$ (b) $\begin{cases} 4x + 7y = -4 \\ -8x - 14y = -4 \end{cases}$

13. (a) $\begin{cases} 2x + \frac{1}{2}y = 1 \\ 2x - \frac{1}{2}y = 0 \end{cases}$ (b) $\begin{cases} 3x - \frac{1}{3}y = 1 \\ 3x + \frac{1}{3}y = 0 \end{cases}$

14. (a) $\begin{cases} 2x + 11y = 2 \\ -2x - 3y = 4 \end{cases}$ (b) $\begin{cases} -5x - y = 1 \\ 4x + 5y = 3 \end{cases}$

15. (a) $\begin{cases} 3x + 8y = 2 \\ 8x - 3y = 2 \end{cases}$ (b) $\begin{cases} 4x - 7y = 1 \\ 7x - 4y = 1 \end{cases}$

8-10 SYSTEMS OF THREE LINEAR EQUATIONS IN THREE VARIABLES

One way to generalize the material in this chapter is to consider systems of three linear equations in two unknowns. For example, you might try to solve a system such as the following.

$$\begin{cases} x + 2y - 5 = 0 \\ 2x + y + 1 = 0 \\ -x + 4y + 7 = 0 \end{cases}$$

You can do this because it does not involve anything new. All you need to do is find the solution of the system

$$\begin{cases} x + 2y - 5 = 0, \\ 2x + y + 1 = 0, \end{cases}$$

and then check to see whether it also is a solution of the third equation.

Another way to generalize the previous work is to consider three variables instead of two. For example, you might try to solve a system such as the one in the following problem.

Problem 1. Solve the system of three equations in three variables

$$\begin{cases} 3x + 4y - z = -3, \\ \quad 4y + 3z = 4, \\ \quad\quad z = 4. \end{cases}$$

Solution. You are asked to find an ordered number triple of values for x, y, and z which makes all three equations true. Looking at the last equation, you see that you must let $z = 4$ to make it a true equation. If you let $z = 4$ in the second equation, you get

$$\begin{aligned} 4y + 12 &= 4, \\ 4y &= -8, \\ y &= -2. \end{aligned}$$

Therefore, you must let $y = -2$ to make the second equation also true. If you let $y = -2$ and $z = 4$ in the first equation, you get

$$\begin{aligned} 3x - 8 - 4 &= -3, \\ 3x &= 9, \\ x &= 3. \end{aligned}$$

Thus, x must equal 3 to make the first equation also true. Therefore, the only solution of the system is

$$(3, -2, 4).$$

Check.

$$\begin{aligned} 3x + 4y - z &= -3 \\ 3 \cdot 3 + 4 \cdot (-2) - 4 &\stackrel{?}{=} -3 \\ 9 - 8 - 4 &\stackrel{?}{=} -3 \\ 9 - 12 &\stackrel{\checkmark}{=} -3 \end{aligned} \qquad \begin{aligned} 4y + 3z &= 4 \\ 4 \cdot (-2) + 3 \cdot 4 &\stackrel{?}{=} 4 \\ -8 + 12 &\stackrel{\checkmark}{=} 4 \end{aligned} \qquad \begin{aligned} z &= 4 \\ 4 &\stackrel{\checkmark}{=} 4 \end{aligned}$$

Every system of three linear equations in three unknowns is equivalent to one in the form of Problem 1, as is illustrated in the next problem.

Problem 2. Michael's bank contains only nickels, dimes and quarters. There are 30 coins in all and their value is \$4.30. Twice the number of dimes added to the number of quarters is five times the number of nickels. How many coins of each kind are there in Michael's bank?

Solution. The fact that there are 30 coins in all means that

$$N + D + Q = 30.$$

Here N is the number of nickels, D of dimes, and Q of quarters. The value of the nickels (in dollars) is $.05N$; of the dimes, $.10D$; and of the quarters, $.25Q$. The second fact given is that

$$.05N + .10D + .25Q = 4.30.$$

Finally, it is stated that

$$2D + Q = 5N.$$

Hence, you are asked to solve the system of three equations in three variables

$$\begin{cases} N + D + Q = 30, \\ .05N + .10D + .25Q = 4.30, \\ 2D + Q = 5N. \end{cases}$$

Your first step in solving this system is as before; replace each equation by an equivalent one having a linear form set equal to zero.

$$\begin{cases} N + D + Q - 30 = 0 \\ .05N + .10D + .25Q - 4.30 = 0 \\ -5N + 2D + Q = 0 \end{cases}$$

Next, multiply each side of the first equation by 5 and of the second equation by 100, obtaining the equivalent system

$$\begin{cases} 5N + 5D + 5Q - 150 = 0, \\ 5N + 10D + 25Q - 430 = 0, \\ -5N + 2D + Q = 0. \end{cases}$$

The equivalent system

$$\begin{cases} 5N + 5D + 5Q - 150 = 0, \\ 5D + 20Q - 280 = 0, \\ 7D + 6Q - 150 = 0 \end{cases}$$

is obtained from the preceding one by keeping the first equation, subtracting the first equation from the second to get the new second equation, and adding the first equation to the third to get the new third equation.

Now divide each side of the first two equations by 5 to obtain the equivalent system

$$\begin{cases} N + D + Q - 30 = 0, \\ D + 4Q - 56 = 0, \\ 7D + 6Q - 150 = 0. \end{cases}$$

You can make the coefficients of D equal in the last two equations by multiplying each side of the second equation by 7. This gives you the equivalent system

$$\begin{cases} N + D + Q - 30 = 0, \\ 7D + 28Q - 392 = 0, \\ 7D + 6Q - 150 = 0. \end{cases}$$

Keeping the first two equations of this system and subtracting the third equation from the second, you get the equivalent system

$$\begin{cases} N + D + Q - 30 = 0, \\ 7D + 28Q - 392 = 0, \\ 22Q - 242 = 0. \end{cases}$$

It is easy to see that this system is equivalent to the system

$$\begin{cases} N + D + Q - 30 = 0, \\ D + 4Q - 56 = 0, \\ Q - 11 = 0. \end{cases}$$

The system now looks like that given in Problem 1, and you can solve the third equation to get $Q = 11$. Then from the second equation, you get $D = 12$, and finally, from the first equation, $N = 7$.

Check.

$$N + D + Q = 30$$
$$7 + 12 + 11 \stackrel{\checkmark}{=} 30$$

$$.05N + .10D + .25Q = 4.30$$
$$.05 \cdot 7 + .10 \cdot 12 + .25 \cdot 11 \stackrel{?}{=} 4.30$$
$$.35 + 1.20 + 2.75 \stackrel{\checkmark}{=} 4.30$$

$$2D + Q = 5N$$
$$2 \cdot 12 + 11 \stackrel{?}{=} 5 \cdot 7$$
$$24 + 11 \stackrel{\checkmark}{=} 35$$

Exercises

Solve each of the following systems of linear equations. Check your answers.

1. (a) $\begin{cases} 3x + 4y + 2z = 5 \\ 2x + 3z = -1 \\ x = 1 \end{cases}$ (b) $\begin{cases} x + 2y - z = 4 \\ y = 4 \\ 2y - z = 5 \end{cases}$

2. (a) $\begin{cases} 2a - b + c = 8 \\ 3a + b = 5 \\ a - 2b = 4 \end{cases}$ (b) $\begin{cases} 3a - c = 2 \\ 2a + 3c = 5 \\ a + b - c = 1 \end{cases}$

3. (a) $\begin{cases} x + y = 4 \\ 2y + z = 6 \\ 3x - z = 4 \end{cases}$ (b) $\begin{cases} 2x - 3y = -3 \\ 3y + 2z = 15 \\ 2x - z = 3 \end{cases}$

4. (a) $\begin{cases} p + 2q + 3r = 9 \\ 3p - q + 2r = -1 \\ -3p + 5q - 4r = -7 \end{cases}$ (b) $\begin{cases} x + 2y + z = 0 \\ 2x - 3y - 4z = 5 \\ 3x + 4y - z = 2 \end{cases}$

5. (a) $\begin{cases} 2r - 3s + 4t = 6 \\ 5r + s + 3t = 14 \\ 3r + 2s - t = 0 \end{cases}$ (b) $\begin{cases} 2x + 3y - z = 5 \\ 7x - 2y - 3z = -5 \\ 5x + 4y - 2z = 8 \end{cases}$

6. (a) $\begin{cases} 3a - 2c + b = -1 \\ 2a - b + 3c = 9 \\ b + 3a - c = 2 \end{cases}$ (b) $\begin{cases} 2x - 3y + 8z = 2 \\ 3x - 3y + 2z = 1 \\ 5x + 6y - 6z = 3 \end{cases}$

---

7. A certain number consists of three digits. If the digits in the units' and tens' positions are interchanged, the number is increased by 18. If the digits in the hundreds' and tens' positions are interchanged, the number is diminished by 90. The sum of the digits is 12. Find the number.

8. The sum of three angles of a triangle is 180°. The largest angle is 5 times as large as the smallest one, and equal to the sum of the two smaller angles. Find the three angles.

9. A grain dealer sold to one customer 5 bushels of wheat, 2 of corn, and 3 of rye, for \$6.60; to another, 2 of wheat, 3 of corn, and 5 of rye, for \$5.80; and to a third, 3 of wheat, 5 of corn, and 2 of rye, for \$5.60. What was the price per bushel of each kind of grain?

---

10. A man goes into a store and borrows as much money as he has and then spends \$10. After doing exactly the same thing in two other stores, he finds that he has no money left. How much money did he initially have?

11. If, in the preceding problem, the man can perform n such borrowing and spending operations before his money is depleted, how much money did he initially have?

*8-11 A LINEAR PROGRAM

Inequalities are used extensively in describing economic problems. The following example illustrates a problem of this type.

Problem. A farmer has at his disposal 100 acres of land for planting, \$450 to spend for seed, and 2400 man-hours of labor to work the land. He decides that he would like to plant two different crops on his land: crop A for which the seed costs \$5 an acre and which requires 8 man-hours of labor per acre, and crop B for which the seed costs \$3 an acre, but which requires 30 man-hours of labor per acre. How many acres of crop A and how many of crop B should he plant?

Solution. If you let

A designate the number of acres of crop A

and

B designate the number of acres of crop B,

then, by the nature of the problem,

$$A \geqq 0 \quad \text{and} \quad B \geqq 0.$$

It seems logical to write the system of equations

$$\begin{cases} A + B = 100, \\ 5A + 3B = 450, \\ 8A + 30B = 2400 \end{cases}$$

to describe the various conditions imposed by the problem. However, this system has no solution. That it has no solution is clear from its graph shown in Fig. 8–7; notice that the graphs of the three equations have no point in common.

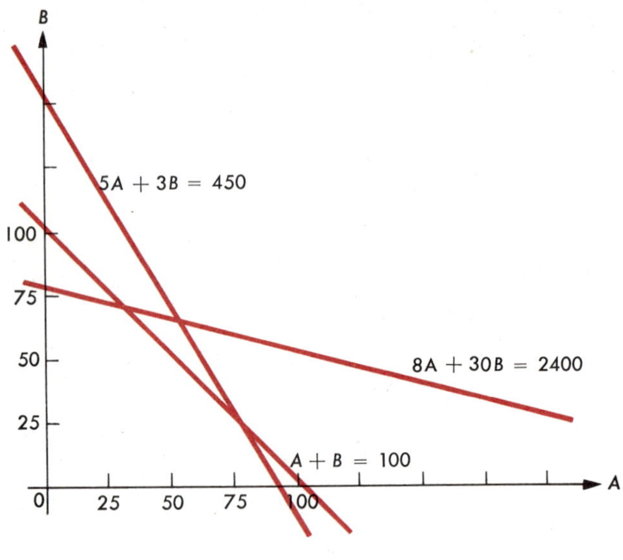

FIGURE 8–7

On second thought, you realize that you should have used inequalities instead of equations. What is really given is that the farmer can plant *no more* than 100 acres, can spend *no more* than \$450, and can use *no more* than 2400 man-hours of labor. Thus, you should have considered this system of five inequalities:

$$\begin{cases} A \geqq 0, & \\ B \geqq 0, & \\ A + B \leqq 100, & \text{(acreage)} \\ 5A + 3B \leqq 450, & \text{(seed cost)} \\ 8A + 30B \leqq 2400. & \text{(labor hours)} \end{cases}$$

The graph of the solution set of this system is the shaded pentagon of Fig. 8–8. Be sure you see which inequality is associated with each of the five sides of the pentagon. The coordinates of each point inside (and on) the pentagon give possible replacements for A and B.

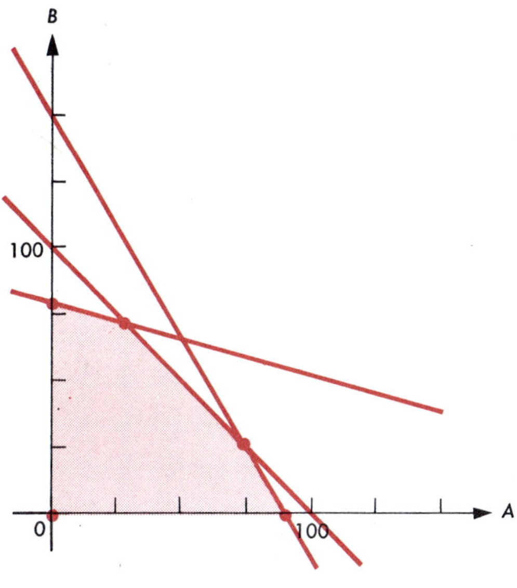

FIGURE 8–8

You do not have enough information to advise the farmer which of the many possible values of A and B he should use. It is probable that the farmer wants to choose values of A and B so that he receives maximum income from his land. You would have to know how much income he receives from one acre each of crop A and of crop B and how much he paid for labor costs before you could possibly advise him.

The problem of choosing the point of the pentagon that gives the desired number pair may be solved by a branch of modern mathematics known as *linear programming.* Linear inequalities are used a great deal in this branch of mathematics.

If a linear form in the variables A and B is evaluated at each point inside and on the boundary of the pentagon in Fig. 8–8, then, according to a result proved in the theory of linear programming, the maximum value of the linear form occurs at a vertex of the pentagon, and the minimum value occurs at another vertex. You may use this result in working the following exercises.

Exercises

1. Find the coordinates of each of the vertices of the pentagon in Fig. 8–8, and explain what each of these coordinates means in terms of land to be planted, labor to be used, and money to be spent on seed.
2. If labor costs the farmer \$1.50 per man-hour and he can sell his harvest at \$28 for an acre's yield of crop A and \$70 for an acre's yield of crop B, write his profit in terms of the number of acres planted to each crop.
3. Using the values given in Exercise 2, find the value of the profit at each vertex of the pentagon of Fig. 8–8.

KEY IDEAS AND KEY WORDS

The **solution set** of a system of two linear equations in two variables is the set of all ordered pairs which are solutions of both equations. If S is the solution set of one equation and T of the other, then

$$S \cap T$$

is the solution set of the system of two equations.

Two systems of equations are said to be **equivalent** if they have the same solution set. Two methods of obtaining equivalent systems from a given system are the **substitution method** and the **addition or subtraction method.** A system can also be solved by **graphing.**

The **solution set** of a system of inequalities is the intersection of the solution sets of the individual inequalities.

If

$$A = \begin{pmatrix} a_1 & b_1 \\ a_2 & b_2 \end{pmatrix}$$

is a 2×2 matrix, then the **determinant** of A is denoted by $|A|$ and is defined by

$$|A| = a_1b_2 - a_2b_1.$$

Linear programming is a branch of modern mathematics by which particular solutions of systems of linear inequalities are found.

CHAPTER REVIEW

1. Solve graphically each of the following systems of linear equations. Give the slope and y-intercept of each line.

   (a) $\begin{cases} x + 2y = 4 \\ y = 4 \end{cases}$

   (b) $\begin{cases} y = 3x + 1 \\ x + 2y = 2 \end{cases}$

2. If $A = \{(x, y) \mid 3x = y\}$ and $B = \{(x, y) \mid 3x + y + 6 = 0\}$, find $A \cap B$ graphically.

3. Use the substitution method to find the solution set of each system of linear equations. Check each answer by substituting into both equations of the given system.

   (a) $\begin{cases} 2x - y + 1 = 0 \\ 3y - 4x = 4 \end{cases}$ (b) $\begin{cases} 3x - 7y = 43 \\ 2x + 3y = -2 \end{cases}$

4. Solve by the addition or subtraction method. Check your answers.

   (a) $\begin{cases} 2x - 6y + 1 = 0 \\ 6x + 6y - 5 = 0 \end{cases}$ (b) $\begin{cases} 3x - 4y = 14 \\ 2x + 3y + 19 = 0 \end{cases}$

5. Use determinants to solve each of the following systems.

   (a) $\begin{cases} 2x + 5y = 1 \\ 5x - y = 2 \end{cases}$ (b) $\begin{cases} 5x + 3y = -2 \\ -3x - y = 4 \end{cases}$

6. Solve the following systems by any method.

   (a) $\begin{cases} 3x - 15y = 51 \\ 4x + 15y = 12 \end{cases}$

   (b) $\begin{cases} \frac{1}{6}x + 4y = 9 \\ 4x + \dfrac{y}{2} = 25 \end{cases}$

   (c) $\begin{cases} 2x + 3y + z = 7 \\ 3x - y - z = 5 \\ x + 2y - 2z = 4 \end{cases}$

7. If twice one number is subtracted from three times another number, the difference is 7. Find the numbers, given that their sum is -1.

8. Separate 42 into two parts so that the smaller part is 3 more than half the larger part.

9. A grocer has one type of coffee which he sells for 80¢ per pound and another type which he sells for \$1.04 per pound. How many pounds of each should he use in 24 pounds of a coffee blend which is to sell for 90¢ a pound?

10. A jet plane made the 3000-mile trip to Europe in 5 hours and returned in 6 hours. If the speed of the wind was constant throughout the round trip, what was the speed of the wind?

11. A motorist drove 120 miles at a certain rate of speed. The return trip, made over the same route at twice his original rate, took 3 hours less. Find his rate on the first trip.

12. The tens' digit of a two-digit number is two more than the units' digit. If the digits are reversed, the number is decreased by 18. Find the number.

13. (a) A box contains 31 coins worth \$4.15. If the coins are all nickels and quarters, how many of each are there?
    (b) Could a box of 31 coins, consisting of pennies and quarters only, be worth \$4.15? If so, how many of each kind would there be?

14. Graph each system of inequalities. Give three elements of each solution set.

(a) $\begin{cases} 6x - y \leqq 0 \\ 6y - x \leqq 0 \end{cases}$ (b) $\begin{cases} x + y + 5 > 0 \\ x - y + 5 < 0 \end{cases}$

15. Graph each of the following systems of inequalities. Describe each graph, giving its shape and the coordinates of the vertices. Find three elements of the solution set of each system.

(a) $\begin{cases} y \geqq -x + 1 \\ y \geqq x - 1 \\ y \leqq 3 \end{cases}$ (b) $\begin{cases} y \leqq 2 \\ y \leqq 2x + 4 \\ y \geqq -2 \\ y \leqq 4 - 2x \end{cases}$

CHAPTER TEST

1. Given $A = \{(x, y) \mid 2x + y + 4 = 0\}$,
   $B = \{(x, y) \mid x - 2y + 7 = 0\}$.

   Find $A \cap B$ graphically.

2. Solve the following systems of linear equations by substitution.

(a) $\begin{cases} x + 2y = 5 \\ x = y - 3 \end{cases}$ (b) $\begin{cases} 2x - 3y = 10 \\ 5x + y = 8 \end{cases}$

3. Use the addition or subtraction method to find the solution set of the following systems of equations.

(a) $\begin{cases} 3x + 4y = 24 \\ 2x - 4y = -20 \end{cases}$ (b) $\begin{cases} 2x + 5y = 14 \\ 4x - 2y + 8 = 0 \end{cases}$

4. Use determinants to solve the system of equations

$$\begin{cases} 3x - 6y + 1 = 0, \\ 2x + 5y = -3. \end{cases}$$

5. One day a contractor has 11 men and 6 boys on his payroll, and the next day he has 7 men and 8 boys. If his payroll is \$181 the first day and \$134 the second day, how much does he pay each man and each boy per day?

6. If the digits of a two-digit number are interchanged, the new number is seven times the tens' digit of the original number. The sum of the digits is eight. Find the original number.

7. Sketch the graph of the system of three inequalities, and give an element of the solution set.

$$\begin{cases} x - 4 \geqq 0 \\ y - 3 \leqq 0 \\ x + y - 2 \geqq 0 \end{cases}$$

CHAPTER 9

Polynominals in Two Variables

Objectives . . .

- To understand the basic arithmetic operations on polynomials.
- To apply the third and fourth laws of exponents to monomials expressions.
- To increase your ability to find products of binomials.
- To increase your ability to factor polynomials.

9–1 ADDITION AND SUBTRACTION OF POLYNOMIALS

The simplest mathematical expression involving two or more variables is the polynomial. Polynomials in two variables will be studied in this chapter. As in Chapter 5, polynomials will be considered to be basic elements of a new algebraic system. Sums and products of polynomials are also polynomials, and the fundamental properties of these operations given in Chapter 2 are valid in this new system. All in all, the algebraic system of polynomials in two variables is quite similar to the system of polynomials in one variable.

A linear form in two variables is an example of a polynomial in two variables. Other examples of polynomials are

$$x^2 - 6xy + 9y^2 + 7,$$
$$a^2 - 3a + 7ab + 5ab^2 - 3,$$
$$uv^3.$$

Thus, a polynomial in two variables consists of one term or a sum of terms, where each term is either a number or a product of a number and powers of the two variables. The terms of a polynomial are called *monomials.* For example,

$$x^2, \quad -6xy, \quad 9y^2, \quad 7$$

are monomials in x and y. The numbers 9 in $9y^2$ and -6 in $-6xy$ are called the *coefficients* of the monomials. The monomial 7 in the polynomial $x^2 - 6xy + 9y^2 + 7$ is called the *constant term* of the polynomial.

Polynomials in two variables are added or subtracted in much the same way as polynomials in one variable. Thus, monomials containing the same powers of the same variables are combined by use of the rearrangement property and the distributive axiom.

Problem. Perform the indicated operations.

$$(3x^2 - 6xy + 5y^2) + (7x^2 + 4xy - 12y^2)$$

Solution.

$$\begin{aligned}(3x^2 - 6xy + 5y^2) &+ (7x^2 + 4xy - 12y^2) \\ &= (3x^2 + 7x^2) + (-6xy + 4xy) + (5y^2 - 12y^2) && \text{(R-A)} \\ &= (3 + 7)x^2 + (-6 + 4)xy + (5 - 12)y^2 && \text{(D)} \\ &= 10x^2 - 2xy - 7y^2\end{aligned}$$

Exercises

Perform the additions or subtractions, as indicated.

1. (a) $(x^2 + 3xy + y^2) + (-x^2 - 3xy - y^2)$
(b) $(3x^2 - 5xy^2 + y^3) + (-3x^2 + 5xy^2 - y^3)$

2. (a) $(2x^2 + 5xy + 4y^2) - (2x^2 + 5xy + 4y^2)$
(b) $(3x^4 - 4x^3y + xy^2) - (3x^4 - 4x^3y + xy^2)$

3. (a) $(7a^2 - 3ab + 2b^2) - (-7a^2 + 3ab - 2b^2)$
(b) $(3x^2 + 2xy - y^2) - (-3x^2 - 2xy + y^2)$

4. (a) $(x^2 + y^2) - (-x^2 + y^2)$
(b) $(3x^3 - y^3) - (-3x^3 - y^3)$

5. (a) $(2a^2 - ab + b^2) + (3a^2 + 5ab - 7b^2)$
(b) $(x^3 + 4x^2y - 2y^3) + (2x^3 - 5x^2y + 5y^3)$

6. (a) $(3x^2 - xy - 8y^2) + (5x^2 - xy - 2y^2) + (7x^2 - 3xy)$
(b) $(a^2 - b^2) + (12b^2 - 2ab + 7a^2) + (12ab - 3a^2 - 5b^2)$

7. (a) $(4x^2 - 3xy + y^2) - (2xy + y^2) - (x^2 + y^2)$
(b) $(5x^2 + 4xy - 3y^2) - (4xy - y^2) - (2x^2 + 3y^2)$

8. (a) $(a^3 - a^2b + 2ab^2 - b^3) - (a^3 - 2a^2b - 3ab^2 - b^3)$
(b) $(10x^2 + 9xy + y^2) + (-7x^2 - 2xy - 3y^2)$

9. (a) $(15x - 3xy) + (7x - 4xy) - (10x - 11xy)$
(b) $(7x - 4x^2y) + (18x + 2x^2y) - (5x - x^2y)$

10. (a) $(2a^3 - 3ab^2 + 4b^3) - (a^3 + 4ab^3 - 2b^3) + 4(b^3 - ab^2)$
(b) $(4x^3 - 2xy^2 + 5y^3) - (x^3 - 3xy^3 + 6y^3) + 2(x^3 - xy^2)$

---

11. What number must be added to $5x - 4y$ to yield 0?

12. What number must be added to $x^2 + y^2$ to yield $-y^2$?

13. Subtract $x^2 - 3y + 4$ from the sum of $2x^2 + 5y - 3$ and $3x^2 - 8y + 2$.

14. Subtract the sum of $2r - 3s + 4$ and $r + 4s - 2$ from zero.

15. Subtract the sum of $2a - 5b + 3$ and $-3a + 2b - 5$ from the sum of $a + 3b + 7$ and $a - 2b - 5$.

16. Find the perimeter of a triangle whose sides are $3x^2 + xy - y^2$, $x^2 - y^2$, and $2xy + y^2$.

Preparation for Section 9–2

Use the first law of exponents to find each of the following products.

1. $x^7 \cdot x^{12}$

2. $m^4 \cdot m^5 \cdot m$

3. $(ab)(ab)$

4. $(x^2y)(-x^3y^2)$

5. Fill in the blanks with the reasons for the key steps in the solution of the following problem.

$$3x^2y(x^2 - xy + y^2)$$
$$= (3x^2y)(x^2) - (3x^2y)(xy) + (3x^2y)(y^2) \qquad \underline{\ ?\ }$$
$$= (3 \cdot x^2 \cdot x^2 \cdot y) - (3 \cdot x^2 \cdot x \cdot y \cdot y) + (3 \cdot x^2 \cdot y \cdot y^2) \qquad \underline{\ ?\ }$$
$$= 3x^4y - 3x^3y^2 + 3x^2y^3 \qquad \underline{\ ?\ }$$

6. Find the product.

$$(2x^2 - x + 3)(x + 2)$$

9–2 MULTIPLICATION OF POLYNOMIALS

You may use the properties of real numbers to find products of polynomials in two variables. For example,

$$3x^2(-4x^2 + 2xy - 5y^2)$$
$$= (3x^2)(-4x^2) + (3x^2)(2xy) + (3x^2)(-5y^2) \qquad \text{(D)}$$
$$= [3 \cdot (-4) \cdot x^2 \cdot x^2] + (3 \cdot 2 \cdot x^2 \cdot x \cdot y) + [3 \cdot (-5) \cdot x^2 \cdot y^2] \qquad \text{(R-M)}$$
$$= -12x^4 + 6x^3y - 15x^2y^2.$$

Problem 1. Find $(2a + b)(4a^2 - 2ab + b^2)$.

Solution. A systematic arrangement of the work will make this problem easier to solve.

$$\begin{array}{lll} 4a^2 - 2ab + b^2 & & \\ \quad\quad 2a + b & & \\ \hline 8a^3 - 4a^2b + 2ab^2 & & = 2a(4a^2 - 2ab + b^2) \\ \quad\quad 4a^2b - 2ab^2 + b^3 & & = b(4a^2 - 2ab + b^2) \\ \hline 8a^3 + \quad 0 + \quad 0 + b^3 & & = \text{sum of the two lines above} \end{array}$$

Therefore, $(2a + b)(4a^2 - 2ab + b^2) = 8a^3 + b^3$.

Problem 2. Find $(3x^2 - 6xy)(2xy + 4y^2)$.

Solution.

$$\begin{array}{rrl} 3x^2 - 6xy & & \\ \underline{2xy + 4y^2} & & \\ 6x^3y - 12x^2y^2 & & = 2xy(3x^2 - 6xy) \\ & 12x^2y^2 - 24xy^3 & = 4y^2(3x^2 - 6xy) \\ \hline 6x^3y + & 0 - 24xy^3 & = \text{sum of two lines above} \end{array}$$

Thus,

$$(3x^2 - 6xy)(2xy + 4y^2) = 6x^3y - 24xy^3.$$

Problem 3. Find $(a^2 - 3a)(b^2 + 4a)$.

Solution.

$$\begin{array}{rrl} a^2 - 3a & & \\ \underline{b^2 + 4a} & & \\ a^2b^2 - 3ab^2 & & = b^2(a^2 - 3a) \\ & + 4a^3 - 12a^2 & = 4a(a^2 - 3a) \\ \hline a^2b^2 - 3ab^2 & + 4a^3 - 12a^2 & = \text{sum of two lines above} \end{array}$$

Thus,

$$(a^2 - 3a)(b^2 + 4a) = a^2b^2 - 3ab^2 + 4a^3 - 12a^2.$$

Exercises

Perform the indicated operations.

1. (a) $\frac{1}{3}xy \cdot \frac{1}{3}xy \cdot \frac{1}{3}xy$ (b) $7ab \cdot 7ab \cdot 7ab$
2. (a) $(-2uv)^3$ (b) $(-\frac{1}{2}x^2y^2)^3$
3. (a) $10xy^2 \cdot 10xy^2 \cdot 10xy^2 \cdot 10xy^2$ (b) $(-3a^8b)^4$
4. (a) $(a^2b^2)(-ab^3)(-a^3b)$ (b) $(-3a^4b^4)(-3a^4b)(3ab^4)$
5. (a) $7xy(x^2 - xy - y)$ (b) $3xy(7x^2 - 3xy - 9y^2)$
6. (a) $-5ab(-6ab + b - 1)$ (b) $-3xy(-2xy - x + 3)$
7. (a) $(x - 3y)(x + 5y)$ (b) $(x + 2y)(x - 4y)$
8. (a) $(3b + 4a)(3b - 4a)$ (b) $(3x - 5y)(3x + 5y)$
9. (a) $(x + y)(x - y)$ (b) $(x - y)^2$
10. (a) $(x - 6y)(x - 4y)$ (b) $(a - 2b)(a - 7b)$
11. (a) $(xy + 7)(xy - 9)$ (b) $(xy - 1)(xy + 8)$
12. (a) $(5a^2 - 4b)(4a - 5b^2)$ (b) $(4a^2 - 5b)(5a - 4b^2)$
13. (a) $3x(2x^2 + xy - 2y^2) - 4y(y^2 - 3xy - x^2)$
    (b) $3ab(6a^2 + 3ab - 5b^2) - 2ab(-2a^2 - 7ab)$

14. (a) $5ab(3a^2 - 4ab + b^2) - 7a^2b(5a - 2b)$
(b) $xy(x^3y - 4x^2y^2 - xy^3) - xy^2(2x^3 + 3x^2y - 4xy^2)$

15. (a) $(49x^2 + 7xy + y^2)(7x - y)$
(b) $(9x^2 - 15xy + 25y^2)(3x + 5y)$

16. (a) $(2x^2 - 3xy + 5y^2)(x + 7y)$
(b) $(4b^2 + 2ab + a^2)(2b - a)$

17. (a) $(9a^2 - 12ab + 4b^2)(3a + 2b)$
(b) $(9x^2 - 6xy + y^2)(3x - y)$

18. (a) $(x + 3xy + y)(x - 3xy + y)$
(b) $(a^2 - ab + b^2)(a^2 + ab + b^2)$

19. (a) $(x - 3y)(x + 2y)(6x - y)$
(b) $(2x - y)(x + 4y)(3x - y)$

20. (a) $(-5xy)[2x(x^3 - y^3) - 3y(xy^2 - y^3)]$
(b) $(-3x^2y^3)[3xy(x^2 - y^2) - 2x^2(x^2 + xy)]$

Perform the indicated operations.

21. $(x^3 + x^2y + xy^2 + y^3)(x - y)$

22. $(a^4 - a^3b + a^2b^2 - ab^3 + b^4)(a - b)$

23. $(x^2 - 4xy + 4y^2)(x^2 + 4xy - 4y^2)$

24. $(20x^2y^2 + 7)(20x^2y^2 - 7)^2$

25. $(2x - 3y)^2 - (2x + 3y)^2$

26. $(7a - b)(7a + b) + (7a + b)^2$

27. $(3a - 5b)(3a + 5b)(3a - 5b)$

28. $(5x + y)^2(5x + y)$

29. $(5x - y)^2(5x - y)$

30. $(a + 2b)(a - 2b)(a + 2b)$

31. $(4a + b)^3$

32. $(6x - y)^2 - (6x + y)(6x - y)$

9-3 POWERS OF MONOMIALS

In Chapter 5, you were given the laws of exponents for monomials in one variable. Now, to find powers of a monomial in two variables, consider the monomial xy. By definition,

$$(xy)^2 = xy \cdot xy,$$
$$(xy)^3 = xy \cdot xy \cdot xy,$$
$$(xy)^4 = xy \cdot xy \cdot xy \cdot xy,$$

and so on. After rearranging and combining the x's and y's in each of these expressions, you obtain

$$(xy)^2 = x^2y^2,$$
$$(xy)^3 = x^3y^3,$$
$$(xy)^4 = x^4y^4.$$

These examples should suggest to you the general rule given below.

THIRD LAW OF EXPONENTS

The equation

$$(xy)^n = x^ny^n \qquad \text{(LE-3)}$$

is true for every positive integer n and all real numbers x and y.

If you wish to find a power of a monomial other than xy, you can use the laws of exponents together with the other laws and axioms of algebra, such as commutative, associative, and so on. For example, to find

$$(3xy)^4,$$

proceed as follows:

$$\begin{aligned} (3xy)^4 &= [3(xy)]^4 && \text{(R-M)} \\ &= 3^4(xy)^4 && \text{(LE-3)} \\ &= 3^4x^4y^4 && \text{(LE-3)} \\ &= 81x^4y^4. \end{aligned}$$

As a second example,

$$\begin{aligned} (-2xy^3)^5 &= [(-2)(xy^3)]^5 && \text{(R-M)} \\ &= (-2)^5(xy^3)^5 && \text{(LE-3)} \\ &= -32x^5(y^3)^5. && \text{(LE-3)} \end{aligned}$$

What power of y is $(y^3)^5$? You can figure this out by finding the product of five y^3's and then using the first law of exponents:

$$\begin{aligned} (y^3)^5 &= y^3 \cdot y^3 \cdot y^3 \cdot y^3 \cdot y^3 \\ &= y^{3+3+3+3+3} && \text{(LE-1)} \\ &= y^{15}. \end{aligned}$$

Hence, $-32x^5(y^3)^5 = -32x^5y^{15}$.

Try a similar problem; find

$$(x^6)^3.$$

You could solve this as follows:

$$\begin{aligned}(x^6)^3 &= x^6 \cdot x^6 \cdot x^6 \\ &= x^{6+6+6} \qquad \text{(LE-1)} \\ &= x^{18}.\end{aligned}$$

Therefore,

$$(x^6)^3 = x^{18}.$$

The preceding problems illustrate this principle: If m and n designate any positive integers, then

$$\begin{aligned}(x^m)^n &= \overbrace{x^m \cdot \ldots \cdot x^m}^{n \text{ factors}} \\ &= x^{\overbrace{m + \cdots + m}^{n \text{ terms}}}.\end{aligned}$$

If each integer is m, the sum of n integers is $m \cdot n$. This leads to the following law of exponents.

FOURTH LAW OF EXPONENTS

The equation

$$(x^m)^n = x^{m \cdot n} \qquad \textit{(LE-4)}$$

is true for every real number x and all positive integers m and n.

Problem 1. Find $(5x^3y^7)^4$.

Solution.

$$\begin{aligned}(5x^3y^7)^4 &= [5(x^3y^7)]^4 && \text{(R-M)} \\ &= 5^4(x^3y^7)^4 && \text{(LE-3)} \\ &= 5^4(x^3)^4(y^7)^4 && \text{(LE-3)} \\ &= 625x^{12}y^{28} && \text{(LE-4)}\end{aligned}$$

Problem 2. Find $(-\frac{3}{2}x^5y^2)^6$.

Solution.

$$\begin{aligned}(-\tfrac{3}{2}x^5y^2)^6 &= [(-\tfrac{3}{2})(x^5y^2)]^6 && \text{(R-M)} \\ &= (-\tfrac{3}{2})^6(x^5y^2)^6 && \text{(LE-3)} \\ &= \tfrac{729}{64}(x^5)^6(y^2)^6 && \text{(LE-3)} \\ &= \tfrac{729}{64}x^{30}y^{12} && \text{(LE-4)}\end{aligned}$$

Exercises

Find the following powers and products of powers of monomials.

1. (a) $(xy)^{100}$ (b) $(ab)^{14}$
2. (a) $(\frac{1}{2}xy)^3$ (b) $(3ab)^2$
3. (a) $(-3xy)^6$ (b) $(-5xy)^4$
4. (a) $(2x^2y^3)^5$ (b) $(\frac{1}{5}x^4y^3)^2$
5. (a) $(-2xy)^3$ (b) $(-3xy)^2$
6. (a) $-(\frac{1}{2}xy)^5$ (b) $-(\frac{1}{3}xy)^4$
7. (a) $x^2 \cdot x^5$ (b) $x^3 \cdot x^4$
8. (a) $(x^2)^5$ (b) $(x^3)^4$
9. (a) $(-\frac{2}{5}x^7y^9)^4$ (b) $(-\frac{2}{3}x^6y^{10})^4$
10. (a) $-(-7x^4y^2)^3$ (b) $-(-\frac{1}{3}x^3y^2)^3$
11. (a) $(7x^3y)^0$, $xy \neq 0$ (b) $(10x^0y^0)^2$, $xy \neq 0$
12. (a) $-(-6ab)^2$ (b) $-(-7a^2b)^2$
13. (a) $(-2ab^3)^2$ (b) $(-7a^2b^2)^3$
14. (a) $x^2y^2 \cdot x^3y^3$ (b) $ab^2 \cdot a^2b^3$
15. (a) $(x^2y^2)^3$ (b) $(ab^2)^3$
16. (a) $(3x^2y^2)^3$ (b) $(10^5x^{10}y^{15})^2$
17. (a) $(-a^8b^9)^5$ (b) $(-x^7y^{12})^4$
18. (a) $(.1ab^2)^3$ (b) $(.2x^2y)^3$
19. (a) $-(-10^2x^2y^2)^3$ (b) $-(-8^2x^2y^2)^2$
20. (a) $-(xy^3)^4$ (b) $-(x^4y)^3$
21. (a) $xy^3 \cdot x^4y^4$ (b) $x^2y \cdot x^4y^5$
22. (a) $(-3xy^3)^4(2x^2y)^3$ (b) $(-2xy)^3(3x^3y)^2$
23. (a) $(\frac{1}{2}a^5b^6)^7(2a^3b^2)^7$ (b) $(\frac{1}{6}x^{12}y^{14})^4(3xy)^3$
24. (a) $(-\frac{1}{5}x^9y^{10})^4(-5x^2)^6$ (b) $(-\frac{1}{4}ab^4)^5(-2a)^{10}$
25. (a) $(-\frac{1}{3}a^4b^5)^6(-3a)^5$ (b) $(-\frac{1}{27}ab^2)^2(-3a^2b)^6$
26. (a) $(-2^0x^0y^2)^5$, $x \neq 0$ (b) $ab^0 \cdot a^3b^2 \cdot a^0b$, $ab \neq 0$

Preparation for Section 9–4

Find the following products.

1. $(x + 4)^2$
2. $(3x - 1)^2$
3. $(9 - 2x)^2$
4. $(2x + 3)(2x - 3)$

9-4 POWERS OF BINOMIALS

In Chapter 5, you learned the pattern for the square of a binomial:

(First term + Second term)^2
= (First term)^2 + 2 (First term)(Second term) + (Second term)^2.

For example,

$$(x + y)^2 = x^2 + 2xy + y^2,$$
$$(x - y)^2 = x^2 - 2xy + y^2.$$

Now consider the cube of $(x + y)$ and of $(x - y)$. You can first obtain

$$\begin{aligned}(x + y)^3 &= (x + y)(x + y)^2 \\ &= (x + y)(x^2 + 2xy + y^2).\end{aligned}$$

Then perform the indicated multiplication.

$$\begin{array}{r} x^2 + 2xy + y^2 \\ x + y \\ \hline x^3 + 2x^2y + xy^2 \\ x^2y + 2xy^2 + y^3 \\ \hline x^3 + 3x^2y + 3xy^2 + y^3 \end{array}$$

Therefore,

$$(x + y)^3 = x^3 + 3x^2y + 3xy^2 + y^3.$$

Observe the pattern of the exponents and coefficients in the cube of $(x + y)$. If you had performed the multiplication for the cube of $(x - y)$, you would have found that

$$(x - y)^3 = x^3 - 3x^2y + 3xy^2 - y^3.$$

Problem 1. Find $(x^2 - 2y)^2$.

Solution. $$\begin{aligned}(x^2 - 2y)^2 &= (x^2)^2 - 2(x^2 \cdot 2y) + (2y)^2 \\ &= x^4 - 4x^2y + 4y^2\end{aligned}$$

Problem 2. Find $(x^2 - 2y)^3$.

Solution. $$\begin{aligned}(x^2 - 2y)^3 &= (x^2 - 2y)(x^2 - 2y)^2 \\ &= (x^2 - 2y)(x^4 - 4x^2y + 4y^2)\end{aligned}$$

The multiplication is performed below.

$$\begin{array}{r} x^4 - 4x^2y + 4y^2 \phantom{{}- 8y^3} \\ x^2 - 2y \phantom{{}- 8y^3} \\ \hline x^6 - 4x^4y + 4x^2y^2 \phantom{{}- 8y^3} \\ - 2x^4y + 8x^2y^2 - 8y^3 \\ \hline x^6 - 6x^4y + 12x^2y^2 - 8y^3 \end{array}$$

Thus,

$$(x^2 - 2y)^3 = x^6 - 6x^4y + 12x^2y^2 - 8y^3.$$

Instead of carrying out the multiplication in Problem 2, you could have used the *pattern for cubing binomials* that you discovered earlier in this section. Variables a and b are used in the following statement of this pattern.

$$(a - b)^3 = a^3 - 3a^2b + 3ab^2 - b^3$$

For example,

$$\begin{aligned} (x^2 - 2y)^3 &= (x^2)^3 - 3(x^2)^2(2y) + 3(x^2)(2y)^2 - (2y)^3, \\ &= x^6 - 6x^4y + 12x^2y^2 - 8y^3. \end{aligned}$$

Notice that a and b are replaced in the pattern by x^2 and $2y$, respectively.

Exercises

Find the following products of binomials.

1. (a) $(8x + 5y)(8x - 5y)$
   (b) $(3x + 10y)(3x - 10y)$

2. (a) $(9a^2 + b^2)(9a^2 - b^2)$
   (b) $(7x^3 + y^3)(7x^3 - y^3)$

3. (a) $(3x^2 + y)(4x^2 + 5y)$
   (b) $(2x + y^2)(4x + 3y^2)$

Use the patterns

$$\begin{aligned} (a + b)^3 &= a^3 + 3a^2b + 3ab^2 + b^3 \\ (a - b)^3 &= a^3 - 3a^2b + 3ab^2 - b^3 \end{aligned}$$

to find the products in Exercises 4–6.

4. (a) $(4x + 3y)^3$
(b) $(3x + 2y)^3$

5. (a) $(\frac{1}{3}x - 4y)^3$
(b) $(2x - \frac{1}{3}y)^3$

6. (a) $(3x^2 - y^3)^3$
(b) $(x^3 - 2y^2)^3$

Find the following powers of binomials.

7. (a) $(2x + 5y)^2$ (b) $(3x + 4y)^2$

8. (a) $(2x - y)^2$ (b) $(4x - y)^2$

9. (a) $(3x^2 - y)^2$ (b) $(3x^2 - 7y)^2$

10. (a) $(x + 4y)^2$ (b) $(5x + y)^2$

11. (a) $(x + 4y)^3$ (b) $(5x + y)^3$

12. (a) $(4y^3 + 9x^2)^2$ (b) $(3x^2 + 5y^3)^2$

13. (a) $(\frac{1}{3}a - b)^2$ (b) $(x - \frac{1}{2}y)^2$

14. (a) $(\frac{1}{3}a - b)^3$ (b) $(x - \frac{1}{2}y)^3$

15. (a) $(2x - y)^3$ (b) $(5x - y)^3$

16. (a) $(2x + 5y)^3$ (b) $(5x + 3y)^3$

17. (a) $(3x^2 - y)^3$ (b) $(4x - 3y)^3$

---

18. (a) Find a pattern for the fourth power of the binomial

$$(x + y)^4$$

by finding the product

$$(x + y)^2(x + y)^2.$$

(b) Check your work by finding the product

$$(x + y)^3(x + y).$$

19. Use the pattern that you discovered in Exercise 18 to find the following powers of binomials.

(a) $(a^2 + b)^4$
(b) $(2a + b)^4$
(c) $(a^2 + 3b)^4$
(d) $(3a + 2b)^4$

20. Find a pattern for

$$(a - b)^4$$

by finding $[(a - b)^2]^2$.

21. Use the pattern that you discovered in Exercise 20 to find
(a) $(3x - y)^4$. (b) $(2x - y^2)^4$.

---

22. Find a pattern for squaring a trinomial by squaring

$$x + y + 1.$$

Hint: First write $(x + y + 1)^2$ as $[(x + y) + 1]^2$.

23. Use the pattern that you discovered in Exercise 22 to find
(a) $(2x + 3y + 1)^2$. (b) $(4x + 5y + 1)^2$.

24. (a) Find a pattern for the square of any trinomial by squaring $x + y + z$.
(b) Describe the pattern that you found for $(x + y + z)^2$ in part (a). How many terms were in the product? How many of the terms were perfect squares? How many terms have even coefficients?

25. Use the pattern that you discovered in Exercise 24 to find each of the following.
(a) $(3x^2 + 2y + 5z)^2$ (b) $(4x + y^3 + 7z)^2$

26. Make a guess about the pattern for $(x + y + z + w)^2$. Now multiply $x + y + z + w$ by itself to determine if you are right. Describe this pattern in your own words.

27. Use the pattern that you found in Exercise 26 to square

$$3x + 2y + 5z + 4w.$$

Review for Sections 9–1 through 9–4

Find the sums, differences, products, or powers as indicated.

1. $(3x^2y + 2xy + 3y^2) - (2x^2y + xy + y^2)$

2. $4mn \cdot 4mn$

3. $(4ab^2 + 3a^2b - 2ab) + (2ab^2 - a^2b + ab)$

4. $(2ab)^3$

5. $3xy^2(4x^2 + 2x - 3y)$

6. $(5 + 3y)(2x + 3xy - 6y)$

7. $(3m^2)^4$

8. $(3a^2 + 2b)(4a^2 + 3a^2b - 5b^2)$
9. $(x^2y^3)^3$
10. $2x(3 - 5xy + 4y^2) + (2x + 3)(4 - 3xy + y^2 - x^2y)$
11. $(3m^2n + 2n)(4mn + 3n^2 - 2n) + 6mn^2$
12. $(14r^3 - 3r^2y + 6ry^2 - y^3) + (r^3 + 2r^2y - 3ry^2)$
13. $(a^3 + 4b)(3a + 2ab + 6b) - (a^3 - 5b)(3a + 4ab - b)$
14. $(x^2y + 5x^2 - 6y^2) - (3x^2y - 2x^2 + y^2)$
15. $(2x + y)^3$
16. $(4m - 3n)^3$
17. $(4m^2 + 3mn + 6n^2) + (2m^2 - 8mn - 6n^2)$
18. $(2x^2 - 4y^3)^3$
19. $(3r^2t + 4t - 3rt^2) - (3t + 8r^2t + 2rt^2)$
20. $(3x^3)^2(4x^2y + 3)$

Answers to Review for Sections 9–1 through 9–4

1. $x^2y + xy + 2y^2$
2. $16m^2n^2$
3. $6ab^2 + 2a^2b - ab$
4. $8a^3b^3$
5. $12x^3y^2 + 6x^2y^2 - 9xy^3$
6. $10x + 21xy - 30y + 9xy^2 - 18y^2$
7. $81m^8$
8. $12a^4 + 9a^4b - 9a^2b^2 + 8a^2b - 10b^3$
9. x^6y^9
10. $-2x^3y - 19x^2y + 10xy^2 - 9xy + 14x + 3y^2 + 12$
11. $12m^3n^2 + 9m^2n^3 - 6m^2n^2 + 14mn^2 + 6n^3 - 4n^2$
12. $15r^3 - r^2y + 3ry^2 - y^3$
13. $-2a^4b + 7a^3b + 27ab + 28ab^2 + 19b^2$
14. $-2x^2y + 7x^2 - 7y^2$
15. $8x^3 + 12x^2y + 6xy^2 + y^3$
16. $64m^3 - 144m^2n + 108mn^2 - 27n^3$
17. $6m^2 - 5mn$
18. $8x^6 - 48x^4y^3 + 96x^2y^6 - 64y^9$
19. $-5r^2t - 5rt^2 + t$
20. $36x^8y + 27x^6$

9-5 FACTORING

To find a monomial factor of a polynomial in two variables, look at the powers of the variables in each term of the polynomial. For example, if you wish to find a monomial factor of highest degree in x and y of the polynomial

$$4x^3y^5 - 17x^7y^4 + 35x^{12}y^7,$$

you should note that the terms of the polynomial have degrees of 3, 7, and 12 in the variable x. The least of these degrees is 3; therefore, x^3 is a factor of each term. Similarly, the terms of the polynomial have degrees of 5, 4, and 7 in y. The least of these degrees is 4; therefore, y^4 is a factor of each term. By the distributive axiom, the given trinomial can be expressed as a product of the monomial x^3y^4 and a trinomial. Thus,

$$\begin{aligned} 4x^3y^5 - 17x^7y^4 + 35x^{12}y^7 & \\ &= (x^3y^4 \cdot 4y) - (x^3y^4 \cdot 17x^4) + (x^3y^4 \cdot 35x^9y^3) \\ &= x^3y^4(4y - 17x^4 + 35x^9y^3). \end{aligned}$$

Problem 1. In the following polynomial, find a monomial factor of highest degree in x and y and factor the polynomial.

$$12x^8y^2 + 9x^2y - 15x^3y^6$$

Solution. The terms have degrees of 8, 2, and 3 in x; of 2, 1, and 6 in y. Therefore x^2y is the monomial factor of highest degree in x and y that occurs in all the terms. You can also factor one 3 from each term:

$$\begin{aligned} 12x^8y^2 + 9x^2y - 15x^3y^6 &= (3x^2y \cdot 4x^6y) + (3x^2y \cdot 3) - (3x^2y \cdot 5xy^5) \\ &= 3x^2y(4x^6y + 3 - 5xy^5). \end{aligned}$$

Some polynomials have binomial, trinomial, and other polynomial factors. For example, the polynomial

$$x(3x - y) + 2(3x - y)$$

has $(3x - y)$ as a factor. By the distributive axiom,

$$x(3x - y) + 2(3x - y) = (x + 2)(3x - y).$$

Problem 2. Factor $x^3 + 2y + 2x^2 + xy$.

Solution. Because there is no monomial factor, you should look for possible binomial factors. By rearranging the terms, you get

$$\begin{aligned} x^3 + 2y + 2x^2 + xy &= (x^3 + 2x^2) + (2y + xy) && \text{(R-A)} \\ &= x^2(x + 2) + y(2 + x) && \text{(D)} \\ &= (x^2 + y)(x + 2). && \text{(D)} \end{aligned}$$

You could have rearranged them as follows:

$$\begin{aligned} x^3 + 2y + 2x^2 + xy &= (x^3 + xy) + (2y + 2x^2) && \text{(R-A)} \\ &= x(x^2 + y) + 2(y + x^2) && \text{(D)} \\ &= (x + 2)(x^2 + y). && \text{(D)} \end{aligned}$$

The two answers that you obtained by rearranging terms differently are equal by the commutative law of multiplication.

Exercises

Find a monomial factor of highest degree in x and y, and factor each of the following polynomials.

1. (a) $7x^2 - 14x^2y^2$
(b) $6x^3 - 18x^3y^3$

2. (a) $3x^2y^2 + 2x^2y^3$
(b) $5x^3y^2 + 4x^2y^3$

3. (a) $6xy - 9xy^2 - 3x^2$
(b) $15a^2b - 3ab^2 - 5b$

4. (a) $8x^2y - 6x^2y^2 + 12x^2$
(b) $24xy^2 - 8x^2y + 6x^2y^2$

5. (a) $28x^2y - 16xy^2 + 64y^3$
(b) $15xy^2 - 25x^3y + 10x^2y^3$

6. (a) $12x^2y^2 - 20x^3y^3 + 28x^4y^4$
(b) $9x^4y^4 - 12x^5y^5 + 18x^2y^2$

7. (a) $4x^5y - 2x^4y^2 + 8x^3y^3$
(b) $33x^3y^4 - 22x^2y^2 + 11x^2y$

8. (a) $-36x^3y^2 + 6x^2y^3 - 18x^5y^2$
(b) $-2x^3y + 4x^2y^2 - 8xy^3$

9. (a) $24x^2y^2 + 6xy^2 + 30x^2y$
(b) $30x^2y^4 + 15x^4y^2 + 45x^4y^4$

10. (a) $32x^5 - 6x^4y + 24x^3y^2$
(b) $16y^7 - 24x^3y^2 + 32x^7y$

Use the distributive axiom to factor the following without removing the parentheses.

11. $y(2x - 3) + 4(2x - 3)$

12. $6x(x + y) - y(x + y)$

13. $x(4x - y) - y(4x - y)$

14. $y(x + 1) + 3(x + 1)$

15. $2x(3y - x) + (3y - x)$

16. $(2x - y) + y(2x - y)$

17. $(x + y)^2 - 2(x + y)$

18. $(a + b)(a - b) + (a + b)$

19. $13x^2(x + 5y) - 7xy(x + 5y) + 4y^2(x + 5y)$

20. $x^2(y - x) - xy(y - x) - (y - x)$

21. $(2x + y)(2x - y) + (2x - y)$

Use the rearrangement property and distributive axiom to factor the following.

22. $2xy + 10x + 3y + 15$

23. $y^3 + 3y^2 + xy + 3x$

24. $2x^2 + 3x + 2xy + 3y$

25. $30a^2 + 35a + 54ab + 63b$

26. $15xy + 28x + 12x^2 + 35y$

27. $6y^2 + 2xy + 3y + x$

28. $3x^2 + 3x + xy + y$

9–6 MORE FACTORING

In Chapter 4, you learned that the difference of two squares is factorable. For example,

$$x^2 - 2^2 = (x + 2)(x - 2).$$

Any binomial that is the difference of two monomial squares is factorable by the same pattern:

$$\begin{aligned}&(\text{First term})^2 - (\text{Second term})^2\\ &\qquad = (\text{First term} + \text{Second term})(\text{First term} - \text{Second term}).\end{aligned}$$

This pattern can be expressed in symbols:

$$x^2 - y^2 = (x + y)(x - y)$$

for all numbers x and y.

The following problems illustrate the procedure for factoring a binomial that is the difference of two monomial squares.

Problem 1. Factor $4x^2 - 25y^2$.

Solution.

$$\begin{aligned}4x^2 - 25y^2 &= (2x)^2 - (5y)^2\\ &= (2x + 5y)(2x - 5y)\end{aligned}$$

Problem 2. Factor $a^6 - 49b^8$.

Solution.
$$a^6 - 49b^8 = (a^3)^2 - (7b^4)^2 = (a^3 + 7b^4)(a^3 - 7b^4)$$

Why is the polynomial

$$x^3 - y^2$$

not factorable by the same method?

Trinomials in two variables can be factored as a product of two binomials in much the same way that was used in factoring trinomials in one variable. Consider the following examples.

Problem 3. If possible, factor the following trinomial as a product of two binomials with integral coefficients.

$$x^2 - xy - 6y^2$$

Solution. The problem is to factor the trinomial

$$x^2 - xy - 6y^2 = (x + \underline{\ ?\ }y)(x + \underline{\ ?\ }y)$$

with integers placed in the blank spaces. Therefore,

$$x^2 - xy - 6y^2 = (x + 2y)(x - 3y).$$

Check.

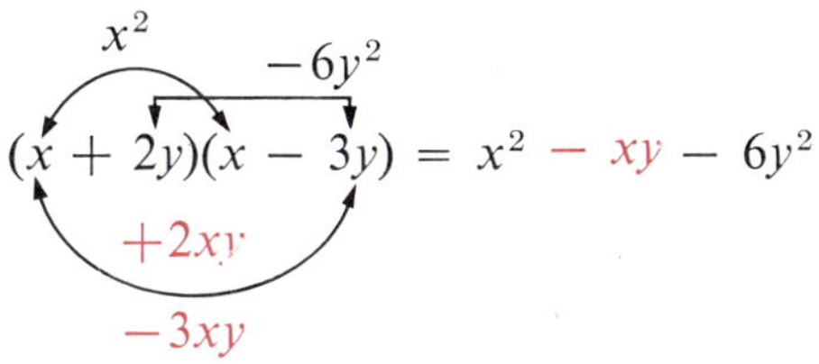

x^2 $-6y^2$

$$(x + 2y)(x - 3y) = x^2 - xy - 6y^2$$

$+2xy$

$-3xy$

Problem 4. Factor $4x^2 - 20xy + 9y^2$.

Solution. If possible, this trinomial is to be factored as follows:

$$4x^2 - 20xy + 9y^2 = (\underline{\ ?\ }x + \underline{\ ?\ }y)(\underline{\ ?\ }x + \underline{\ ?\ }y).$$

The two blank spaces in front of the x's must be filled with integers whose product is 4; those in front of the y's must be filled with integers whose product is 9. Because the middle term, $-20xy$, has a negative coefficient, while the other two terms have positive coefficients, both factors of either 4 or 9 must be chosen to be negative numbers.

By trial and error, you can find that the factors of 4 are 2 and 2, and the factors of 9 are -1 and -9:

$$4x^2 - 20xy + 9y^2 = (2x - y)(2x - 9y).$$

Check.

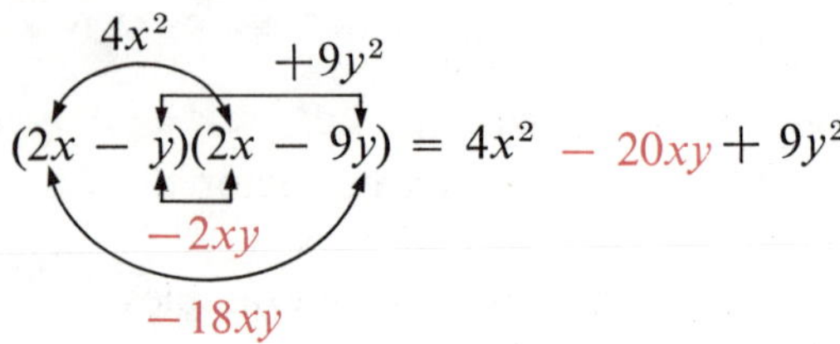

Problem 5. Factor $6a^2 - 7ab - 5b^2$.

Solution. If

$$6a^2 - 7ab - 5b^2 = (\underline{\ ?\ }a + \underline{\ ?\ }b)(\underline{\ ?\ }a + \underline{\ ?\ }b),$$

which integers can be used to fill the blank spaces in front of the a's? in front of the b's?

Exercises

Factor each of the following polynomials. Remember to look for monomial factors.

1. (a) $x^2 - 4y^2$ (b) $9x^2 - y^2$
2. (a) $x^2 + 2xy + y^2$ (b) $x^2 - 2xy + y^2$
3. (a) $x^2 + 2xy - 3y^2$ (b) $x^2 - 2xy - 3y^2$
4. (a) $x^2 + 13xy + 36y^2$ (b) $x^2 + 6xy + 8y^2$
5. (a) $a^2 - 2ab - 35b^2$ (b) $a^2 - ab - 12b^2$
6. (a) $x^2 - 14xy + 40y^2$ (b) $x^2 - 16xy + 60y^2$
7. (a) $a^2 + 5ab - 14b^2$ (b) $a^2 + 7ab - 18b^2$
8. (a) $x^2 - 34xy - 35y^2$ (b) $a^2 - 11ab - 12b^2$
9. (a) $36x^2 - 12xy + y^2$ (b) $25x^2 - 10xy + y^2$
10. (a) $16x^2 - 8xy + y^2$ (b) $9x^2 - 6xy + y^2$
11. (a) $9a^2 + 30ab + 25b^2$ (b) $4a^2 + 28ab + 49b^2$
12. (a) $5b^2 + 7ab + 2a^2$ (b) $12y^2 + 7xy + x^2$
13. (a) $3x^2 + xy - 2y^2$ (b) $8x^2 + 2xy - 3y^2$
14. (a) $16x^4 - y^4$ (b) $81x^4 - y^4$
15. (a) $25x^6 - y^4$ (b) $49x^4 - y^2$

16. (a) $12a^2 - 19ab - 18b^2$ (b) $30a^2 - 7ab - 15b^2$

17. (a) $56a^2 + 15ab - 56b^2$ (b) $40x^2 + 39xy - 40y^2$

18. (a) $3a^4 - 8a^2b^2 - 3b^4$ (b) $4x^4 - 15x^2y^2 - 4y^4$

19. (a) $6x^6 + 13x^3y^3 + 6y^6$ (b) $12x^5 - 25x^3y^3 + 12y^6$

20. (a) $4a^4 - 10a^2b + 6b^2$ (b) $6a^2 - 10ab^2 + 4b^4$

In the exercises below, first look for monomial factors. Then write each polynomial as the product of as many factors as possible.

21. $7a^2 - 14ab - 105b^2$

22. $x^3 - x^2y - 20xy^2$

23. $x^3y^3 - xy$

24. $2x^2 - 128y^2$

25. $64x^8 - y^8$

26. $15a^2b^2 - 5ab - 10$

27. $x^3y^4 - 6x^3y^2 - 16x^3$

28. $x^4 - 17x^3y + 16x^2y^2$

29. $.01x^8 - y^{10}$

30. $5x^2 - 80y^2$

31. $x^4y^4 - 81$

32. $98a^2 + 140ab + 50b^2$

If possible, factor each of the polynomials in Exercises 33–42. (Do not remove any parentheses before factoring.)

33. $x^2 + xy + y^2$

34. $x^2 + y^2$

35. $(x - y)^2 - 3^2$

36. $(2a - b)^2 - 5^2$

37. $9(x + y)^2 - 6(x + y) + 1$

38. $(2x - y)^2 + 5(2x - y) + 6$

39. $x^2 - xy + y^2$

40. $x^2 + 4y^2$

41. $3y(x - 2y) + 2x(x - 2y)$

42. $4a^2(5a + b) - 9b^2(5a + b)$

43. You can easily recognize that each of the following polynomials in two variables is factorable if you first write it as shown in the example below. Use the method suggested in this example to complete the factoring of the others.

$$\begin{aligned} -9x^4 - 78x^2y^2 - 169y^4 &= -1(9x^4 + 78x^2y^2 + 169y^4) \\ &= -1(3x^2 + 13y^2)^2 \end{aligned}$$

(a) $-9a^2 - 96ab - 256b^2$ (b) $40xy - 16x^2 - 25y^2$

(c) $56a^2b^2 - 8a^4 - 98b^4$ (d) $-121a^2b^2 - 132ab^2 - 36b^2$

EXTRA!

A difference of two squares can be factored:

$$x^2 - y^2 = (x + y)(x - y).$$

Can a difference of two cubes,

$$x^3 - y^3,$$

be factored? A mathematical "trick" answers this question.

$$\begin{aligned} x^3 - y^3 &= x^3 + (-x^2y + x^2y) - y^3 \quad (\text{because } -x^2y + x^2y = 0) \\ &= (x^3 - x^2y) + (x^2y - y^3) \\ &= x^2(x - y) + y(x^2 - y^2) \\ &= x^2(x - y) + y(x + y)(x - y) \\ &= [x^2 + y(x + y)](x - y) \\ &= (x^2 + xy + y^2)(x - y) \end{aligned}$$

So the answer is "yes, $x^3 - y^3$ can be factored."

Continuing, can a difference of two fourth powers,

$$x^4 - y^4,$$

be factored? The answer is easily seen to be "yes":

$$\begin{aligned} x^4 - y^4 &= (x^2 + y^2)(x^2 - y^2) \\ &= (x^2 + y^2)(x + y)(x - y) \\ &= (x^3 + x^2y + xy^2 + y^3)(x - y). \quad (\text{on multiplying } x^2 + y^2 \text{ by } x + y) \end{aligned}$$

Using the same "trick" as for the difference of two cubes above, you should be able to factor $x^5 - y^5$:

$$x^5 - y^5 = x^5 - x^4y + x^4y - y^5 \qquad (\text{because } -x^4y + x^4y = 0)$$

and so on. Eventually, you should get

$$x^5 - y^5 = (x^4 + x^3y + x^2y^2 + xy^3 + y^4)(x - y).$$

Similarly, you can show that

$$x^6 - y^6 = (x^5 + x^4y + x^3y^2 + x^2y^3 + xy^4 + y^5)(x - y).$$

Do you see the pattern in factoring $x^n - y^n$ for every positive integer n? Always, one factor is $x - y$. How would you describe the other factor?

A sum of two squares cannot be factored. However, a sum of two cubes can be factored from results above.

$$x^3 + y^3 = x^3 - (-y)^3 \qquad \text{[because } (-y)^3 = -y^3\text{]}$$
$$= [x^2 + x \cdot (-y) + (-y)^2][x - (-y)]$$

Thus,

$$x^3 + y^3 = (x^2 - xy + y^2)(x + y).$$

A sum of two fourth powers can be factored by a "trick" method similar to one above.

$$x^4 + y^4 = x^4 + 2x^2y^2 + y^4 - 2x^2y^2$$
$$\text{(because } 2x^2y^2 - 2x^2y^2 = 0\text{)}$$
$$= (x^2 + y^2)^2 - (\sqrt{2}\,xy)^2$$
$$= [(x^2 + y^2) + \sqrt{2}\,xy][(x^2 + y^2) - \sqrt{2}\,xy]$$
$$\text{(difference of two squares)}$$

Thus,

$$x^4 + y^4 = (x^2 + \sqrt{2}\,xy + y^2)(x^2 - \sqrt{2}\,xy + y^2).$$

Try factoring the following polynomials by the methods above.

1. $x^5 + y^5$

2. $x^7 - y^7$

3. $x^7 + y^7$

4. $x^4 + x^2y^2 + y^4$

5. $x^4 - 2x^2y^2 + y^4$

6. $x^9 - y^9$ [Hint: $a^9 = (a^3)^3$]

7. $x^6 + y^6$ [Hint: $a^6 = (a^2)^3$]

8. $x^8 + y^8$

9. $x^8 - y^8$

10. $x^4 + 4y^4$

*9–7 NUMBER THEORY II

Although there are many ways to factor 24, there is only one way if you use prime divisors.

$$24 = 2 \times 2 \times 2 \times 3$$

Similarly, you can factor 42 in only one way if you use prime divisors: $42 = 2 \times 3 \times 7$. If you look at the prime factorizations of 24 and 42, you see that 2×3, or 6, is the greatest common divisor.

$$24 = (2 \times 3) \times (2 \times 2)$$
$$42 = (2 \times 3) \times 7$$

You can also see that the least common multiple of 24 and 42 is

$$2 \times 2 \times 2 \times 3 \times 7, \quad \text{or } 168.$$

The greatest common divisor (g.c.d.) of a pair of integers is the greatest integer which is a divisor of both. The least common multiple (l.c.m.) of a pair of integers is the least integer which is a multiple of both.

The only prime divisor of 4 is 2, and the only prime divisors of 15 are 3 and 5. Therefore, 4 and 15 have no common prime divisor. This means that the g.c.d. of 4 and 15 is 1, and their l.c.m. is 4×15, or 60.

Two integers are said to be *relatively prime* if their g.c.d. is 1. Relatively prime integers do not have to be prime integers, as the example above shows. Thus, 4 and 15 are relatively prime but neither one is a prime. However, any two different primes are relatively prime.

Problem. Find the g.c.d and l.c.m. of the pair of integers

$$858 \quad \text{and} \quad 910.$$

Solution. Since 2 is a divisor of 858,

$$858 = 2 \times 429.$$

The other divisors of 858 are the divisors of 429. Although 2 is not a divisor of 429, 3 is such a divisor, and

$$858 = 2 \times 3 \times 143.$$

Further divisors of 858 are found by factoring 143. If you try the primes in order, you see that 3, 5, and 7 are not divisors of 143, but that 11 is, and

$$858 = 2 \times 3 \times 11 \times 13.$$

The prime factors of 858 are 2, 3, 11, and 13.

You can factor 910 in a similar way,

$$\begin{aligned} 910 &= 2 \times 455 \\ &= 2 \times 5 \times 91 \\ &= 2 \times 5 \times 7 \times 13. \end{aligned}$$

The prime factors of 910 are 2, 5, 7 and 13.

You may conclude that 2×13, or 26, is the g.c.d. of 858 and 910, since one 2 and one 13 are the only common prime factors of 858 and 910.

By inspecting the prime factors of 858 and 910, you can determine the l.c.m. of 858 and 910: $2 \times 3 \times 5 \times 7 \times 11 \times 13$, or 30,030.

There is another way of finding the g.c.d. of two numbers. To find the g.c.d. of 858 and 910, first divide 910 by 858.

$$\begin{array}{r} 1 \\ 858\overline{)910} \\ \underline{858} \\ 52 \end{array} \qquad 910 = 1 \times 858 + 52$$

Next, divide the previous divisor, 858, by the remainder, 52, and obtain the following.

$$\begin{array}{r} 16 \\ 52\overline{)858} \\ \underline{52} \\ 338 \\ \underline{312} \\ 26 \end{array} \qquad 858 = 16 \times 52 + 26$$

Finally, divide the last divisor, 52, by the last remainder, 26, and get the following.

$$\begin{array}{r} 2 \\ 26\overline{)52} \\ \underline{52} \\ 0 \end{array} \qquad 52 = 2 \times 26 + 0$$

The last nonzero remainder encountered in your successive divisions was 26 and, therefore, 26 is the g.c.d. of 858 and 910. Can you see why this is so?

Let us use the same process on 24 and 42.

$$\begin{array}{r} 1 \\ 24\overline{)42} \\ \underline{24} \\ 18 \end{array} \qquad 42 = 1 \times 24 + 18$$

$$\begin{array}{r} 1 \\ 18\overline{)24} \\ \underline{18} \\ 6 \end{array} \qquad 24 = 1 \times 18 + 6$$

$$\begin{array}{r} 3 \\ 6\overline{)18} \\ \underline{18} \\ 0 \end{array} \qquad 18 = 3 \times 6 + 0$$

The last nonzero remainder is 6, and 6 is the g.c.d. of 24 and 42.

The process of repeatedly dividing to find the g.c.d. of two numbers is called *Euclid's algorithm;* it appeared in Euclid's *Elements* about 2300 years ago! It is effective chiefly for this reason: *A divisor of each of two numbers is also a divisor of the remainder obtained from dividing one of the numbers by the other.*

Exercises

Use Euclid's algorithm to find the g.c.d. of each of the following pairs of integers.

1. 8 and 18

2. 22 and 36

3. 144 and 132

4. 72 and 125

5. 319 and 583

6. 273 and 429

7. 129 and 838

8. 187 and 408

9. 437 and 551

10. 1032 and 2196

In Exercises 11–20, give the g.c.d. and l.c.m. in factored form.

11. $(5 \cdot 13 \cdot 17)$ and $(13 \cdot 41)$

12. $(7 \cdot 19 \cdot 31)$ and $(19 \cdot 43)$

13. 2^5 and $(2^2 \cdot 3^4)$

14. 3^6 and $(2^2 \cdot 5^4)$

15. $(3 \cdot 11 \cdot 19)$ and $(2 \cdot 13 \cdot 17)$

16. $(5 \cdot 7 \cdot 11)$ and $(2 \cdot 13 \cdot 23)$

17. $(11^2 \cdot 13^3)$ and $(3 \cdot 11 \cdot 13 \cdot 17)$

18. $(3^3 \cdot 5^2 \cdot 7^5)$ and $(2 \cdot 3 \cdot 5^2 \cdot 7^2)$

19. $(17^3 \cdot 29)$ and $(17^5 \cdot 31)$

20. $(101 \cdot 107)$ and $(3 \cdot 101^2 \cdot 109)$

21. If each of two given integers is expressed as a product of primes, describe how to find the g.c.d. and l.c.m. of the two integers.

22. The product of two integers is 1260 and their g.c.d. is 6. Find the l.c.m. of the two integers.

23. The l.c.m. of two integers is 4752 and their g.c.d. is 18. Find the product of the two integers.

24. The product of two integers is 24,500 and their l.c.m. is 700. Find the g.c.d. of the two integers.

25. How are the product, l.c.m., and g.c.d. of two integers related to each other?

26. An integer n has a remainder of 1 when divided by 2, of 2 when divided by 3, of 3 when divided by 4, of 4 when divided by 5, of 5 when divided by 6, and of 6 when divided by 7. Find the smallest such integer n.

KEY IDEAS AND KEY WORDS

A **monomial** in variables x and y is a number or a product of a number by powers of x and y. For example, $8x^3y^5$ is a monomial in x and y. The number 8 is called the **coefficient** of the monomial. A **polynomial** in x and y is a sum of monomials in x and y. An example of a polynomial in x and y is $8x^3y^5 - 23xy^4 + 17y^7 - 8x + 13$. The monomial that has no variable is called the **constant term** of the polynomial.

The **third** and **fourth laws of exponents** are as follows:

$$(xy)^n = x^n y^n, \qquad \text{(LE-3)}$$
$$(x^m)^n = x^{m \cdot n}. \qquad \text{(LE-4)}$$

These are true for all numbers x and y and all positive integers m and n.

The patterns for **squaring and cubing a binomial** are as follows:

$$(a + b)^2 = a^2 + 2ab + b^2,$$
$$(a + b)^3 = a^3 + 3a^2b + 3ab^2 + b^3.$$

The pattern for **factoring a difference of two squares** is as follows:

$$a^2 - b^2 = (a + b)(a - b).$$

The **greatest common divisor (g.c.d)** of a pair of integers is the greatest integer which is a divisor of both. The **least common multiple (l.c.m.)** of a pair of integers is the least integer which is a multiple of both.

CHAPTER REVIEW

1. Give an example of

(a) two polynomials in x and y that are additive inverses of each other.
(b) a trinomial in x and y that is the square of the difference of two monomials.
(c) a binomial in x and y that is the product of two binomials.
(d) a trinomial in x and y that is the product of two binomials.

Complete the computation in Exercises 2–19.

2. $5x(2x^2 - 3y^2) - 7y^2(4x - y)$

3. $(2x - 3y)(x^2 - 5xy - 4y^2)$

4. $(9x + 4y)(9x - 4y)$

5. $(7x^2 - y)^2$

6. $(5y + 2x)^3$

7. $(-3x^2y^5)(x^3y)(-\frac{1}{6}xy^4)$

8. $(2x^3y)^4(-3xy^3)^2$
9. $(\frac{1}{3}x^3 + 2y^3)^2$
10. $(3x^3 + 4x^2y - 2y^3) - (3x^2y + 4y^3)$
11. $(2a^3 - 3a^2b^2 + b^3) - 2(a^3 + a^2b^2 - 3b^3)$
12. $2(x^2 + 3y^2) - 3(2xy + y^2) + 4(3x^2 - xy)$
13. $(-\frac{1}{2}ab)^5$
14. $(7a^3b^2)(-ab)^3$
15. $(-4a^2b^3)^4$
16. $(3x - 5y)^2$
17. $(3x - 5y)^3$
18. $(3a^3b^2)^5 \cdot (2ab^4)^3$
19. $a^2b(a^2 + ab - 2) - ab(a^3 - a^2 + a)$

Factor each of the following polynomials.

20. $3x^4 + 5xy - 7xy^3$
21. $x^2y^2 + x^3y^3 - x^4y^4$
22. $x(x + 2) - y(x + 2)$
23. $2x(x - 1) + (x - 1)$
24. $3x(4x + 1) - (4x + 1)$
25. $64x^2 - 9y^2$
26. $x^2y^2 - 25$
27. $x^2 - 17xy + 16y^2$
28. $21x^3y^5 - 35x^5y^4 + 14x^7y^3$
29. $7x(5x + y) - 3y(5x + y)$
30. $x^3 + 3x^2 + 4xy + 12y$
31. $x^2 + 36xy + 324y^2$
32. $196x^2 - 225y^2$
33. $5x^2 - 14xy - 3y^2$
34. $25x^4 - 10x^2y^2 + y^4$
35. $11x^2 - 19xy - 6y^2$
36. $8x^4 - 32y^4$

Write each of the following as a product of three or more factors.

37. $100x^3 + 36x^3y^4 - 120x^3y^2$
38. $64x^2 - 4y^6$
39. $5x^2 - 80y^2$
40. $81x^4 - y^8$

CHAPTER TEST

Simplify each of the following.

1. $(-\frac{2}{3}x^2y)(-\frac{15}{4}x^3y^2)$
2. $(-4xy^2)^3(-\frac{1}{2}x^2y)$
3. $-5xy(3x^2y^3 - \frac{1}{15}xy)$
4. $7x(4x - 5y) - 3y(2x + 7y)$

Find the following products.

5. $(x - y)(x^2 + xy + y^2)$
6. $(6x - 7y)^2$
7. $(\frac{1}{3}x^2 + y^2)(\frac{1}{3}x^2 - y^2)$
8. $(x + y)^3$

Factor each of the following.

9. $30x^2y^2 - 45xy^4$

10. $7r(3r - s) - 5s(3r - s)$

11. $x^2y^2 - 49$

12. $4x^4 + 4x^2y^2 + y^4$

13. $9y^2 - 6xy + x^2$

14. $35x^2 + xy - 6y^2$

Write each as a product of three factors.

15. $3x^3y - 27xy^3$

16. $5x^4 - 5000xy^3$

Group and factor each of the following.

17. $14xy + 35y + 6x^2 + 15x$

18. $16x^2 + 8x + 1 - 9y^2$

CHAPTER 10

Rational Algebraic Expressions

Objectives . . .

- To express rational expressions in reduced form.
- To apply the four basic arithmetic operations to rational expressions.
- To find reduced forms of complex fractions.
- To work with negative exponents and scientific notation.
- To solve equations involving rational algebraic expressions.

10–1 REDUCED EXPRESSIONS

An algebraic expression which can be represented as a quotient of two polynomials is called a *rational algebraic expression.* The word *rational* is used in a similar way in the definition of a *rational number* as one which can be represented as a quotient of two integers.

For example, each of the following is a rational algebraic expression in one variable:

$$\frac{2x + 5}{3x^2 - 5x + 1}$$

and

$$\frac{y^3 + 1}{y^3 - 1}.$$

Similarly,

$$\frac{3x + 2y}{x^2 + 4xy + y^2}$$

and

$$\frac{m^3 + n^3}{m + n}$$

are rational algebraic expressions in two variables. In this chapter, all coefficients of variables will be rational numbers.

To find products of rational algebraic expressions, you can use the rule for multiplication of fractions which was given previously.

$$\frac{a}{b} \cdot \frac{c}{d} = \frac{a \cdot c}{b \cdot d} \qquad \text{(Fr-M)}$$

For example,

$$\frac{3x}{4 - x} \cdot \frac{x + 2}{5x^2} = \frac{3x \cdot (x + 2)}{(4 - x) \cdot 5x^2}.$$

Rational algebraic expressions which have the same numerator and denominator are equal to 1. For example, consider

$$\frac{5}{5}, \quad \frac{x^2 + 2x}{x^2 + 2x}, \quad \frac{5xy^3}{5xy^3}.$$

In other words, their values are 1 for all values of the variables except for those which make the denominator equal zero.

Every polynomial can be expressed as a rational algebraic expression with denominator 1. For example,

$$3x + 7 = \frac{3x + 7}{1}.$$

Also, every rational algebraic expression with rational coefficients is equal to a quotient of two polynomials with integral coefficients. For example,

$$\frac{\frac{1}{2}x - 3y}{x + \frac{1}{3}y} = \frac{\frac{1}{2}x - 3y}{x + \frac{1}{3}y} \cdot \frac{6}{6}, \quad \text{or} \quad \frac{3x - 18y}{6x + 2y}.$$

Just as the rational number $\frac{9}{12}$ can be expressed in a simpler form, $\frac{3}{4}$, some algebraic expressions can be expressed in simpler forms. For example,

$$\frac{3x(x + 2)}{5x^2(4 - x)},$$

may be simplified, or *reduced*, as follows:

$$\frac{3x(x + 2)}{5x^2(4 - x)} = \frac{3(x + 2)}{5x(4 - x)} \cdot \frac{x}{x}, \quad \text{or} \quad \frac{3(x + 2)}{5x(4 - x)}$$

(since $x/x = 1$ if $x \neq 0$).

A rational algebraic expression is said to be in reduced form if it is a quotient of two polynomials having integral coefficients and having no common polynomial factors other than 1 and −1.

A rational algebraic expression and its reduced form have the same value for every choice of values of the variables. The only restriction imposed on the values of the variables is that the denominator of the rational algebraic expression can never have value zero. Some examples of rational algebraic expressions and their reduced forms are given in the problems below.

Problem 1. Find the reduced form of

$$\frac{(x-1)^2(x+1)}{(x-1)(x+2)}.$$

Solution.

$$\frac{(x-1)^2(x+1)}{(x-1)(x+2)} = \frac{x-1}{x-1} \cdot \frac{(x-1)(x+1)}{x+2}$$

$$= \frac{(x-1)(x+1)}{x+2}, \quad x \neq 1 \text{ and } x \neq -2$$

The values of x for which the denominator $(x-1)(x+2)$ is zero are 1 and -2. For any other value of x, the original expression has the same value as the reduced expression. For example, if $x = 3$, then

$$\frac{(x-1)(x+1)}{x+2} = \frac{(3-1)(3+1)}{3+2}, \quad \text{or } \frac{8}{5}.$$

This is also the value of the original expression when $x = 3$.

Problem 2. Find the reduced form of

$$\frac{\frac{1}{3}a^5(a+2b)}{\frac{2}{5}a^2b(a-2b)}.$$

Solution.

$$\frac{\frac{1}{3}a^5(a+2b)}{\frac{2}{5}a^2b(a-2b)} = \frac{15}{15} \cdot \frac{\frac{1}{3}a^5(a+2b)}{\frac{2}{5}a^2b(a-2b)}, \quad \text{or} \quad \frac{5a^5(a+2b)}{6a^2b(a-2b)}$$

$$= \frac{a^2}{a^2} \cdot \frac{5a^3(a+2b)}{6b(a-2b)}, \quad \text{or} \quad \frac{5a^3(a+2b)}{6b(a-2b)}$$

For all values of a and b other than $a = 0$, $b = 0$, and $a = 2b$, the given expression has the same value as the reduced expression. For example, if $a = 1$ and $b = 2$, then

$$\frac{5a^3(a+2b)}{6b(a-2b)} = \frac{(5 \cdot 1^3)(1+4)}{(6 \cdot 2)(1-4)}, \quad \text{or } -\frac{25}{36}.$$

In the remainder of this chapter, the domains of the variables in a rational algebraic expression will not usually be discussed. You need only remember that you may use any values of the variables except those which will make some denominator have value zero.

Exercises

Find the reduced form of each of the following.

1. (a) $\frac{45}{48}$ (b) $\frac{60}{64}$

2. (a) $\dfrac{7a}{21a}$ (b) $\dfrac{28x}{4x}$

3. (a) $\dfrac{12x}{48}$ (b) $\dfrac{15x}{75}$

4. (a) $\dfrac{15x}{24y}$ (b) $\dfrac{20a}{65b}$

5. (a) $-\dfrac{3s}{9s}$ (b) $-\dfrac{6x}{42x}$

6. (a) $\dfrac{5(a+2)}{10(a+2)}$ (b) $\dfrac{6(x-4)}{18(x-4)}$

7. (a) $\dfrac{2y(x+y)}{6y}$ (b) $\dfrac{4a(x-1)}{2a}$

8. (a) $\dfrac{x^2y}{xy}$ (b) $\dfrac{a^2b}{ab^2}$

9. (a) $\dfrac{15p^2q}{48pq}$ (b) $\dfrac{x^3y^5}{x^4y^2}$

10. (a) $\dfrac{\pi r^2}{2\pi r}$ (b) $\dfrac{6xy^2}{30xy}$

11. (a) $\dfrac{4(x-y)^2}{3(x-y)}$ (b) $\dfrac{3(x-2)^2}{6(x-2)^4}$

12. (a) $\dfrac{\dfrac{x}{5}-\dfrac{y}{4}}{\dfrac{x}{10}+\dfrac{y}{2}}$ (b) $\dfrac{\frac{1}{4}x-y}{x+\frac{1}{12}y}$

13. (a) $\dfrac{\frac{2}{3}x^5(x+y)}{\frac{14}{15}x^2y(x-y)}$ (b) $\dfrac{\frac{4}{5}x^3y(2x-1)}{\frac{8}{25}x^4y^2(2x+1)}$

Factor each numerator and denominator. Then find the reduced form for each rational algebraic expression.

14. (a) $\dfrac{3x-3}{3x+6}$ (b) $\dfrac{15x-5}{5x+25}$

15. (a) $\dfrac{4x^2+6}{8x+6}$ (b) $\dfrac{9+12a}{6a-9}$

16. (a) $\dfrac{x^3+x}{x^2-x}$ (b) $\dfrac{x-x^4}{x+x^3}$

17. (a) $\dfrac{7x-14x^2}{21x^3-56x}$ (b) $\dfrac{36x^2+6x}{18x^3-12x}$

18. (a) $\dfrac{3a^5 - 2a^4}{a^3 + 5a^2}$ (b) $\dfrac{4x^3 - 5x^5}{6x^4 + 5x^6}$

19. (a) $\dfrac{x - 9}{7x - 63}$ (b) $\dfrac{12x - 6}{2x - 1}$

20. (a) $\dfrac{3x - 3}{4x - 4}$ (b) $\dfrac{5x + 5}{7x + 7}$

21. (a) $\dfrac{3x^2 - 27}{3 - x}$ (b) $\dfrac{4y + 4}{1 - y^2}$

22. (a) $\dfrac{4x^2 - y^2}{2x - y}$ (b) $\dfrac{3c + 5d}{25d^2 - 9c^2}$

23. (a) $\dfrac{(a - b)^2}{a^2 - b^2}$ (b) $\dfrac{(a + b)^2}{a^2 - b^2}$

24. (a) $\dfrac{x^2 + 3x - 10}{x^2 - 4x + 4}$ (b) $\dfrac{y^2 - 9y - 22}{y^2 - 13y + 22}$

25. (a) $\dfrac{5x^2 - 45x + 90}{180 - 5x^2}$ (b) $\dfrac{ax^3 - 20ax^2 + 99ax}{ax^3 - 2ax^2 - 99ax}$

In each of the following algebraic expressions, what real number or numbers must be excluded from the domain of x?

26. (a) $\dfrac{15}{x}$

(b) $\dfrac{15x}{x - 2}$

27. (a) $\dfrac{x + 3}{(x - 1)^2(x + 3)}$

(b) $\dfrac{x + 4}{(x^2 - 1)(x + 4)}$

28. (a) $\dfrac{2x}{x^2 + x - 12}$

(b) $\dfrac{3x^3}{x^2 + x - 2}$

Find the reduced form for each of the following rational algebraic expressions. Give the restrictions on the domains of the variables.

29. $\dfrac{x^2 - 7x + 10}{10 - 3x - x^2}$

30. $\dfrac{5x^2 - 20}{2 - x}$

31. $\dfrac{15 + x - 6x^2}{6x^2 - 19x + 15}$

32. In each set of rational algebraic expressions listed below, compare the resulting values when $x = 8$ and $y = 5$.

(a) $\dfrac{10}{x-4}$ and $\dfrac{10}{4-x}$

(b) $\dfrac{1}{y-x}$ and $-\left(\dfrac{1}{x-y}\right)$

(c) $\dfrac{12}{x-y}$, $\dfrac{12}{y-x}$, and $\dfrac{-12}{x-y}$

(d) $\dfrac{-18}{7-x}$, $\dfrac{18}{-(7-x)}$, $\dfrac{18}{7-x}$, and $\dfrac{18}{x-7}$

(e) Why is 4 not a permissible value of x in the fraction

$$\frac{10}{x-4}?$$

For $x \neq 4$, how does the value of

$$\frac{10}{x-4}$$

compare with the value of

$$\frac{10}{4-x}?$$

33. (a) Why is 5 not a permissible value of x in the rational expression

$$\frac{x^2 - x - 12}{x^2 - 5x}?$$

(b) In the same expression, can you find another value of x that would not be permissible?

34. (a) Find the value of each of the rational expressions

$$\frac{x^2 + x - 2}{x^2 + 3x + 2} \quad \text{and} \quad \frac{x-1}{x+1}$$

when x is given the values -2, 0, 1, 4, and 5.

(b) For what values of x are the values of the two expressions *not* the same?

(c) Write a general statement about the values of a rational algebraic expression and the values of its reduced form.

10-2 PRODUCTS OF EXPRESSIONS

Products of rational algebraic expressions can sometimes be reduced, as the following examples show.

Problem 1. Express the product in reduced form.

$$\frac{2x}{3y} \cdot \frac{6y^2}{4x^2}$$

Solution.

$$\frac{2x}{3y} \cdot \frac{6y^2}{4x^2} = \frac{2x \cdot 6y^2}{3y \cdot 4x^2} = \frac{12xy^2}{12x^2y}$$

$$= \frac{12xy}{12xy} \cdot \frac{y}{x} = 1 \cdot \frac{y}{x}, \quad \text{or } \frac{y}{x}$$

Problem 2. Reduce

$$\frac{3x - 3}{x + 3} \cdot \frac{x^2 - 2x - 15}{x^2 - 6x + 5}.$$

Solution. You should factor each polynomial *before multiplying* in order to see all possible common factors of the numerator and denominator.

$$\frac{3x - 3}{x + 3} \cdot \frac{x^2 - 2x - 15}{x^2 - 6x + 5} = \frac{3(x - 1) \cdot (x - 5)(x + 3)}{(x + 3) \cdot (x - 5)(x - 1)}$$

$$= \frac{(x - 1)(x - 5)(x + 3)}{(x - 1)(x - 5)(x + 3)} \cdot \frac{3}{1} = \frac{3}{1}, \quad \text{or } 3$$

Problem 3. Reduce

$$\frac{9a^2 - b^2}{6ab + 2b^2} \times \frac{2ab + 4b^2}{4a^2 - 8a}.$$

Solution. Again, you should first try to factor each polynomial.

$$\frac{9a^2 - b^2}{6ab + 2b^2} \times \frac{2ab + 4b^2}{4a^2 - 8a} = \frac{(3a - b)(3a + b) \times 2b(a + 2b)}{2b(3a + b) \times 4a(a - 2)}$$

$$= \frac{(3a - b)(a + 2b)}{4a(a - 2)} \times \frac{2b(3a + b)}{2b(3a + b)}$$

$$= \frac{(3a - b)(a + 2b)}{4a(a - 2)} \times 1 = \frac{(3a - b)(a + 2b)}{4a(a - 2)}$$

Exercises

Perform the indicated operations and give your answers in reduced form.

1. (a) $\frac{4}{5} \cdot \frac{15}{28}$ (b) $\frac{7}{6} \cdot \frac{42}{70}$
2. (a) $\frac{3x}{4} \cdot \frac{8}{9x}$ (b) $\frac{5}{7x} \cdot \frac{28x}{45}$
3. (a) $\frac{3}{a} \cdot \frac{7}{b}$ (b) $\frac{x}{4} \cdot \frac{5}{y}$
4. (a) $\frac{4a}{5b} \cdot \frac{15b}{16a}$ (b) $\frac{3x}{10y} \cdot \frac{20y}{9x}$
5. (a) $\frac{a}{b} \cdot \frac{b}{c} \cdot \frac{c}{a^2}$ (b) $\frac{x}{y^2} \cdot \frac{y^2}{z} \cdot \frac{z}{x^3}$
6. (a) $x \cdot \frac{x}{10}$ (b) $\frac{m}{7} \cdot m$
7. (a) $\frac{7x}{66y} \cdot \frac{11y}{14}$ (b) $\frac{30x}{13y} \cdot \frac{65}{90x}$
8. (a) $\frac{2x}{y} \cdot \frac{x}{4}$ (b) $\frac{3x}{7} \cdot \frac{2x}{y}$
9. (a) $\frac{1}{2x} \cdot \frac{-1}{2x}$ (b) $\frac{-4}{3x} \cdot \frac{-4}{3x}$
10. (a) $\frac{5x^2}{3y} \cdot \frac{2x}{y^3}$ (b) $\frac{7x^2}{9y^2} \cdot \frac{2x}{3y}$
11. (a) $\frac{1}{x - y} \cdot \frac{1}{x + y}$ (b) $\frac{1}{x - 4y} \cdot \frac{1}{x + 4y}$
12. (a) $\frac{2x + 3y}{5} \cdot \frac{10}{2x + 3y}$ (b) $\frac{17}{3x - 5y} \cdot \frac{3x - 5y}{34}$
13. (a) $\frac{2a}{7b^2} \times \frac{3ab}{8} \times \frac{4b}{9a^2}$ (b) $\frac{5x^2}{3y} \times \frac{9xy}{20} \times \frac{4y^2}{3x^2}$
14. (a) $\frac{2}{15x^4} \times \frac{5x^2}{14} \times \frac{7x^3}{10}$ (b) $\frac{3}{22x} \times \frac{7x^2}{24} \times \frac{33}{14x}$
15. (a) $\frac{1 - x}{5} \cdot \frac{10}{x - 1}$ (b) $\frac{4 - 3y}{6} \cdot \frac{3}{3y - 4}$

In Exercises 16–25, factor numerators and denominators before multiplying and give each product in reduced form.

16. (a) $\frac{6a + 12}{5} \times \frac{15a}{7a + 14}$ (b) $\frac{3a - 15}{9a} \cdot \frac{a}{6a - 30}$
17. (a) $\frac{25 - x^2}{12} \cdot \frac{6}{5 - x}$ (b) $\frac{3}{1 - x^2} \cdot \frac{1 + x}{3}$
18. (a) $\frac{x^2 + 2x + 1}{xy^2} \cdot \frac{y}{x + 1}$ (b) $\frac{x^2 + x}{x^2} \cdot \frac{3x - 3}{x^2 - 1}$

19. (a) $\frac{9x^2 - 1}{6x + 2} \cdot \frac{2}{3x - 1}$ (b) $\frac{a^2 - 1}{2a - 4} \cdot \frac{a^2 - 4}{a^2 + a - 2}$

20. (a) $\frac{3ay}{a^2 - y^2} \times \frac{ay - a^2}{12y^3}$ (b) $\frac{(p - q)^2}{p^2} \times \frac{p^3}{p^2 - q^2}$

21. (a) $\frac{x^2 + 5x + 6}{x + 3} \cdot \frac{x + 1}{x^2 + 3x + 2}$ (b) $\frac{x + 3}{x^2 + 7x + 12} \cdot \frac{x^2 + 9x + 20}{x + 5}$

22. (a) $\frac{x^2 - 5x - 6}{x^2 + 3x} \cdot \frac{x + 3}{6 - x}$ (b) $\frac{x^2 + 10x + 25}{2x - 10} \cdot \frac{6}{x + 5}$

23. (a) $\frac{64x^2 - 1}{64x^2 + 16x + 1} \cdot \frac{8x + 1}{16x - 2}$ (b) $\frac{4x + x^2}{x^3} \cdot \frac{12 - 3x}{16 - x^2}$

24. (a) $\frac{12x^2 - 31x + 20}{9x - 12} \cdot \frac{6x + 12}{4x^2 - 13x + 10}$

(b) $\frac{x^2 - 4}{x^2 + 4x + 4} \times \frac{3x + 6}{x^2 + 2x - 8}$

25. (a) $\frac{a^2 - ab - 6b^2}{a^2 + 6ab + 8b^2} \times \frac{a^2 - ab - 20b^2}{a^2 - 9ab + 18b^2}$

(b) $\frac{x^2 - 9x - 10}{x^2 - 14x + 49} \cdot \frac{x^2 - 49}{x^2 - 4}$

Express each product in reduced form.

26. $\frac{12 + 12x}{x} \cdot \frac{x^3}{4x^2 - 4} \cdot \frac{x^2 + 6x + 5}{3x^2 + 15x}$

27. $\frac{3x^2 - 75}{2x^2 - 32} \cdot \frac{5x + 20}{7x - 35}$

28. $\frac{2x^2 - 5x - 3}{1 - 4x^2} \cdot \frac{2x^2 + 5x - 3}{9 - x^2}$

29. $\frac{x^2 + xy}{y - x} \cdot \frac{x^2 - xy}{y + x}$

30. $\frac{25 - x^2}{3x - 15} \cdot \frac{7}{x + 5}$

31. $\frac{7 - x}{7 + x} \cdot \frac{x + 7}{x - 7}$

Express each product in reduced form.

32. $\frac{v^9 - v^7}{u^2v} \cdot \frac{u^4}{uv^3 - uv}$

33. $\frac{x^3 + 8y^3}{x^2 - 4y^2} \cdot \frac{(x - 2y)^2}{x^2 - 2xy + y^2}$

34. $\frac{x^3 - y^3}{x + y} \cdot \frac{x^3 + y^3}{(x^2 - y^2)(x^2 + xy + y^2)}$

35. $\frac{a^6 - b^6}{a^3 - b^3} \cdot \frac{5a^2}{a^3 + b^3}$

10–3 QUOTIENTS OF EXPRESSIONS

Each nonzero rational algebraic expression has a reciprocal, which is obtained by interchanging the numerator and denominator. For example,

$$\frac{2x}{x+1} \quad \text{has reciprocal} \quad \frac{x+1}{2x} \quad \text{because} \quad \frac{2x}{x+1}\cdot\frac{x+1}{2x} = 1.$$

$$\frac{a^2-b^2}{a^2+b^2} \quad \text{has reciprocal} \quad \frac{a^2+b^2}{a^2-b^2} \quad \text{because} \quad \frac{a^2-b^2}{a^2+b^2}\cdot\frac{a^2+b^2}{a^2-b^2} = 1.$$

The quotient of two expressions is found by multiplying the first expression by the reciprocal of the second.

DIVISION OF FRACTIONS

$$\frac{a}{b} \div \frac{c}{d} = \frac{a}{b}\cdot\frac{1}{c/d}, \quad \text{or} \quad \frac{a}{b}\cdot\frac{d}{c} \qquad \textit{(Fr-D)}$$

Problem 1. Express the quotient in reduced form.

$$\frac{3}{10x^2} \div \frac{4}{15x}$$

Solution.

$$\begin{aligned} \frac{3}{10x^2} \div \frac{4}{15x} &= \frac{3}{10x^2}\cdot\frac{15x}{4} \\ &= \frac{45x}{40x^2} \\ &= \frac{5x}{5x}\cdot\frac{9}{8x} \\ &= \frac{9}{8x} \end{aligned}$$

Problem 2. Find the quotient of

$$\frac{2xy}{x^2-4xy+3y^2} \div \frac{xy-6y^2}{x^2+xy-2y^2}.$$

Solution.

$$\frac{2xy}{x^2 - 4xy + 3y^2} \div \frac{xy - 6y^2}{x^2 + xy - 2y^2}$$

$$= \frac{2xy}{x^2 - 4xy + 3y^2} \cdot \frac{x^2 + xy - 2y^2}{xy - 6y^2}$$

$$= \frac{2xy}{(x - y)(x - 3y)} \cdot \frac{(x + 2y)(x - y)}{y(x - 6y)}$$

$$= \frac{y(x - y)}{y(x - y)} \cdot \frac{2x(x + 2y)}{(x - 3y)(x - 6y)}$$

$$= 1 \cdot \frac{2x(x + 2y)}{(x - 3y)(x - 6y)}, \quad \text{or} \quad \frac{2x(x + 2y)}{(x - 3y)(x - 6y)}$$

Thus,

$$\frac{2xy}{x^2 - 4xy + 3y^2} \div \frac{xy - 6y^2}{x^2 + xy - 2y^2} = \frac{2x(x + 2y)}{(x - 3y)(x - 6y)}.$$

Exercises

Give the reciprocal of each rational algebraic expression.

1. (a) $\dfrac{3x}{x + 3}$ (b) $\dfrac{x - y}{x + y}$

2. (a) $\dfrac{x^2 + 1}{x^3 - 1}$ (b) $\dfrac{1 - x^3}{1 + x^2}$

3. (a) $x^2 + xy + y^2$ (b) $\dfrac{1}{x^3 - x^2y + x}$

Express each quotient in reduced form.

4. (a) $\frac{6}{35} \div \frac{3}{14}$ (b) $\frac{22}{49} \div \frac{55}{14}$

5. (a) $\dfrac{6}{11} \div \dfrac{12}{7x}$ (b) $\dfrac{4}{7x} \div \dfrac{16}{21}$

6. (a) $\dfrac{7x}{15} \div \dfrac{5}{4x}$ (b) $\dfrac{5}{9x} \div \dfrac{18x}{25}$

7. (a) $\dfrac{4}{x} \div \dfrac{6}{x}$ (b) $\left(\dfrac{9}{x}\right) \div \left(\dfrac{-x}{9}\right)$

8. (a) $6x \div \left(\dfrac{3x}{4}\right)$ (b) $2x^2 \div \left(\dfrac{8x^2}{3}\right)$

9. (a) $\dfrac{9x^2y}{13} \div \dfrac{3xy^2}{26}$ (b) $\dfrac{3x^2y^3}{15} \div \dfrac{9x^3y^2}{30}$

10. (a) $\dfrac{3x^2y}{8} \div 6xy^2$ (b) $\dfrac{9x^2}{5} \div 3x$

11. (a) $-\left(\frac{x^2}{5}\right) \div \frac{x}{15}$ (b) $-\left(\frac{x^3}{20}\right) \div \frac{x^2}{4}$

12. (a) $\frac{3a}{4b^2} \div \frac{9a^2}{16b}$ (b) $\frac{26x^2}{15y^2} \div \frac{39x^3}{20y}$

13. (a) $\frac{125ab}{16} \div \frac{15b^2}{8a}$ (b) $\frac{22x^3}{35y^4} \div \frac{11x}{7y^5}$

Find each quotient and give your answer in reduced and factored form.

14. (a) $(x - y) \div (y - x)$ (b) $(x + y)^3 \div (x + y)^2$

15. (a) $\frac{4x - 6}{5} \div \frac{6x - 9}{25}$ (b) $\frac{6x + 2}{7} \div \frac{15x + 5}{56}$

16. (a) $\frac{x^2 - y^2}{xy^2} \div \frac{xy + y^2}{x}$ (b) $\frac{x^2 - 4y^2}{x^2y^2} \div \frac{xy + 2y^2}{x^3y^3}$

17. (a) $\frac{b - 2}{a^2 + a} \div \frac{b^2 - 4}{a + 1}$ (b) $\frac{1 - c}{d + d^2} \div \frac{1 - c^2}{1 - d^2}$

18. (a) $\frac{3x - 9y}{8} \div (3y - x)$ (b) $(5x - 3y) \div \frac{9y - 15x}{2}$

19. (a) $\frac{8a - 32}{a + b} \div \frac{16}{3a + 3b}$ (b) $\frac{17}{5x - 10y} \div \frac{34x + 51y}{3x - 6y}$

20. (a) $\frac{2x + x^2}{4x - 5} \div \frac{4x^2 + 2x^3}{16x - 20}$ (b) $\frac{5x^3 + 5x}{8x - 20} \div \frac{10x^3 + 10x^2}{2x - 5}$

21. (a) $\frac{a^2 - 5a}{2a^3 - a^2} \div \frac{a^2 - 7a + 10}{2a^2 - 5a + 2}$ (b) $\frac{3a^2 + 6a - 24}{a^2 - 7a + 10} \div \frac{3a^2 + 4a}{a^3 - 5a^2}$

22. (a) $\frac{x^2 + 13xy + 12y^2}{3xy^2} \div (x + y)$

(b) $(x^2 - 100y^2) \div \frac{2x}{x^2 - 9xy - 10y^2}$

Express each quotient in reduced form.

23. $\frac{a - 3b}{a^3b} \div \frac{3b - a}{ab^3}$

24. $\frac{x^2 + 3x}{9 - x^2} \div \frac{x^2 - 5x}{x^2 - 8x + 15}$

25. $\frac{c^2 - d^2}{c - d} \div \frac{(c - d)^2}{(c + d)^2}$

26. $\frac{x^4 - 4x^3}{4y - xy} \div \frac{x^3}{y^2}$

27. $\left(\frac{p^2 + 6pq + 5q^2}{p + q}\right)\left(\frac{p^2 + 4pq + 4q^2}{p^2 + 7pq + 10q^2}\right) \div (p + 2q)$

28. $\frac{8x^3 - y^3}{6x} \div \frac{2x - y}{12x^2}$

29. $\frac{x^3 + 64y^3}{x - 4y} \div \frac{x^2 - 4xy + 16y^2}{x^2 - 16y^2}$

30. $\frac{4y}{x^2 - x - 56} \div \frac{y^4}{8 - x}$

31. $\frac{a^2 - 1}{15 - 5a} \div \frac{6a + 6}{6 - 6a}$

32. $(x^6 - y^6) \div (x^4 + x^2y^2 + y^4)$

Preparation for Section 10–4

Complete the following problems.

1. $\frac{3}{10} + \frac{9}{14} = (\frac{3}{10} \cdot \frac{7}{7}) + (\frac{9}{14} \cdot \frac{5}{5})$
$= \frac{21}{70} + \frac{45}{70}$
$= (21 \cdot \frac{1}{70}) + (45 \cdot \frac{1}{70})$
$=$ _?_

2. $\frac{6}{7x} - \frac{3}{14x^2} + \frac{5}{2x} = \left(\frac{6}{7x} \cdot \frac{2x}{2x}\right) - \frac{3}{14x^2} + \left(\frac{5}{2x} \cdot \frac{7x}{7x}\right)$
$= \frac{12x}{14x^2} - \frac{3}{14x^2} + \frac{35x}{14x^2}$
$= \frac{1}{14x^2}$(_?_)
$=$ _?_

3. $\frac{x + 2}{x - 3} + \frac{7}{3 - x} = \frac{x + 2}{x - 3} + \frac{7}{-(x - 3)} = \frac{x + 2}{x - 3} - \frac{7}{x - 3}$
$= \frac{1}{x - 3}$(_?_) = _?_

10–4 ADDITION AND SUBTRACTION OF EXPRESSIONS

The distributive axiom is of great help when you add and subtract rational numbers. For example,

$$\frac{8}{13} + \frac{3}{13} = (8 \cdot \frac{1}{13}) + (3 \cdot \frac{1}{13})$$
$$= (8 + 3) \cdot \frac{1}{13}, \quad \text{or } \frac{11}{13}.$$

The two rational numbers above have the same denominator. If this had not been the case, you would have had to find a common denominator before performing the addition. For example,

$$\frac{5}{6} + \frac{3}{8} = \frac{5 \cdot 4}{6 \cdot 4} + \frac{3 \cdot 3}{8 \cdot 3}$$
$$= \frac{20}{24} + \frac{9}{24}$$
$$= (20 \cdot \frac{1}{24}) + (9 \cdot \frac{1}{24})$$
$$= (20 + 9) \cdot \frac{1}{24}, \quad \text{or } \frac{29}{24}.$$

Similarly,

$$\tfrac{5}{6} - \tfrac{3}{8} = (20 \cdot \tfrac{1}{24}) - (9 \cdot \tfrac{1}{24})$$
$$= (20 - 9) \cdot \tfrac{1}{24}) \quad \text{or } \tfrac{11}{24}.$$

You add and subtract rational algebraic expressions in the same way; that is, put each numerator over a common denominator. (In Chapter 9, you learned about the least common multiple (l.c.m.) of two numbers. The least common multiple of the denominators of two rational algebraic expressions is called their *least common denominator* (l.c.d.).)

Problem 1. Express the sum in reduced form.

$$\frac{2}{3y} + \frac{9}{4y}$$

Solution. The l.c.d. of the fractions, or the l.c.m. of $3y$ and $4y$, is $12y$. Therefore,

$$\frac{2}{3y} + \frac{9}{4y} = \left(\frac{2}{3y} \cdot \frac{4}{4}\right) + \left(\frac{9}{4y} \cdot \frac{3}{3}\right)$$
$$= \frac{8}{12y} + \frac{27}{12y}$$
$$= \left(8 \cdot \frac{1}{12y}\right) + \left(27 \cdot \frac{1}{12y}\right)$$
$$= (8 + 27) \cdot \frac{1}{12y}, \quad \text{or } \frac{35}{12y}.$$

Problem 2. Express the difference in reduced form.

$$\frac{x}{x - 2y} - \frac{3y}{x - 2y}$$

Solution. The denominators are the same. Therefore,

$$\frac{x}{x - 2y} - \frac{3y}{x - 2y} = \left(x \cdot \frac{1}{x - 2y}\right) - \left(3y \cdot \frac{1}{x - 2y}\right)$$
$$= (x - 3y) \cdot \frac{1}{x - 2y}, \quad \text{or } \frac{x - 3y}{x - 2y}.$$

Problem 3. Express the sum in reduced form.

$$\frac{b + 1}{2b - 1} + \frac{3}{1 - 2b}$$

Solution. What is the l.c.d.? Note that

$$1 - 2b = -(2b - 1).$$

Since the l.c.m. of a polynomial and its negative is the polynomial itself, you have

$$\begin{aligned}\frac{b+1}{2b-1} + \frac{3}{1-2b} &= \frac{b+1}{2b-1} + \frac{3}{1-2b} \cdot \frac{-1}{-1} \\ &= \frac{b+1}{2b-1} + \frac{-3}{2b-1} \\ &= (b+1)\left(\frac{1}{2b-1}\right) + (-3)\left(\frac{1}{2b-1}\right) \\ &= [(b+1) - 3] \cdot \frac{1}{2b-1}, \quad \text{or } \frac{b-2}{2b-1}.\end{aligned}$$

Exercises

Find each indicated sum or difference. Give your answers in reduced form.

1. (a) $\frac{7}{15} + \frac{5}{6}$ (b) $\frac{8}{21} + \frac{3}{14}$
2. (a) $\dfrac{5}{12t} - \dfrac{1}{12t}$ (b) $\dfrac{14}{13x} - \dfrac{1}{13x}$
3. (a) $\dfrac{4x}{9y} - \dfrac{x}{9y}$ (b) $\dfrac{5x}{12y} - \dfrac{x}{12y}$
4. (a) $\dfrac{3}{a} + \dfrac{a-3}{a}$ (b) $\dfrac{x-4}{x} + \dfrac{4}{x}$
5. (a) $\dfrac{7a}{4b} - \dfrac{5a^2}{4b}$ (b) $\dfrac{2y}{3x^2} - \dfrac{5y^2}{3x^2}$
6. (a) $\dfrac{x}{2} + \dfrac{y}{2}$ (b) $\dfrac{x}{7} + \dfrac{y}{7}$
7. (a) $\dfrac{x}{3} + \dfrac{y}{7}$ (b) $\dfrac{x}{7} + \dfrac{y}{5}$
8. (a) $\dfrac{5}{3x} - \dfrac{7}{4x}$ (b) $\dfrac{1}{2x} - \dfrac{3}{5x}$
9. (a) $\dfrac{6x}{13} - \dfrac{5x}{13} + \dfrac{12x}{13}$ (b) $\dfrac{4x}{9} - \dfrac{5x}{9} + \dfrac{10x}{9}$
10. (a) $\dfrac{7}{10x} - \dfrac{3}{5x} - \dfrac{5}{2x}$ (b) $\dfrac{5}{6x} - \dfrac{2}{3x} - \dfrac{4}{5x}$

Find the indicated sum or difference, and if possible, reduce.

11. (a) $\frac{3x^2 - x}{2x + 1} - \frac{x^2 - 2x}{2x + 1}$ (b) $\frac{x^2 + 3x}{1 - 2x} - \frac{3x^2 + 2x}{1 - 2x}$

12. (a) $\frac{x - 5}{x^2 - 1} + \frac{5}{x^2 - 1}$ (b) $\frac{2x}{x^2 + 1} + \frac{2x - 1}{x^2 + 1}$

13. (a) $\frac{y}{y - x} + \frac{x}{y - x}$ (b) $\frac{x}{x + y} - \frac{y}{x + y}$

14. (a) $\frac{a^2}{a - b} - \frac{b^2}{a - b}$ (b) $\frac{x^2}{x + 4y} - \frac{16y^2}{x + 4y}$

15. (a) $\frac{x}{2x - 4} + \frac{x}{2x - 4}$ (b) $\frac{2x}{3x - 9} + \frac{x}{3x - 9}$

16. (a) $\frac{x^2 + y^2}{x + y} + \frac{2xy}{x + y}$ (b) $\frac{x^2 + y^2}{x - y} - \frac{2xy}{x - y}$

17. (a) $\frac{6a}{a + 3} + \frac{a^2 + 9}{a + 3}$ (b) $\frac{x^2 + 16}{x + 4} + \frac{8x}{x + 4}$

18. (a) $\frac{x^2}{x - 3} + \frac{9}{x - 3} + \frac{6x}{3 - x}$ (b) $\frac{x^2}{x - 4} + \frac{8x}{4 - x} + \frac{16}{x - 4}$

19. (a) $\frac{4x + 10}{2x^2 + x - 3} - \frac{2x + 7}{2x^2 + x - 3}$

(b) $\frac{5x + 7}{2x^2 + 3x - 5} - \frac{3x + 2}{2x^2 + 3x - 5}$

20. (a) $\frac{8}{a - 3} + \frac{5}{3 - a}$ (b) $\frac{10}{x - 4} + \frac{4}{4 - x}$

21. (a) $\frac{7}{x - y} - \frac{2}{y - x}$ (b) $\frac{12}{a^2 - b^2} - \frac{2}{b^2 - a^2}$

Find the indicated sum or difference, and if possible, reduce.

22. $\frac{7a - 2}{a(b + c)} - \frac{2}{b + c}$

23. $\frac{4}{5} + \frac{2}{x + 5}$

24. $\frac{3}{x^2 + xy} + \frac{2}{x + y}$

25. $\frac{10}{x^2 + x} + \frac{2}{3x^2 + 3x}$

Preparation for Section 10–5

Each addition or subtraction of fractions is followed by a set of suggestions for a common denominator. In each case, select the correct l.c.d.

1. $\frac{3}{ab^2} - \frac{4}{a^2b} + \frac{7}{ab}$; $(ab^2)(a^2b)$ or ab^2 or a^2b^2

2. $\frac{x}{x^2 + 2x} - \frac{5}{x^2 + 3x}$; $(x^2 + 2x)(x^2 + 3x)$ or $x(x + 2)(x + 3)$

3. $5x + \frac{x - y}{7x + 3y}$; $5x(7x + 3y)$ or $7x + 3y$

4. $\frac{x}{2x - y} + \frac{y}{x + y} + \frac{3x}{y - 2x}$; $(2x - y)(x + y)$ or $(2x - y)(x + y)(y - 2x)$

5. $\frac{a}{9 - a^2} - \frac{3}{12 + 4a}$; $(9 - a^2)(12 + 4a)$ or $4(3 - a)(3 + a)$

6. $\frac{3}{x^2 + 5x + 4} - \frac{x}{x^2 + 11x + 28}$; $(x + 4)(x + 1)(x + 7)$ or $(x^2 + 5x + 4)(x^2 + 11x + 28)$

10–5 MORE ADDITION AND SUBTRACTION

The following problems are some more examples of sums and differences of expressions.

Problem 1. Express the difference in a reduced form.

$$\frac{3}{ab} - \frac{4b}{a^2}$$

Solution. The least common denominator is a^2b. Thus,

$$\begin{aligned}
\frac{3}{ab} - \frac{4b}{a^2} &= \left(\frac{3}{ab} \cdot \frac{a}{a}\right) - \left(\frac{4b}{a^2} \cdot \frac{b}{b}\right) \\
&= \frac{3a}{a^2b} - \frac{4b^2}{a^2b} \\
&= \left(3a \cdot \frac{1}{a^2b}\right) - \left(4b^2 \cdot \frac{1}{a^2b}\right) \\
&= (3a - 4b^2) \cdot \frac{1}{a^2b}, \quad \text{or} \quad \frac{3a - 4b^2}{a^2b}.
\end{aligned}$$

Problem 2. Express the sum in reduced form.

$$\frac{x+y}{x^2y^2}+\frac{1-y}{xy^3}$$

Solution. The l.c.d. is x^2y^3. Thus,

$$\begin{aligned}\frac{x+y}{x^2y^2}+\frac{1-y}{xy^3} &= \left(\frac{x+y}{x^2y^2}\cdot\frac{y}{y}\right)+\left(\frac{1-y}{xy^3}\cdot\frac{x}{x}\right)\\ &= \frac{y(x+y)}{x^2y^3}+\frac{x(1-y)}{x^2y^3}\\ &= \left[y(x+y)\cdot\frac{1}{x^2y^3}\right]+\left[x(1-y)\cdot\frac{1}{x^2y^3}\right]\\ &= [y(x+y)+x(1-y)]\cdot\frac{1}{x^2y^3}\\ &= \frac{xy+y^2+x-xy}{x^2y^3}, \quad \text{or} \quad \frac{y^2+x}{x^2y^3}.\end{aligned}$$

Problem 3. Express the difference in reduced form.

$$\frac{x}{x-2}-\frac{3}{x-3}$$

Solution. The l.c.d. is $(x-2)(x-3)$, and

$$\begin{aligned}&\frac{x}{x-2}-\frac{3}{x-3}\\ &\quad = \frac{x(x-3)}{(x-2)(x-3)}-\frac{3(x-2)}{(x-3)(x-2)}\\ &\quad = \left[x(x-3)\cdot\frac{1}{(x-2)(x-3)}\right]-\left[3(x-2)\cdot\frac{1}{(x-2)(x-3)}\right]\\ &\quad = \frac{x(x-3)-3(x-2)}{(x-2)(x-3)}\\ &\quad = \frac{x^2-3x-3x+6}{(x-2)(x-3)}, \quad \text{or} \quad \frac{x^2-6x+6}{x^2-5x+6}.\end{aligned}$$

Problem 4. Express the sum in reduced form.

$$4a+\frac{a-b}{2a+7b}$$

Solution. First find that

$$4a + \frac{a-b}{2a+7b} = \frac{4a}{1} + \frac{a-b}{2a+7b}.$$

Hence, the l.c.d. is $(2a + 7b)$, and

$$\begin{aligned}\frac{4a}{1} + \frac{a-b}{2a+7b} &= \left(\frac{4a}{1} \cdot \frac{2a+7b}{2a+7b}\right) + \left(\frac{a-b}{2a+7b}\right)\\ &= \frac{4a(2a+7b)}{2a+7b} + \frac{a-b}{2a+7b}\\ &= \frac{4a(2a+7b)+(a-b)}{2a+7b}\\ &= \frac{8a^2+28ab+a-b}{2a+7b}.\end{aligned}$$

Problem 5. Express the difference in reduced form.

$$\frac{x}{2x^2 - xy - 10y^2} - \frac{x+y}{15y^2 - xy - 2x^2}$$

Solution. First factor each denominator.

$$\begin{aligned}&\frac{x}{2x^2 - xy - 10y^2} - \frac{x+y}{15y^2 - xy - 2x^2}\\ &\qquad = \frac{x}{(2x-5y)(x+2y)} - \frac{x+y}{(5y-2x)(3y+x)}\end{aligned}$$

Since $5y - 2x$ is the negative of $2x - 5y$, the least common denominator is $(2x - 5y)(x + 2y)(x + 3y)$. Hence,

$$\begin{aligned}&\frac{x}{2x^2 - xy - 10y^2} - \frac{x+y}{15y^2 - xy - 2x^2}\\ &= \frac{x(x+3y)}{(2x-5y)(x+2y)(x+3y)} - \frac{-1(x+y)(x+2y)}{(2x-5y)(x+3y)(x+2y)}\\ &= \frac{x(x+3y)-(-1)(x+y)(x+2y)}{(2x-5y)(x+2y)(x+3y)}\\ &= \frac{(x^2+3xy)+(x^2+3xy+2y^2)}{(2x-5y)(x+2y)(x+3y)}\\ &= \frac{2x^2+6xy+2y^2}{(2x-5y)(x+2y)(x+3y)}.\end{aligned}$$

Exercises

Express each sum or difference in reduced form.

1. (a) $\frac{2b}{a} - \frac{3}{2}$ (b) $\frac{3x}{4} - \frac{2}{y}$

2. (a) $\frac{1}{8a} - \frac{5b}{2a}$ (b) $\frac{5}{12x} - \frac{2y}{3x}$

3. (a) $\frac{6x-1}{4} + \frac{2x+1}{6}$ (b) $\frac{3x+5}{3} + \frac{2x-1}{4}$

4. (a) $\frac{3x-7}{5} - \frac{2x+1}{6}$ (b) $\frac{2a-5}{7} - \frac{3a+1}{2}$

5. (a) $\frac{4}{xy} - \frac{5y}{x^2}$ (b) $\frac{3}{ab} - \frac{6b}{a^2}$

6. (a) $\frac{2}{x} - \frac{3}{x^2} + \frac{5}{x^3}$ (b) $\frac{3}{x^4} - \frac{1}{x^5} + \frac{2}{x^6}$

7. (a) $\frac{7}{ab^2} - \frac{6}{a^2b}$ (b) $\frac{13}{xy^2} - \frac{11}{x^2y}$

8. (a) $\frac{3}{x^2} + \frac{5}{y^2} - \frac{2}{xy}$ (b) $\frac{4a}{5b^3} - \frac{2}{ab} + \frac{5b}{6a^2}$

Express each sum or difference in reduced form.

9. (a) $\frac{8}{a+2} - \frac{3}{a-4}$ (b) $\frac{7}{a+5} - \frac{2}{a-2}$

10. (a) $\frac{2x}{x+2} + \frac{5}{x-3}$ (b) $\frac{3x^2}{x+1} + \frac{1}{x-2}$

11. (a) $\frac{3}{a-4} + \frac{6}{a+3}$ (b) $\frac{a}{a-5} + \frac{2}{a+2}$

12. (a) $\frac{3x}{2x-3} - \frac{2x}{x-1}$ (b) $\frac{4x}{2x+1} - \frac{x}{x-2}$

13. (a) $\frac{3}{4x-1} - \frac{x}{4x+1}$ (b) $\frac{2}{3x-2} - \frac{x}{3x+2}$

14. (a) $\frac{x-2}{x+3} - \frac{x+5}{x-3}$ (b) $\frac{x-4}{x+5} - \frac{x-1}{x-5}$

15. (a) $\frac{3}{a} + \frac{7}{a+5}$ (b) $\frac{2}{x} - \frac{7}{x+1}$

16. (a) $\frac{b-1}{2b-5} - \frac{b-2}{2b+5}$ (b) $\frac{x+3}{2x-3} - \frac{x-3}{2x+3}$

Find the sum or difference. Express each answer in reduced form.

17. (a) $\frac{3x-2}{5x+7} + 9$ (b) $\frac{4x-1}{2x+7} + 2$

18. (a) $a + \frac{3}{a+1}$ (b) $x + \frac{4}{x+4}$

19. (a) $x + \frac{7}{x}$ (b) $\frac{3}{a} - a$

20. (a) $20x^2 - \frac{x^2 - 1}{x - 2}$ (b) $3x^2 - \frac{x^2 + 1}{x - 1}$

21. (a) $5 + \frac{x + y}{x - y}$ (b) $4 + \frac{x - y}{x + y}$

22. (a) $\frac{3a + 5}{2a - 7} - 10a$ (b) $\frac{5a + 3}{7a - 2} - a$

23. (a) $7 - \frac{2a - b}{2a + b}$ (b) $11 - \frac{3x + 2y}{3x - 2y}$

24. (a) $\frac{a}{b} + 3 + \frac{b}{a}$ (b) $\frac{x}{2y} + 1 + \frac{2y}{x}$

Find each of the following differences.

25. $\frac{3a}{(a + 2)^2} - \frac{2a - 1}{a + 2}$

26. $\frac{y}{2y + 3} - \frac{5}{(2y + 3)^2}$

27. $\frac{4}{(x + 4)(x + 1)} - \frac{3}{(x + 7)(x + 4)}$

28. $\frac{5}{(2x + 1)(x - 4)} - \frac{2}{(2x + 1)(x + 4)}$

First factor each denominator. Then find the sums or differences, and express them in reduced form.

29. $\frac{4}{x - 2} - \frac{3x + 2}{x^2 - 4}$

30. $\frac{3}{2a + 18} + \frac{27}{a^2 - 81}$

31. $\frac{7}{x^2 - y^2} - \frac{5}{3x + 3y}$

32. $\frac{5}{b^2 - 4} - \frac{7}{3b - 6}$

33. $\frac{6}{9 - a^2} - \frac{3}{12 + 4a}$

34. $\frac{3}{x^2 - 2x - 8} + \frac{2}{x^2 - 4}$

35. $\frac{2}{x + y} - \frac{5}{x^2 - y^2} + \frac{1}{x - y}$

36. $\frac{5}{xy + y^2} + \frac{2}{x^2 + xy}$

Review for Sections 10–1 through 10–5

Find the reduced form of each of the following rational expressions.

1. $\frac{4ab}{8b^2}$

2. $\frac{a^2 - 4}{2(a + 2)}$

3. $\frac{6(x + y)}{3x^2 + 3xy}$

4. $\frac{a^4 - 16}{3(a^2 + 4)(a - 2)}$

5. $\frac{x^2 + x - 6}{x^2 + 8x + 15}$

6. $\frac{b^2 - ab - 2a^2}{b^2 + 2ab + a^2}$

Perform the indicated operations and give your answers in reduced form.

7. $\frac{3x}{2} \cdot \frac{4xy}{3x}$

8. $\frac{3m}{2} \div \frac{m^2}{6}$

9. $\frac{b}{a + 2b} + \frac{a}{b}$

10. $\frac{m}{m^2 - n^2} - \frac{n}{m^2 - n^2}$

11. $4t^2 \div \left(\frac{8t}{3}\right)$

12. $\frac{4x}{y} \cdot \frac{3xy}{x^2} \cdot \frac{y^2}{6x^2}$

13. $\frac{r^2 + t}{r^2 - t^2} + \frac{r - r^2}{r^2 - t^2}$

14. $\frac{2}{2m} - \frac{1 + 2m}{m + 3}$

15. $\frac{4a - 8}{3a} \times \frac{a}{2a - 4}$

16. $\frac{13x}{2y^2} \div \frac{26xy}{4}$

17. $\frac{3n}{m + 2n} + 2$

18. $\frac{a - 1}{6} - \frac{1}{a}$

19. $(x^2 - 1) \div (x + 1)$

20. $\frac{m^2 - 2m - 15}{m^2 + m - 2} \times \frac{3m + 6}{6m - 30}$

21. $\frac{x + 4y}{y - x} - 1$

22. $\frac{3a + b}{2(a - b)} \div \frac{6a + 2b}{a^2 - b^2}$

23. $\frac{4 - m}{5m + 20} - \frac{m + 12}{5(m + 4)}$

24. $\frac{2x}{3(x - y)} + \frac{2y}{3y - 3x}$

25. $\frac{2t + 1}{2t - 1} + \frac{1 - 4t^2}{4t^2 - 1}$

26. $\frac{15ab}{a^2 + 2b} \div \frac{5b^2}{a^4 - 4b^2}$

27. $\frac{5y^3}{3x} \times \frac{2y}{x}$

28. $\frac{xy}{2x + y} - \frac{3xy}{y - 2x}$

29. $\frac{m + n}{m^2 - 4mn - 12n^2} + \frac{n}{m^2 - 4mn - 12n^2}$

30. $\frac{t^2 - 1}{4t - 4} \cdot \frac{t^2}{t^2 + t}$

Answers to Review for Sections 10–1 through 10–5

1. $\frac{a}{2b}$

2. $\frac{a - 2}{2}$

3. $\frac{2}{x}$

4. $\frac{a + 2}{3}$

5. $\frac{x - 2}{x + 5}$

6. $\frac{b - 2a}{b + a}$

7. $2xy$

8. $\frac{9}{m}$

9. $\frac{(a + b)^2}{b(a + 2b)}$

10. $\frac{1}{m + n}$

11. $\frac{3t}{2}$

12. $\frac{2y^2}{x^2}$

13. $\dfrac{1}{r-t}$ 14. $\dfrac{3-2m^2}{m(m+3)}$ 15. $\frac{2}{3}$

16. $\dfrac{1}{y^3}$ 17. $\dfrac{7n+2m}{m+2n}$ 18. $\dfrac{(a-3)(a+2)}{6a}$

19. $x-1$ 20. $\dfrac{m+3}{2(m-1)}$ 21. $\dfrac{2x+3y}{y-x}$

22. $\dfrac{a+b}{4}$ 23. $-\frac{2}{5}$ 24. $\frac{2}{3}$

25. $\dfrac{2}{2t-1}$ 26. $\dfrac{3a(a^2-2b)}{b}$ 27. $\dfrac{10y^4}{3x^2}$

28. $\dfrac{-2xy(4x+y)}{(2x+y)(y-2x)}$ 29. $\dfrac{1}{m-6n}$ 30. $\dfrac{t}{4}$

10–6 COMPLEX FRACTIONS

An algebraic expression which is made up of sums, differences, products, and quotients of rational algebraic expressions is often called a *complex fraction.* The following are examples of complex fractions.

Problem 1. Find a reduced form of

$$\left(3-\frac{3}{x}\right)\left(\frac{5}{x}+\frac{6}{x^2}\right).$$

Solution. First express each factor as a quotient of two polynomials; then multiply in the usual way.

$$\begin{aligned}\left(3-\frac{3}{x}\right)\left(\frac{5}{x}+\frac{6}{x^2}\right) &= \left(\frac{3x}{x}-\frac{3}{x}\right)\left(\frac{5x}{x^2}+\frac{6}{x^2}\right)\\ &= \left(\frac{3x-3}{x}\right)\left(\frac{5x+6}{x^2}\right)\\ &= \frac{(3x-3)(5x+6)}{x^3}, \quad \text{or} \quad \frac{3(x-1)(5x+6)}{x^3}\end{aligned}$$

Problem 2. Find a reduced form of

$$\left(\frac{x}{y}-\frac{y}{x}\right)\div\left(\frac{2x}{y}+\frac{y}{2x}\right).$$

Solution. First express the dividend and divisor as quotients of polynomials, then divide in the usual way.

$$\left(\frac{x}{y} - \frac{y}{x}\right) \div \left(\frac{2x}{y} + \frac{y}{2x}\right) = \left(\frac{x^2}{xy} - \frac{y^2}{xy}\right) \div \left(\frac{4x^2}{2xy} + \frac{y^2}{2xy}\right)$$

$$= \frac{x^2 - y^2}{xy} \div \frac{4x^2 + y^2}{2xy}$$

$$= \frac{x^2 - y^2}{xy} \cdot \frac{2xy}{4x^2 + y^2}$$

$$= \frac{2(x^2 - y^2)}{4x^2 + y^2} \cdot \frac{xy}{xy}$$

$$= \frac{2(x^2 - y^2)}{4x^2 + y^2}$$

Problem 3. Find a reduced form of

$$\frac{1 + \frac{1}{a}}{1 - \frac{1}{a^2}}.$$

Solution. Because you can consider the fraction bar as a division sign, you could write this complex fraction as

$$\left(1 + \frac{1}{a}\right) \div \left(1 - \frac{1}{a^2}\right)$$

and proceed as in Problem 2. Another procedure is shown below. Be sure that you understand why a^2 was selected.

$$\frac{1 + \frac{1}{a}}{1 - \frac{1}{a^2}} = \frac{\left(1 + \frac{1}{a}\right)}{\left(1 - \frac{1}{a^2}\right)} \cdot \frac{a^2}{a^2}$$

$$= \frac{\left(1 + \frac{1}{a}\right)\left(\frac{a^2}{1}\right)}{\left(1 - \frac{1}{a^2}\right)\left(\frac{a^2}{1}\right)}$$

$$= \frac{a^2 + a}{a^2 - 1}$$

$$= \frac{a(a + 1)}{(a + 1)(a - 1)}, \quad \text{or} \quad \frac{a}{a - 1}$$

Exercises

Perform the indicated operations and give your answers in reduced form.

1. (a) $(\frac{3}{4}) \div (\frac{9}{16})$ (b) $(\frac{10}{11}) \div (\frac{100}{33})$

2. (a) $\left(\frac{x}{y}\right) \div \left(\frac{x^2}{y^2}\right)$ (b) $\left(\frac{a^2}{b^2}\right) \div \left(\frac{a}{b}\right)$

3. (a) $\dfrac{x+y}{\frac{a}{b}}$ (b) $\dfrac{a+b}{\frac{x}{y}}$

4. (a) $(\frac{3}{5} + \frac{1}{2}) \div (\frac{1}{4} - 3)$ (b) $(\frac{4}{7} + \frac{1}{4}) \div (\frac{1}{2} - 4)$

5. (a) $(x^2 - 9) \div \left(\frac{x+3}{x}\right)$ (b) $\left(\frac{x-4}{x}\right) \div (x^2 - 16)$

6. (a) $\left(x + \frac{5}{y}\right) \div \left(\frac{10}{xy}\right)$ (b) $\left(x - \frac{4}{y}\right) \div \left(\frac{8}{xy}\right)$

7. (a) $\left(a + \frac{1}{b}\right) \div \left(b + \frac{1}{a}\right)$ (b) $\left(\frac{1}{x} + \frac{1}{y}\right) \div \left(x + \frac{1}{y}\right)$

8. (a) $\left(\frac{a}{b} - 4\right) \div \left(\frac{a}{b} + 4\right)$ (b) $\left(\frac{x}{y} + 4\right) \div \left(\frac{x}{y} - 4\right)$

9. (a) $\left(\frac{x}{4} - \frac{4}{x}\right) \cdot \left(\frac{8x}{x-4}\right)$ (b) $\left(\frac{3}{x} + \frac{x}{3}\right) \cdot \left(\frac{6x}{6+x}\right)$

10. (a) $\left(9 - \frac{16}{x^2}\right) \div \left(3 - \frac{4}{x}\right)$ (b) $\left(25 - \frac{49}{y^2}\right) \div \left(5 - \frac{7}{y}\right)$

11. (a) $\left(\frac{2}{a} + \frac{3}{b}\right) \div \left(\frac{5}{a} - \frac{2}{b}\right)$ (b) $\left(\frac{3}{x} - \frac{2}{y}\right) \div \left(\frac{2}{x} - \frac{3}{y}\right)$

12. (a) $\left(\frac{1}{x^2} - \frac{1}{y^2}\right) \div \left(\frac{y}{x} - \frac{x}{y}\right)$ (b) $\left(\frac{4}{x^2} - \frac{1}{y^2}\right) \div \left(\frac{2}{x} + \frac{1}{y}\right)$

13. (a) $\left(\frac{x}{y} + 1\right) \div \left(\frac{x}{y} - 1\right)$ (b) $\left(4 - \frac{x}{y}\right) \div \left(4 + \frac{x}{y}\right)$

14. (a) $\left(2x + \frac{3}{y^2}\right) \div \left(3x + \frac{2}{y^3}\right)$ (b) $\left(3x + \frac{2}{y^2}\right) \div \left(2x + \frac{3}{y^3}\right)$

15. (a) $1 \div \left(x + 1 - \frac{1}{x}\right)$ (b) $4 \div \left(\frac{x}{4} - 4 + x\right)$

Find a reduced form for the expressions in Exercises 16–20.

16. (a) $\dfrac{\frac{1}{2} - \frac{1}{3}}{1 + (\frac{1}{2})(\frac{1}{3})}$ (b) $\dfrac{\frac{1}{4} - \frac{1}{3}}{1 + (\frac{1}{4})(\frac{1}{3})}$

17. (a) $\dfrac{3 - \frac{x}{y}}{9 - \frac{x^2}{y^2}}$ (b) $\dfrac{4 + \frac{x}{y}}{16 - \frac{x^2}{y^2}}$

18. (a) $\dfrac{2-\dfrac{a}{b}}{2+\dfrac{a}{b}}$ (b) $\dfrac{1+\dfrac{x}{y}}{1-\dfrac{x}{y}}$

19. (a) $\dfrac{x-\dfrac{4}{x}}{x-5+\dfrac{6}{x}}$ (b) $\dfrac{x+\dfrac{2}{x}}{x-2+\dfrac{4}{x}}$

20. (a) $\left(1-\dfrac{3}{x-2}\right)\div\left(1+\dfrac{3}{2-x}\right)$

(b) $\left(\dfrac{4}{3-x}-2\right)\div\left(-2-\dfrac{4}{x-3}\right)$

---

21. Do the indicated operations. Give your results in reduced form.

(a) $\left(2x-\dfrac{1}{2x}\right)\left(6x-\dfrac{x+3}{2x+1}\right)\left(\dfrac{8x^2+2}{16x^4-1}\right)$

(b) $\left[\left(\dfrac{y}{x}-\dfrac{4x}{y}\right)\div\left(\dfrac{y^2}{4x^2}-\dfrac{4x^2}{y^2}\right)\right]\cdot\left(\dfrac{y^2+4x^2}{4xy}\right)$

(c) $\left(1-\dfrac{6}{x}+\dfrac{9}{x^2}\right)\left(2-\dfrac{6}{x+3}\right)\div\dfrac{9-x^2}{x}$

(d) $\dfrac{\dfrac{x}{x-1}-\dfrac{2x}{x-2}}{\dfrac{2x}{x-2}-\dfrac{3x}{x-3}}$

(e) $\dfrac{\dfrac{1}{x}-\dfrac{3}{y}+\dfrac{1}{z}}{\dfrac{5}{xy}-\dfrac{2}{yz}+\dfrac{1}{xz}}$

Preparation for Section 10–7

Simplify each of the following expressions.

1. $x^{12}\cdot x^9$ **2.** $\dfrac{x^{12}}{x^9}$ **3.** $(xy)^{12}$

4. $(x^{12})^9$ **5.** $(x^4)^2\cdot x^3$ **6.** $(a^2b)^2\cdot(a^2)^2\cdot a$

7. $\dfrac{(a^2b^2)^3}{ab}$

10-7 THE FIFTH LAW OF EXPONENTS

The following laws of exponents are helpful when working with powers of polynomials.

$$x^m \cdot x^n = x^{m+n} \qquad \text{(LE-1)}$$

$$\frac{x^m}{x^n} = x^{m-n} \quad \text{if } m \geqq n \text{ and } x \neq 0 \qquad \text{(LE-2)}$$

$$(xy)^n = x^n y^n \qquad \text{(LE-3)}$$

$$(x^m)^n = x^{m \cdot n} \qquad \text{(LE-4)}$$

These equations are true for all numbers x and y and all non-negative integers m and n. Recall that, by definition,

$$x^0 = 1 \quad \text{for every } x \neq 0.$$

It is not possible to give a useful definition of zero to the zero power, so it is left undefined.

An additional law of exponents is needed to work with powers of rational algebraic expressions. The law is suggested by the equations below:

$$\left(\frac{x}{y}\right)^1 = \frac{x}{y},$$

$$\left(\frac{x}{y}\right)^2 = \frac{x}{y} \cdot \frac{x}{y}, \quad \text{or } \frac{x^2}{y^2},$$

$$\left(\frac{x}{y}\right)^3 = \frac{x}{y} \cdot \frac{x}{y} \cdot \frac{x}{y}, \quad \text{or } \frac{x^3}{y^3},$$

and so on.

FIFTH LAW OF EXPONENTS

The equation

$$\left(\frac{x}{y}\right)^n = \frac{x^n}{y^n}, \quad y \neq 0, \qquad \text{(LE-5)}$$

is true for all numbers x and y (y not equal to zero) and every non-negative integer n.

For example,

$$\left(\frac{2x}{3y^2}\right)^4 = \frac{(2x)^4}{(3y^2)^4} = \frac{2^4x^4}{3^4(y^2)^4}, \quad \text{or} \quad \frac{16x^4}{81y^8}.$$

Exercises

Simplify each of the following expressions.

1. (a) $\left(\frac{x}{4}\right)^2$ (b) $\left(\frac{3}{y}\right)^2$

2. (a) $\left(\frac{2x}{3}\right)^3$ (b) $\left(\frac{3x}{2}\right)^3$

3. (a) $\left(-\frac{2x}{5y}\right)^3$ (b) $\left(-\frac{x}{3y}\right)^3$

4. (a) $-\left(\frac{2x}{y}\right)^3$ (b) $-\left(\frac{3x}{2y}\right)^3$

5. (a) $\left(\frac{-2x}{y}\right)^5$ (b) $\left(\frac{-3x}{2y}\right)^5$

6. (a) $\left(\frac{c^2}{ad^3}\right)^5$ (b) $\left(\frac{b^4}{a^3c^2}\right)^4$

7. (a) $\left(\frac{b^2c}{a^5}\right)^3$ (b) $\left(\frac{m^4n}{p^3}\right)^2$

8. (a) $\left(\frac{2m^5}{3}\right)^4$ (b) $\left(\frac{m^8}{3n^3}\right)^4$

9. (a) $\left(\frac{3c^2}{-5}\right)^3$ (b) $\left(\frac{-5a^5}{b^4}\right)^3$

10. (a) $\left(\frac{2a^3}{b^4}\right)^2$ (b) $\left(\frac{3a^4}{b^3}\right)^2$

11. (a) $\left(\frac{a^5}{2b}\right)^3$ (b) $\left(\frac{a^7}{5b}\right)^2$

12. (a) $\left(\frac{-3n^3}{m^6}\right)^2$ (b) $\left(\frac{-3x^2}{y^3}\right)^4$

13. (a) $-\left(\frac{a^2b^5}{5c^3}\right)^2$ (b) $-\left(\frac{-4x^2}{3y}\right)^2$

14. (a) $\left(\frac{a^3b^2}{x^5y}\right)^6$ (b) $-\left(\frac{3a^2}{2b}\right)^6$

15. (a) $\left(\frac{m^6n^8}{-x^4y^3}\right)^2$ (b) $\left(\frac{-x^4y^{10}}{2ab^5}\right)^2$

16. (a) $-\left(\frac{-x^3y^5}{4z^2}\right)^3$ (b) $-\left(\frac{-5x^4}{2y^3}\right)^3$

17. (a) $\left(\frac{-2x^2}{a^3b^3}\right)^5$ (b) $\left(\frac{-3x^3}{m^4y^5}\right)^3$

18. (a) $\left(\frac{2n}{3a^2}\right)\left(\frac{ax}{ny}\right)^3$ (b) $-\left(\frac{3x}{2y}\right)^2\left(\frac{2x^4}{3y^2}\right)$

19. (a) $\left(\frac{8x^2y^3}{10xy^2w}\right)\left(\frac{6w^2}{3wy}\right)^2$ (b) $\left(\frac{15a^4}{14b^7}\right)\cdot\left(\frac{2b^4}{3a^2}\right)^4$

20. (a) $\left(\frac{12a^3}{8b^2}\right)^3\left(\frac{10a^2b^4}{27a^2b^2}\right)$ (b) $\left(\frac{5a^3}{8b^{10}}\right)\cdot\left(\frac{12b^4}{10a^2}\right)^3$

21. (a) $\left(\frac{a^2}{ab}\right)^3 \div \left(\frac{c}{d}\right)^4$ (b) $\left(\frac{a}{x}\right)^3 \div \left(\frac{5}{3}\right)^2$

22. (a) $\frac{2x^2}{7y^2} \div \left(\frac{4xy}{21}\right)^2$ (b) $\left(\frac{2b}{a}\right)^3 \div \frac{6ab^2}{18a^2b}$

23. (a) $\left(-\frac{7^2x^4}{5^3y}\right)^5 \div \left[-\left(\frac{7^9x^2}{5^{12}y}\right)\right]$ (b) $\left(-\frac{6a^7}{7b^4}\right)^3 \div \left[-\left(\frac{6a^9}{7b^3}\right)^4\right]$

24. (a) $\left(-\frac{4x^5}{3y^6}\right)^3\left(\frac{3y^2}{2x^2}\right)^6$ (b) $\left(-\frac{5x^5}{2y^7}\right)^5\left(\frac{2y^3}{10x^4}\right)^4$

10–8 NEGATIVE EXPONENTS

By the second law of exponents, (LE-2),

$$\frac{x^5}{x^3} = x^{5-3}, \quad \text{or } x^2.$$

What is $x^3 \div x^5$? If you use (LE-2) and ignore the condition $m \geqq n$, you get

$$\frac{x^3}{x^5} = x^{3-5}, \quad \text{or } x^{-2}.$$

Of course, $x^3 \div x^5$ can be found as follows:

$$\frac{x^3}{x^5} = \frac{x^3}{x^3}\cdot\frac{1}{x^2}, \quad \text{or } \frac{1}{x^2}.$$

This suggests that x^{-2} be defined as

$$x^{-2} = \frac{1}{x^2}.$$

As another example,

$$\frac{x^7}{x^8} = \frac{x^7}{x^7}\cdot\frac{1}{x}, \quad \text{or } \frac{1}{x},$$

but if you use (LE-2), you obtain

$$\frac{x^7}{x^8} = x^{7-8}, \quad \text{or } x^{-1}.$$

Again, this suggests that you *define* x^{-1} as

$$x^{-1} = \frac{1}{x}.$$

With these examples in mind, the following definition is made.

Definition of a negative exponent

The equation

$$x^{-n} = \frac{1}{x^n} \qquad \textit{(Def-Neg. Exp.)}$$

is true for every positive integer n and every nonzero number x.

Now x^n has been defined for every integer n, positive, negative, or zero.

Problem 1. Find reduced forms of the following.

(a) $x^6 \cdot x^{-4}$

(b) $x^5 \cdot x^{-5}$

(c) $(y^2)^{-3}$

(d) $(y^{-4})^{-2}$

Solution.

(a) $x^6 \cdot x^{-4} = x^6 \cdot \dfrac{1}{x^4}$

$= x^2 \cdot x^4 \cdot \dfrac{1}{x^4}$

$= x^2 \cdot 1, \quad \text{or } x^2$

If you assume that (LE-1) is also true for negative exponents, then

$$x^6 \cdot x^{-4} = x^{6-4}, \quad \text{or } x^2.$$

(b) $x^5 \cdot x^{-5} = x^5 \cdot \dfrac{1}{x^5}, \quad \text{or } 1$

If (LE-1) is true for negative as well as positive exponents, then

$$x^5 \cdot x^{-5} = x^{5-5} = x^0, \quad \text{or } 1.$$

(c) $(y^2)^{-3} = \dfrac{1}{(y^2)^3} = \dfrac{1}{y^6}, \quad \text{or } y^{-6}$

If (LE-4) is true for positive and negative exponents, then

$$(y^2)^{-3} = y^{2(-3)}, \quad \text{or } y^{-6}.$$

(d) $(y^{-4})^{-2} = \dfrac{1}{(y^{-4})^2} = \dfrac{1}{\left(\dfrac{1}{y^4}\right)^2} = \dfrac{1}{\dfrac{1}{y^8}} = 1 \cdot \dfrac{y^8}{1}, \quad \text{or } y^8$

If (LE-4) is true for both positive and negative exponents, then

$$(y^{-4})^{-2} = y^{(-4)(-2)}, \quad \text{or } y^8.$$

In each example above you saw that the laws of exponents give the correct result for a negative, zero, or positive exponent. These examples suggest the following true statement.

The five laws of exponents are valid for any integral exponents—positive, negative, or zero.

Problem 2. Find the reduced form of $(\frac{3}{4})^{-3}$.

Solution. By definition,

$$\left(\frac{3}{4}\right)^{-3} = \frac{1}{(\frac{3}{4})^3} = \frac{1}{\frac{27}{64}}, \quad \text{or } \frac{64}{27}.$$

Thus, $(\frac{3}{4})^{-3} = (\frac{4}{3})^3$. This leads to the conclusion that the equation

$$\left(\frac{x}{y}\right)^{-n} = \left(\frac{y}{x}\right)^n$$

is true for all nonzero numbers x and y and every integer n.

Exercises

For each of the following, find an equivalent expression which has only positive exponents.

1. (a) x^{-7} (b) x^{-4}
2. (a) $\dfrac{1}{y^{-2}}$ (b) $\dfrac{1}{x^{-3}}$

3. (a) 3^{-2} (b) 4^{-3}
4. (a) $(\frac{4}{5})^{-1}$ (b) $(\frac{3}{4})^{-1}$
5. (a) $(x+y)^{-3}$ (b) $(x-1)^{-2}$
6. (a) $(x^{-1}y^2)^0$ (b) $(x^{-3}y^{-4})^0$
7. (a) $[(\frac{5}{6})^{-1}]^{-1}$ (b) $[(\frac{3}{4})^{-1}]^{-1}$
8. (a) $(4x^{-5})^{-2}$ (b) $(5x^{-4})^{-2}$
9. (a) $(3x^2y)^{-2}$ (b) $(2x^{-3}y)^{-2}$
10. (a) $(x^{-2})(x^0)$ (b) $(x^{-3})(x^0)$

Perform each of the indicated operations and give your answers with positive exponents.

11. (a) $x^{12} \cdot x^{-4}$ (b) $x^9 \cdot x^{-5}$
12. (a) $(x^{-4})^3$ (b) $(x^{-3})^4$
13. (a) $(x^2)^{-5}$ (b) $(x^3)^{-5}$
14. (a) $(x^{-3})^{-2}$ (b) $(x^{-4})^{-5}$
15. (a) $5^{-7} \cdot 5^8$ (b) $3^{12} \cdot 3^{-9}$
16. (a) $2^{-3} \cdot 2^3$ (b) $12^{-7} \cdot 12^7$
17. (a) $3^{-9} \cdot 3^8$ (b) $10^{-11} \cdot 10^{10}$
18. (a) $(7^{-3})^{-2}$ (b) $(7^{-4})^{-2}$
19. (a) $(\frac{2}{3})^{-2}$ (b) $(\frac{3}{5})^{-3}$
20. (a) $(\frac{3}{4})^{-4}(\frac{3}{4})^6$ (b) $(\frac{7}{8})^{-10}(\frac{7}{8})^{13}$
21. (a) $2xy^{-3} \cdot 3x^{-2}y$ (b) $a^{-6}b^2 \cdot ab^{-1}$
22. (a) $(2x^{-3})^2$ (b) $(3x^{-2})^3$
23. (a) $\dfrac{x^8}{x^{-3}}$ (b) $\dfrac{y^{10}}{3y^{-4}}$
24. (a) $\dfrac{3x^{-4}}{6y^{-2}}$ (b) $\dfrac{5x^{-7}}{10y^{-2}}$
25. (a) $\dfrac{x^0}{x^{-3}}$ (b) $\dfrac{x^{-4}}{x^0}$

Perform each of the indicated operations and give your answer with positive exponents.

26. $(2xy^{-2})^3$
27. $(3ab^2)^{-2}$
28. $(2x^{-1}y^{-2})^{-3}$
29. $3a^{-2}b$
30. $2x^3y^{-4}$
31. $3a^{-2}b^{-2}$
32. $(x+y)^{-1}$
33. $(x+y)^{-2}$

34. $(a^3 \cdot a^{-4} \cdot a^{-5})^{-1}$

35. $\dfrac{xy^6}{x^{-2}y^{-3}}$

36. $\dfrac{2a^{-3}b^{-4}}{a^{-4}b^{-5}}$

37. $(x^3y^{-2})^4$

38. $(4a^3b^{-4})^3$

39. $(x - 2y)^{-2}$

40. $(\frac{21}{121})^{-5}(\frac{36}{121})^4$

41. $\left(\dfrac{x^{-2}}{y^{-3}}\right)\left(\dfrac{y^{-3}}{x^{-2}}\right)$

Preparation for Section 10–9

1. Express each of the following in scientific notation.

(a) 186,000

(b) 33

(c) 193,000,000

2. Complete each of the following statements.

(a) $.0456 = \dfrac{4.56}{100} = 4.56 \times \dfrac{1}{10^?} = 4.56 \times 10^?$

(b) $.389 = \dfrac{3.89}{10} = 3.89 \times \dfrac{1}{10} = 3.89 \times 10^?$

10–9 SCIENTIFIC NOTATION

In Chapter 1, you used scientific notation to express a very large number as a product of a number between 1 and 10 and a positive integral power of 10. For instance, 330,000,000,000 can be expressed in scientific notation as

$$3.3 \times 10^{11}.$$

It is also possible to express very small numbers in scientific notation if you use negative powers of 10. By definition,

$$10^{-1} = \tfrac{1}{10}, \quad \text{or } .1;$$

$$10^{-2} = \frac{1}{10^2}, \quad \text{or } .01;$$

$$10^{-3} = \frac{1}{10^3}, \quad \text{or } .001;$$

and so on. If you wish to express

$$.0000238$$

in scientific notation, the number between one and ten that you must use is 2.38. Since

$$.0000238 = \frac{2.38}{100{,}000} = \frac{2.38}{10^5},$$

you can write .0000238 in scientific notation as follows:

$$.0000238 = 2.38 \times 10^{-5}.$$

A physicist uses scientific notation in describing the small quantities encountered in atomic physics. For example, he uses

$$9.108 \times 10^{-31} \text{ kilograms}$$

to express the mass of an electron. Try writing this number without using scientific notation!

Exercises

Represent each of the following numbers in scientific notation by writing it as a number between 1 and 10 multiplied by a power of 10.

1. (a) $\frac{65}{100{,}000{,}000}$ (b) $\frac{9}{10{,}000}$

2. (a) .00089 (b) .000987

3. (a) .00000678 (b) .0000000123

4. (a) .000000706 (b) .00002003

5. (a) .0642 (b) .00745

6. (a) .000456 (b) .0000273

7. (a) .4981 (b) .36

8. (a) .000000894 (b) .000000000275

Express each of the following in scientific notation.

9. (a) The wave length of the longest x-rays: .0000001 cm
(b) The wave length of the γ-rays: .0000000002 cm

10. (a) The photon energy of radar waves:

.00000000000000000002 erg

(One erg is the amount of work required to raise $\frac{1}{981}$ of a gram vertically through one centimeter.)

(b) The electronic charge: .00000000000000000016 coulomb

11. (a) The diameter of the sun: 130,000,000,000 cm
(b) The mass of the sun: 2,000,000,000,000,000,000,000,000,000 tons

Complete the computation and express each answer in scientific notation.

12. (a) $(65 \times 10^{-7})(80 \times 10^{-6})$ (b) $(7 \times 10^{-6})(8 \times 10^{-9})$

13. (a) $(.0009)^2$ (b) $(.003)^4$

14. (a) $(5 \times 10^{-4})^3$ (b) $(6 \times 10^{-5})^4$

15. (a) $\dfrac{18 \times 109 \times 10^{-3}}{10^{20} \times 10^{30}}$ (b) $\dfrac{7 \times 16 \times 10^{-10}}{10^{12} \times 10^{22}}$

16. (a) $\dfrac{(52 \times 10^2)(21 \times 10^5)}{(350 \times 10^4)(65 \times 10^8)}$ (b) $\dfrac{(49 \times 10^4)(63 \times 10^8)}{(54 \times 10^9)(35 \times 10^5)}$

Preparation for Section 10–10

Complete each of the following solutions of the equation

$$\frac{y-6}{y+1} = \frac{21}{(y+1)^2} + 1.$$

Give reasons for the key steps and check the solution.

1. First solution:

$$\frac{y-6}{y+1} = \frac{21}{(y+1)^2} + 1,$$

$$\frac{(y-6)(y+1)}{(y+1)^2} = \frac{21 + (y+1)^2}{(y+1)^2},$$

$$(y-6)(y+1) = 21 + (y+1)^2. \qquad \underline{\quad ? \quad}$$

The rest of the problem is left for you to complete.

2. Second solution:

$$\frac{y-6}{y+1} = \frac{21}{(y+1)^2} + 1,$$

$$\frac{y-6}{y+1} - \frac{21}{(y+1)^2} - 1 = 0,$$

$$\frac{(y-6)(y+1)}{(y+1)^2} - \frac{21}{(y+1)^2} - \frac{(y+1)^2}{(y+1)^2} = 0,$$

$$\frac{(y-6)(y+1) - 21 - (y+1)^2}{(y+1)^2} = 0,$$

$$(y-6)(y+1) - 21 - (y+1)^2 = 0. \qquad \underline{\quad ? \quad}$$

The rest of the problem is left for you to complete.

10–10 EQUATIONS

The properties of the real number system that were given in Chapter 2 may be used to solve equations involving rational algebraic expressions. The following is a special property which will be used extensively.

QUOTIENT EQUALS ZERO

$$\frac{a}{b} = 0 \quad \textit{if, and only if,} \quad a = 0 \qquad (\textit{assuming } b \neq 0) \qquad (Q)$$

In other words, a quotient of two numbers is zero if, and only if, the numerator is zero and the denominator is nonzero.

Problem 1. Solve the equation

$$\frac{2x + 1}{3x - 5} = 0.$$

Solution. Proceed as follows:

$$\frac{2x + 1}{3x - 5} = 0,$$

$$2x + 1 = 0, \qquad \text{(Q)}$$

$$x = -\tfrac{1}{2}.$$

Thus, $\{-\frac{1}{2}\}$ is the solution set provided $3x - 5$ does not have the value 0 when $x = -\frac{1}{2}$. The following check shows that $3x - 5 \neq 0$ when $x = -\frac{1}{2}$.

Check.

$$3x - 5 \stackrel{?}{=} 0$$

$$3(-\tfrac{1}{2}) - 5 \stackrel{?}{=} 0$$

$$-\tfrac{3}{2} - 5 \neq 0$$

Problem 2. Solve the equation

$$\frac{3}{x} + \frac{4}{9} = \frac{5}{12}.$$

Solution. Proceed as follows:

$$\frac{3}{x} + \frac{4}{9} = \frac{5}{12},$$

$$\frac{3}{x} + \frac{4}{9} - \frac{5}{12} = \frac{5}{12} - \frac{5}{12},$$

$$\frac{108 + 16x - 15x}{36x} = 0,$$

$$108 + x = 0, \qquad \text{(Q)}$$

$$x = -108.$$

Check.

$$\frac{3}{-108} + \frac{4}{9} \stackrel{?}{=} \frac{5}{12}$$

$$\frac{1}{-36} + \frac{16}{36} \stackrel{?}{=} \frac{5}{12}$$

$$\frac{15}{36} \stackrel{\checkmark}{=} \frac{5}{12}$$

Thus, $\{-108\}$ is the solution set of the given equation.

If

$$\frac{a}{7} = \frac{c}{7},$$

then

$$a \cdot \frac{1}{7} = c \cdot \frac{1}{7} \quad \text{and} \quad a = c. \qquad \text{(Can-M)}$$

Conversely, if $a = c$ then

$$a \cdot \frac{1}{7} = c \cdot \frac{1}{7} \quad \text{and} \quad \frac{a}{7} = \frac{c}{7}. \qquad \text{(Multi-P)}$$

This illustrates the following property.

UNIQUENESS OF QUOTIENTS

$$\frac{a}{b} = \frac{c}{b} \quad \textit{if, and only if,} \quad a = c \qquad (\textit{assuming } b \neq 0) \qquad \text{(U-Q)}$$

Problem 3. Solve the equation

$$\frac{3}{x-3} - \frac{2}{3+x} = \frac{5}{9-x^2}.$$

Solution.

$$\frac{3}{x-3} - \frac{2}{3+x} = \frac{5}{9-x^2}$$

$$\frac{-3}{3-x} - \frac{2}{3+x} = \frac{5}{9-x^2}$$

$$\frac{-3(3+x)}{9-x^2} - \frac{2(3-x)}{9-x^2} = \frac{5}{9-x^2}$$

$$\frac{-3(3+x) - 2(3-x)}{9-x^2} = \frac{5}{9-x^2}$$

By uniqueness of quotients, you may proceed as follows:

$$-3(x+3) - 2(3-x) = 5,$$
$$-3x - 9 - 6 + 2x = 5,$$
$$-x - 15 = 5,$$
$$-x = 20,$$
$$x = -20.$$

Check.

$$\frac{3}{-20-3} - \frac{2}{3-20} \stackrel{?}{=} \frac{5}{9-(-20)^2}$$

$$\frac{3}{-23} - \frac{2}{-17} \stackrel{?}{=} \frac{5}{9-400}$$

$$\frac{3(-17) - 2(-23)}{(-23)\cdot(-17)} \stackrel{?}{=} \frac{5}{-391}$$

$$\frac{-51+46}{391} \stackrel{\checkmark}{=} \frac{5}{-391}$$

Thus, $\{-20\}$ is the solution set of the given equation.

Exercises

Solve each of the following equations and check your solutions.

1. (a) $\frac{9x}{11} + \frac{3}{11} = 0$ (b) $\frac{4x}{7} - \frac{2}{7} = 0$

2. (a) $\frac{5x-3}{7x+1} = 0$ (b) $\frac{6x+12}{3x-4} = 0$

3. (a) $\frac{2x+5}{3x}=0$ (b) $\frac{3x}{2x+5}=0$

4. (a) $\frac{a-2}{a}=\frac{5}{a}$ (b) $\frac{x+6}{x}=\frac{4}{x}$

5. (a) $\frac{x}{6}+\frac{3x-1}{6}=\frac{5}{6}$ (b) $\frac{2a}{3}+\frac{a-2}{3}=\frac{2}{3}$

6. (a) $\frac{(3x+8)-(x+3)}{2x+5}=\frac{20(2x+5)}{2x+5}$

(b) $\frac{(6x+4)-(3x+5)}{3x-1}=\frac{6(3x-1)}{3x-1}$

7. (a) $\frac{2}{x}+3=0$ (b) $\frac{6}{x}+2=0$

8. (a) $\frac{3}{x}-\frac{2}{5}=\frac{3}{10}$ (b) $\frac{4}{x}-\frac{3}{4}=\frac{1}{8}$

9. (a) $\frac{5}{y}+\frac{3}{8}=\frac{7}{16}$ (b) $\frac{4}{7}+\frac{2}{y}=\frac{11}{14}$

10. (a) $\frac{2}{y}+\frac{5}{y}=\frac{14}{9}$ (b) $\frac{3}{2x}+\frac{5}{2x}=\frac{5}{6}$

11. (a) $\frac{3}{4a}-\frac{3}{5a}+\frac{1}{10}=0$ (b) $\frac{1}{4y}-\frac{2}{3y}+\frac{1}{12}=0$

12. (a) $\frac{2}{y}-\frac{1}{y}+6=0$ (b) $\frac{1}{2x}+\frac{3}{8}=\frac{2}{x}$

13. (a) $\frac{1}{2x}-\frac{2}{3x}=7-\frac{5}{x}$ (b) $\frac{3}{5x}-\frac{1}{2x}=10-\frac{5}{x}$

14. (a) $\frac{y-7}{y}=\frac{21}{y^2}+1$ (b) $\frac{6}{5-x}-\frac{7}{5+x}=\frac{21}{25-x^2}$

15. (a) $\frac{8-x}{15x}+\frac{1}{5}=\frac{8x+10}{5x}$ (b) $\frac{3x-4}{2x}-\frac{x+1}{3x}+\frac{x+2}{5x}=\frac{2}{5}$

Solve each of the following equations and check your solutions.

16. $\frac{28}{x^2-9}+\frac{x}{3-x}+\frac{7}{3+x}=1$

17. $\frac{2}{x+2}+\frac{x}{x-2}=\frac{x^2+4}{x^2-4}$

18. $\frac{x}{x-1}-\frac{2}{1-x^2}=\frac{8}{x+1}$

19. $\frac{2}{y+1}+\frac{1-y}{y}=\frac{1}{y^2+y}$

20. $\frac{y+2}{2y-6}+\frac{3}{3-y}=\frac{y}{2}$

21. $\frac{12}{y}=\frac{12}{y+1}+1$

22. $\frac{x}{x+4}+\frac{4}{4-x}=\frac{x^2+16}{x^2-16}$

10–11 WORD PROBLEMS

If a person working steadily completes a job in 2 hours, then he or she could complete $\frac{1}{2}$ of the job in 1 hour. If the job takes 3 hours of steady work, then $\frac{1}{3}$ of the job can be done in 1 hour. More generally, if a job requires n hours of steady work, then $1/n$ of the job can be done in 1 hour. The following problems illustrate the use of this principle.

Problem 1. Rosa and Kim are painting rooms in an office building. Rosa can paint a large room in 4 hours and Kim can paint a large room in 6 hours. How long would it take them to paint a large room together?

Solution: Let x denote the number of hours that Rosa and Kim, working together, need to paint the room. Then

$$\frac{1}{4} = \text{part that Rosa paints in 1 hour.}$$

$$\frac{1}{6} = \text{part that Kim paints in 1 hour.}$$

$$\frac{1}{x} = \text{part that Rosa and Kim paint in 1 hour.}$$

However, the part they paint together in 1 hour is the sum of the parts they paint individually in 1 hour:

$$\frac{1}{x} = \frac{1}{4} + \frac{1}{6}.$$

Since

$$\tfrac{1}{4} + \tfrac{1}{6} = \tfrac{3}{12} + \tfrac{2}{12}, \text{ or } \tfrac{5}{12},$$

you have

$$\frac{1}{x} = \frac{5}{12},$$

$$x = \tfrac{12}{5}, \text{ or } 2\tfrac{2}{5}.$$

Thus, Rosa and Kim should be able to paint a large room together in $2\frac{2}{5}$ hours, or 2 hours and 24 minutes.

Problem 2. It takes two computers working together 10 hours to solve problems about a wing design for an airplane. One computer working alone can do the job in 16 hours. How long does it take the other computer to do the job alone?

Solution.

x = number of hours it takes the other computer to complete the job

$\frac{1}{x}$ = part of job other computer does in 1 hour

$\frac{1}{16}$ = part of job first computer does in 1 hour

$\frac{1}{10}$ = part of job they do in 1 hour working together

Hence,

$$\frac{1}{x} + \frac{1}{16} = \frac{1}{10},$$

$$\frac{1}{x} = \frac{1}{10} - \frac{1}{16},$$

$$\frac{1}{x} = \frac{3}{80},$$

$$x = \frac{80}{3}, \quad \text{or} \quad 26\tfrac{2}{3} \text{ hours.}$$

Exercises

In each of the following exercises, choose a variable, state what it denotes, and write an equation using the information given. Solve the equation and check the correctness by using the answer in the original problem.

1. (a) John can shovel the snow from the sidewalk in 12 minutes, and James can do it in 18 minutes. How long will it take them to shovel the snow from the sidewalk if they work together?

(b) Amy can build a table in 6 hours. It takes Nan 12 hours to build the same table. How long will it take Amy and Nan working together to build the table?

2. (a) Jill can paper a room in 8 hours. She and Diane working together can paper the room in 6 hours. How long would it take Diane working alone?

(b) A bank teller counts a certain amount of money in 20 minutes, but if she is helped by another teller, they count the money in 15 minutes. How long would it take the second teller to count the money alone?

3. (a) One pipe can fill a tank in 5 hours. A second can fill it in 4 hours. How long will it take both pipes together to fill the tank?

(b) The hot water faucet fills a tub in 10 minutes, and the cold water faucet in 7 minutes. How long will it take to fill the tub if both faucets are turned on?

4. (a) The denominator of a certain fraction is 14 more than its numerator. If numerator and denominator are both increased by 5, the new fraction is equal to $\frac{11}{18}$. Find the original fraction.
   (b) The numerator of a certain fraction is 8 less than the denominator. If both are increased by 9, the resulting fraction is equal to $\frac{2}{3}$. Find the original fraction.
5. (a) Find a number which when added to both the numerator and denominator of the fraction $\frac{11}{13}$ produces a fraction equal to $\frac{18}{19}$.
   (b) Find a number which when subtracted from both the numerator and denominator of the fraction $\frac{33}{43}$ produces a fraction equal to $\frac{5}{7}$.
6. (a) A dump truck can haul enough gravel to fill a pit in 8 days. A larger dump truck can fill the same pit in 6 days. How long will it take two of the smaller trucks and one of the larger trucks together to fill the pit?
   (b) One bulldozer clears a certain area in 5 days. Another takes 3 days to clear the same area. How long will it take two of the first and three of the second kind to clear the area?

---

7. (a) It takes one man 8 hours to do a certain job. After the first man has been at work for 3 hours, another man is put to work. The two men then complete the job in 2 more hours. How long would it take the second man to do the job working by himself?
   (b) It takes Mr. Smith 10 hours to paint a room. If he works for 1 hour and then asks Mrs. Smith to help him, they finish in 2 more hours. In what time can Mrs. Smith do the whole job?
8. (a) One pump fills a tank twice as fast as another. If together they fill the tank in 20 minutes, how long does the larger pump take?
   (b) One painter works twice as fast as his partner. If together they paint a house in 24 hours, how long does the second painter take to do the job working by himself?
9. (a) One pipe can fill a tank in 3 hours; a second also can fill it in 3 hours, but a third needs 5 hours. How long will it take to fill the tank if all three pipes are open?
   (b) Three computers can prepare a payroll in 2 hours. The two faster computers can do the job by themselves in 3 hours. In what time can the slowest computer do the job?

---

10. (a) A swimming pool has two inlet pipes. One fills the pool in 4 hours, the other in 6 hours. The outlet pipe empties the pool in 5 hours. Once the outlet pipe was left open when the pool was being filled. In how many hours was the pool full?

(b) The hot water faucet fills a tub in 30 minutes, and the cold water faucet, in 20 minutes. The tub can be drained in 15 minutes. If both faucets are open while the drain is open, how soon will the tub be full?

11. Two bus drivers traveled the same route for a distance of 315 miles. The average speed of one driver was 10 miles per hour faster than that of the other, enabling him to make the trip in 2 hours less time. Find the average rate at which each man drove.

12. An outlet pipe takes 1 hour longer to empty a tank than an inlet pipe takes to fill it. With both pipes open, $\frac{1}{10}$ of the tank is full at the end of 2 hours. If the outlet pipe is now turned off, how long will it take the inlet pipe alone to finish the tank?

13. On a fishing trip Mr. Smith goes 15 miles upstream and returns; he spends eight hours in all. If the average rate of the current is $2\frac{1}{2}$ miles per hour, what is the average speed of Mr. Smith's boat in still water?

14. If 12 skilled men and 15 unskilled men work together, they can complete a construction job in 10 days. The same job can be done in 8 days by 25 skilled men. How long should it take one unskilled man alone to complete the job?

KEY IDEAS AND KEY WORDS

A **rational algebraic expression** can be expressed as the quotient of two polynomials. If the two polynomials have integral coefficients and have no common factors other than 1 and -1, the rational algebraic expression is said to be in **reduced form.**

The least common multiple of the denominators of two rational algebraic expressions is called the **least common denominator,** (l.c.d.), of these expressions. For example, the expressions

$$\frac{7x}{y(2y + x)} \quad \text{and} \quad \frac{4x - 3y}{y(2y - x)}$$

have as their least common denominator

$$y(2y + x)(2y - x), \quad \text{or } y(4y^2 - x^2).$$

Positive exponents were previously defined. **Negative exponents** are defined as follows:

$$x^{-n} = \frac{1}{x^n}$$

for every positive integer n and every nonzero number x.

The five **laws of exponents** are given below.

$$x^m \cdot x^n = x^{m+n} \qquad \text{(LE-1)}$$

$$\frac{x^m}{x^n} = x^{m-n}, \quad x \neq 0 \qquad \text{(LE-2)}$$

$$(xy)^n = x^n y^n \qquad \text{(LE-3)}$$

$$(x^m)^n = x^{m \cdot n} \qquad \text{(LE-4)}$$

$$\left(\frac{x}{y}\right)^n = \frac{x^n}{y^n}, \quad y \neq 0 \qquad \text{(LE-5)}$$

Each is true for all real values of x and y (except 0 for y in LE-5) and for all integral values of n.

Every number can be expressed in **scientific notation** as a product of a number between 1 and 10 and a power of 10.

The properties of **quotient equals zero,** (Q), and **uniqueness of quotients,** (U-Q), are used in solving equations involving rational algebraic expressions.

$$\frac{a}{b} = 0 \text{ if, and only if, } a = 0 \text{ (assuming } b \neq 0) \qquad \text{(Q)}$$

$$\frac{a}{b} = \frac{c}{b} \text{ if, and only if, } a = c \text{ (assuming } b \neq 0) \qquad \text{(U-Q)}$$

An algebraic expression which is made up of sums, differences, products, and quotients of rational algebraic expressions is often called a **complex fraction.**

CHAPTER REVIEW

Find the reduced form of each of the following rational algebraic expressions.

1. $\dfrac{21x^3y^4}{70x^5y}$

2. $\dfrac{11x^2y}{132x^2y^3}$

3. $\dfrac{18x - 18}{15 - 15x}$

4. $\dfrac{x - 2}{x^2 - 3x + 2}$

5. $\dfrac{x^2 - 25y^2}{25y^2 - x^2}$

6. $\dfrac{35 + 2x - x^2}{x^2 - 10x + 21}$

7. Give the following products and quotients in reduced form.

(a) $\dfrac{9x^2}{10y} \cdot \dfrac{55x}{24y^3}$

(b) $\dfrac{9x^3}{13y^3} \div \dfrac{3x}{26y^2}$

(c) $\dfrac{9x^2 - y^2}{5x} \cdot \dfrac{10x^2}{x^2 + 6xy + 9y^2}$ (d) $\dfrac{8x + 20}{6x^2 - 5x - 50} \div \dfrac{36}{30 - 9x}$

Do the indicated operations and give your answers in reduced form.

8. $\dfrac{9}{ab} - \dfrac{7b}{a^2}$

9. $\dfrac{4}{b} - \dfrac{5}{b - 2}$

10. $\dfrac{3x + 2}{15} + \dfrac{x - 3}{10}$

11. $\dfrac{4a}{a + 1} - \dfrac{1}{2a - 5}$

12. $\dfrac{3a - 1}{4a} - \dfrac{5a - 2}{6a}$

13. $\dfrac{3}{x - y} - 1$

14. $\dfrac{3x - 2}{2x} - \dfrac{7x - 6}{11x^2}$

15. $\dfrac{x}{x^2 - y^2} + \dfrac{y}{y^2 - x^2}$

16. $\dfrac{9}{y^2 - 3y + 2} + \dfrac{3}{y^2 - 5y + 6} + \dfrac{2}{y^2 - 4}$

17. $\dfrac{\dfrac{x}{12} + 4}{\dfrac{x}{12} - \dfrac{1}{3}}$

18. $\dfrac{\dfrac{1}{14y} - \dfrac{1}{2y^2}}{\dfrac{1}{7} - \dfrac{6}{7y} - \dfrac{1}{y^2}}$

19. $\left(25 - \dfrac{9}{x^2}\right) \div \left(5 + \dfrac{3}{x}\right)$

20. $\dfrac{\frac{2}{7} - \frac{3}{5}}{(1 + \frac{2}{7}) \cdot \frac{3}{5}}$

21. $\left(\dfrac{2y}{3} - \dfrac{3}{2y}\right) \cdot \left(\dfrac{12y}{3 - 2y}\right)$

22. $\left(2 - \dfrac{4}{x - 5}\right) \div \left(1 + \dfrac{x}{5 - x}\right)$

23. $-\left(\dfrac{-6x^3}{9y^2}\right)^4$

24. $\left(\dfrac{-2x^7}{5y^3}\right)^3 \div \left(\dfrac{-4x^3}{15y^4}\right)^2$

25. $2x^{-5} \cdot 6x^{-3}$

26. $7x^{-8} \div 14x^{-4}$

27. $(-\frac{4}{7})^{-3}$

28. $8^{-10} \div 8^{-15}$

29. $(x^{-5})^2$

30. $5x^{-4}y \cdot (10xy^{-2})^2$

31. $(5x^6y^{-3})^{-2}$

32. $\dfrac{17x^{-5}y^{-6}}{51x^{-4}y^{-7}}$

33. Express in scientific notation.

(a) $\dfrac{267}{1{,}000{,}000}$ (b) .00173

(c) .0000903 (d) $\dfrac{(21 \times 10^{-4})(13 \times 10^5)}{(39 \times 10^8)(35 \times 10)}$

(e) $(13 \times 10^{-5})^3$ (f) $(.0000015)^2$

34. Write each rational algebraic expression as the sum of a polynomial and a rational algebraic expression with the degree of the numerator smaller than that of the denominator.

(a) $\dfrac{a^3 - 16}{a - 2}$ (b) $\dfrac{5x^5 - 3x^3 - x + 1}{x^3 - 2x^2 + 1}$

In Exercises 35–39, find the solution of each equation. Check your answers.

35. $\frac{4}{x} + \frac{3}{2x} = \frac{11}{6}$

36. $1 + \frac{x}{2x - 1} = \frac{2x^2}{2x - 1} - 3x$

37. $\frac{2}{x + 4} - \frac{5}{3 - x} = \frac{6}{x - 3}$

38. $4 - \frac{3x - 5}{x} = \frac{5}{x}$

39. $\frac{x + 1}{x - 1} - \frac{3x + 9}{6 - x} = \frac{2x^2}{x^2 - 7x + 6}$

40. An airplane travels 3500 miles in the same time that a passenger train travels 600 miles. If the plane goes 50 miles an hour more than 5 times the speed of the train, what is the rate of each?

41. Working alone, Fred can paint a certain house in 60 hours. If he and Tom work together, the same house can be painted in 40 hours. How long would it take Tom to do the painting by himself?

CHAPTER TEST

Find the reduced form for each of the following expressions.

1. $\frac{3 - x}{2x - 6}$

2. $\frac{11x^4}{12y^3} \div \frac{3x^7}{44y}$

3. $\frac{8y - xy}{y^4} \cdot \frac{4y}{x^2 - x - 56}$

4. $\frac{3x + 9}{x^2 - 2x - 8} \div \frac{6x + 18}{3x - 12}$

Do the indicated arithmetic and give your answers in reduced form.

5. $\frac{5x - 3}{12} - \frac{2x - 5}{18}$

6. $\frac{8}{xy} - \frac{3x}{y^2}$

7. $(y - 5) - \frac{7}{y + 5}$

8. $\frac{a + 1}{1 - a} + \frac{a - 1}{1 + a}$

9. $\frac{2}{x^2 + 7x + 10} - \frac{1}{x^2 + 3x - 10}$

Simplify each of the expressions in Exercises 10–16.

10. $\dfrac{\frac{y}{9} - 1}{\frac{y}{2} - \frac{1}{3}}$

11. $\left(49 - \frac{4}{x^2}\right) \div \left(7 + \frac{2}{x}\right)$

12. $-\left(\dfrac{-15x^2}{35y^4}\right)^3$

13. $(\frac{3}{7})^{-2}$

14. $9^{-7} \div 9^{-5}$

15. $(2x^{-3}y)^{-2} \div (17x^{-2}y^{-3})^0$

16. $\dfrac{16y^4 - 1}{4y^2 + 1}$

17. Find the solution set and check your answer.

$$\frac{18}{x-1} - 8 = x + \frac{x+6}{x-1}$$

18. The head of the billing department can process a batch of bills in 16 hours. Her assistant requires 24 hours to do the same job. If the two of them work together, how long will it take to complete the job?

CUMULATIVE REVIEW III

Do the indicated operations.

1. $(x^2 - 2x - 35) \div (x + 5)$
2. $(x^4 - x^3 + x^2 - x) \div (x^2 + 1)$
3. $(18a^3b - 12a^2b^2 + 30a^4b^4) \div 6a^2b$
4. $(5x - 9)(2x - 1)$
5. $(5x - 3y)^2$
6. $(a + b - c) + (a - b + c) - [(b - a + c) + (b + a + c)]$

Factor each of the following, if possible.

7. $6 + 9x + 3x^2$

8. $a(x - 2) + 3(x - 2)$

9. $m^2 - 2m + 2$

10. $4x^2 - 11x + 9$

11. $x^2 - 3x - 28$

12. $x^8 - y^8$

13. $x^2 + 6xy - 16y^2$

14. $27x^3 - 8y^3$

In Exercises 15 and 16, graph the linear equations. Give the slope, x-intercept, and y-intercept.

15. $y + 3x = 5$

16. $2x + 3y = 9$

17. Write an equation of the line passing through the points $(2, 5)$ and $(-1, -4)$.

18. Write an equation of the line passing through the point $(1, -1)$ and having a slope of 2.

19. Write an equation of the line that has a slope of $\frac{1}{2}$ and y-intercept of 2.

20. Graph $x + 2y > 8$.

21. Which of the following points are to the right, on, and to the left of the line $x - y = 2$?

(a) $(5, 3)$ (b) $(-3, -1)$ (c) $(2, 4)$ (d) $(1, -1)$ (e) $(6, -8)$

22. Reduce the following expressions.

(a) $\dfrac{x^2y}{yx^2}$ (b) $\dfrac{2x + 10}{x^2 - 25}$ (c) $\dfrac{a^2 - 9}{a^2 - a - 6}$ (d) $\dfrac{y - 5}{5 - y}$

23. Express each sum or difference in reduced form.

(a) $\dfrac{3x}{4} + \dfrac{9x}{8}$ (b) $\dfrac{x}{x + y} - \dfrac{y}{x + y}$

(c) $\dfrac{x^2}{x - 2} + \dfrac{x}{x - 3}$ (d) $\dfrac{y^2}{y - 1} - \dfrac{y}{y + 1}$

24. Simplify the expressions.

(a) $\dfrac{\dfrac{x + y}{a}}{\dfrac{x - y}{a}}$ (b) $\dfrac{\dfrac{1}{y} - \dfrac{1}{x}}{1 - \dfrac{x}{y}}$

(c) $\dfrac{x^2 - 16y^2}{xy} \div (x - 4y)$ (d) $\dfrac{2y - 2x}{y^2 - 9x^2} \div \dfrac{4y - 4x}{y + 3x}$

Do the indicated arithmetic and use only positive exponents to give your results.

25. $y^9 \cdot y^{-5}$ **26.** $x^{12} \cdot x^3$ **27.** $x^7 \div x^{-3}$

28. $x^{-8} \div x^{-5}$ **29.** $6^{-2} \cdot 6^3$ **30.** $\dfrac{5^{-4}}{5^{-7}}$

31. $(a^{-3})^2$ **32.** $(b^4)^{-5}$ **33.** $(c^{-6})^{-2}$

34. $(3x)^{-1}$ **35.** $3x^{-1}$ **36.** $(\frac{2}{3})^{-3}$

37. $(\frac{1}{2})^{-1}$ **38.** 6^{-2} **39.** $\left(\dfrac{2}{x}\right)^{-3}$

40. $\left(\dfrac{2x}{y}\right)^{-3}$ **41.** $(3x^{-2}y^{-3})^{-4}$ **42.** $(2x^{-1}y^{-2})^{-3}$

Solve the following systems of equations.

43. $\begin{cases} 2x - y = -1 \\ 4x + 3y = -12 \end{cases}$ **44.** $\begin{cases} 3x + 2y = -17 \\ x - 3y = 9 \end{cases}$

45. $\begin{cases} 3x - 5y = 63 \\ 2x + 3y = -15 \end{cases}$ **46.** $\begin{cases} \dfrac{2}{x} + \dfrac{1}{y} = 3 \\ \dfrac{3}{x} - \dfrac{2}{y} = 8 \end{cases}$

47. Use the graphical method to solve the following system.

$$\begin{cases} x + 3y = 4 \\ 2x - y = 1 \\ x + y = 2 \end{cases}$$

48. Graph the system of inequalities

$$\begin{cases} x + 2y \geq 4, \\ 2x + y \leq 8. \end{cases}$$

49. In which of the following equations is y a function of x? In which is x a function of y?

(a) $y = x + 2$ (b) $y = x^2 + x$
(c) $y^2 = x^2 + 5$ (d) $y^2 = 2x + 3$

CHAPTER 11

Roots

Objectives . . .

- To apply the laws of radicals to algebraic expressions involving roots.
- To apply the basic arithmetic operations to expressions involving radicals.
- To use the notation and terminology of rational exponents to simplify expressions and solve equations.
- To apply your knowledge of squares and square roots to the Pythagorean Theorem.

11–1 A SECOND LOOK AT THE REAL NUMBER SYSTEM

The real number system consists of rational and irrational numbers. One way to describe the set of real numbers is to say that it is the set of all infinite decimals. Then the set of repeating decimals is the set of rational numbers, and the set of nonrepeating decimals is the set of irrational numbers. Even a terminating decimal can be considered a repeating decimal. For example, you can think of $\frac{3}{4}$ as either

$$.75000\ldots \quad \text{or} \quad .74999\ldots.$$

Repeating decimals are indicated by using the convenient underlining notation that was introduced in Chapter 3. Thus, you can represent

$$\begin{aligned} 5.16666\ldots &\quad \text{by} \quad 5.1\underline{6}, \\ -1.030303\ldots &\quad \text{by} \quad -1.\underline{03}, \\ .74999\ldots &\quad \text{by} \quad .74\underline{9}, \\ 12.876876876\ldots &\quad \text{by} \quad 12.\underline{876}. \end{aligned}$$

You have learned that you can geometrically represent the set of real numbers as the set of all points on a number line. First, represent the integers as equally spaced points on the line. Then the rational numbers are fitted between the integers proportionally. The rest of the points are associated with irrational numbers. This illustrates a property which can be stated in the following way.

The graph of each real number is a unique point on a number line, and each point on the line is the graph of a unique real number.

Each irrational number can be approximated as accurately as you wish by a rational number. For example, let r be the irrational number

$$2.44222444422222\ldots.$$

Then the decimal representation of r consists of one 2, two 4's, three 2's, four 4's, five 2's, and so on. Therefore, r is between

$$\begin{array}{rcl} 2 & \text{and} & 3, \\ 2.4 & \text{and} & 2.5, \\ 2.44 & \text{and} & 2.45, \\ 2.442 & \text{and} & 2.443, \\ 2.4422 & \text{and} & 2.4423, \end{array}$$

and so on. Thus, you have found successively better rational approximations of r: 2, 2.4, 2.44, 2.442, 2.4422, and so on indefinitely. If you magnify the number line between 2 and 3, as shown in Fig. 11–1, it is clear that the intervals between the pairs of numbers above get smaller and smaller. Furthermore, each interval fits inside the previous one. It is reasonable to assume, therefore, that this chain of intervals will eventually trap one, and only one, point on the line. This point is the graph of r.

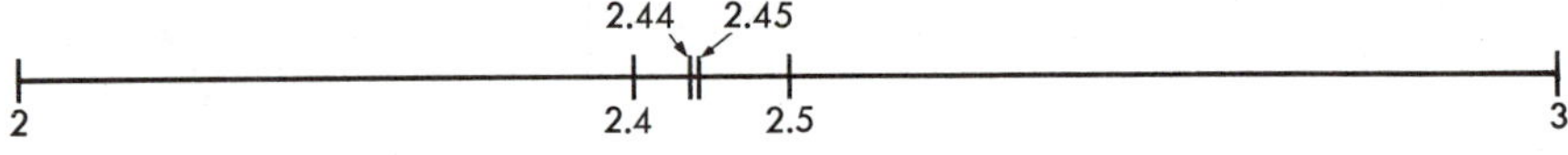

FIGURE 11–1

Exercises

Find the repeating decimal representation of each of the following rational numbers.

1. (a) $\frac{3}{8}$ (b) $-1\frac{5}{8}$

2. (a) $\frac{7}{9}$ (b) $\frac{4}{9}$

3. (a) $\frac{14}{13}$ (b) $\frac{9}{13}$

4. (a) $\frac{11}{12}$ (b) $\frac{7}{12}$

5. (a) $\frac{2}{11}$ (b) $\frac{4}{11}$

6. (a) $-2\frac{4}{7}$ (b) $-3\frac{2}{7}$

Express the rational number represented by each of the following infinite repeating decimals as a fraction.

7. (a) $4.\underline{61}$ (b) $-1.\underline{03}$

8. (a) $.34\underline{9}$ (b) $.74\underline{9}$

9. (a) $.75\underline{0}$ (b) $.3125\underline{0}$

10. (a) $.\underline{27}$ (b) $.\underline{41}$

11. (a) $12.\underline{876}$ (b) $13.\underline{135}$

12. (a) $.\underline{076923}$ (b) $.\underline{153846}$

---

13. (a) The irrational number

$$r = 2.44222444422222\ldots$$

lies between 2 and 3. What is the length of this interval?

(b) Suppose that you have found that r is located between 2.4 and 2.5. What is the length of the interval? What is the length of the interval from 2.44 to 2.45? from 2.442 to 2.443?

(c) Can r ever be located on an interval of length 10^{-4}? of length 10^{-5}? of length 10^{-1000}?

14. How many rational numbers are there between 0 and .1?

11–2 SQUARE ROOTS

The product of two positive numbers or two negative numbers is always positive. Therefore,

$$x^2 > 0 \text{ for every nonzero real number } x.$$

Definition of square root

A number a is called a square root of a number b if

$$a^2 = b. \qquad \textit{(Def-Sq. Rt.)}$$

For example, 5 is a square root of 25 because

$$5^2 = 25.$$

Also, -5 is a square root of 25 because

$$(-5)^2 = 25.$$

The following is the statement of an important property of the system of real numbers, which is illustrated above.

Every positive real number has two square roots, each of which is the negative of the other.

For example, 49 has the two square roots: 7 and -7.

The number 0 has one square root: 0. There is no real number a such that $a^2 = -4$, because $a^2 > 0$ for every real number a. This illustrates the following statement.

Negative real numbers have no real square roots.

The symbol $\sqrt{\ }$, called a *radical sign,* is used to designate the *non-negative square root* of a number. Thus,

$$\sqrt{4} = 2,$$
$$\sqrt{\tfrac{9}{16}} = \tfrac{3}{4},$$
$$\sqrt{0} = 0.$$

To designate the negative square root of a number, a negative sign is put in front of the symbol $\sqrt{\ }$. Thus,

$$-\sqrt{4} = -2,$$
$$-\sqrt{\tfrac{9}{16}} = -\tfrac{3}{4}.$$

The squares of consecutive integers from 0 to 13 are listed below.

0, 1, 4, 9, 16, 25, 36, 49, 64, 81, 100, 121, 144, 169

Notice that if you take the differences between these consecutive squares, you obtain

1, 3, 5, 7, 9, 11, 13, 15, 17, 19, 21, 23, 25,

which are the first 13 positive odd integers. This suggests the following property of squares of integers:

The sum of the first n positive odd integers is n^2.

For example,

$$1 + 3 = 2^2,$$
$$1 + 3 + 5 = 3^2,$$
$$1 + 3 + 5 + 7 = 4^2.$$

The nonsquare positive integers,

$$2,\ 3,\ 5,\ 6,\ 7,\ 8,\ 10,\ 11,\ 12,\ 13,\ 14,\ 15,\ 17,$$

and so on, have irrational square roots. It has been known that $\sqrt{2}$ is an irrational number since Euclid gave a geometric proof of it in his *Elements*.

Exercises

Find the integer denoted by each of the following expressions.

1. (a) $\sqrt{81}$ (b) $\sqrt{144}$

2. (a) $-\sqrt{25}$ (b) $-\sqrt{64}$

3. (a) $\sqrt{10^2}$ (b) $\sqrt{21^2}$

4. (a) $\sqrt{400}$ (b) $\sqrt{324}$

5. (a) $-\sqrt{21^2}$ (b) $-\sqrt{95^2}$

6. (a) $\sqrt{3600}$ (b) $\sqrt{4900}$

7. (a) $\sqrt{(910)^6}$ (b) $\sqrt{(212)^4}$

Find the rational number denoted by each of the following expressions.

8. (a) $\sqrt{.0081}$ (b) $\sqrt{.01}$

9. (a) $\sqrt{(\frac{7}{17})^2}$ (b) $\sqrt{(\frac{3}{23})^2}$

10. (a) $-\sqrt{\frac{16}{169}}$ (b) $-\sqrt{\frac{121}{100}}$

11. (a) $\sqrt{1.44}$ (b) $\sqrt{1.21}$

12. (a) $-\sqrt{\frac{9}{49}}$ (b) $-\sqrt{\frac{36}{49}}$

13. (a) $\sqrt{\frac{64}{225}}$ (b) $\sqrt{\frac{81}{256}}$

14. (a) $\sqrt{\dfrac{289}{10^6}}$ (b) $\sqrt{\dfrac{169}{5^8}}$

15. (a) $-\sqrt{\frac{16}{25}}$ (b) $-\sqrt{\frac{49}{81}}$

16. (a) $-\sqrt{.49}$ (b) $-\sqrt{.64}$

17. Find the following sums.

(a) $1 + 3 + 5 + 7 + 9 + 11 + 13 + 15 + 17 + 19$
(b) $1 + 3 + 5 + 7 + 9 + 11 + 13 + 15 + 17 + 19 + 21 + 23 + 25 + 27 + 29 + 31$

18. (a) What is the sum of the first 15 positive odd integers?
(b) What is the sum of the first 22 positive odd integers?

19. If you begin with $1 + 3$ and continue to add odd integers in order, will you eventually obtain a sum of 26? of 220? of 223? Explain your answers.

20. Prove that the square of the nth positive integer plus the $(n + 1)$st positive odd integer is the square of the $(n + 1)$st integer.

11-3 OTHER ROOTS

The number 4 is a cube root of 64 because

$$4^3 = 64,$$

and -3 is a cube root of -27 because

$$(-3)^3 = -27.$$

Definition of cube root

A number a is called a cube root of a number c if

$$a^3 = c. \qquad \textit{(Def-Cu. Rt.)}$$

Fourth roots, fifth roots, and so on, are defined in a similar way. For example, 2 is a fourth root of 16 and it is a fifth root of 32, because

$$2^4 = 16 \quad \text{and} \quad 2^5 = 32.$$

Just as every positive real number has a positive square root, every positive real number has a positive cube root, fourth root, fifth root, and so on. Also, each negative number has a negative cube root, fifth root, and so on.

For each integer $n > 1$, every positive real number has a unique positive nth root. If n is an odd integer, then every negative real number has a unique negative nth root.

For example,

$$\sqrt[3]{125} = 5, \quad \sqrt[3]{-125} = -5, \quad \sqrt[4]{81} = 3.$$

Definition of the *n*th root of a number

For every integer $n > 1$ and all real numbers a and b, the number b is called an nth root of a if

$$b^n = a. \qquad (Def\text{-}nth\ Rt.)$$

If $a > 0$, $\sqrt[n]{a}$ denotes the positive nth root of a. If n is odd and $a < 0$, $\sqrt[n]{a}$ denotes the negative nth root of a. Of course,

$$\sqrt[n]{0} = 0 \text{ for every integer } n > 1.$$

The integer n in $\sqrt[n]{a}$ is called the *index* and the real number a the *radicand* of $\sqrt[n]{a}$. The index 2 is usually omitted, and you write $\sqrt{a}$ for the square root of a.

The positive integers,

$$1,\ 8,\ 27,\ 64,\ 125,\ 216,\ 343,$$

and so on, are perfect cubes. In other words, they are cubes of integers. The negative integers,

$$-1,\ -8,\ -27,\ -64,\ -125,\ -216,\ -343,$$

and so on, are also perfect cubes. Many other integers have irrational cube roots. Thus,

$$\sqrt[3]{2},\ \sqrt[3]{9},\ \sqrt[3]{32},\ \sqrt[3]{-5},\ \sqrt[3]{-13}$$

are irrational numbers. Some rational numbers which are perfect cubes are

$$\tfrac{64}{27},\ -(\tfrac{8}{125}),\ \tfrac{343}{1000}.$$

Exercises

In Exercises 1–12, find the integer equal to the following expressions.

1. (a) $\sqrt{64}$ (b) $\sqrt{81}$

2. (a) $\sqrt[3]{64}$ (b) $\sqrt[3]{27}$

3. (a) $-\sqrt[6]{64}$ (b) $-\sqrt[4]{81}$

4. (a) $\sqrt[3]{-64}$ (b) $\sqrt[3]{-27}$

5. (a) $\sqrt[3]{-125}$ (b) $\sqrt[3]{-216}$

6. (a) $\sqrt[5]{-32}$ (b) $\sqrt[5]{-243}$

7. (a) $-\sqrt[4]{625}$ (b) $-\sqrt[4]{2401}$

8. (a) $\sqrt[5]{243}$ (b) $\sqrt[10]{1024}$
9. (a) $\sqrt[3]{7^3}$ (b) $\sqrt[8]{9^8}$
10. (a) $\sqrt[7]{12^7}$ (b) $\sqrt[6]{6^6}$
11. (a) $\sqrt[3]{125{,}000}$ (b) $\sqrt[3]{64000}$
12. (a) $-\sqrt[4]{1296}$ (b) $-\sqrt[4]{256}$

Find the rational number denoted by each of the following expressions.

13. (a) $\sqrt[3]{\frac{27}{64}}$ (b) $\sqrt[3]{\frac{64}{125}}$
14. (a) $\sqrt[4]{\frac{1}{81}}$ (b) $\sqrt[4]{\frac{1}{256}}$
15. (a) $\sqrt[3]{-.064}$ (b) $\sqrt[3]{-.008}$
16. (a) $\sqrt[5]{\frac{-243}{32}}$ (b) $\sqrt[5]{-\frac{32}{243}}$
17. (a) $\sqrt[3]{-\frac{8}{27}}$ (b) $\sqrt[3]{-\frac{216}{343}}$
18. (a) $\sqrt[4]{-\frac{16}{81}}$ (b) $\sqrt[4]{-\frac{81}{256}}$
19. (a) $\sqrt[7]{(.1)^7}$ (b) $\sqrt[11]{(.1)^{11}}$
20. (a) $-\sqrt[5]{\frac{-1}{243}}$ (b) $-\sqrt[5]{\frac{-1}{32}}$
21. (a) $\sqrt[6]{(.6)^6}$ (b) $\sqrt[20]{(.2)^{20}}$

If $S = \{x \mid x^2 = 4\}$, or $S = \{2, -2\}$, then S is the set of square roots of 4. Similarly, each set below is the set of roots of some number. List the elements of each set.

22. (a) $A = \{x \mid x^2 = 144\}$
    (b) $A' = \{x \mid x^2 = 156\}$
23. (a) $B = \{x \mid x^3 = -27\}$
    (b) $B' = \{x \mid x^3 = -729\}$
24. (a) $C = \{x \mid x^5 = 32\}$
    (b) $C' = \{x \mid x^5 = -\frac{243}{32}\}$
25. (a) $D = \{y \mid y^4 = 256\}$
    (b) $D' = \{r \mid r^4 = 2401\}$
26. (a) $E = \{x \mid x^6 = 10^6\}$
    (b) $E' = \{x \mid x^{14} = 5^{14}\}$
27. (a) $F = \{n \mid n^3 = .008\}$
    (b) $F' = \{y \mid y^3 = .027\}$
28. (a) $G = \{y \mid y^2 = -\frac{1}{4}\}$
    (b) $G' = \{x \mid x^2 = -\frac{1}{9}\}$

11-4 ROOTS OF PRODUCTS

Since

$$(\sqrt{3}\cdot\sqrt{7})^2 = (\sqrt{3})^2(\sqrt{7})^2 = 3\cdot 7 = (\sqrt{3\cdot 7})^2,$$

evidently $\sqrt{3}\cdot\sqrt{7} = \sqrt{21}$. Similarly,

$$\sqrt{xy} = \sqrt{x}\cdot\sqrt{y} \text{ for all positive numbers } x \text{ and } y.$$

It may also be shown that this is true for every index n.

FIRST LAW OF RADICALS

The equation

$$\sqrt[n]{xy} = \sqrt[n]{x}\,\sqrt[n]{y} \qquad \textbf{(LR-1)}$$

is true for all positive numbers x and y.

The above property is true for all real numbers x and y if n is an odd integer.

The first law of radicals can be used to simplify roots of numbers. For example,

$$\sqrt{300} = \sqrt{100\cdot 3} = \sqrt{100}\cdot\sqrt{3} = 10\sqrt{3}.$$

The square root of 300 is expressed in *simplest form* above, because 3 has no perfect-square integer greater than 1 as a factor. For this reason, 3 is called a *square-free integer.* Other square-free integers are 10, 14, 37, and 55.

Problem 1. Find the simplest form of the following square roots.

(a) $\sqrt{171}$

(b) $\sqrt{2520}$

Solution.

(a) You see that $171 = 9 \times 19$, and hence

$$\sqrt{171} = \sqrt{9\times 19} = \sqrt{9}\times\sqrt{19} = 3\sqrt{19}.$$

This is in simplest form, since 19 is a square-free integer.

(b) By trial and error, $2520 = 4 \times 630 = 4 \times 9 \times 70$ and 70 is square-free. Thus

$$\sqrt{2520} = \sqrt{4 \times 9 \times 70} = \sqrt{4} \times \sqrt{9} \times \sqrt{70}$$
$$= 2 \times 3 \times \sqrt{70} = 6\sqrt{70}$$

is in simplest form.

While it is true that

$$\sqrt{x^2} = x \text{ for every number } x \geq 0,$$

it is not true that $\sqrt{x^2} = x$ if x is negative. For example,

$$\sqrt{(-3)^2} = \sqrt{9} = 3;$$

and hence $\sqrt{x^2} = -x$ if $x = -3$. As this example illustrates,

$$\sqrt{x^2} = |x| \quad \text{for every real number } x.$$

If the variable x is raised to an even power such as $2n$ where n is any integer, then you can extract the square root of this power as follows:

$$\sqrt{x^{2n}} = \sqrt{(x^n)^2} = |x^n|.$$

For example, $\sqrt{x^{12}} = \sqrt{(x^6)^2} = |x^6| = x^6$.

Problem 2. Simplify.

(a) $\sqrt{25a^4b^6}$ (b) $\sqrt{8x^3y^2}$

Solution.

(a) $\sqrt{25a^4b^6} = \sqrt{25} \cdot \sqrt{a^4} \cdot \sqrt{b^6} = |5a^2b^3|$

(b) $\sqrt{8x^3y^2} = \sqrt{4} \cdot \sqrt{x^2} \cdot \sqrt{y^2} \cdot \sqrt{2x} = |2xy| \times \sqrt{2x}$

(Note that you must have $x \geq 0$ in order that $\sqrt{2x}$ exists.)

Exercises

Use (LR-1) to find the simplest form of each of the following roots.

1. (a) $\sqrt{12}$ (b) $\sqrt{20}$

2. (a) $\sqrt{18}$ (b) $\sqrt{27}$

3. (a) $\sqrt{50}$ (b) $\sqrt{32}$

4. (a) $\sqrt{8}$ (b) $\sqrt{24}$

5. (a) $\sqrt{45}$ (b) $\sqrt{54}$

6. (a) $\sqrt{112}$ (b) $\sqrt{75}$

7. (a) $\sqrt{72}$ (b) $\sqrt{98}$
8. (a) $\sqrt{540}$ (b) $\sqrt{325}$
9. (a) $\sqrt{882}$ (b) $\sqrt{279}$
10. (a) $\sqrt{328}$ (b) $\sqrt{147}$
11. (a) $\sqrt{16800}$ (b) $\sqrt{1296}$
12. (a) $\sqrt{10^5}$ (b) $\sqrt{15^7}$
13. (a) $\sqrt{2 \cdot 6^3}$ (b) $\sqrt{3 \cdot 7^5}$
14. (a) $\sqrt[3]{-56}$ (b) $\sqrt[3]{40}$
15. (a) $\sqrt[4]{80}$ (b) $\sqrt[4]{48}$
16. (a) $\sqrt[5]{224}$ (b) $\sqrt[3]{250}$

Simplify each of the algebraic expressions below. The domain of each variable is assumed to be the set of positive real numbers.

17. (a) $\sqrt{81x^6}$ (b) $\sqrt{121y^8}$
18. (a) $-\sqrt{64a^8}$ (b) $-\sqrt{100y^{10}}$
19. (a) $\sqrt{98x^4}$ (b) $\sqrt{180y^5}$
20. (a) $\sqrt{\frac{4}{9}x^4y^6}$ (b) $\sqrt{\frac{25}{36}a^{10}b^{12}}$
21. (a) $\sqrt{50x^9y^{11}}$ (b) $\sqrt{72x^{13}y^{15}}$
22. (a) $\sqrt{\frac{1}{4}(x+y)^2}$ (b) $\sqrt{\frac{1}{9}(a-b)^2}$
23. (a) $\sqrt{128x^6y^6}$ (b) $\sqrt{80x^8y^{10}}$
24. (a) $-\sqrt{10^3x^7y}$ (b) $-\sqrt{9^9x^9y}$
25. (a) $\sqrt{121x^8y^{10}}$ (b) $\sqrt{81x^{14}y^{16}}$
26. (a) $\sqrt[3]{125x^3y^9}$ (b) $\sqrt[3]{64x^{12}y^{21}}$
27. (a) $\sqrt[4]{256x^{20}}$ (b) $\sqrt[4]{243y^{24}}$
28. (a) $\sqrt{.01(a+2b)^4}$ (b) $\sqrt{.09(x-3y)^4}$
29. (a) $\sqrt[3]{2x^5y^6}$ (b) $\sqrt[3]{3x^4y^9}$
30. (a) $\sqrt[3]{16x^4y^5}$ (b) $\sqrt[3]{54x^7y^{10}}$

---

31. Tell which of the following statements are true, and correct the statements that you marked false.
    (a) $\sqrt{36x^2y^8} = 6xy^4$ for every x and y.
    (b) $\sqrt{49x^4y^6} = 7x^2y^3$ for every $y \geqq 0$ and every x.
    (c) $\sqrt{125x^{10}y^{14}} = 5x^5y^7\sqrt{5}$ for every $x \geqq 0$ and every y.
    (d) $\sqrt[3]{-8x^3y^{12}} = -2xy^4$ for every x and y.

Simplify.

32. $-\sqrt[5]{96}$ **33.** $-\sqrt[4]{1250}$ **34.** $-\sqrt[5]{-64}$

35. $\sqrt[7]{2^7 \cdot 3^9}$ **36.** $\sqrt[12]{8^{12} \cdot 9^{13} \cdot x^{14}}$ **37.** $\sqrt[3]{-.027x^4y^7}$

38. $\sqrt[6]{10^6 \cdot 11^{12} \cdot x^{18}}$ **39.** $\sqrt[7]{4^7 \cdot 3^{21} \cdot x^{15}}$ **40.** $\sqrt[5]{3^6 \cdot 4^7 \cdot a^{11}}$

11-5 PRODUCTS OF ROOTS

Certain products of roots of numbers can be simplified by use of the first law of radicals, (LR-1), as illustrated below.

Problem 1. Express $\sqrt{15} \times \sqrt{12}$ in simplest form.

Solution.

$$\begin{aligned}
\sqrt{15} \times \sqrt{12} &= \sqrt{15 \times 12} && \text{(LR-1)}\\
&= \sqrt{(3 \times 5) \times (3 \times 4)}\\
&= \sqrt{9 \times 4 \times 5} && \text{(R-M)}\\
&= \sqrt{9} \times \sqrt{4} \times \sqrt{5} && \text{(LR-1)}\\
&= 3 \times 2 \times \sqrt{5}\\
&= 6\sqrt{5}
\end{aligned}$$

In other words,

$$\sqrt{15} \times \sqrt{12} = 6\sqrt{5}.$$

This is the simplest form of the product because 5 is square-free.

Problem 2. Express $\sqrt{6} \times \sqrt{21} \times \sqrt{70}$ in simplest form.

Solution. You have

$$\begin{aligned}
\sqrt{6} \times \sqrt{21} \times \sqrt{70} &= \sqrt{6 \times 21 \times 70} && \text{(LR-1)}\\
&= \sqrt{(2 \times 3) \times (3 \times 7) \times (2 \times 5 \times 7)}\\
&= \sqrt{2^2 \times 3^2 \times 7^2 \times 5} && \text{(R-M)}\\
&= \sqrt{2^2 \times 3^2 \times 7^2} \times \sqrt{5} && \text{(LR-1)}\\
&= 2 \times 3 \times 7 \times \sqrt{5}, \quad \text{or } 42\sqrt{5}.
\end{aligned}$$

Problem 3. Simplify $\sqrt{3a} \times \sqrt{6ab^2} \times \sqrt{2ab}$

Solution.

$$\begin{aligned}
\sqrt{3a} \times \sqrt{6ab^2} \times \sqrt{2ab} &= \sqrt{(3a) \cdot (6ab^2) \cdot (2ab)}\\
&= \sqrt{36a^3b^3}\\
&= \sqrt{36(ab)^2(ab)}\\
&= |6ab|\sqrt{ab}
\end{aligned}$$

Exercises

Express each product of square roots in simplest form.

1. (a) $\sqrt{3} \times \sqrt{6}$ (b) $\sqrt{2} \times \sqrt{10}$
2. (a) $\sqrt{5} \times \sqrt{20}$ (b) $\sqrt{7} \times \sqrt{28}$
3. (a) $\sqrt{14} \times \sqrt{7}$ (b) $\sqrt{15} \times \sqrt{5}$
4. (a) $\sqrt{2} \times \sqrt{6} \times \sqrt{3}$ (b) $\sqrt{7} \times \sqrt{11} \times \sqrt{77}$
5. (a) $\sqrt{7} \times \sqrt{21} \times \sqrt{3}$ (b) $\sqrt{5} \times \sqrt{30} \times \sqrt{6}$
6. (a) $\sqrt{12} \cdot \sqrt{15}$ (b) $\sqrt{21} \cdot \sqrt{28}$
7. (a) $\sqrt{18} \cdot \sqrt{6}$ (b) $\sqrt{24} \cdot \sqrt{8}$
8. (a) $\sqrt{21a} \cdot \sqrt{15}$ (b) $\sqrt{13x^2} \cdot \sqrt{26}$
9. (a) $\sqrt{39} \cdot \sqrt{26}$ (b) $\sqrt{52} \cdot \sqrt{78}$
10. (a) $\sqrt{17} \cdot \sqrt{51}$ (b) $\sqrt{18} \cdot \sqrt{54}$
11. (a) $\sqrt{33} \cdot \sqrt{55}$ (b) $\sqrt{48} \cdot \sqrt{60}$
12. (a) $\sqrt{30} \cdot \sqrt{6} \cdot \sqrt{60}$ (b) $\sqrt{15} \cdot \sqrt{10} \cdot \sqrt{75}$
13. (a) $\sqrt{97a} \cdot \sqrt{97a}$ (b) $\sqrt{93x^2} \cdot \sqrt{93x^2}$
14. (a) $-\sqrt{80} \cdot \sqrt{40x^2}$ (b) $-\sqrt{105} \cdot \sqrt{5a^2}$
15. (a) $\sqrt{5a} \cdot \sqrt{20ab}$ (b) $\sqrt{6xy} \cdot \sqrt{24x}$
16. (a) $\sqrt{7a^2b} \cdot \sqrt{42a^3b^2}$ (b) $\sqrt{56x^2y^7} \cdot \sqrt{8xy}$
17. (a) $\sqrt{12x^3} \cdot \sqrt{5x} \cdot \sqrt{45}$ (b) $\sqrt{20a^5} \cdot \sqrt{7a^2} \cdot \sqrt{42}$
18. (a) $(-\sqrt{8a^5b}) \cdot (-\sqrt{2ab^5})$ (b) $(-\sqrt{18x^3y^{10}}) \cdot (-\sqrt{2xy^2})$
19. (a) $\sqrt{10x^6y^3} \cdot \sqrt{2x^5y}$ (b) $\sqrt{15xy^{12}} \cdot \sqrt{3x^3y^5}$
20. (a) $\sqrt{12x^3y^5z} \cdot \sqrt{5xy^2z}$ (b) $\sqrt{8xyz^3} \cdot \sqrt{10x^3y^2z}$
21. (a) $\sqrt{6x^3} \cdot \sqrt{5x^5} \cdot \sqrt{10x^6}$ (b) $\sqrt{12x^6} \cdot \sqrt{7x^3} \cdot \sqrt{42x}$
22. (a) $\sqrt{3ab^2c} \cdot \sqrt{12a^3bc} \cdot \sqrt{ab^7}$ (b) $\sqrt{4a^3bc} \cdot \sqrt{8ab^3c} \cdot \sqrt{2abc^3}$
23. (a) $\sqrt{20x^3y^4} \cdot \sqrt{5x^5y} \cdot \sqrt{8x^4y}$ (b) $\sqrt{50xy^5} \cdot \sqrt{10x^2y} \cdot \sqrt{20x^2y^2}$

In Exercises 24–41, express each product in simplest form.

24. $\sqrt{128} \cdot \sqrt{243}$
25. $(\sqrt{5})^3$
26. $\sqrt[3]{5} \cdot \sqrt[3]{25}$
27. $\sqrt[3]{21} \cdot \sqrt[3]{18}$
28. $\sqrt[3]{-12} \cdot \sqrt[3]{-10}$
29. $\sqrt[4]{24} \cdot \sqrt[4]{10}$
30. $\sqrt[5]{54} \cdot \sqrt[5]{63}$
31. $\sqrt[6]{2^4 \cdot 3^2} \cdot \sqrt[6]{2^3 \cdot 3^{10}}$
32. $\sqrt[3]{7a} \cdot \sqrt[3]{7a} \cdot \sqrt[3]{7a}$
33. $\sqrt[4]{17x} \cdot \sqrt[4]{17x^3}$

34. $(\sqrt[8]{5x})^8$

35. $\sqrt[3]{-50} \cdot \sqrt[3]{65}$

36. $\sqrt[4]{36x^2} \cdot \sqrt[4]{180x^2}$

37. $\sqrt[3]{30a^3} \cdot \sqrt[3]{-25a^6}$

38. $\sqrt[7]{10^3x^7} \cdot \sqrt[7]{10^5}$

39. $\sqrt[8]{2^7x^3} \cdot \sqrt[8]{2^9x^{13}}$

40. $\sqrt[10]{15^3x^4} \cdot \sqrt[10]{15^8x^9}$

41. $\sqrt[4]{36x^3} \cdot \sqrt[4]{72x^5}$

11–6 ROOTS OF QUOTIENTS

The nth root of a quotient of two numbers is the quotient of the nth roots of the numbers according to the second law of radicals.

SECOND LAW OF RADICALS

The equation

$$\sqrt[n]{\frac{x}{y}} = \frac{\sqrt[n]{x}}{\sqrt[n]{y}} \qquad \textit{(LR-2)}$$

is true for all positive numbers x and y.

This property is also true if x or y is negative and n is an odd integer. An example for $n = 2$ is given below.

$$\sqrt{\frac{16}{49}} = \frac{\sqrt{16}}{\sqrt{49}}, \quad \text{or} \quad \frac{4}{7}$$

In expressing a rational number as a quotient of two integers, you may always choose a perfect square for one of the integers. For example,

$$\frac{3}{2} = \frac{3}{2} \cdot \frac{2}{2}, \quad \text{or} \quad \frac{6}{2^2};$$

$$.7 = \frac{7}{10} \cdot \frac{10}{10}, \quad \text{or} \quad \frac{70}{10^2}.$$

You may use this fact to simplify square roots of rational numbers. (A number is not considered to be in simplified form if there is a radical sign in the denominator.) Thus,

$$\sqrt{\frac{3}{2}} = \sqrt{\frac{6}{2^2}}$$

$$= \frac{\sqrt{6}}{\sqrt{2^2}}, \quad \text{or} \quad \frac{\sqrt{6}}{2}, \quad \text{or} \quad \frac{1}{2}\sqrt{6}. \qquad \text{(LR-2)}$$

Also,

$$\sqrt{.7} = \sqrt{\frac{70}{10^2}}$$

$$= \frac{\sqrt{70}}{\sqrt{10^2}}, \quad \text{or} \quad \frac{\sqrt{70}}{10}, \quad \text{or} \quad \frac{1}{10}\sqrt{70}. \qquad \text{(LR-2)}$$

Problem 1. Simplify $\sqrt{\frac{8}{27}}$.

Solution. What is the *least* multiple of 27 that is a perfect square? If you express $27 = 3^3$, then $3^3 \cdot 3$, or 3^4, or 9^2, is the least perfect square which is a multiple of 27. You should also notice that $8 = 2^2 \cdot 2$. Hence,

$$\sqrt{\frac{8}{27}} = \sqrt{\frac{8}{27} \cdot \frac{3}{3}}$$

$$= \sqrt{\frac{2^2 \cdot 2 \cdot 3}{3^4}}$$

$$= \frac{\sqrt{2^2 \cdot 6}}{\sqrt{3^4}} \qquad \text{(LR-2)}$$

$$= \frac{2\sqrt{6}}{3^2}, \quad \text{or} \quad \frac{2}{9}\sqrt{6}. \qquad \text{(LR-1)}$$

Problem 2. Simplify $\sqrt{\frac{3a}{5b}}$.

Solution. You can proceed as follows (assuming $a > 0$ and $b > 0$):

$$\sqrt{\frac{3a}{5b}} = \sqrt{\frac{3a \cdot 5b}{(5b)^2}}$$

$$= \frac{\sqrt{15ab}}{5b}.$$

Problem 3. Simplify $\dfrac{\sqrt{14x^3y^5}}{\sqrt{20x^4y}}$.

Solution. You might argue as follows ($x > 0$ and $y > 0$):

$$\begin{aligned}\frac{\sqrt{14x^3y^5}}{\sqrt{20x^4y}} &= \frac{\sqrt{14xy(xy^2)^2}}{\sqrt{5y(2x^2)^2}} \\ &= \frac{xy^2}{2x^2}\,\frac{\sqrt{14xy}}{\sqrt{5y}} \\ &= \frac{y^2}{2x}\sqrt{\frac{14xy}{5y}} \\ &= \frac{y^2}{2x}\sqrt{\frac{70x}{25}} \\ &= \frac{y^2}{10x}\sqrt{70x}.\end{aligned}$$

Exercises

Simplify.

1. (a) $\sqrt{\frac{2}{25}}$ (b) $\sqrt{\frac{3}{64}}$

2. (a) $\sqrt{\frac{1}{16}}$ (b) $\sqrt{\frac{1}{36}}$

3. (a) $\sqrt{\frac{1}{2}}$ (b) $\sqrt{\frac{1}{3}}$

4. (a) $\sqrt{\frac{4}{7}}$ (b) $\sqrt{\frac{5}{11}}$

5. (a) $\sqrt{\frac{27}{125}}$ (b) $\sqrt{\frac{8}{75}}$

6. (a) $\sqrt{.3}$ (b) $\sqrt{.7}$

7. (a) $\sqrt{.9}$ (b) $\sqrt{.1}$

8. (a) $\sqrt{\dfrac{5x}{9y^2}}$ (b) $\sqrt{\dfrac{3x}{16y^4}}$

9. (a) $\sqrt{\dfrac{9a^4}{16}}$ (b) $\sqrt{\dfrac{36x^{10}}{49}}$

10. (a) $\sqrt{.24x^6}$ (b) $\sqrt{.48x^{10}}$

11. (a) $\sqrt{\dfrac{75a}{28b}}$ (b) $\sqrt{\dfrac{150x}{63y}}$

12. (a) $\sqrt{\dfrac{20x^2}{3y^4}}$ (b) $\sqrt{\dfrac{44x^8}{5y^6}}$

13. (a) $\sqrt{\dfrac{15x}{2y}}$ (b) $\sqrt{\dfrac{17x}{3y}}$

14. (a) $\sqrt{\frac{7x}{27y^2}}$ (b) $\sqrt{\frac{13x^3}{216y^4}}$

15. (a) $\sqrt{\frac{8x^3y}{125}}$ (b) $\sqrt{\frac{175xy^3}{27}}$

16. (a) $\sqrt{\frac{7a^5}{8b^5}}$ (b) $\sqrt{\frac{10x^5}{7y^7}}$

17. (a) $\sqrt{\frac{a^3b^7}{x^4y^9}}$ (b) $\sqrt{\frac{x^9y^3}{a^6b^{11}}}$

18. (a) $\sqrt{\frac{ab^6}{10x^5y}}$ (b) $\sqrt{\frac{x^{13}y}{3a^5b^3}}$

19. (a) $\sqrt{\frac{3ab}{2x^5y^9}}$ (b) $\sqrt{\frac{2x^5y^{11}}{3a^7b^5}}$

20. (a) $\frac{\sqrt{30x^5y^2}}{\sqrt{10x^3y^4}}$ (b) $\frac{\sqrt{28x^5y^5}}{\sqrt{7xy^2}}$

21. (a) $\frac{\sqrt{40x^4y^5}}{\sqrt{32xy}}$ (b) $\frac{\sqrt{48x^9y^{20}}}{\sqrt{75x^3y^3}}$

22. (a) $\sqrt{\frac{11x^3y^{11}}{2ab}}$ (b) $\sqrt{\frac{21x^2y^9}{5a^5b}}$

23. (a) $\sqrt{\frac{3x^5y^6}{5a^3b^2}}$ (b) $\sqrt{\frac{13x^2y^7}{3m^3n^9}}$

24. (a) $\sqrt{\frac{abc}{2xyz}}$ (b) $\sqrt{\frac{x^2y^3z^4}{5a^5b^4c^3}}$

25. (a) $\sqrt{\frac{28x^4y^5z}{75ab^7c}}$ (b) $\sqrt{\frac{55a^2b^3c^6}{32xyz^3}}$

Simplify each of the following roots.

26. $\sqrt[3]{\frac{1}{2}}$

27. $\sqrt[3]{\frac{5}{16}}$

28. $\sqrt[3]{\frac{3}{32}}$

29. $-\sqrt[4]{\frac{2}{27}}$

30. $\sqrt[4]{\frac{5}{8}}$

31. $\sqrt[3]{\frac{2a}{9b^2}}$

32. $\sqrt[3]{\frac{1}{5x^4}}$

33. $\sqrt[5]{-\frac{3}{16x^4}}$

34. $\sqrt[3]{\frac{-x}{10^5y}}$

35. $-\sqrt[5]{-\frac{8x}{27y^6}}$

11–7 SUMS OF ROOTS

There is no simpler numeral that can denote

$$\sqrt{2} + \sqrt{3}.$$

That is, you cannot express this number as a rational number times the square root of an integer. However, the irrational numbers

$$3\sqrt{2} + 7\sqrt{2} \quad \text{and} \quad 10\sqrt{2}$$

are equal according to the distributive axiom.

$$3\sqrt{2} + 7\sqrt{2} = (3 + 7)\sqrt{2}, \quad \text{or } 10\sqrt{2}.$$

An expression consisting of a sum or difference of radicals having the same index can sometimes be simplified by combining terms with the same radicand. This process is illustrated in the following problems.

Problem 1. Simplify $5\sqrt{2} - \sqrt{3} + 9\sqrt{2} + 8\sqrt{3}$.

Solution. On successive lines, expressions equal to the given one are written.

$$\begin{array}{ll} 5\sqrt{2} - \sqrt{3} + 9\sqrt{2} + 8\sqrt{3} & \\ 5\sqrt{2} + 9\sqrt{2} - \sqrt{3} + 8\sqrt{3} & \text{(R-A)} \\ (5 + 9)\sqrt{2} + (-1 + 8)\sqrt{3} & \text{(D)} \end{array}$$

Thus,

$$14\sqrt{2} + 7\sqrt{3}$$

is the desired simplification.

Problem 2. Simplify $12\sqrt{\frac{2}{3}} + \sqrt{\frac{24}{25}}$.

Solution. You might proceed as follows:

$$\begin{aligned} 12\sqrt{\frac{2}{3}} + \sqrt{\frac{24}{25}} &= 12\sqrt{\frac{2 \cdot 3}{9}} + \sqrt{\frac{4 \cdot 6}{25}} \\ &= 12\frac{\sqrt{6}}{3} + 2\frac{\sqrt{6}}{5} \\ &= 4\sqrt{6} + \frac{2}{5}\sqrt{6} \\ &= \frac{22}{5}\sqrt{6}. \end{aligned}$$

Problem 3. Find the product and simplify.

$$(2\sqrt{5} - \sqrt{3})(3\sqrt{5} + 4\sqrt{3})$$

Solution. Using the distributive axiom first and then simplifying, you obtain:

$$\begin{aligned}(2\sqrt{5} - \sqrt{3})&(3\sqrt{5} + 4\sqrt{3})\\ &= 2\sqrt{5}(3\sqrt{5} + 4\sqrt{3}) - \sqrt{3}(3\sqrt{5} + 4\sqrt{3})\\ &= 2\sqrt{5}\cdot 3\sqrt{5} + 2\sqrt{5}\cdot 4\sqrt{3} - \sqrt{3}\cdot 3\sqrt{5} - \sqrt{3}\cdot 4\sqrt{3}\\ &= 6\sqrt{5\cdot 5} + 8\sqrt{5\cdot 3} - 3\sqrt{3\cdot 5} - 4\sqrt{3\cdot 3}\\ &= 30 + 8\sqrt{15} - 3\sqrt{15} - 12\\ &= 18 + 5\sqrt{15}\end{aligned}$$

Problem 4. Simplify $(\sqrt{ab} - 2\sqrt{bc})^2$, assuming $a > 0$, $b > 0$, and $c > 0$.

Solution. You have

$$\begin{aligned}(\sqrt{ab} - 2\sqrt{bc})^2 &= (\sqrt{ab})^2 + 2(\sqrt{ab})(-2\sqrt{bc}) + (-2\sqrt{bc})^2\\ &= ab - 4\sqrt{ab\cdot bc} + 4bc\\ &= ab - 4\sqrt{ab^2c} + 4bc\\ &= ab + 4bc - 4b\sqrt{ac}.\end{aligned}$$

Exercises

In Exercises 1–10 simplify and name the property used in the simplification.

1. (a) $8\sqrt{6} - \sqrt{6}$ (b) $9\sqrt{7} - 2\sqrt{7}$

2. (a) $9\sqrt{7} + 6\sqrt{7} + \sqrt{7}$ (b) $\sqrt{11} + 3\sqrt{11} + 5\sqrt{11}$

3. (a) $6\sqrt{3} - 4\sqrt{3} + 7\sqrt{3}$ (b) $7\sqrt{7} - 10\sqrt{7} + 11\sqrt{7}$

4. (a) $9\sqrt{7} + 5\sqrt{7} + \sqrt{7}$ (b) $3\sqrt{10} + \sqrt{10} + 5\sqrt{10}$

5. (a) $\frac{1}{2}\sqrt{6} - \frac{1}{3}\sqrt{6} + \frac{1}{12}\sqrt{6}$ (b) $\frac{1}{4}\sqrt{3} - \frac{1}{2}\sqrt{3} + \frac{1}{12}\sqrt{3}$

6. (a) $7 + \sqrt{5} - 3 + 4\sqrt{5}$
(b) $-5 + 2\sqrt{13} - 2 + \sqrt{13}$

7. (a) $\sqrt{5} + 2\sqrt{6} - 3\sqrt{5} - \sqrt{6}$
(b) $3\sqrt{14} + 5\sqrt{13} - \sqrt{14} - \sqrt{13}$

8. (a) $\sqrt{10} + 1 + 12\sqrt{10} - 10$
(b) $17 + 3\sqrt{11} - 18 + \sqrt{11}$

9. (a) $\frac{1}{5}\sqrt{5} - \sqrt{13} - \sqrt{5} + 8\sqrt{13}$
   (b) $\sqrt{17} - \frac{1}{2}\sqrt{7} + 5\sqrt{17} - \sqrt{7}$
10. (a) $8\sqrt{2} - \sqrt{14} + \sqrt{2} - 5\sqrt{14} + \sqrt{3}$
   (b) $9\sqrt{3} - 2\sqrt{5} + \sqrt{3} - \sqrt{5} + \sqrt{3}$

Simplify each radical and then simplify the complete expression. What properties of radicals are involved?

11. (a) $3\sqrt{5} - 5\sqrt{20}$ (b) $2\sqrt{2} - 3\sqrt{32}$
12. (a) $8\sqrt{72} + 2\sqrt{20} - 3\sqrt{5}$ (b) $2\sqrt{27} - 4\sqrt{48} + 5\sqrt{7}$
13. (a) $2\sqrt{7} - \sqrt{28} + \sqrt{63}$ (b) $3\sqrt{44} + \sqrt{99} - \sqrt{55}$
14. (a) $\sqrt{18} + \sqrt{108} - \sqrt{50}$ (b) $\sqrt{48} - \sqrt{12} + \sqrt{300}$
15. (a) $\sqrt{20} - \sqrt{24} - \sqrt{180} + \sqrt{54}$
   (b) $\sqrt{18} + \sqrt{30} - \sqrt{80} + \sqrt{63}$
16. (a) $2\sqrt{108} - \sqrt{27} + \sqrt{363}$ (b) $3\sqrt{72} + \sqrt{98} - \sqrt{40}$
17. (a) $\sqrt{\frac{9}{2}} + \sqrt{\frac{1}{2}}$ (b) $\sqrt{\frac{10}{3}} - \sqrt{\frac{1}{3}}$
18. (a) $3\sqrt{\frac{1}{7}} + 4\sqrt{\frac{1}{28}}$ (b) $2\sqrt{\frac{1}{5}} + 3\sqrt{\frac{1}{20}}$
19. (a) $4\sqrt{\frac{2}{5}} + \sqrt{\frac{5}{2}}$ (b) $4\sqrt{\frac{2}{3}} + 2\sqrt{\frac{3}{2}}$
20. (a) $12\sqrt{\frac{3}{2}} + 18\sqrt{\frac{2}{3}}$ (b) $5\sqrt{\frac{7}{2}} + \sqrt{\frac{2}{7}}$
21. (a) $7\sqrt{\frac{8}{15}} - 6\sqrt{\frac{6}{5}}$ (b) $3\sqrt{\frac{8}{21}} - 2\sqrt{\frac{14}{3}}$
22. (a) $5\sqrt{\frac{9}{6}} - 8\sqrt{\frac{75}{2}}$ (b) $3\sqrt{\frac{16}{10}} - 4\sqrt{\frac{50}{2}}$

Find the following products and give each in simplified form.

23. (a) $\sqrt{10}(2\sqrt{10} + 3)$ (b) $\sqrt{5}(3\sqrt{5} + 2)$
24. (a) $\sqrt{3}(\sqrt{18} - \sqrt{2})$ (b) $\sqrt{6}(\sqrt{18} - \sqrt{5})$
25. (a) $\sqrt{7}(\sqrt{7} - \sqrt{28} + \sqrt{14})$ (b) $\sqrt{5}(\sqrt{5} - \sqrt{45} + \sqrt{10})$
26. (a) $(4\sqrt{6a} + 5)(\sqrt{6a} - 3)$ (b) $(2\sqrt{5a} + 6)(\sqrt{5a} - 1)$
27. (a) $(4\sqrt{13b} + \sqrt{6})(2\sqrt{13b} - 3\sqrt{6})$
   (b) $(3\sqrt{11x} + \sqrt{2})(2\sqrt{11x} - 3\sqrt{2})$
28. (a) $(\sqrt{8} - \sqrt{5})^2$ (b) $(\sqrt{7} - \sqrt{11})^2$
29. (a) $(2\sqrt{11} - 3\sqrt{6})^2$ (b) $(3\sqrt{15} - 2\sqrt{7})^2$
30. (a) $(2\sqrt{5x} - 3)(2\sqrt{5x} + 3)$ (b) $(\sqrt{5y} - 2\sqrt{7})(\sqrt{5y} + 2\sqrt{7})$
31. (a) $(6 + \sqrt{3y})(6 - \sqrt{3y})$ (b) $(8 + \sqrt{5x})(8 - \sqrt{5x})$
32. (a) $(\sqrt{27} + \sqrt{2})^2$ (b) $(\sqrt{32} + \sqrt{3})^2$

Simplify.

33. $\sqrt{\frac{5}{6}} + \sqrt{120} - \frac{2}{3}\sqrt{30}$
34. $\sqrt{5} - 5\sqrt{\frac{1}{5}} + \sqrt{45}$
35. $(\sqrt{6} + \sqrt{7} + \sqrt{8})^2$
36. $(\sqrt{xy} + 3\sqrt{yz})^2$
37. $(\sqrt{2x} - \sqrt{3y})(\sqrt{3x} + \sqrt{2y})$
38. $(\sqrt{x} - \sqrt{y})(\sqrt{x} + \sqrt{y})$

39. $(\sqrt{6x} - \sqrt{2y})(\sqrt{6x} + \sqrt{2y})$
40. $(\sqrt{3xy} - 3\sqrt{yz})^2$
41. $(\sqrt{a} + \sqrt{b})^3$
42. $(\sqrt{27a} + \sqrt{2ab})^2$
43. $(\sqrt{63x^3} - \sqrt{7x})^2$
44. $3\sqrt[3]{16} - 2\sqrt[3]{54}$
45. $\sqrt[4]{32} - 3\sqrt[4]{2} + \sqrt[4]{162}$
46. $(\sqrt{11} - 3)(\sqrt{11} + 3)(2\sqrt{7} - 3\sqrt{5})$
47. $(\sqrt{7xy} + \sqrt{3yz})(\sqrt{3xy} - \sqrt{7yz})$
48. $(1 + \sqrt{2} + \sqrt{3})(1 + \sqrt{2} - \sqrt{3})$

Review for Sections 11–1 through 11–7

Find the repeating decimal representation of each of the following rational numbers.

1. $2\frac{7}{8}$
2. $-3\frac{1}{4}$
3. $-1\frac{3}{11}$
4. $2\frac{3}{7}$

What integer or rational number is denoted by each of the following expressions?

5. $\sqrt[3]{8}$
6. $-\sqrt{\frac{4}{9}}$
7. $\sqrt{.25}$
8. $\sqrt[5]{\frac{1}{32}}$
9. $-\sqrt{.0025}$
10. $\sqrt{51^2}$
11. $\sqrt[8]{91^8}$
12. $\sqrt{\frac{9}{100}}$
13. $\sqrt{49}$
14. $\sqrt[3]{-27}$
15. $\sqrt[5]{(-26)^5}$
16. $-\sqrt{36}$
17. $\sqrt[4]{\frac{16}{81}}$
18. $-\sqrt[4]{29^4}$
19. $-\sqrt{42^2}$
20. $-\sqrt[3]{-1}$

Simplify each of the following algebraic expressions. The domain of each variable is assumed to be the set of positive real numbers.

21. $\sqrt{24}$
22. $\sqrt{3} \cdot \sqrt{8}$
23. $\sqrt{\frac{3}{8}}$
24. $2\sqrt{5} + 3\sqrt{5}$
25. $\sqrt{5xy} \cdot \sqrt{20x}$
26. $\sqrt{\dfrac{15r^{10}}{12rt}}$
27. $2\sqrt{\frac{3}{5}} + \frac{1}{5}\sqrt{15}$
28. $\sqrt{72x^3}$
29. $\sqrt{.32a^4}$
30. $(\sqrt{rt} + 3\sqrt{ts})^2$
31. $-\sqrt{75m^2n^5}$
32. $\sqrt{18a^3b} \cdot \sqrt{2a}$
33. $\sqrt{\frac{2}{5}} + 3\sqrt{\frac{5}{2}}$
34. $\sqrt{\dfrac{2a}{3b}}$
35. $\sqrt{15rt} \cdot \sqrt{5r^3}$
36. $(4\sqrt{3x} - \sqrt{6x})(2\sqrt{3x} + 3\sqrt{6x})$
37. $\sqrt{56}$
38. $\sqrt{\dfrac{5m^3}{3n}}$
39. $\sqrt{14m^3n} \cdot \sqrt{2mn^5}$
40. $\sqrt{60a^4b}$

Answers to Review for Sections 11–1 through 11–7

1. $2.875\underline{0}$ **2.** $-3.25\underline{0}$ **3.** $-1.\underline{27}$ **4.** $2.\underline{428571}$

5. 2 **6.** $-\frac{2}{3}$ **7.** .5 **8.** $\frac{1}{2}$

9. $-.05$ **10.** 51 **11.** 91 **12.** $\frac{3}{10}$

13. 7 **14.** -3 **15.** -26 **16.** -6

17. $\frac{2}{3}$ **18.** -29 **19.** -42 **20.** 1

21. $2\sqrt{6}$ **22.** $2\sqrt{6}$ **23.** $\frac{1}{4}\sqrt{6}$ **24.** $5\sqrt{5}$

25. $10x\sqrt{y}$ **26.** $\frac{r^4}{2t}\sqrt{5rt}$ **27.** $\frac{3}{5}\sqrt{15}$ **28.** $6x\sqrt{2x}$

29. $\frac{4a^2}{10}\sqrt{2}$, or $\frac{2a^2}{5}\sqrt{2}$ **30.** $rt + 6t\sqrt{rs} + 9ts$

31. $-5mn^2\sqrt{3n}$ **32.** $6a^2\sqrt{b}$ **33.** $\frac{17}{10}\sqrt{10}$

34. $\frac{1}{3b}\sqrt{6ab}$ **35.** $5r^2\sqrt{3t}$ **36.** $6x + 30x\sqrt{2}$

37. $2\sqrt{14}$ **38.** $\frac{m}{3n}\sqrt{15mn}$ **39.** $2m^2n^3\sqrt{7}$

40. $2a^2\sqrt{15b}$

11–8 QUOTIENTS OF RADICAL EXPRESSIONS

Quotients of radical expressions can be simplified as illustrated below.

Problem 1. Simplify $\dfrac{3\sqrt{12} - 8\sqrt{24}}{2\sqrt{2}}$.

Solution. You can eliminate the radical in the denominator by multiplying the fraction by $\sqrt{2}/\sqrt{2}$ as shown below.

$$\frac{3\sqrt{12} - 8\sqrt{24}}{2\sqrt{2}} = \frac{(3\sqrt{12} - 8\sqrt{24})\sqrt{2}}{(2\sqrt{2})\sqrt{2}}$$

$$= \frac{3\sqrt{12}\cdot\sqrt{2} - 8\sqrt{24}\cdot\sqrt{2}}{2(\sqrt{2})^2}$$

$$= \frac{3\sqrt{24} - 8\sqrt{48}}{2\cdot 2}$$

$$= \frac{3\sqrt{4\cdot 6} - 8\sqrt{16\cdot 3}}{4}$$

$$= \frac{3\cdot 2\sqrt{6} - 8\cdot 4\sqrt{3}}{4}$$

$$= \tfrac{3}{2}\sqrt{6} - 8\sqrt{3}$$

The observation that products such as

$$(\sqrt{7} - 2\sqrt{2})(\sqrt{7} + 2\sqrt{2}) = (\sqrt{7})^2 - (2\sqrt{2})^2 = -1$$

are rational numbers allows you to simplify radical expressions of the following form.

Problem 2. Simplify $\dfrac{\sqrt{5} - 2\sqrt{3}}{4\sqrt{5} - 5\sqrt{3}}$.

Solution. You can *rationalize the denominator* as follows.

$$\begin{aligned}
\frac{\sqrt{5} - 2\sqrt{3}}{4\sqrt{5} - 5\sqrt{3}} &= \frac{(\sqrt{5} - 2\sqrt{3})(4\sqrt{5} + 5\sqrt{3})}{(4\sqrt{5} - 5\sqrt{3})(4\sqrt{5} + 5\sqrt{3})} \\
&= \frac{\sqrt{5} \cdot 4\sqrt{5} + \sqrt{5} \cdot 5\sqrt{3} - 2\sqrt{3} \cdot 4\sqrt{5} - 2\sqrt{3} \cdot 5\sqrt{3}}{(4\sqrt{5})^2 - (5\sqrt{3})^2} \\
&= \frac{4 \cdot 5 + 5\sqrt{15} - 8\sqrt{15} - 10 \cdot 3}{16 \cdot 5 - 25 \cdot 3} \\
&= \frac{-10 - 3\sqrt{15}}{5} \\
&= -2 - \tfrac{3}{5}\sqrt{15}
\end{aligned}$$

Exercises

In Exercises 1–24 simplify each of the following quotients.

1. (a) $15\sqrt{3} \div 5\sqrt{3}$ (b) $21\sqrt{7} \div 3\sqrt{7}$
2. (a) $-2\sqrt{20} \div \sqrt{2}$ (b) $-5\sqrt{72} \div \sqrt{3}$
3. (a) $9\sqrt{3} \div \sqrt{18}$ (b) $6\sqrt{2} \div \sqrt{27}$
4. (a) $\sqrt{50} \div 5\sqrt{9}$ (b) $\sqrt{45} \div 4\sqrt{25}$
5. (a) $3\sqrt{5} \div 6\sqrt{35}$ (b) $6\sqrt{6} \div 5\sqrt{54}$
6. (a) $5\sqrt{19} \div 20\sqrt{38}$ (b) $2\sqrt{17} \div 10\sqrt{34}$
7. (a) $\dfrac{5\sqrt{6} - 3\sqrt{21}}{\sqrt{3}}$ (b) $\dfrac{3\sqrt{10} - 2\sqrt{15}}{\sqrt{5}}$
8. (a) $\dfrac{6\sqrt{15} + 4\sqrt{30}}{\sqrt{5}}$ (b) $\dfrac{2\sqrt{14} + 5\sqrt{35}}{\sqrt{7}}$
9. (a) $\dfrac{3\sqrt{24} - \sqrt{56} + 7\sqrt{150}}{\sqrt{3}}$ (b) $\dfrac{6\sqrt{12} - \sqrt{18} + \sqrt{32}}{\sqrt{2}}$
10. (a) $\dfrac{\sqrt{90} - \sqrt{45} + 3\sqrt{15}}{\sqrt{15}}$ (b) $\dfrac{2\sqrt{65} - \sqrt{78} + \sqrt{117}}{\sqrt{13}}$

11. (a) $\dfrac{1}{\sqrt{2}-3}$ (b) $\dfrac{1}{\sqrt{3}-2}$

12. (a) $\dfrac{3}{\sqrt{2}+1}$ (b) $\dfrac{2}{\sqrt{3}+1}$

13. (a) $\dfrac{1}{\sqrt{5}-2}$ (b) $\dfrac{1}{\sqrt{6}-3}$

14. (a) $\dfrac{1}{2\sqrt{2}-5}$ (b) $\dfrac{1}{3\sqrt{2}-7}$

15. (a) $\dfrac{4}{5-2\sqrt{6}}$ (b) $\dfrac{5}{6-3\sqrt{5}}$

16. (a) $\dfrac{7}{\sqrt{6}-\sqrt{5}}$ (b) $\dfrac{5}{\sqrt{6}-\sqrt{7}}$

17. (a) $\dfrac{\sqrt{5}-\sqrt{3}}{2+\sqrt{3}}$ (b) $\dfrac{\sqrt{11}-\sqrt{2}}{3+\sqrt{2}}$

18. (a) $\dfrac{-8+\sqrt{6}}{2-\sqrt{6}}$ (b) $\dfrac{-3+\sqrt{7}}{5-\sqrt{7}}$

19. (a) $\dfrac{1-4\sqrt{5}}{3+2\sqrt{5}}$ (b) $\dfrac{3-2\sqrt{11}}{2+3\sqrt{11}}$

20. (a) $\dfrac{\sqrt{7}+\sqrt{2}}{3\sqrt{7}-5\sqrt{2}}$ (b) $\dfrac{\sqrt{10}+\sqrt{3}}{2\sqrt{10}-3\sqrt{3}}$

21. (a) $\dfrac{\sqrt{10}+2\sqrt{5}}{3\sqrt{10}-\sqrt{5}}$ (b) $\dfrac{\sqrt{6}-3\sqrt{3}}{2\sqrt{6}-\sqrt{3}}$

22. (a) $\dfrac{\sqrt{10}-2\sqrt{5}}{\sqrt{10}+2\sqrt{5}}$ (b) $\dfrac{\sqrt{15}}{\sqrt{5}-3\sqrt{3}}$

23. (a) $\dfrac{\sqrt{7}-\sqrt{3}}{\sqrt{7}+\sqrt{3}}$ (b) $\dfrac{\sqrt{5}-\sqrt{13}}{\sqrt{5}+\sqrt{13}}$

24. (a) $\dfrac{2\sqrt{6}-\sqrt{2}}{\sqrt{6}-\sqrt{2}}$ (b) $\dfrac{5\sqrt{8}-\sqrt{3}}{\sqrt{8}-\sqrt{3}}$

---

25. In each of the following, rationalize the denominator.

(a) $\dfrac{1}{\sqrt{3}}$ (b) $\dfrac{1}{\sqrt{5}}$ (c) $\dfrac{1}{\sqrt{17}}$ (d) $\dfrac{1}{\sqrt{23}}$

26. Use the results that you found in Exercise 25 to make a statement about

$$\frac{1}{\sqrt{n}}, \quad n > 0.$$

27. In each of the following, rationalize the denominator.

(a) $\dfrac{6}{\sqrt{6}}$ (b) $\dfrac{10}{\sqrt{10}}$

(c) $\dfrac{21}{\sqrt{21}}$ (d) $\dfrac{80}{\sqrt{80}}$

28. Use the results that you found in Exercise 27 to make a statement about

$$\frac{n}{\sqrt{n}}, \quad n > 0.$$

In each of the following, rationalize the denominator.

29. $\dfrac{1}{\sqrt[3]{2}}$

30. $\dfrac{1}{\sqrt[3]{3}}$

31. $\dfrac{1}{\sqrt[3]{5}}$

32. $\dfrac{1}{\sqrt[3]{n}}, \quad n \neq 0$

11–9 RATIONAL EXPONENTS

You have seen that exponents are a valuable part of the language of mathematics. In previous chapters, x^n was defined for every integer n and every nonzero real number x. For example,

$$\begin{aligned} 5^4 &= 5 \cdot 5 \cdot 5 \cdot 5, \quad \text{or } 625, \\ 5^1 &= 5, \\ 5^0 &= 1, \\ 5^{-3} &= \frac{1}{5^3}, \quad \text{or } \frac{1}{125}. \end{aligned}$$

Soon after exponents were introduced into mathematical language, mathematicians discovered that they could be used to represent square roots, cube roots, and so on. In this section, you will learn how to use exponents in this way.

Since $x^{12} = (x^6)^2$ by the fourth law of exponents, evidently

$$\sqrt{x^{12}} = x^{\frac{12}{2}}, \quad \text{or } x^6.$$

Similarly, the equation

$$\sqrt{x^n} = x^{\frac{n}{2}}$$

is true for every even integer n and every positive number x.

It seems natural, therefore, to define the $n/2$ power of x to be the square root of x^n for every integer n, even or odd. Thus,

$$x^{\frac{1}{2}} = \sqrt{x}, \quad x^{\frac{3}{2}} = \sqrt{x^3}, \quad x^{\frac{5}{2}} = \sqrt{x^5},$$

and so on. In general, a rational power of x is defined in the following manner.

Definition of rational exponents

The equation

$$x^{\frac{m}{n}} = \sqrt[n]{x^m} \qquad \textbf{(Def-Rat. Exp.)}$$

is true for every integer m, every integer n greater than 1, and every positive real number x.

In particular, if $m = 1$, you have

$$x^{\frac{1}{n}} = \sqrt[n]{x}.$$

If m is a positive integer, then

$$x^m = \overbrace{x \cdot \ldots \cdot x}^{m \text{ factors}},$$

$$\sqrt[n]{x^m} = \sqrt[n]{\overbrace{x \cdot \ldots \cdot x}^{m \text{ factors}}},$$

or

$$\sqrt[n]{x^m} = \overbrace{\sqrt[n]{x} \cdot \ldots \cdot \sqrt[n]{x}}^{m \text{ factors}}, \qquad \text{(LR-1)}$$

$$\sqrt[n]{x^m} = (\sqrt[n]{x})^m.$$

This argument suggests the rational power law.

RATIONAL POWER LAW

The equation

$$x^{\frac{m}{n}} = (x^m)^{\frac{1}{n}} \quad \textit{or} \quad x^{\frac{m}{n}} = (x^{\frac{1}{n}})^m \qquad \textbf{(R-P)}$$

is true for all integers m and n, with n > 1 and every positive real number x. If n is an odd integer and x is a negative number, the equation is still true.

For example, $4^{\frac{3}{2}}$ can be computed in either of the following two ways:

$$4^{\frac{3}{2}} = (4^3)^{\frac{1}{2}} = \sqrt{64} = 8,$$

or

$$4^{\frac{3}{2}} = (4^{\frac{1}{2}})^3 = 2^3 = 8.$$

The expression x^a has been defined for every positive number x and every rational number a. The laws of exponents, which were stated in previous chapters, are true for rational, as well as integral, exponents. For convenience, these laws are restated below.

$$x^a \cdot x^b = x^{a+b} \qquad \text{(LE-1)}$$

$$\frac{x^a}{x^b} = x^{a-b} \qquad \text{(LE-2)}$$

$$(x \cdot y)^a = x^a \cdot y^a \qquad \text{(LE-3)}$$

$$(x^a)^b = x^{a \cdot b} \qquad \text{(LE-4)}$$

$$\left(\frac{x}{y}\right)^a = \frac{x^a}{y^a} \qquad \text{(LE-5)}$$

The above equations are true for all rational numbers a and b and all positive real numbers x and y. For certain rational numbers a and b, the equations are also true for negative values of x or y.

The laws of exponents make it easy for you to multiply or divide roots having different indexes. You can do this by changing all roots to rational exponents, as illustrated below.

Problem. Perform the indicated operation, and express the answer as a single root of a rational number.

(a) $\sqrt[3]{2} \cdot \sqrt{2}$ (b) $\dfrac{\sqrt{7}}{\sqrt[5]{49}}$

Solution. First change each radical to exponential form and then use the laws of exponents as indicated.

$$\text{(a)}\ \sqrt[3]{2} \cdot \sqrt{2} = 2^{\frac{1}{3}} \cdot 2^{\frac{1}{2}}$$

$$= 2^{\frac{1}{3}+\frac{1}{2}} \qquad \text{(LE-1)}$$

$$= 2^{\frac{5}{6}}$$

$$= \sqrt[6]{2^5}, \quad \text{or} \quad \sqrt[6]{32}$$

(b) $\dfrac{\sqrt{7}}{\sqrt[5]{49}} = \dfrac{7^{\frac{1}{2}}}{(7^2)^{\frac{1}{5}}}$

$= \dfrac{7^{\frac{1}{2}}}{7^{\frac{2}{5}}}$

$= 7^{\frac{1}{2}-\frac{2}{5}}$ (LE-2)

$= 7^{\frac{1}{10}}$, or $\sqrt[10]{7}$

Exercises

Use the definition of rational exponents to simplify each of the following expressions.

1. (a) $9^{\frac{1}{2}}$ (b) $16^{\frac{1}{2}}$
2. (a) $8^{\frac{1}{3}}$ (b) $27^{\frac{1}{3}}$
3. (a) $32^{\frac{1}{5}}$ (b) $64^{\frac{1}{6}}$
4. (a) $27^{\frac{2}{3}}$ (b) $100^{\frac{3}{2}}$
5. (a) $81^{\frac{3}{4}}$ (b) $125^{\frac{4}{3}}$
6. (a) $4^{\frac{1}{2}} - 4^{-\frac{1}{2}}$ (b) $49^{-\frac{1}{2}}$
7. (a) $(-64)^{\frac{2}{3}}$ (b) $(-32)^{\frac{3}{5}}$
8. (a) $8^{-\frac{2}{3}}$ (b) $64^{-\frac{2}{3}}$
9. (a) $9^0 \cdot 25^{\frac{1}{2}}$ (b) $(100^{\frac{3}{2}})^0$
10. (a) $(81)^{\frac{1}{4}}(81)^{-\frac{1}{4}}$ (b) $(216)^{\frac{2}{3}}(216)^{-\frac{2}{3}}$
11. (a) $(\frac{49}{81})^{-\frac{3}{2}}$ (b) $(\frac{8}{27})^{-\frac{4}{3}}$
12. (a) $(\frac{1}{64})^{-\frac{1}{3}}$ (b) $(\frac{1}{125})^{-\frac{2}{3}}$

Simplify each of the following expressions.

13. (a) $(36x^2)^{\frac{1}{2}}$ (b) $(64x^4)^{\frac{1}{2}}$
14. (a) $(.001a^3)^{\frac{1}{3}}$ (b) $(.008x^3)^{\frac{1}{3}}$
15. (a) $(4x^4y^6)^{\frac{3}{2}}$ (b) $(27x^6y^9)^{\frac{2}{3}}$
16. (a) $-(16x^4y^8)^{\frac{1}{4}}$ (b) $-(81x^{12}y^{16})^{\frac{1}{4}}$
17. (a) $(25x^6)^{-\frac{1}{2}}$ (b) $(49x^2y^2)^{-\frac{3}{2}}$
18. (a) $(81a^{12}b^{20})^{-\frac{3}{4}}$ (b) $(5^7x^{14}y^{21})^{-\frac{1}{7}}$
19. (a) $11^{\frac{2}{3}} \div 11^{\frac{3}{2}}$ (b) $7^{\frac{3}{4}} \div 7^{\frac{1}{4}}$
20. (a) $(3x^{\frac{1}{3}}y^{\frac{4}{3}})^{-3}$ (b) $(2x^{\frac{1}{4}}y^{\frac{3}{4}})^{-4}$

First express the roots with fractional exponents; then perform the indicated multiplications and divisions and, where possible, simplify.

21. (a) $\sqrt{2} \cdot \sqrt[3]{2}$ (b) $\sqrt[3]{3} \cdot \sqrt{3}$
22. (a) $\sqrt[3]{6} \div \sqrt[4]{2}$ (b) $\sqrt[4]{8} \div \sqrt[3]{2}$

23. (a) $\sqrt{8} \div \sqrt[3]{2}$ (b) $\sqrt{10} \div \sqrt[3]{2}$
24. (a) $\sqrt{2} \cdot \sqrt[3]{4}$ (b) $\sqrt[3]{3} \cdot \sqrt{2}$
25. (a) $\sqrt{x} \cdot \sqrt[5]{x}$ (b) $\sqrt[3]{x} \cdot \sqrt[5]{x}$
26. (a) $\sqrt[4]{x^3} \cdot \sqrt[5]{x^4}$ (b) $\sqrt[3]{x^2} \cdot \sqrt[7]{x^6}$
27. (a) $\sqrt[3]{7} \cdot \sqrt[4]{7}$ (b) $\sqrt[3]{6} \cdot \sqrt[5]{6}$
28. (a) $\sqrt[4]{7^3} \div \sqrt[4]{7}$ (b) $\sqrt[3]{6^2} \div \sqrt[4]{6}$
29. (a) $\sqrt[4]{9} \div \sqrt[6]{27}$ (b) $\sqrt[4]{25} \div \sqrt[6]{125}$
30. (a) $\sqrt{10} \cdot \sqrt[3]{10^2}$ (b) $\sqrt[4]{5} \cdot \sqrt[4]{5^3}$
31. (a) $\sqrt[3]{8x^3} \div \sqrt[3]{y^6}$ (b) $\sqrt[4]{16x^8} \div \sqrt[4]{y^4}$
32. (a) $3\sqrt{x^{\frac{1}{3}}} \div \sqrt[4]{x}$ (b) $6\sqrt{x^{\frac{1}{2}}} \div \sqrt[3]{x}$

Preparation for Section 11-10

Which of the following statements are true? Correct the statements that you marked false.

1. If two numbers are equal, their squares are equal.
2. If two numbers are equal, their square roots are equal.
3. If $\sqrt{x} = 7$, $x = 49$.
4. If $x = 64$, $-\sqrt{x} = -8$.
5. If $-\sqrt{x} = -6$, then $x = 36$.
6. If $x = 36$, then $\sqrt{x} = 6$.
7. If $x = 36$, then $\sqrt{x} = -6$.
8. (a) Is 13 a solution of the equation $\sqrt{x - 4} + 1 = 4$?
(b) Are the equations $x = 13$ and $\sqrt{x - 4} + 1 = 4$ equivalent?
9. Are the equations $x = 2$ and $\sqrt{x^2 + 2x - 7} + 4 = x + 1$ equivalent?
10. Find the solution set of the equation $\sqrt{x} = -1$.

11-10 RADICAL EQUATIONS

Equations containing radical expressions occur frequently in mathematics. For example, you might have seen the following equations. The first gives the radius of a circle with area A, the second the edge of a cube with volume V, and the third the time it takes an object to fall a distance s.

$$r = \sqrt{\frac{A}{\pi}}$$

$$e = \sqrt[3]{V}$$

$$t = \tfrac{1}{4}\sqrt{s}$$

In solving radical equations, you make use of the fact that if two numbers are equal, then their squares, cubes, and, in general, their nth powers are equal. This property might be stated as follows:

If $x = y$, then $x^n = y^n$ for every positive integer n.

Is the converse of this statement true? In other words, if $x^n = y^n$ for some integer n, then does $x = y$? Consider, for example, the true equation

$$(-2)^2 = 2^2.$$

However, -2 is not equal to 2. Thus, the converse of the statement above is not true for all real numbers x and y, but it is true when x and y are positive.

Problem 1. Solve the equation $\sqrt{3x - 2} = 4$.

Solution. If you square the expression on each side of this equation, every solution of the given equation will also be a solution of the new equation. You solve this new equation as follows:

$$(\sqrt{3x - 2})^2 = 4^2,$$
$$3x - 2 = 16,$$
$$3x = 18,$$
$$x = 6.$$

The only solution of this equation is 6, and therefore, the only possible solution of the given equation is 6.

Check.
$$\sqrt{(3 \cdot 6) - 2} \overset{?}{=} 4$$
$$\sqrt{16} \overset{\checkmark}{=} 4$$

Thus, $\{6\}$ is the solution set of the given equation.

Problem 2. Solve the equation $3\sqrt{y} + 1 = 11$.

Solution. If you square the expression on each side of this equation, the resulting equation,

$$(3\sqrt{y} + 1)^2 = 11^2,$$

or

$$9y + 6\sqrt{y} + 1 = 121,$$

is no easier to solve than the given one. However, if you first replace the given equation by the equivalent equation

$$3\sqrt{y} = 10$$

and square the expression on each side of this equation, you may solve the resulting equation as follows:

$$(3\sqrt{y})^2 = 10^2,$$
$$9y = 100,$$
$$y = \tfrac{100}{9}.$$

The only possible solution of the given equation is $\frac{100}{9}$. You should check the solution in the following way.

Check.

$$3\sqrt{\tfrac{100}{9}} + 1 \stackrel{?}{=} 11$$
$$(3 \cdot \tfrac{10}{3}) + 1 \stackrel{?}{=} 11$$
$$10 + 1 \stackrel{\checkmark}{=} 11$$

Exercises

In Exercises 1–18 solve each equation and check each result.

1. (a) $4\sqrt{x} = 20$ (b) $5\sqrt{x} = 30$

2. (a) $\sqrt{x - 7} = 3$ (b) $\sqrt{x - 9} = 5$

3. (a) $\sqrt{x + 4} = 5$ (b) $\sqrt{x + 3} = 2$

4. (a) $\sqrt{2x + 1} = 1$ (b) $\sqrt{2x - 1} = 1$

5. (a) $\sqrt{x} - 9 = 7$ (b) $\sqrt{x} - 10 = 5$

6. (a) $\sqrt{y + 5} = 6$ (b) $\sqrt{x + 3} = 7$

7. (a) $\sqrt{x + 30} = 5$ (b) $\sqrt{x + 40} = 2$

8. (a) $5\sqrt{3b} = 45$ (b) $3\sqrt{5b} = 45$

9. (a) $\sqrt{3x - 2} - 5 = 0$ (b) $\sqrt{2x + 5} - 7 = 0$

10. (a) $\sqrt{a - 3} = 2\sqrt{5}$ (b) $\sqrt{b - 7} = 3\sqrt{2}$

11. (a) $\sqrt{\dfrac{5x}{3}} = 10$ (b) $\sqrt{\dfrac{2x}{3}} = 6$

12. (a) $\sqrt{4x - 5} = 3\sqrt{7}$ (b) $\sqrt{5a - 2} = 2\sqrt{5}$

13. (a) $\frac{1}{6}\sqrt{10m} = \frac{1}{3}\sqrt{20}$ (b) $\frac{1}{4}\sqrt{5x} = \frac{1}{2}\sqrt{10}$

14. (a) $\sqrt{5 - x} = -3$ (b) $\sqrt{6x + 1} = -5$

15. (a) $\sqrt{x} = \dfrac{9}{\sqrt{x}}$ (b) $\sqrt{2x} = \dfrac{3}{\sqrt{2x}}$

16. (a) $\sqrt{x^2 + 21} = x + 3$ (b) $\sqrt{x^2 + 9} = x + 4$

17. (a) $5 + \sqrt{x + 4} = 12$ (b) $12 + \sqrt{x + 4} = 5$

18. (a) $\sqrt[3]{11x + 26} = 5$ (b) $\sqrt[3]{2x + 5} = 2$

In Exercises 19–24, solve each equation and check each result.

19. $\sqrt{x} - \frac{1}{3} = 4$

20. $\frac{3}{5} - \sqrt{3x} = ^{-}\frac{21}{5}$

21. $4\sqrt{x} + 5 = 12$

22. $\frac{1}{3}\sqrt{y} - 8 = 2$

23. $\sqrt{\frac{5x - 1}{6}} = 2$

24. $\sqrt{\frac{8x - 1}{7}} - 1 = 2$

25. Find a number such that the square root of five less than three times the number is seven.

26. If possible, find a number such that the square root of seven more than three times the number is equal to the square root of seven less than the number.

---

27. Find the solution set of each of the following equations, where $x \geqq 0$ and $y \geqq 0$.

(a) $\sqrt{x} + \sqrt{y} = \sqrt{x + y}$
(b) $\sqrt[3]{x} + \sqrt[3]{y} = \sqrt[3]{x + y}$

28. For each replacement of x by a positive real number, which of the three algebraic expressions

$$x + x^{-1}, \quad x^{\frac{1}{2}} + x^{-\frac{1}{2}}, \quad x^{\frac{1}{3}} + x^{-\frac{1}{3}}$$

has the smallest value and which has the largest?

11–11 APPROXIMATIONS

You should recall that each irrational number may be approximated as accurately as you wish by a rational number. In particular, each positive root of a positive number may be approximated by a positive rational number.

If a is a real number and b is an integer such that

$$b < a < b + 1,$$

then either b or $(b + 1)$ is called an *integral approximation* of a. If c is a rational number such that

$$c < a < c + .1,$$

then either c or $(c + .1)$ is called a *one-decimal-place approximation* of a. If d is a rational number such that

$$d < a < d + .01,$$

then either d or $(d + .01)$ is called a *two-decimal-place approximation* of a, and so on.

Problem 1. Find a two-decimal-place approximation of $\sqrt{14}$.

Solution. Since 14 is between the perfect squares 9 and 16, then

$$3 < \sqrt{14} < 4,$$

and either 3 or 4 is an integral approximation of $\sqrt{14}$. To find a one-decimal-place approximation of $\sqrt{14}$, you would consider the squares of the numbers 3.1, 3.2, 3.3, . . . , 3.9. Since 14 is closer to 16 than to 9, it would seem that $\sqrt{14}$ is closer to 3.9 than to 3.1. By trial and error, you obtain

$$3.7^2 = 13.69 \quad \text{and} \quad 3.8^2 = 14.44.$$

Therefore,

$$3.7^2 < 14 < 3.8^2$$

and

$$3.7 < \sqrt{14} < 3.8.$$

Thus, either 3.7, or 3.8 is a one-decimal-place approximation of $\sqrt{14}$. In turn, you find that $3.74^2 = 13.9876$ and $3.75^2 = 14.0625$. Hence,

$$3.74^2 < 14 < 3.75^2,$$
$$3.74 < \sqrt{14} < 3.75.$$

Either 3.74 or 3.75 is a two-decimal-place approximation of $\sqrt{14}$.

Problem 2. Find a one-decimal-place approximation of $\sqrt[3]{14}$.

Solution. Since $2^3 < 14 < 3^3$, you have

$$2 < \sqrt[3]{14} < 3.$$

In turn,

$$(2.4)^3 = 13.824 \quad \text{and} \quad (2.5)^3 = 15.625,$$

and therefore,

$$(2.4)^3 < 14 < (2.5)^3.$$

Thus,

$$2.4 < \sqrt[3]{14} < 2.5.$$

Either 2.4 or 2.5 is a one-decimal-place approximation of $\sqrt[3]{14}$.

For convenience, the symbol

$$\doteq$$

is used to denote *is approximately equal to.* Then the statement

$$\pi \doteq 3.14$$

means that 3.14 is a two-decimal-place approximation of π, and

$$\pi \doteq 3.141$$

means that 3.141 is a three-decimal-place approximation of π.

For many centuries, mathematicians have been fascinated by the problem of finding rational approximations of π. That π is actually irrational was first proved by the German mathematician Lambert in 1767. In 1873, an Englishman named Shanks gave a 707-decimal-place approximation of π after 15 years of work. Later, it was discovered that his approximation was correct to only 528 places. In 1949, a 2037-decimal-place approximation of π was found after 70 hours of work on an ENIAC computer. More recently, a 100,000-decimal-place approximation of π was found in 9 hours on an IBM 7090 computer. The methods used to compute π are too advanced to be given here. A 40-decimal-place approximation of π follows:

$$3.1415926535897932384626433832795028841971.$$

A table of three-decimal-place approximations of square roots of integers is given in the Appendix. Although this table is only for integers between 1 and 100, it may be used for other numbers which can be factored as products of integers less than 100.

For example,

$$\sqrt{300} = 10\sqrt{3}$$

and therefore,

$$\sqrt{300} \doteq 10 \times 1.732, \quad \text{or } 17.32,$$

As another example,

$$\begin{aligned} \sqrt{497} &= \sqrt{7} \times \sqrt{71} \\ &\doteq 2.646 \times 8.426 \\ &\doteq 22.295. \end{aligned}$$

You will note that the answer has been "rounded off" to three decimal places. Thus, although

$$2.646 \times 8.426 = 22.295196,$$

you cannot possibly expect the approximation of $\sqrt{497}$ to be more accurate than those of $\sqrt{7}$ and $\sqrt{71}$, which were approximations to three decimal places.

You may use the Table of Square Roots in working problems like the following.

Problem 3. Approximate

$$\frac{6}{\sqrt{7} - \sqrt{3}}.$$

Solution. You could proceed by finding $\sqrt{7} - \sqrt{3} \doteq 2.646 - 1.732$, or .914, and then computing $6 \div .914$. However, another approach is to first rationalize the denominator and then proceed as follows:

$$\begin{aligned}\frac{6}{\sqrt{7} - \sqrt{3}} &= \frac{6}{\sqrt{7} - \sqrt{3}} \cdot \frac{\sqrt{7} + \sqrt{3}}{\sqrt{7} + \sqrt{3}} \\ &= \frac{6(\sqrt{7} + \sqrt{3})}{(\sqrt{7})^2 - (\sqrt{3})^2} \\ &= \frac{6(\sqrt{7} + \sqrt{3})}{7 - 3}, \quad \text{or} \quad \frac{3}{2}(\sqrt{7} + \sqrt{3}).\end{aligned}$$

Then

$$\begin{aligned}\frac{6}{\sqrt{7} - \sqrt{3}} &\doteq \frac{3}{2}(2.646 + 1.732) \\ &\doteq \frac{3}{2}(4.378), \quad \text{or} \quad 6.567.\end{aligned}$$

Exercises

In Exercises 1–8 find a three-decimal-place approximation of each expression.

1. (a) $\sqrt{15}$ (b) $\sqrt{21}$

2. (a) $\sqrt{60}$ (b) $\sqrt{84}$

3. (a) $\frac{1}{\sqrt{15}}$ (b) $\frac{1}{\sqrt{21}}$

4. (a) $\sqrt{312}$ (b) $\sqrt{138}$

5. (a) $\sqrt{549}$ (b) $\sqrt{546}$

6. (a) $\dfrac{1}{\sqrt{5}}$ (b) $\dfrac{1}{\sqrt{3}}$

7. (a) $\dfrac{2}{\sqrt{6}}$ (b) $\dfrac{3}{\sqrt{10}}$

8. (a) $\sqrt{.6}$ (b) $\sqrt{.7}$

Find a one-decimal-place approximation of each of the following.

9. (a) $\sqrt[3]{23}$ (b) $\sqrt[3]{34}$

10. (a) $\sqrt[3]{2}$ (b) $\sqrt[3]{3}$

Find a three-decimal-place approximation of each of the following.

11. (a) $\dfrac{1}{\sqrt{5}-\sqrt{3}}$ (b) $\dfrac{1}{\sqrt{11}-\sqrt{5}}$

12. (a) $\dfrac{6}{\sqrt{7}+\sqrt{2}}$ (b) $\dfrac{7}{\sqrt{6}+\sqrt{3}}$

13. (a) $\sqrt{738}$ (b) $\sqrt{325}$

14. (a) $\dfrac{1}{2\sqrt{2}-3}$ (b) $\dfrac{1}{3\sqrt{3}-2}$

15. (a) $\dfrac{4}{5-3\sqrt{3}}$ (b) $\dfrac{5}{4-2\sqrt{2}}$

16. (a) $\dfrac{3}{\sqrt{12}+\sqrt{24}}$ (b) $\dfrac{4}{\sqrt{7}+\sqrt{14}}$

17. (a) $\dfrac{2}{5-5\sqrt{5}}$ (b) $\dfrac{3}{2-2\sqrt{2}}$

18. (a) $\dfrac{4}{2\sqrt{3}+3\sqrt{2}}$ (b) $\dfrac{3}{4\sqrt{2}+3\sqrt{3}}$

---

19. If the positive rational number r is an approximation of $\sqrt{2}$, show that $\dfrac{2+r}{1+r}$ is always a better approximation.

20. If the positive rational number r is an approximation of $\sqrt{3}$, show how to find a better approximation. (Hint: See Exercise 19.)

21. Show that the two numbers

$$\sqrt{100.002}-10 \quad \text{and} \quad .002 \div (\sqrt{100.002}+10)$$

are equal. Then obtain a four-decimal-place approximation for the number $\sqrt{100.002}-10$.

11–12 LINEAR INTERPOLATION

Linear interpolation is a useful method for approximating roots of numbers and also values of other functions. This method is illustrated below in finding a rational approximation of $\sqrt{45}$.

One way of approximating $\sqrt{45}$ is to graph the set

$$S = \{(x, y) \mid y = x^2, x \geqq 0\}.$$

This is done in Fig. 11–2 by plotting the points

$$(0, 0),\ (1, 1),\ (2, 4),\ (5, 25),\ (6, 36),\ (7, 49),\ (8, 64)$$

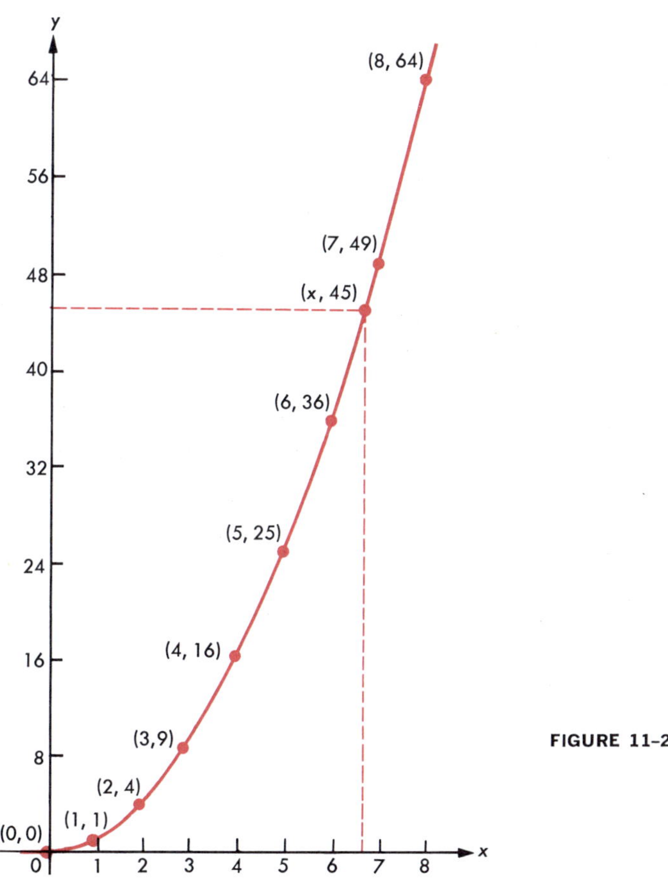

FIGURE 11–2

and then joining them by a smooth curve. You can find an approximation of $\sqrt{45}$ by drawing the line $y = 45$ and finding the approximate x-coordinate of the point at which it crosses the graph of S. Since the exact x-coordinate of this point is $\sqrt{45}$, the number that you find is only an approximation of $\sqrt{45}$.

There is an indirect way of using the graph of S to approximate $\sqrt{45}$. Note that in Fig. 11–2, the graph of S appears to be almost a straight line segment between the points (5, 25) and (6, 36), between the points (6, 36) and (7, 49), and so on. Therefore, the straight-line segment joining the points (6, 36) and (7, 49) is an approximation of the graph of the set S. This line segment is shown in Fig. 11–3. You should observe that for this line,

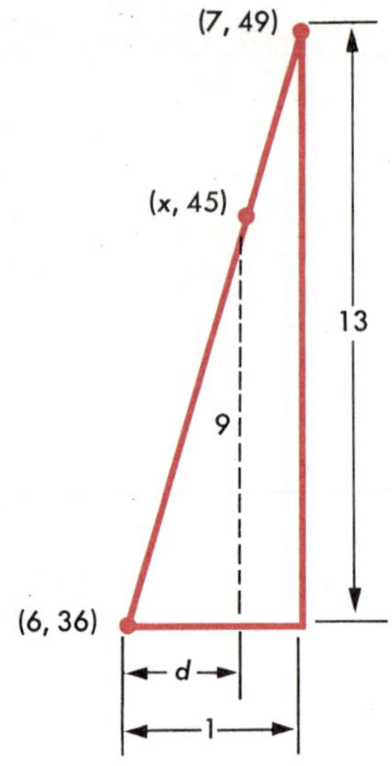

FIGURE 11-3

$$\frac{\text{rise}}{\text{run}} = \frac{13}{1},$$

and hence the slope of the line is 13.

The point on this segment with y-coordinate 45 has an x-coordinate approximately equal to $\sqrt{45}$. You can find the x-coordinate of this point with very little trouble. The rise from the point (6, 36) to the point $(x, 45)$ is 9. If you let d designate the run between these points, then $d = x - 6$.

$$\frac{9}{d} = \frac{13}{1}$$

You can solve this equation in the following way.

$$9 = 13d$$
$$d = \tfrac{9}{13}, \quad \text{or approximately .7}$$

Thus, the x-coordinate of the point $(x, 45)$ is approximately .7 more than that of (6, 36). Therefore, its x-coordinate is approximately 6.7, and

$$\sqrt{45} \doteq 6.7.$$

You may use the method of linear interpolation without drawing graphs. This fact is illustrated below by approximating $\sqrt{177}$.

First, find out which two perfect squares 177 lies between. Since

$$13^2 = 169 \quad \text{and} \quad 14^2 = 196,$$

you see that

$$13^2 < 177 < 14^2.$$

The straight line joining the points (13, 169) and (14, 196) has a slope of

$$\frac{196 - 169}{14 - 13}, \quad \text{or } \frac{27}{1}.$$

The line joining (13, 169) to (x, 177) has a rise of 8. If you designate its run by d, then $d = x - 13$.

$$\frac{8}{d} = \frac{27}{1}$$

$$8 = 27d$$

$$d = \frac{8}{27}, \quad \text{or approximately .3}$$

Therefore, $x = 13 + d$ is approximately $13 + .3$, or

$$x = \sqrt{177}$$
$$\doteq 13.3.$$

Exercises

Use the method of linear interpolation to find one-decimal-place approximations of each of the following square roots.

1. (a) $\sqrt{54}$ (b) $\sqrt{63}$

2. (a) $\sqrt{70}$ (b) $\sqrt{110}$

3. (a) $\sqrt{74}$ (b) $\sqrt{46}$

4. (a) $\sqrt{30}$ (b) $\sqrt{42}$

5. (a) $\sqrt{34}$ (b) $\sqrt{51}$

6. (a) $\sqrt{420}$ (b) $\sqrt{600}$

7. (a) $\sqrt{85}$ (b) $\sqrt{95}$

8. (a) $\sqrt{68}$ (b) $\sqrt{52}$

---

9. Use $x = 0, \frac{1}{2}, 1, \frac{3}{2}, 3, \frac{7}{2}, 4$ to sketch the graph of the set

$$C = \{(x, y) \mid y = x^3, x \geqq 0\}.$$

Use linear interpolation to find one-decimal-place approximations of each of the following cube roots.

10. $\sqrt[3]{25}$ **11.** $\sqrt[3]{30}$ **12.** $\sqrt[3]{40}$

13. $\sqrt[3]{70}$ **14.** $\sqrt[3]{80}$ **15.** $\sqrt[3]{100}$

---

16. The *method of errors* is a more effective way of approximating square roots than linear interpolation. For example, you can approximate $\sqrt{177}$ by the method of errors as follows. Take 13 as a first approximation, and let e be the error of this approximation. Thus, $13 + e = \sqrt{177}$, $(13 + e)^2 = 177$, and $26e + e^2 = 8$. If 13 is a close enough approximation of $\sqrt{177}$, then you can neglect e^2 in the equation above and obtain $26e \doteq 8$, $e = \frac{8}{26}$, or $e \doteq .308$. Consequently, $\sqrt{177} \doteq 13.308$. Since $\sqrt{177} = 13.3041\ldots$, this approximation is off by about .004. Start with 12 as your first approximation, and use the method of errors to approximate $\sqrt{151}$ to three decimal places.

In Exercises 17–20, use the method of errors to find three-decimal-place approximations for each of the numbers.

17. $\sqrt{45}$ **18.** $\sqrt{70}$ **19.** $\sqrt{230}$ **20.** $\sqrt{907}$

21. The method of errors may be used to approximate cube roots of numbers. Show how you think it might be used to approximate $\sqrt[3]{66}$.

11–13 PYTHAGOREAN THEOREM

Squares and square roots occur in problems involving right triangles, that is, triangles having a right angle. The two sides that form the right angle are called *legs;* the third and longest side is called the *hypotenuse* of the right triangle.

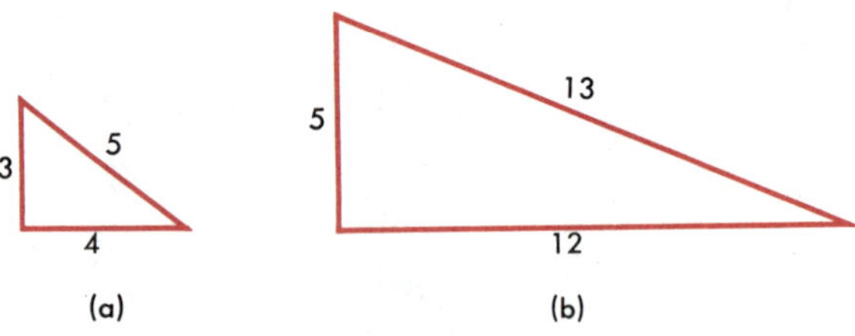

FIGURE 11–4

It has been known for thousands of years that a triangle with sides measuring 3, 4, and 5 units is a right triangle. Similarly, a triangle with sides measuring 5, 12, and 13 units is a right triangle. See Fig.

11–4(a) and 11–4(b). Note that the numbers 3, 4, and 5 are related by the equation

$$3^2 + 4^2 = 5^2,$$

and that the numbers 5, 12, and 13 are related by the similar equation

$$5^2 + 12^2 = 13^2.$$

The problems above illustrate the following theorem. See Fig. 11–5.

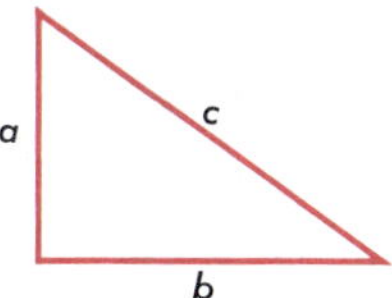

FIGURE 11-5

PYTHAGOREAN THEOREM

If a and b are the lengths of the legs and c is the length of the hypotenuse of a right triangle, then

$$a^2 + b^2 = c^2.$$

Credit for the first proof of this theorem is given to the Greek mathematician Pythagoras, who lived about 500 B.C. However, there is historical evidence that the Babylonians were aware of this property of right triangles at least a thousand years before the time of Pythagoras.

Problem 1. Find the length of the hypotenuse of a right triangle, if the legs have lengths 14 inches and 48 inches, respectively.

Solution. Proceed as follows:

$$\begin{aligned} c^2 &= a^2 + b^2, \\ c^2 &= 14^2 + 48^2, \\ c^2 &= 196 + 2304, \\ c^2 &= 2500, \\ c &= 50. \end{aligned}$$

Therefore, the hypotenuse has a length of 50 inches.

Problem 2. If the hypotenuse of a right triangle is 18 inches long and if one of the legs is 14 inches long, how long is the other leg?

Solution. Proceed as follows:

$$c^2 = a^2 + b^2,$$
$$18^2 = a^2 + 14^2,$$
$$324 = a^2 + 196,$$
$$a^2 = 128,$$
$$a = \sqrt{128}, \quad \text{or } 8\sqrt{2},$$
$$a \doteq 8 \times 1.414,$$
$$a \doteq 11.3.$$

Thus, the other leg is exactly $8\sqrt{2}$, or approximately 11.3 inches long.

The converse of the pythagorean theorem is also true.

CONVERSE OF THE PYTHAGOREAN THEOREM

If a, b, and c are the lengths of the sides of a triangle and if

$$a^2 + b^2 = c^2,$$

then the triangle is a right triangle with right angle opposite the side of length c.

Problem 3. Is a triangle with sides of lengths 20, 21, and 29 a right triangle?

Solution. It is a right triangle only if 29 is the length of the hypotenuse and

$$20^2 + 21^2 = 29^2.$$

Since

$$20^2 = 400, \quad 21^2 = 441, \quad 29^2 = 841,$$

and

$$400 + 441 = 841,$$

it is a right triangle.

Exercises

In the exercises below, a and b stand for the lengths of the legs of a right triangle, and c stands for the length of the hypotenuse. Find the missing length for each right triangle. Give a one-decimal-place approximation of each irrational number.

1. (a) $a = 6, b = 8$ (b) $a = 10, b = 24$

2. (a) $a = 18, b = 24$ (b) $a = 35, b = 84$

3. (a) $a = 36, c = 39$ (b) $a = 15, c = 25$

4. (a) $a = 7, b = 7$ (b) $a = 10, b = 10$

5. (a) $b = 8, c = 16$ (b) $a = 8, c = 9$

6. (a) $b = 9, c = 9\sqrt{2}$ (b) $b = 6, c = 6\sqrt{3}$

Test each of the following number triples to see which has three numbers that might be used as the lengths of the sides of a right triangle; that is, which triple (in some order) is a solution of $a^2 + b^2 = c^2$?

7. (a) 21, 72, 75 (b) 72, 54, 90

8. (a) 20, 30, 40 (b) 40, 96, 104

9. (a) 42, 144, 150 (b) 180, 112, 106

10. (a) 14, 48, 52 (b) 7.5, 10, 12.5

Make a sketch of each of the following problems. Give a one-decimal-place approximation of each irrational answer.

11. (a) A wire from the top of a telephone pole to a point on the ground 20 feet from the pole is 40 feet long. How high is the pole?
(b) If the bottom of a 10-foot ladder is 6 feet from a wall, how high on the wall does it reach?

12. (a) A rectangular lot is 50 feet long and 30 feet wide. How long is a straight line from one corner to the corner diagonally opposite?
(b) A square lot has one side 25 feet long. How long is a straight line from one corner to the corner diagonally opposite?

In about 1900 B.C., the Babylonians inscribed a tablet with many of the following number triples, supposedly solutions of $a^2 + b^2 = c^2$. Errors have been found in some of their work. Test the triples to see which are right and which are wrong.

13. 120, 119, 169 **14.** 600, 541, 769

15. 72, 65, 97 **16.** 56, 90, 53

17. The side of a square is x. Use the pythagorean theorem to show that the length of the diagonal of a square is always $x\sqrt{2}$.

18. The side of an equilateral triangle is x. Use the pythagorean theorem to show that the height is $\frac{x}{2}\sqrt{3}$.

11–14 PROOF IN ALGEBRA IV

For each positive integer n, every positive real number x has a unique positive nth root $\sqrt[n]{x}$, *(U-Pos. Rad.)*. If n is odd, then every negative real number x has a unique negative nth root which is again denoted by $\sqrt[n]{x}$. In this section, some of the properties of nth roots will be proved.

Problem 1. Prove the second law of radicals:

$$\sqrt[n]{\frac{x}{y}} = \frac{\sqrt[n]{x}}{\sqrt[n]{y}}. \qquad \text{(LR-2)}$$

Solution. Restrict x and y to nonzero numbers having nth roots. Let

$$a = \sqrt[n]{x}, \quad b = \sqrt[n]{y}.$$

Then

$$a^n = x, \quad b^n = y, \qquad \text{(Def-}n\text{th Rt.)}$$

$$\frac{a^n}{b^n} = \frac{x}{y},$$

$$\left(\frac{a}{b}\right)^n = \frac{x}{y}, \qquad \text{(LE-5)}$$

$$\frac{a}{b} = \sqrt[n]{\frac{x}{y}}, \qquad \text{(Def-}n\text{th Rt.)}$$

$$\frac{\sqrt[n]{x}}{\sqrt[n]{y}} = \sqrt[n]{\frac{x}{y}}.$$

This proves the second law of radicals.

You recall that for every integer m and every integer $n > 1$, the m/nth power of a number is defined in the following way.

$$x^{\frac{m}{n}} = \sqrt[n]{x^m} \qquad \text{(Def-Rat. Exp.)}$$

The rational number m/n is assumed to be in lowest form. This definition is valid for all nonzero numbers having an nth root.

Problem 2. Prove that $\sqrt[n]{x^m} = (\sqrt[n]{x})^m$.

Solution. Let

$$a = \sqrt[n]{x}.$$

Then

$$\begin{aligned} a^n &= x, && \text{(Def-}n\text{th Rt.)} \\ a^{nm} &= x^m, && \\ (a^m)^n &= x^m, && \text{(LE-4)} \\ a^m &= \sqrt[n]{x^m}, && \text{(Def-}n\text{th Rt.)} \\ (\sqrt[n]{x})^m &= \sqrt[n]{x^m}. && \end{aligned}$$

This is what you wished to prove.

Exercises

1. Assume that the laws of exponents are true for integral exponents. Then prove each of the following.

   (a) $x^{\frac{m}{n}} \cdot y^{\frac{m}{n}} = (x \cdot y)^{\frac{m}{n}}$ (b) $\dfrac{x^{\frac{m}{n}}}{y^{\frac{m}{n}}} = \left(\dfrac{x}{y}\right)^{\frac{m}{n}}$

2. Prove $\sqrt[n]{xy} = \sqrt[n]{x}\sqrt[n]{y}$.

3. Prove (LR-1),

   $$\sqrt[n]{xy} = \sqrt[n]{x}\sqrt[n]{y} \quad \text{for } x > 0 \text{ and } y > 0,$$

   by showing that the nth power of

   $$\sqrt[n]{x} \cdot \sqrt[n]{y}$$

   is equal to xy.

4. What is the product of $\sqrt[n]{x}$ and $\sqrt[m]{x}$?

5. Prove that $\sqrt{x^2 + y^2} \neq x + y$ for any positive numbers x and y.

6. Prove

   $$\frac{a}{b} = 0$$

   if, and only if, $a = 0$ and $b \neq 0$.

KEY IDEAS AND KEY WORDS

Every positive real number has two **square roots,** one of which is the negative of the other. The positive square root of a positive number b is denoted by $\sqrt{b}$; the negative square root of b is denoted by $-\sqrt{b}$. By definition,

$$(\sqrt{b})^2 = b.$$

Every positive real number b has a unique **positive nth root,** denoted by $\sqrt[n]{b}$, for every integer $n > 1$. By definition, $\sqrt[n]{b}$ is a number such that its nth power is b:

$$(\sqrt[n]{b})^n = b.$$

A negative number b has a unique negative nth root, again denoted by $\sqrt[n]{b}$, if n is an odd integer.

A positive integer which has no perfect-square factor greater than 1 is called a **square-free integer.** A square root of a positive integer can always be expressed in **simplest form** as the product of an integer and the square root of a square-free integer. For example, $\sqrt{12}$ has simplest form $2\sqrt{3}$.

The first two **laws of radicals** are as follows:

$$\sqrt[n]{x \cdot y} = \sqrt[n]{x} \cdot \sqrt[n]{y}, \qquad \text{(LR-1)}$$

$$\sqrt[n]{\frac{x}{y}} = \frac{\sqrt[n]{x}}{\sqrt[n]{y}}. \qquad \text{(LR-2)}$$

They are true for all positive real numbers x and y if n is an even integer, and all nonzero real numbers x and y if n is an odd integer.

An algebraic expression involving roots of other algebraic expressions is called a **radical expression.**

Rational exponents are defined by

$$x^{\frac{m}{n}} = \sqrt[n]{x^m}$$

for every integer m, every integer $n > 1$, and every positive real number x. The five laws of exponents are true for rational exponents as well as integral exponents.

The **rational power law** states that the equations

$$x^{\frac{m}{n}} = (x^m)^{\frac{1}{n}} \quad \text{and} \quad x^{\frac{m}{n}} = (x^{\frac{1}{n}})^m$$

are true for all integers m and n, with $n > 1$, and every positive real number x. If n is an odd integer and x is a negative number, the equations are still true.

The symbol

$$\doteq$$

means **approximately equal to.**

For example, $\sqrt{2} \doteq 1.414$. Therefore 1.414 is called a three-decimal-place approximation of $\sqrt{2}$.

The **pythagorean theorem** states that if a and b are the lengths of the legs and c is the length of the hypotenuse of a right triangle, then

$$a^2 + b^2 = c^2.$$

CHAPTER REVIEW

1. Express some irrational number between 3 and 4 as

(a) an infinite nonrepeating decimal.

(b) a root of a rational number.

2. Find the decimal representation of each of the following rational numbers.

(a) $-(\frac{8}{37})$ (b) $\frac{22}{23}$

3. Express the rational number denoted by each of the following radical expressions.

(a) $-.\underline{4}$ (b) $-.9\underline{6}$ (c) $.\underline{54321}$

Name the rational number denoted by each of the following radical expressions.

4. $\sqrt[3]{125}$ **5.** $\sqrt[4]{16}$ **6.** $\sqrt{10^6}$ **7.** $\sqrt{\frac{16}{25}}$ **8.** $\sqrt{.36}$

List the elements of each set.

9. $\{x \mid x^2 = 9\}$ **10.** $\{x \mid x^3 = -8\}$

Find the simplest form of each of the following roots.

11. $\sqrt[3]{12} \cdot \sqrt[3]{10}$ **12.** $\sqrt{356}$ **13.** $\sqrt{960}$

14. $\sqrt{405}$ **15.** $\sqrt[5]{-2430}$ **16.** $\sqrt[3]{-500}$

17. $\sqrt[4]{240}$ **18.** $-2\sqrt{19} \cdot 3\sqrt{19}$ **19.** $\sqrt{\frac{7}{9}} \cdot \sqrt{\frac{9}{7}}$

20. $\sqrt[5]{-\frac{2}{27}}$ **21.** $\sqrt[4]{\frac{3}{8}}$ **22.** $\sqrt{\frac{5}{7}}$

23. $\sqrt[3]{\frac{2}{5}}$ **24.** $\sqrt{\frac{45}{8}}$ **25.** $\sqrt{\frac{20}{3}}$

26. $\sqrt{\frac{18}{14}}$ **27.** $\sqrt{\frac{54}{30}}$ **28.** $\sqrt{\frac{300}{6}}$

29. $\sqrt[3]{40} \div \sqrt[3]{5}$ **30.** $\sqrt{15^{10}} \div \sqrt{15^{11}}$ **31.** $\sqrt[3]{-2} \div \sqrt[3]{-6}$

32. $\sqrt{289 - 225}$ **33.** $\sqrt{289} - \sqrt{225}$ **34.** $\sqrt{289 \cdot 225}$

35. $2\sqrt{6} - \sqrt{24} + 3\sqrt{54}$

36. $3\sqrt{12} - \sqrt{75} + 2\sqrt{3} - \sqrt{\frac{1}{3}}$

37. $3\sqrt{8} - \sqrt{50} - 2\sqrt{48}$

38. $\sqrt[3]{\frac{1}{3}} - \sqrt[3]{2}$

39. $\sqrt[3]{\frac{1}{2}} + \sqrt[3]{\frac{1}{3}}$

Simplify each of the following radical expressions. The domain of each variable is the set of positive real numbers.

40. $-\sqrt{.25x^4y^5}$ **41.** $\sqrt{49a^2b^6}$

42. $\sqrt{245x^3y^3}$ **43.** $\sqrt{1.44p^{16}q^{36}}$

44. $\sqrt[3]{\frac{125}{27}x^{12}y^{15}}$ **45.** $-\sqrt[4]{625c^{16}}$

46. $\sqrt{\dfrac{2}{x}}$ **47.** $\sqrt{\dfrac{4y^2}{x^4}}$

48. $\frac{1}{5}\sqrt{x^3y^3} \cdot \frac{5}{7}\sqrt{xy}$

49. $\sqrt{14x} \div \sqrt{2x^5}$

50. $(-\sqrt{5xy})(-\sqrt[6]{x^5y^7})$

51. $\sqrt{8xy} \div \sqrt{18x^3y^5}$

52. $(.16)^{\frac{1}{2}}$

53. $27^{-\frac{2}{3}}$

54. $-\sqrt[6]{64}$

55. $(\sqrt{\frac{7}{19}})^2$

56. $(\frac{3}{5})^{-2}$

57. $(\frac{4}{49})^{-\frac{1}{2}}$

58. $\sqrt{\frac{7}{30}} \div \sqrt{\frac{2}{5}}$

59. $\sqrt{\frac{2}{3}} \div \sqrt{\frac{4}{15}}$

60. $(-2\sqrt[3]{50})(\sqrt[3]{-105})$

61. $-3\sqrt{5}(2\sqrt{15} - 4\sqrt{5})$

62. $(3\sqrt{13} - 2\sqrt{7})(3\sqrt{13} - 2\sqrt{7})$

63. $(2\sqrt{17} - 3\sqrt{7})^2$

64. $(2\sqrt{5} - \sqrt{3})(3\sqrt{5} + \sqrt{3})$

65. $\dfrac{10}{3 + \sqrt{2}}$

66. $\dfrac{20 + \sqrt{5}}{3 - \sqrt{5}}$

67. $\dfrac{8}{\sqrt{7} - \sqrt{6}}$

68. $\dfrac{4}{3\sqrt{6} - 4\sqrt{3}}$

69. $\dfrac{\sqrt{5} + \sqrt{7}}{2\sqrt{5} + \sqrt{7}}$

70. $64^{\frac{2}{3}} - 64^0$

71. $7^{\frac{2}{3}} \cdot 7^{\frac{1}{2}}$

72. $10^{\frac{4}{5}} \div 10^{\frac{1}{2}}$

73. $(-3x^{\frac{1}{2}}y^{\frac{3}{4}})^2$

74. $\sqrt{2x} \times \sqrt{3x^2} \times \sqrt{6x}$

75. $\sqrt{5xy} \times \sqrt{10y} \times \sqrt{2x}$

76. $\dfrac{6\sqrt{12} - 4\sqrt{3}}{\sqrt{3}}$

77. $\dfrac{5\sqrt{7} - 2\sqrt{14}}{\sqrt{2}}$

78. $\sqrt{\dfrac{13x}{7y}}$

79. $\sqrt{\dfrac{15x^3}{2y}}$

80. $\sqrt{\dfrac{20x^5}{3y^3}}$

81. $\dfrac{(\sqrt{ab} - 1)(\sqrt{ab} + 1)}{\sqrt{ab}}$

82. $\sqrt{7} \cdot \sqrt[3]{7} \cdot \sqrt[4]{7}$

In Exercises 83–88, solve each equation. Check to see if the final equation is equivalent to the given one.

83. $3\sqrt{x} = 15$

84. $4\sqrt{x} - 52 = 0$

85. $\sqrt{x - 7} + 10 = 21$

86. $\sqrt{x + 5} = \sqrt{2x + 5}$

87. $\sqrt{x^2 + 6x + 1} + 4 = x + 3$

88. $\sqrt{x^2 + 3} + x = 3$

89. Use linear interpolation to find a one-decimal-place approximation for each of the following.

(a) $\sqrt{11}$ (b) $\sqrt{33}$ (c) $\sqrt{110}$

90. (a) Find the length of the diagonal of a rectangle if the legs have lengths 16 inches and 30 inches.

(b) Find the missing side of a right triangle if the hypotenuse has length 10 inches and one side has length 5 inches.

CHAPTER TEST

Find the simplest form of each of the following.

1. $\sqrt{180}$
2. $\sqrt{.81}$
3. $\sqrt{\frac{3}{13}}$
4. $(-4\sqrt{42})(-2\sqrt{63})$
5. $\sqrt{\frac{54}{10}}$
6. $\dfrac{3}{\sqrt{10}}$
7. $\dfrac{5}{\sqrt{125}}$
8. $\sqrt{150} - 3\sqrt{24} + 4\sqrt{54}$
9. $\sqrt{27} - \sqrt{9} + \sqrt{81}$
10. $\sqrt{\frac{3}{5}} - \sqrt{\frac{5}{3}} - \sqrt{60}$

Simplify each of the following radical expressions. The domain of each variable is the set of positive real numbers.

11. $\sqrt{x^4y^5}$
12. $\sqrt{.9x^2}$
13. $\sqrt[3]{\frac{2}{9}}$
14. $(-5\sqrt{56})(-3\sqrt{182})$
15. $\sqrt{\frac{5}{13}} \div \sqrt{\frac{15}{52}}$
16. $\sqrt[3]{9} \cdot \sqrt[3]{3}$
17. $\sqrt{6x^3y^5} \cdot \sqrt{10x^7y^7}$
18. $\dfrac{2}{\sqrt{2} + \sqrt{3}}$
19. $\dfrac{\sqrt{15} - \sqrt{5}}{\sqrt{15} + \sqrt{5}}$
20. $(2\sqrt{3} + 5)(3\sqrt{3} - 8)$

Write each of the following with positive integral exponent or radical signs, whenever necessary.

21. $100^{-\frac{1}{2}}$
22. $\left(\dfrac{x^2}{y^4}\right)^{-\frac{3}{2}}$
23. $\dfrac{(x^{-2})(x^{-3})}{x^{-5}}$
24. $\dfrac{1}{(x)^{-\frac{1}{2}}}$
25. $81^{-\frac{3}{4}}$
26. $\dfrac{1}{(16)^{-\frac{1}{2}}}$
27. $4^{-\frac{2}{3}}$

Find the solution set for each of the equations in Exercises 28 and 29. Check your answers.

28. $5\sqrt{2x} - 3 = 17$
29. $\sqrt{4y^2 + 5y} - 2y = 1$
30. Use linear interpolation to find a one-decimal-place approximation of $\sqrt{13}$.
31. If the hypotenuse of a right triangle is 125 inches long and the shorter leg is 35 inches, how long is the other leg?

CHAPTER 12

Quadratic Equations

Objectives . . .

- To solve quadratic equations by factoring or by the quadratic formula.
- To solve word problems that lead to quadratic equations.
- To evaluate the discriminant of a quadratic equation.
- To graph parabolas by plotting ordered pairs of a quadratic function.

12–1 SOLVING BY FACTORING

Second-degree polynomial equations, such as

$$3x^2 - 5x + 7 = 0,$$
$$x^2 + 2\sqrt{2}x = 10,$$

are commonly called quadratic equations. Because quadratic equations occur frequently in mathematics and in applications of mathematics to the sciences, it is worthwhile to study these equations in detail. In Chapter 5, you learned how to solve some kinds of quadratic equations. In this chapter, you shall learn to solve any quadratic equation which has real solutions.

Every quadratic equation is equivalent to an equation having a second-degree polynomial on one side of the equals sign and zero on the other side. For example, the quadratic equation

$$x^2 = 8x - 15$$

is equivalent to the equation

$$x^2 - 8x + 15 = 0.$$

This particular equation contains a factorable polynomial; therefore, it is equivalent to the equation

$$(x - 3)(x - 5) = 0.$$

When written in the form above, a quadratic equation may be solved by using the property of the real number system which was discussed in Chapter 5, Section 5–13.

$$a \cdot b = 0 \quad \textit{if, and only if,} \quad a = 0 \quad \textit{or} \quad b = 0 \qquad (F\text{-}0)$$

The property of factors of zero means that a product of two numbers is zero if, and only if, at least one of the numbers is zero.

Problem 1. Solve the equation $x^2 = 8x - 15$.

Solution. This equation is equivalent to the given equation by the methods of factoring shown above.

$$(x - 3)(x - 5) = 0$$

However, by factors of zero,

$$(x - 3)(x - 5) = 0$$

if, and only if,

$$x - 3 = 0 \quad \text{or} \quad x - 5 = 0.$$

Then the solution set of the given equation is

$$\begin{aligned} \{x \mid (x - 3)(x - 5) = 0\} &= \{x \mid x - 3 = 0\} \cup \{x \mid x - 5 = 0\} \\ &= \{3\} \cup \{5\} \\ &= \{3, 5\}. \end{aligned}$$

Check.

$$3^2 \stackrel{?}{=} (8 \cdot 3) - 15 \qquad 5^2 \stackrel{?}{=} (8 \cdot 5) - 15$$
$$9 \stackrel{\checkmark}{=} 24 - 15 \qquad 25 \stackrel{\checkmark}{=} 40 - 15$$

Problem 2. Solve the equation $3x^2 - 14 = x$.

Solution. Each of the following equations is equivalent to the given one.

$$\begin{aligned} 3x^2 - x - 14 &= 0 \\ (3x - 7)(x + 2) &= 0 \end{aligned}$$

Thus, the solution set, S, of this equation is

$$S = \{x \mid 3x - 7 = 0\} \cup \{x \mid x + 2 = 0\},$$

or $S = \{\frac{7}{3}, -2\}$.

Exercises

Solve each of the following quadratic equations. Check each solution.

1. (a) $(2x + 1)(3x + 7) = 0$ (b) $(x + \sqrt{6})(x - \sqrt{6}) = 0$
2. (a) $(4x + 5)^2 = 0$ (b) $(2x - 3)^2 = 0$
3. (a) $x(x - 2\sqrt{2}) = 0$ (b) $x(x + 3\sqrt{2}) = 0$

4. (a) $(x + 5)(x + 1) = 0$ (b) $(x + 5)(x + 3) = 0$
5. (a) $(x + 5)(x + 1) = -3$ (b) $(x + 5)(x + 3) = -1$
6. (a) $c^2 - 11c + 18 = 0$ (b) $a^2 - 9a + 20 = 0$
7. (a) $x^2 - 4x + 4 = 0$ (b) $x^2 - 6x + 9 = 0$
8. (a) $x^2 + x - 6 = 0$ (b) $x^2 + 2x - 8 = 0$
9. (a) $x^2 + 2x - 63 = 0$ (b) $x^2 + 3x - 70 = 0$
10. (a) $b^2 + 7b = 60$ (b) $x^2 + 7x = 8$
11. (a) $y^2 - 4y = 21$ (b) $y^2 - 6y = 27$
12. (a) $2p^2 - 14 = 3p$ (b) $4x^2 - 3 = 4x$
13. (a) $6r^2 + r = 1$ (b) $10x^2 - 3x = 1$
14. (a) $16p^2 - 25 = 0$ (b) $25p^2 - 16 = 0$
15. (a) $48t - 16t^2 = 0$ (b) $36x - 24x^2 = 0$
16. (a) $3x^2 + 8x = -4$ (b) $2x^2 + 3x = 5$
17. (a) $35 - 26m + 3m^2 = 0$ (b) $12 - 4x - 5x^2 = 0$

---

18. If you are given the solution set of a quadratic equation, you can find an equation to which the solution set applies. This is done in the following way.

$$S = \{-1, 3\}$$
$$x = -1 \quad \text{or} \quad x = 3$$
$$x + 1 = 0 \quad \text{or} \quad x - 3 = 0$$
$$(x + 1)(x - 3) = 0$$
$$x^2 - 2x - 3 = 0$$

Because your steps are reversible, you can verify that the quadratic equation $x^2 - 2x - 3 = 0$ has the solution set $\{-1, 3\}$. Use this procedure to find a quadratic equation for each of the following solution sets.

(a) $\{3, 2\}$ (b) $\{4, \frac{1}{2}\}$
(c) $\{-2, 0\}$ (d) $\{-1, -3\}$
(e) $\{\frac{1}{2}, \frac{3}{4}\}$ (f) $\{\frac{1}{2}\sqrt{3}, -\frac{1}{2}\sqrt{3}\}$

Simplify each equation and find its solution set. Check your answers.

19. $\frac{n^2}{20} = \frac{13n}{40} - \frac{1}{2}$ 20. $\frac{x}{3} = \frac{16x^2}{9}$ 21. $\frac{x + 3}{x - 3} = \frac{2x - 3}{x + 3}$
22. $\frac{n^2}{5} + \frac{1}{5} = \frac{n}{2}$ 23. $\frac{x^2}{9} = 12 + \frac{x}{3}$ 24. $\frac{5}{3x^2} + \frac{13}{6x} - 1 = 0$
25. $\frac{x^2}{16} + \frac{x}{4} = \frac{35}{4}$ 26. $\frac{x^2}{48} + \frac{x}{24} = 1$

For each of the given equations in Exercises 27–30, find an equation without radicals and solve it. Check each answer.

27. $x + 3 = \sqrt{3 + 11x}$

28. $x + 2 = \sqrt{3x + 34}$

29. $x + 1 = \sqrt{3x + 7}$

30. $3x + \sqrt{2x - 3} = 5$

31. Solve the equation

$$\frac{1}{x^2 + 3x + 2} = \frac{2}{x^2 - 1} - \frac{1}{x - 1}$$

by first finding a related quadratic equation without fractions and then solving it. Are both solutions of the quadratic equation also solutions of the given equation?

Preparation for Section 12–2

Factor each of the following algebraic expressions.

1. $4x^2 - 1$

2. $3x^2 - 5x$

3. $9x^2 - 25$

4. $10x^2 - 9x$

5. $x^3 - 9x$

12–2 INCOMPLETE QUADRATIC EQUATIONS

A quadratic equation is *incomplete* if either its constant term is zero or its first-degree term is missing. Thus,

$$2x^2 + 7x = 0,$$
$$3x^2 - 5 = 0$$

are examples of incomplete quadratic equations. Most incomplete quadratic equations are easily solved, as illustrated below.

Problem 1. Solve the equation $x^2 - 81 = 0$.

Solution. This equation is incomplete because it has no term of the first degree in x. Yet, you can solve it by factoring. The given equation is equivalent to

$$(x + 9)(x - 9) = 0.$$

Its solution set, S, is given by

$$\begin{aligned} S &= \{x \mid x + 9 = 0\} \cup \{x \mid x - 9 = 0\} \\ &= \{-9, 9\}. \end{aligned}$$

Check. $9^2 - 81 \stackrel{?}{=} 0$ $\quad (-9)^2 - 81 \stackrel{?}{=} 0$
$81 - 81 \stackrel{\checkmark}{=} 0$ $\quad 81 - 81 \stackrel{\checkmark}{=} 0$

You might have approached the above problem differently by considering the equivalent equation

$$x^2 = 81.$$

The values of x that make x^2 equal 81 are the two square roots of 81, 9 and -9. Let us solve the next problem by this method.

Problem 2. Solve the equation $3x^2 - 8 = 0$.

Solution. The given incomplete quadratic equation is equivalent to

$$3x^2 = 8,$$

or

$$x^2 = \tfrac{8}{3}.$$

The solutions of this last equation are the two square roots of $\frac{8}{3}$:

$$x = \sqrt{\tfrac{8}{3}} \quad \text{or} \quad x = -\sqrt{\tfrac{8}{3}}.$$

Thus, $\{\frac{2}{3}\sqrt{6}, -\frac{2}{3}\sqrt{6}\}$ is the solution set of the given equation.

Problem 3. Solve the equation $5x^2 - 6x = 0$.

Solution. This equation has a zero constant term and, therefore, is incomplete. Since it is equivalent to the equation

$$x(5x - 6) = 0,$$

its solution set, S, is given by

$$\begin{aligned} S &= \{x \mid x = 0\} \cup \{x \mid 5x - 6 = 0\} \\ &= \{0, \tfrac{6}{5}\}. \end{aligned}$$

Check. $(5 \cdot 0^2) - (6 \cdot 0) \stackrel{?}{=} 0$
$0 - 0 \stackrel{\checkmark}{=} 0$

$[5 \cdot (\frac{6}{5})^2] - (6 \cdot \frac{6}{5}) \stackrel{?}{=} 0$
$(5 \cdot \frac{36}{25}) - (6 \cdot \frac{6}{5}) \stackrel{?}{=} 0$
$\frac{36}{5} - \frac{36}{5} \stackrel{\checkmark}{=} 0$

Problem 4. Solve the equation $(2y - 3)^2 = 16$.

Solution. If you were to square $(2y - 3)$, you would see that this equation is not incomplete. However, its form is like an incomplete quadratic equation in that the square of the expression $(2y - 3)$ equals a number. Therefore, $(2y - 3)$ must equal one of the square roots of the number. Thus,

$$\begin{aligned} 2y - 3 &= 4 \quad \text{or} \quad 2y - 3 = -4, \\ 2y &= 7 \quad \text{or} \quad 2y = -1, \\ y &= \tfrac{7}{2} \quad \text{or} \quad y = -\tfrac{1}{2}. \end{aligned}$$

The solution set of the given equation is

$$\{\tfrac{7}{2}, -\tfrac{1}{2}\}.$$

Check.

$$(2 \cdot \tfrac{7}{2} - 3)^2 \overset{?}{=} 16 \qquad [2(-\tfrac{1}{2}) - 3]^2 \overset{?}{=} 16$$
$$(7 - 3)^2 \overset{?}{=} 16 \qquad (-1 - 3)^2 \overset{?}{=} 16$$
$$4^2 \overset{\checkmark}{=} 16 \qquad (-4)^2 \overset{\checkmark}{=} 16$$

Problem 5. Solve the equation $(3x + 1)^2 = 0$.

Solution. The form of this equation is like that of an incomplete equation. Since the square of a number is zero if, and only if, the number is zero, the solution set of the equation is given by

$$\{x \mid 3x + 1 = 0\}, \quad \text{or } \{-\tfrac{1}{3}\}.$$

Thus, this particular quadratic equation has only one solution.

Check.

$$[3(-\tfrac{1}{3}) + 1]^2 \overset{?}{=} 0$$
$$(-1 + 1)^2 \overset{?}{=} 0$$
$$0^2 \overset{\checkmark}{=} 0$$

Exercises

Solve each of the following incomplete quadratic equations.

1. (a) $x^2 - 49 = 0$ (b) $x^2 - 100 = 0$

2. (a) $x^2 = 25$ (b) $x^2 = 144$

3. (a) $m^2 - 36 = 0$ (b) $y^2 - 64 = 0$

4. (a) $6x^2 - 11x = 0$ (b) $3x^2 - 14x = 0$

5. (a) $7x^2 - 4x = 0$ (b) $6x^2 - 5x = 0$

6. (a) $8x^2 = 9x$ (b) $10x^2 = 7x$

7. (a) $3x^2 = 2x^2$ (b) $15x^2 = 13x^2$

8. (a) $3m^2 = \frac{4}{3}$ (b) $4x^2 = \frac{9}{4}$

9. (a) $x^2 - 5 = 0$ (b) $y^2 - 32 = 0$

10. (a) $4x^2 - 5 = 0$ (b) $9x^2 - 10 = 0$

11. (a) $s^2 = \frac{1}{4}s$ (b) $p^2 = .01p$

12. (a) $3y = \dfrac{36}{y}$ (b) $5x = \dfrac{20}{x}$

Solve each of the following equations without first removing the parentheses.

13. (a) $(x + 6)^2 = 49$ (b) $(x + 3)^2 = 25$

14. (a) $(4x + 3)^2 = 36$ (b) $(2y + 5)^2 = 36$

15. (a) $(2x - 7)^2 = 0$ (b) $(3x - 1)^2 = 0$

16. (a) $(5x - 6)^2 = 196$ (b) $(2x - 9)^2 = 169$

17. (a) $(4x - 5)^2 = (x + 10)^2$ (b) $(3x - 2)^2 = (x + 4)^2$

These equations are the same type as those in Exercises 13–17. Factor the polynomial on the left-hand side before you begin to solve them.

18. $9x^2 - 30x + 25 = 169$ **19.** $49x^2 + 14x + 1 = 144$

20. $4x^2 - 20x + 25 = 9$ **21.** $x^2 - 6x + 9 = 121$

In Exercises 22 and 23, first simplify and then find the solution set of each of the equations. Check your answers.

22. $4(r^2 + 2) - 5(r^2 - 3) = 14$ **23.** $\dfrac{80}{c} = \dfrac{c}{45}$

---

24. Is there a real value for x that will make the equation $x^2 + 4 = 0$ true? What number added to 4 yields 0? What is the solution set of $x^2 + 4 = 0$?

Preparation for Section 12–3

In Exercises 1–4, express each trinomial as the square of a binomial.

1. $x^2 - 8x + 16$ **2.** $c^2 + 22c + 121$

3. $a^2 + \frac{2}{3}a + \frac{1}{9}$ **4.** $c^2 + 4\sqrt{5}c + 20$

12–3 COMPLETING THE SQUARE

The quadratic equation

$$x^2 - 6x + 9 = 2$$

can be solved rather easily if you observe that the polynomial $x^2 - 6x + 9$ is a perfect-square trinomial:

$$x^2 - 6x + 9 = (x - 3)^2.$$

Thus, the given equation is equivalent to

$$(x - 3)^2 = 2.$$

If the square of $(x - 3)$ is 2, then $(x - 3)$ must be one of the square roots of 2:

$$x - 3 = \sqrt{2} \quad \text{or} \quad x - 3 = -\sqrt{2}$$

and

$$x = 3 + \sqrt{2} \quad \text{or} \quad x = 3 - \sqrt{2}.$$

Therefore, the solution set of the given equation is

$$\{3 + \sqrt{2}, 3 - \sqrt{2}\}.$$

Both solutions are irrational numbers.

The equation above is easy to solve, because one side of it is a perfect-square trinomial and the other side is a positive number. Any equation which is equivalent to an equation of this form can be solved in exactly the same manner.

Is every quadratic equation equivalent to an equation of the form

$$(x + m)^2 = n, \quad n \geqq 0,$$

for some numbers m and n? If so, then you can solve every quadratic equation by the above method. Recall from Chapter 5 that

$$(x + m)^2 = x^2 + 2mx + m^2.$$

You can solve each quadratic equation that is equivalent to an equation of the form

$$x^2 + 2mx + m^2 = n, \quad n \geqq 0,$$

for some numbers m and n. $x^2 + 2mx + m^2$ is called a *perfect-square trinomial.*

Problem 1. Make each trinomial a perfect square by filling in the blank space.

(a) $x^2 - 18x + \underline{\ ?\ }$ (b) $x^2 + \underline{\ ?\ }x + 49$

Solution.

(a) If you imagine that

$$x^2 - 18x + \underline{\ ?\ } = x^2 + 2mx + m^2,$$

then $2m = -18$ and $m = -9$. The number in the blank space is m^2, or 81:

$$x^2 - 18x + 81 = (x - 9)^2.$$

(b) In order to have

$$x^2 + \underline{\quad ? \quad} x + 49 = x^2 + 2mx + m^2,$$

you must have

$$m^2 = 49.$$

Thus, m is either 7 or -7, and $2m$, the coefficient of x, is either 14 or -14. You may have either

$$x^2 + 14x + 49 = (x + 7)^2$$

or

$$x^2 - 14x + 49 = (x - 7)^2.$$

In part (a), you started with the polynomial $x^2 - 18x$ and derived the perfect-square trinomial $x^2 - 18x + 81$. This process is called *completing the square.* It is used in solving quadratic equations, as shown in the following problems.

Problem 2. Solve the equation $x^2 - 18x + 74 = 0$.

Solution. The polynomial

$$x^2 - 18x + 74$$

is not a perfect-square trinomial. However, if you replace the given equation by the equivalent equation

$$x^2 - 18x = -74,$$

you can then complete the square of the polynomial $x^2 - 18x$ by adding 81 to it. In order to obtain an equivalent equation, add 81 to each side of the above equation:

$$x^2 - 18x + 81 = -74 + 81.$$

Now the polynomial $x^2 - 18x + 81$ is a perfect-square trinomial, and you may solve the equation above in the following way.

$$(x - 9)^2 = 7$$

$$x - 9 = \sqrt{7} \quad \text{or} \quad x - 9 = -\sqrt{7}$$

$$x = 9 + \sqrt{7} \quad \text{or} \quad x = 9 - \sqrt{7}$$

Hence, the solution set of the given equation is

$$\{9 + \sqrt{7}, 9 - \sqrt{7}\}.$$

Check.

$$(9+\sqrt{7})^2 - 18(9+\sqrt{7}) + 74 \stackrel{?}{=} 0$$
$$81 + 18\sqrt{7} + (\sqrt{7})^2 - 162 - 18\sqrt{7} + 74 \stackrel{?}{=} 0$$
$$81 + 7 - 162 + 74 \stackrel{\checkmark}{=} 0$$

$$(9-\sqrt{7})^2 - 18(9-\sqrt{7}) + 74 \stackrel{?}{=} 0$$
$$81 - 18\sqrt{7} + (\sqrt{7})^2 - 162 + 18\sqrt{7} + 74 \stackrel{?}{=} 0$$
$$81 + 7 - 162 + 74 \stackrel{\checkmark}{=} 0$$

The solutions of the quadratic equation in Problem 2 are irrational numbers. If you use a two-decimal-place approximation for $\sqrt{7}$,

$$\sqrt{7} \doteq 2.65,$$

then the rational numbers

$$9 + 2.65 \quad \text{and} \quad 9 - 2.65,$$

or 11.65 and 6.35, are rational approximations of the irrational solutions of this equation.

Procedure for solving a quadratic equation by completing the square

1. Find an equivalent equation of the form

$$x^2 + cx = d.$$

2. Add $(c/2)^2$ to each side of the above equation to make the left side a perfect-square trinomial,

$$x^2 + cx + \left(\frac{c}{2}\right)^2 = d + \left(\frac{c}{2}\right)^2.$$

3. Express the left side of the equation above as the square of a binomial,

$$\left(x + \frac{c}{2}\right)^2 = d + \left(\frac{c}{2}\right)^2.$$

4. Solve the equation by expressing $x + c/2$ as either the positive or the negative square root of the number on the right side of the above equation. Thus,

$$x + \frac{c}{2} = \sqrt{d + \frac{c^2}{4}} \quad \text{or} \quad x + \frac{c}{2} = -\sqrt{d + \frac{c^2}{4}}.$$

5. Then

$$x = \frac{-c}{2} + \frac{1}{2}\sqrt{c^2 + 4d} \quad \text{or} \quad x = \frac{-c}{2} - \frac{1}{2}\sqrt{c^2 + 4d}.$$

Exercises

Fill each of the following blanks so that the resulting trinomial is a perfect square. Then express the trinomial as the square of a binomial.

1. (a) $x^2 + 6x + _?_$ (b) $x^2 + 8x + _?_$
2. (a) $y^2 + _?_ y + 1$ (b) $y^2 + _?_ y + 25$
3. (a) $z^2 - 5z + _?_$ (b) $w^2 - \frac{1}{2}w + _?_$

Solve each equation by completing the square, and check each solution. Find a two-decimal-place approximation for each irrational solution.

4. (a) $y^2 + 2y - 4 = 0$ (b) $y^2 + 2y - 7 = 0$
5. (a) $a^2 - 12a + 36 = 25$ (b) $x^2 + 14x + 49 = 4$
6. (a) $a^2 - 12a + 36 = 5$ (b) $x^2 + 14x + 49 = 3$
7. (a) $m^2 - 24m + 144 = 0$ (b) $m^2 - 18m + 81 = 0$
8. (a) $m^2 - 24m - 6 = 0$ (b) $m^2 - 18m - 4 = 0$
9. (a) $x^2 = 20x + 8$ (b) $x^2 = 20x - 1$
10. (a) $x^2 - 7x - 8 = 0$ (b) $x^2 - 4x - 5 = 0$
11. (a) $x^2 - 10x - 12 = 0$ (b) $x^2 - 14x - 11 = 0$
12. (a) $22y = -1 - y^2$ (b) $22y = 4 - y^2$
13. (a) $x^2 - 7x + 1 = 0$ (b) $x^2 - 9x + 3 = 0$
14. (a) $m^2 + m - 1 = 0$ (b) $m^2 + m - 4 = 0$
15. (a) $m^2 + m - \frac{1}{2} = 0$ (b) $m^2 + m - \frac{1}{4} = 0$

Fill each of the following blanks so that the resulting trinomial is a perfect square. Then express the trinomial as the square of a binomial.

16. $a^2 - \sqrt{12}a + _?_$
17. $y^6 + 30y^3 + _?_$
18. $x^2 - 2\sqrt{7}x + _?_$

Solve each of the following quadratic equations.

19. $a^2 - 2\sqrt{2}a + 2 = 0$
20. $c^2 + 2\sqrt{3}c + 3 = 0$
21. $a^2 - 2\sqrt{2}a + 1 = 0$
22. $c^2 + 2\sqrt{3}c - 1 = 0$
23. $x^2 - 2\sqrt{5}x + 5 = 0$
24. $x^2 + 2\sqrt{6}x + 6 = 16$

Find four solutions of each of the equations in Exercises 25 and 26.

25. $x^4 - 4x^2 + 3 = 0$
26. $x^4 - 6x^2 + 8 = 0$

27. (a) Solve the equation

$$x^2 - 2x - 2 = 0.$$

(b) Check the positive irrational solution by substitution.
(c) Approximate this solution to one-decimal-place accuracy and substitute it into the original equation as a check.
(d) Substitute into the original equation a two-decimal-place approximation of this same solution.
(e) Will any rational number check in the original equation?

12–4 MORE ON COMPLETING THE SQUARE

How do you solve an equation like

$$2x^2 - 6x = 1?$$

This equation differs from the ones solved in the preceding section. There, the coefficients of x^2 were always 1. In this equation, the coefficient of x^2 is 2. If you try to solve it by completing the square, you do not know which number should be added to each side of it to make the left side a perfect-square trinomial. However, if you replace the given equation by the equivalent one obtained by multiplying each side by $\frac{1}{2}$, $x^2 - 3x = \frac{1}{2}$, then you can proceed as before and solve the new equation by completing the square. Do you understand that any quadratic equation is equivalent to an equation having 1 as the coefficient of x^2?

Problem 1. Solve the quadratic equation $2x^2 - 6x = 1$.

Solution. As stated above, you should solve the equivalent equation $x^2 - 3x = \frac{1}{2}$ instead of the original one. Complete the square and solve as indicated below.

$$x^2 - 3x + (-\tfrac{3}{2})^2 = \tfrac{1}{2} + (-\tfrac{3}{2})^2$$

$$(x - \tfrac{3}{2})^2 = \tfrac{1}{2} + \tfrac{9}{4}$$

$$(x - \tfrac{3}{2})^2 = \tfrac{11}{4}$$

$$x - \frac{3}{2} = \frac{\sqrt{11}}{2} \quad \text{or} \quad x - \frac{3}{2} = \frac{-\sqrt{11}}{2}$$

$$x = \frac{3 + \sqrt{11}}{2} \quad \text{or} \quad x = \frac{3 - \sqrt{11}}{2}$$

Thus,

$$\left\{\frac{3+\sqrt{11}}{2}, \frac{3-\sqrt{11}}{2}\right\}$$

is the solution set of the given equation. The check of one solution is presented here. The other is left for you to check.

Check.

$$2\left(\frac{3-\sqrt{11}}{2}\right)^2 - 6\left(\frac{3-\sqrt{11}}{2}\right) \stackrel{?}{=} 1$$

$$2\left(\frac{9-6\sqrt{11}+11}{4}\right) - 3(3-\sqrt{11}) \stackrel{?}{=} 1$$

$$\frac{20-6\sqrt{11}}{2} - 9 + 3\sqrt{11} \stackrel{?}{=} 1$$

$$10 - 3\sqrt{11} - 9 + 3\sqrt{11} \stackrel{\checkmark}{=} 1$$

Problem 2. Solve the quadratic equation $7x^2 + 2x + 3 = 0$.

Solution. Proceed as before. That is, first replace the given equation by the equivalent one

$$7x^2 + 2x = -3$$

and replace this, in turn, by the equivalent equation

$$x^2 + \tfrac{2}{7}x = -\tfrac{3}{7}.$$

You can now complete the square.

$$\begin{aligned} x^2 + \tfrac{2}{7}x + (\tfrac{1}{7})^2 &= -\tfrac{3}{7} + (\tfrac{1}{7})^2 \\ (x + \tfrac{1}{7})^2 &= -\tfrac{3}{7} + \tfrac{1}{49} \\ (x + \tfrac{1}{7})^2 &= -\tfrac{20}{49} \end{aligned}$$

For what value of x is the square of

$$x + \tfrac{1}{7}$$

a negative number? If you recall that the square of each real number is non-negative, then the answer is "for no value." Thus, the given equation has no real-number solution. You can conclude that the solution set of the given equation is the empty set, $\emptyset$.

Every quadratic equation can be solved by completing the square; that is, all solutions of a quadratic equation may be found by this method. Naturally, if solutions can be found by factoring, there is no need to use the more complicated technique of completing the square.

Exercises

Solve each of the following equations by completing the square. Check your solution.

1. (a) $2x^2 + 2x - 1 = 0$ (b) $2x^2 + 4x - 3 = 0$

2. (a) $3x^2 + 9x - 2 = 0$ (b) $3x^2 - 12x + 5 = 0$

3. (a) $-5x^2 + 10x = -3$ (b) $-6x^2 - 11x = 4$

4. (a) $4x^2 + 12x + 9 = 0$ (b) $3t^2 + 9t + 5 = 0$

5. (a) $-2y^2 = -6y + 1$ (b) $-4y^2 = -12y + 3$

6. (a) $9y^2 = 12y - 2$ (b) $8x^2 = 12x - 3$

7. (a) $3x^2 + 9x - 81 = 0$ (b) $5x^2 + 15x - 25 = 0$

8. (a) $2y^2 - 10y - 10 = 0$ (b) $7x^2 - 14x - 14 = 0$

9. (a) $a^2 + 1 = \sqrt{5}a$ (b) $x^2 - 2 = \sqrt{3}x$

10. (a) $3y^2 + \sqrt{3}y = 2$ (b) $5x^2 + \sqrt{5}x = 3$

Simplify the equations and solve them by the easiest method.

11. $(x + 2)(2x + 3) = 2$

12. $\frac{2}{3}x^2 - 1 = \frac{1}{5}x$

13. $\dfrac{3}{x + 2} = \dfrac{x}{4} - 1$

14. $\dfrac{x - 3}{x} + \dfrac{x - 1}{5x} = \dfrac{19}{10x^2}$

15. $\dfrac{y + 5}{y - 4} + \dfrac{5}{2 - y} = \dfrac{7y - 4}{y^2 - 6y + 8}$

16. $\dfrac{2x + 5}{2x - 2} - \dfrac{3 - x}{x} = \dfrac{7}{3}$

17. $\dfrac{x + 1}{x - 1} + \dfrac{x}{2} = x + 2$

18. $\sqrt{3x + 6} = 2\sqrt{x} - 1$

19. $\dfrac{2 - 3y}{4y^2 + 7y - 15} = \dfrac{6 - y}{6y^2 + 17y - 3}$

20. $\sqrt{3y - 2} = y$

21. $\sqrt{2x + 5} + x = 5$

22. $3y + \sqrt{2y - 3} = 5$

23. $\dfrac{7}{4 + \sqrt{7 + x}} = \dfrac{\sqrt{7 + x}}{3}$

24. $\sqrt{4 - x} + \sqrt{x + 6} = \sqrt{1 - 3x}$

Use the method of completing the square to show that each of the following equations has no real number solution.

25. $c^2 + c + 1 = 0$

26. $2m^2 + 2m + 1 = 0$

27. $4a^2 = a - 1$

28. $a^2 + 1 = \sqrt{3}a$

29. $7c^2 = 3\sqrt{3}c - 1$

30. $\dfrac{3x + 5}{x^2 + 4x + 3} + \dfrac{x}{x + 1} = \dfrac{2}{x + 3}$

Review for Sections 12–1 through 12–4

Solve each of the following quadratic equations. You may use any of the methods discussed in Sections 12–1 through 12–4. Remember to check each solution.

1. $(x - 4)(x + 2) = 0$
2. $2m^2 = 32$
3. $a^2 + 2a = 2$
4. $x^2 = 10x - 25$
5. $4y^2 + 4y + 1 = 0$
6. $4m^2 - 15m = 0$
7. $x^2 - 2x - 24 = 0$
8. $23 = 10a - a^2$
9. $3m^2 - 12m = 0$
10. $a^2 - 4a = 12$
11. $2y^2 - y = 10$
12. $(x + 3)(x - 5) = 0$
13. $(m + 3)^2 = 36$
14. $y^2 + 3y - 2 = 0$
15. $a^2 - 8a + 12 = 0$
16. $y^2 + \sqrt{3}y + \frac{1}{2} = 0$
17. $x^2 - 10x + 21 = 0$
18. $4m^2 = 16$
19. $y^2 - 6y + 4 = 0$
20. $a^2 + 13a + 4 = 0$

Answers to Review for Sections 12–1 through 12–4

1. $\{-2, 4\}$
2. $\{-4, 4\}$
3. $\{-1 - \sqrt{3}, -1 + \sqrt{3}\}$
4. $\{5\}$
5. $\{-\frac{1}{2}\}$
6. $\{0, \frac{15}{4}\}$
7. $\{-4, 6\}$
8. $\{5 - \sqrt{2}, 5 + \sqrt{2}\}$
9. $\{0, 4\}$
10. $\{-2, 6\}$
11. $\{-2, \frac{5}{2}\}$
12. $\{-3, 5\}$
13. $\{-9, 3\}$
14. $\left\{\frac{-3 - \sqrt{17}}{2}, \frac{-3 + \sqrt{17}}{2}\right\}$
15. $\{2, 6\}$
16. $\left\{\frac{-\sqrt{3} - 1}{2}, \frac{-\sqrt{3} + 1}{2}\right\}$
17. $\{3, 7\}$
18. $\{2, -2\}$
19. $\{3 - \sqrt{5}, 3 + \sqrt{5}\}$
20. $\left\{\frac{-13 - 3\sqrt{17}}{2}, \frac{-13 + 3\sqrt{17}}{2}\right\}$

12–5 WORD PROBLEMS

When translated into mathematical language, certain word problems lead to quadratic equations. Several such problems are given below.

Problem 1. A rectangle has a perimeter of 30 inches and an area of 54 square inches. Find its dimensions. (See Fig. 12–1.)

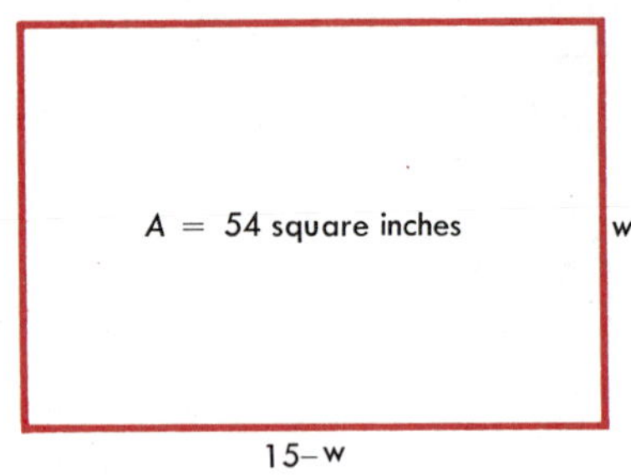

FIGURE 12–1

Solution. The perimeter of a rectangle is twice the sum of its length and width. Therefore, the sum of its length and width is one-half the perimeter, or 15 inches. If you let w denote the width, then $(15 - w)$ denotes the length. Since the area of a rectangle is the product of the length and width, and since the area of this particular rectangle is given as 54 square inches, you have

$$w(15 - w) = 54.$$

This equation may be solved as follows:

$$\begin{aligned} 15w - w^2 &= 54, \\ w^2 - 15w + 54 &= 0, \\ (w - 6)(w - 9) &= 0, \\ w = 6 \quad \text{or} \quad w &= 9. \end{aligned}$$

Thus, the width is either 6 inches or 9 inches. If the width is 6 inches, the length is 15 − 6, or 9 inches. If the width is 9 inches, the length is 6 inches. In other words, the dimensions of the rectangle are 6 inches by 9 inches.

Check.

$$\begin{aligned} 6 + 6 + 9 + 9 &\stackrel{\checkmark}{=} 30 \\ 6 \times 9 &\stackrel{\checkmark}{=} 54 \end{aligned}$$

Problem 2. If you stand on top of a building 240 feet high and throw a rock upward with a speed of 32 feet per second, by the laws of physics, the equation

$$h = 240 + 32t - 16t^2$$

gives the height of the rock above the ground t seconds after it is thrown. How many seconds does it take the rock to reach the ground?

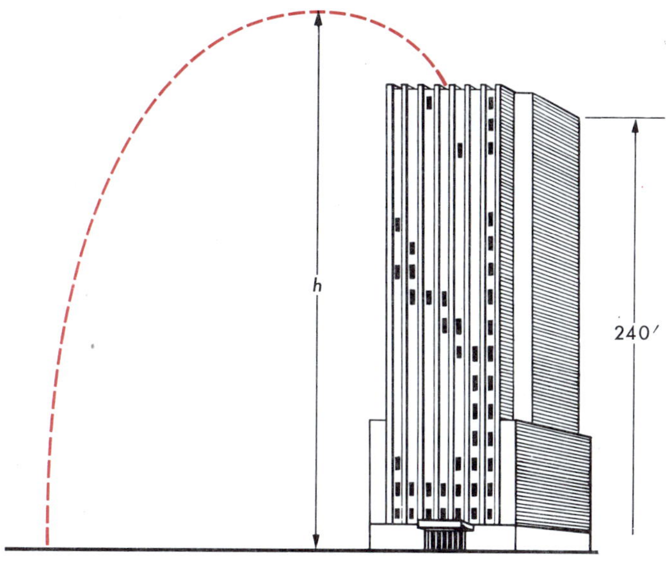

FIGURE 12-2

Solution. See Fig. 12–2. When the rock hits the ground, it is 0 feet above the ground. Thus, you must solve the equation

$$0 = 240 + 32t - 16t^2.$$

This equation is equivalent to each of the following two equations.

$$-16(t^2 - 2t - 15) = 0$$

$$-16(t - 5)(t + 3) = 0$$

The latter equation has the solution set

$$\{5, -3\}.$$

The solution 5 is the only one that is sensible from a physical standpoint. You can not measure intervals of time in negative seconds!

Check. Let $t = 5$ in the given equation to get

$$\begin{aligned} h &= 240 + (32 \times 5) - (16 \times 5^2) \\ &= 240 + 160 - 400 \\ &= 0. \end{aligned}$$

Thus, the rock reaches the ground 5 seconds after it is thrown.

Problem 3. Find a number such that the sum of the number and its reciprocal is 4.

Solution. If N designates the desired number, then $1/N$ designates its reciprocal and

$$N + \frac{1}{N} = 4,$$

according to the given information. You can solve this equation as follows:

$$\begin{aligned} N^2 + 1 &= 4N, \\ N^2 - 4N &= -1, \\ N^2 - 4N + 4 &= -1 + 4, \\ (N - 2)^2 &= 3, \end{aligned}$$

$$N - 2 = \sqrt{3} \quad \text{or} \quad N - 2 = -\sqrt{3},$$
$$N = 2 + \sqrt{3} \quad \text{or} \quad N = 2 - \sqrt{3}.$$

Check. The numbers $(2 + \sqrt{3})$ and $(2 - \sqrt{3})$ are reciprocals of each other since

$$(2 + \sqrt{3})(2 - \sqrt{3}) = 2^2 - (\sqrt{3})^2, \quad \text{or } 1.$$

Thus, either $(2 + \sqrt{3})$ or $(2 - \sqrt{3})$ is a number that satisfies the conditions of the problem.

$$(2 + \sqrt{3}) + (2 - \sqrt{3}) = 4$$

Exercises

For each problem, write an equation, solve it, and reject any solutions that are not appropriate to the question asked. State clearly what is denoted by each variable that you use.

1. (a) Find two consecutive integers such that their product is 182.
(b) Find two consecutive integers such that their product is 272.

2. (a) If a number is added to its square, the result is 42. Find the number.
(b) If twice a number is added to its square, the result is 35. Find the number.

3. (a) The product of two consecutive odd integers is 783. Find the integers.
(b) The product of two consecutive even integers is 528. Find the integers.

4. (a) Find two consecutive even integers that have 580 as the sum of their squares.
   (b) Find two consecutive odd integers that have 290 as the sum of their squares.

5. (a) The difference between two numbers is 4 and their product is 77. Find the numbers.
   (b) The sum of two numbers is 18 and their product is 80. Find the numbers.

6. (a) The sum of a number and four times its reciprocal is $\frac{20}{3}$. Find the number.
   (b) A number less twice its reciprocal is $\frac{23}{5}$. Find the number.

7. (a) A number exceeds its square by $\frac{2}{9}$. What is the number?
   (b) A number is 42 less than its square. What is the number?

8. (a) One number is 3 more than $\frac{1}{2}$ a second number, and the difference between their squares is 495. Find the numbers.
   (b) One number is 2 less than $\frac{1}{2}$ a second number, and the sum of their squares is 221. Find the numbers.

For each problem write an equation, solve it, and reject any solutions that are not appropriate to the question asked. State clearly what is denoted by each variable that you use.

9. What are the dimensions of a rectangle that has a perimeter of 40 inches and an area of 96 square inches?

10. What are the dimensions of a badminton court with a length 2.2 times the width and an area of 880 square feet?

11. A rectangular field is 20 yards longer than it is wide. What are its dimensions if its area is 4800 square yards?

12. A rectangle is 3 feet wider than it is long, and its area is 180 square feet. What are its dimensions?

13. The length of a rectangle exceeds twice its width by 3 feet. Find the dimensions of the rectangle if its area is 65 square feet.

For each problem in Exercises 14–17, write an equation, and when you solve it, reject any solutions that are not appropriate to the question asked. State clearly what is denoted by each variable that you use.

14. The width of a rectangle is 9 inches shorter than its diagonal and 7 inches shorter than its length. How long is the diagonal?

15. The base of a triangle is 9 inches longer than its altitude. If the area of the triangle is 56 square inches, find the altitude.

16. One of two numbers is 4 greater than the other. If the sum of their reciprocals is $\frac{3}{8}$, what are the numbers?

17. The denominator of a fraction is 5 more than the numerator. If 3 is added to both numerator and denominator, the resulting fraction is $\frac{5}{36}$ more than the original fraction. Find the original fraction.

18. If an object is thrown vertically upward with a velocity of v feet per second, the distance above the ground after t seconds is found by using the formula

$$d = vt - 16t^2.$$

If an object is thrown upward with a velocity of 96 feet per second,

(a) how high will it be at the end of 1 second? 2 seconds? 3 seconds? 4 seconds? 5 seconds?

(b) in how many seconds will it be 44 feet above the ground? 140 feet above the ground? 144 feet above the ground?

12–6 QUADRATIC FORMULA

The method of completing the square is applicable to every quadratic equation in one variable. The solution set of every quadratic equation can be found by this method, but this does not mean that there is nothing further you can do about solving quadratic equations. What has been established so far is a *procedure* for replacing a given quadratic equation by successively simpler equivalent equations. There are several steps in this procedure. The ideal method would be to go from the given equation to the solution set in just one step.

You have already observed that every quadratic equation in the variable x is equivalent to one having the form

$$(\text{a number})x^2 + (\text{a number})x + (\text{a number}) = 0,$$

where the coefficient of x^2 is not zero. Thus, every quadratic equation fits the pattern

$$ax^2 + bx + c = 0,$$

where a, b, and c are real numbers, with $a \neq 0$. It is the numbers a, b, and c that change from one quadratic equation to another. This suggests that the solutions of the equation depend solely on the numbers a, b, and c. To find out whether this is so, you might try to use the method of completing the square on this "general" quadratic equation.

Completing the square for the equation above involves the following steps.

$$ax^2 + bx + c = 0$$

$$x^2 + \frac{b}{a}x + \frac{c}{a} = 0$$

$$x^2 + \frac{b}{a}x = \frac{-c}{a}$$

$$x^2 + \frac{b}{a}x + \left(\frac{b}{2a}\right)^2 = \frac{-c}{a} + \left(\frac{b}{2a}\right)^2$$

$$\left(x + \frac{b}{2a}\right)^2 = \frac{-c}{a} + \frac{b^2}{4a^2}$$

$$\left(x + \frac{b}{2a}\right)^2 = \frac{b^2 - 4ac}{4a^2}$$

The usual procedure is to set $x + b/2a$ equal to one of the square roots of the number $(b^2 - 4ac)/4a^2$. However, this number has a square root if, and only if, it is non-negative. The number above is expressed as a quotient of the number $b^2 - 4ac$ divided by the *positive* number $4a^2$. Therefore, it is non-negative if, and only if,

$$b^2 - 4ac \geqq 0.$$

The number $b^2 - 4ac$ is called the *discriminant* of the quadratic equation $ax^2 + bx + c = 0$.

If $b^2 - 4ac \geqq 0$, the two square roots of the number $(b^2 - 4ac)/4a^2$ are

$$\frac{\sqrt{b^2 - 4ac}}{2a} \quad \text{and} \quad \frac{-\sqrt{b^2 - 4ac}}{2a}.$$

In this case, you may proceed with the solution of the given equation as follows:

$$x + \frac{b}{2a} = \frac{\sqrt{b^2 - 4ac}}{2a} \quad \text{or} \quad x + \frac{b}{2a} = \frac{-\sqrt{b^2 - 4ac}}{2a},$$

$$x = \frac{-b}{2a} + \frac{\sqrt{b^2 - 4ac}}{2a} \quad \text{or} \quad x = \frac{-b}{2a} - \frac{\sqrt{b^2 - 4ac}}{2a}.$$

Finally,

$$x = \frac{-b + \sqrt{b^2 - 4ac}}{2a} \quad \text{or} \quad x = \frac{-b - \sqrt{b^2 - 4ac}}{2a}.$$

You now have a formula for solving any quadratic equation.

QUADRATIC FORMULA

The quadratic equation

$$ax^2 + bx + c = 0, \quad a \neq 0,$$

has the real-number solutions

$$x = \frac{-b \pm \sqrt{b^2 - 4ac}}{2a}$$

if its discriminant is non-negative.

The symbol $\pm$ is read *plus or minus*, and indicates that either $+$ or $-$ may be used to obtain a solution. The following problem illustrates the use of this formula.

Problem 1. Solve the equation $2x^2 + x - 2 = 0$.

Solution. This equation has the form

$$ax^2 + bx + c = 0$$

with

$$a = 2, \quad b = 1, \quad c = -2.$$

For this equation, the discriminant is given by

$$\begin{aligned} b^2 - 4ac &= 1^2 - 4 \cdot 2(-2) \\ &= 1 - (-16), \quad \text{or } 17. \end{aligned}$$

Since $b^2 - 4ac > 0$, this equation has solutions. By the quadratic formula, the solutions are

$$x = \frac{-1 \pm \sqrt{17}}{4}.$$

Thus,

$$\frac{-1 + \sqrt{17}}{4} \quad \text{and} \quad \frac{-1 - \sqrt{17}}{4}$$

are the two solutions of the given equation. Since $\sqrt{17} \doteq 4.123$, these irrational solutions may be approximated by the rational numbers

$$.781 \quad \text{and} \quad -1.281.$$

Exercises

For each equation, state the values of a, b, and c and find the value of the discriminant, $b^2 - 4ac$. Then use the quadratic formula to solve the equation.

1. (a) $3x^2 - 4x - 1 = 0$ (b) $3x^2 - 4x + 1 = 0$
2. (a) $9x^2 - 6x + 1 = 0$ (b) $9x^2 - 6x - 1 = 0$

Use the quadratic formula to solve each of the following equations.

3. (a) $2x^2 + 3x + 1 = 0$ (b) $2x^2 + x - 3 = 0$
4. (a) $2x^2 - 3x + 1 = 0$ (b) $2x^2 - x - 3 = 0$
5. (a) $x^2 - 14x + 49 = 0$ (b) $x^2 - 10x + 25 = 0$
6. (a) $x^2 - 14x - 49 = 0$ (b) $x^2 - 10x - 25 = 0$
7. (a) $3x^2 + x - 1 = 0$ (b) $2x^2 + x - 1 = 0$
8. (a) $3x^2 - x - 5 = 0$ (b) $4x^2 - x - 4 = 0$
9. (a) $3x^2 - 2x - 3 = 0$ (b) $2x^2 - 3x - 1 = 0$
10. (a) $15x^2 + 2x - 16 = 0$ (b) $10x^2 + 3x - 5 = 0$
11. (a) $15x^2 - 2x - 16 = 0$ (b) $10x^2 - 3x - 5 = 0$
12. (a) $11x^2 - x - 3 = 0$ (b) $10x^2 - x - 4 = 0$

Simplify each equation and solve it. Then check your solution in the original equation by substitution.

13. $\dfrac{9}{2(y-2)} = \dfrac{5}{y} + \dfrac{7}{2(y+2)}$ **14.** $\dfrac{5}{2(y-2)} - \dfrac{2}{(y-2)^2} = \dfrac{3}{2y}$

15. $\dfrac{3}{y+8} - 2 = \dfrac{4}{y-2}$ **16.** $\dfrac{2x-2}{5} = 1 - \dfrac{x-6}{x-4}$

17. $\dfrac{5x}{x-1} = \dfrac{2x+1}{x-2} - \dfrac{2x-7}{x-1}$ **18.** $\dfrac{x-1}{2x-1} = 1 - \dfrac{2x-1}{x}$

For each equation, find a simpler one without radicals and solve it. Then check each solution.

19. $\sqrt{x-3} = x - 9$ **20.** $x + 5 = 5\sqrt{x-1}$

In Exercises 21–24, use the quadratic formula to solve each of the equations.

21. $x^2 + \sqrt{5}x + 1 = 0$ **22.** $\sqrt{12}x^2 + x - \sqrt{3} = 0$

23. $x^2 + \sqrt{2}x - 3 = 0$ **24.** $\sqrt{3}y^2 = \sqrt{12} - 5y$

25. (a) What was the discriminant for the quadratic equation in Exercise 21? What kind of solutions did the equation have?

(b) What was the discriminant for the quadratic equation in Exercise 22? What kind of solutions did the equation have?

26. Suppose that a, b, and c are integers such that $b^2 - 4ac = 144$. What can you say about the solution set of $ax^2 + bx + c = 0$?

27. If a, b, and c are rational numbers such that $b^2 - 4ac = 0$, what can you predict about the solution set of $ax^2 + bx + c = 0$?

28. If a, b, and c are rational numbers such that $b^2 - 4ac = 15$, what can you say about the solution set of $ax^2 + bx + c = 0$?

12–7 USING THE QUADRATIC FORMULA

The quadratic formula is used to solve the following problems.

Problem 1. Solve the equation $3x^2 - 4x + 2 = 0$.

Solution. When you compare this equation with the formal equation

$$ax^2 + bx + c = 0,$$

you should see that

$$a = 3, \quad b = -4, \quad c = 2.$$

Thus, its discriminant is given by

$$\begin{aligned} b^2 - 4ac &= (-4)^2 - (4 \cdot 3 \cdot 2) \\ &= 16 - 24 \\ &= -8. \end{aligned}$$

Since $b^2 - 4ac < 0$, the given equation has no real-number solution.

Problem 2. Solve the equation $-4x^2 + 4x + 19 = 0$.

Solution. When you compare this equation with the equation

$$ax^2 + bx + c = 0,$$

you see that

$$a = -4, \quad b = 4, \quad c = 19.$$

Thus, its discriminant is given by

$$\begin{aligned} b^2 - 4ac &= 4^2 - 4 \cdot (-4) \cdot 19 \\ &= 4^2 + 4^2 \cdot 19 \\ &= 4^2(1 + 19) \\ &= 4^2 \cdot 20, \quad \text{or } 4^3 \cdot 5, \quad \text{or } 2^6 \cdot 5. \end{aligned}$$

Since the discriminant is positive, the given equation has the following two solutions.

$$\begin{aligned} x &= \frac{-4 \pm \sqrt{2^6 \cdot 5}}{-8} \\ &= \frac{-4 \pm 2^3\sqrt{5}}{-8}, \quad \text{or} \quad \frac{-1 \pm 2\sqrt{5}}{-2}. \end{aligned}$$

In other words,

$$\tfrac{1}{2} - \sqrt{5} \quad \text{and} \quad \tfrac{1}{2} + \sqrt{5}$$

are the two solutions of the given equation. Since $\sqrt{5} \doteq 2.236$,

$$-1.736 \quad \text{and} \quad 2.736$$

are rational approximations of the irrational solutions.

How many real-number solutions does a quadratic equation have? Some equations have two solutions; for example, the equation in Problem 2 has two solutions. The equation

$$x^2 - 4x + 4 = 0$$

has one solution: 2. Finally, some quadratic equations have no real-number solutions. The equation in Problem 1 is of this type.

In summary, if $ax^2 + bx + c = 0$ and

if $b^2 - 4ac > 0$, the solution set of the equation contains two real numbers.

if $b^2 - 4ac = 0$, the solution set of the equation contains one real number.

if $b^2 - 4ac < 0$, the solution set of the equation contains no real numbers. In other words, when the domain of the variable is the set of real numbers, the solution set is the empty set.

In more advanced mathematics, a number system called the *system of complex numbers* is used. The complex number system contains the real number system and the square roots of negative real numbers. In the complex number system, every quadratic equation has a nonempty solution set.

Exercises

Find the value of the discriminant and determine which of the following quadratic equations have no real-number solutions.

1. (a) $3x^2 + x + 1 = 0$ (b) $7x^2 + 7x - 2 = 0$
2. (a) $x^2 = 28x - 196$ (b) $x^2 = 24x - 144$
3. (a) $-7x^2 - 3x + 1 = 0$ (b) $-7x^2 + 7x - 2 = 0$

Find the value of the discriminant of each of the following quadratic equations. Tell which equations have no real-number solutions. Then use whatever method you wish to find the solution set of the equations that have real-number solutions.

4. (a) $-5x^2 + 2x - 3 = 0$ (b) $-5x^2 - 9x + 3 = 0$
5. (a) $4x^2 = 28x - 49$ (b) $9x^2 = 6x - 1$
6. (a) $x^2 - 4x + 1 = 0$ (b) $x^2 - 3x + 1 = 0$
7. (a) $x^2 - 10x + 12 = 0$ (b) $x^2 - 8x + 5 = 0$
8. (a) $x^2 - 2x - 4 = 0$ (b) $x^2 - 6x - 5 = 0$
9. (a) $x^2 + 8x - 5 = 0$ (b) $x^2 + 11x - 1 = 0$
10. (a) $2x^2 - 2x - 1 = 0$ (b) $4x^2 - 3x - 1 = 0$
11. (a) $3x^2 + 7x - 2 = 0$ (b) $5x^2 + 2x + 3 = 0$
12. (a) $3x^2 - 1 = 0$ (b) $5x^2 - 1 = 0$
13. (a) $2x^2 + 8 = 0$ (b) $3x^2 + 9 = 0$
14. (a) $5x^2 - 6x = 0$ (b) $6x^2 - 7x = 0$
15. (a) $x + \frac{1}{4}x - \frac{3}{8} = 0$ (b) $x^2 + .08x + .0015 = 0$

---

16. If a, b, and c are integers and the integer $b^2 - 4ac$ is a perfect square, what kind of solutions do you think the quadratic equation $ax^2 + bx + c = 0$ will have? Test your answer on the following equations.

(a) $x^2 + 4x - 12 = 0$ (b) $3x^2 + 5x - 2 = 0$

Find a simpler form for each of the equations below. Solve it, and then check your solution in the given equation to see if the two equations are equivalent.

17. $\dfrac{x^2}{x^2 - 4} = \dfrac{x - 3}{2 - x} + \dfrac{x + 3}{x + 2}$ **18.** $x + \sqrt{x + 9} = 3$

19. $2x = \sqrt{3 - x} - 4$

20. $x = \sqrt{x^2 - 9} - 1$

---

21. (a) Find a simplified form of the sum

$$\frac{-b + \sqrt{b^2 - 4ac}}{2a} + \frac{-b - \sqrt{b^2 - 4ac}}{2a}.$$

(b) Since

$$\left\{\frac{-b + \sqrt{b^2 - 4ac}}{2a}, \frac{-b - \sqrt{b^2 - 4ac}}{2a}\right\}$$

is the solution set of the quadratic equation, $ax^2 + bx + c = 0$, what was proved in Exercise 21(a) about the sum of the solutions of *any* quadratic equation?

22. Verify that $\{\frac{1}{2}, 1\}$ is the solution set of $2x^2 - 3x + 1 = 0$. Is $\frac{1}{2} + 1 = -(-\frac{3}{2})$? Does this agree with your answer to Exercise 21?

23. Verify that $\left\{\frac{-1 + \sqrt{13}}{6}, \frac{-1 - \sqrt{13}}{6}\right\}$ is the solution set of

$$3x^2 + x - 1 = 0.$$

Is $\frac{-1 + \sqrt{13}}{6} + \frac{-1 - \sqrt{13}}{6} = \frac{-1}{3}$?

In Exercises 24–27, first find the solution set of each equation and then find the sum of the solutions. Is the sum of the solutions of a quadratic equation always equal to $-(b/a)$, where a and b are the coefficients of the variable in $ax^2 + bx + c = 0$?

24. $x^2 - 4x - 5 = 0$

25. $x^2 - 4x - 8 = 0$

26. $x^2 - x - 12 = 0$

27. $5x^2 - 20x + 13 = 0$

28. (a) Find a simplified form of the product

$$\frac{-b + \sqrt{b^2 - 4ac}}{2a} \times \frac{-b - \sqrt{b^2 - 4ac}}{2a}.$$

(b) Since

$$\left\{\frac{-b + \sqrt{b^2 - 4ac}}{2a}, \frac{-b - \sqrt{b^2 - 4ac}}{2a}\right\}$$

is the solution set of the quadratic equation $ax^2 + bx + c = 0$, what was proved in Exercise 28(a) about the product of the solutions of *any* quadratic equation?

29. Is $\{-2 + \sqrt{7}, -2 - \sqrt{7}\}$ the solution set of $x^2 + 4x - 3 = 0$? Is

$$(-2 + \sqrt{7})(-2 - \sqrt{7}) = \frac{-3}{1}?$$

30. Verify that $\{\frac{3}{2}, \frac{1}{5}\}$ is the solution set of $10x^2 - 17x + 3 = 0$. Is $\frac{3}{2} \cdot \frac{1}{5} = \frac{3}{10}$?

Find the solution set of each equation below. Is the product of the solutions of each quadratic equation equal to c/a, where a is the coefficient of x^2 and c is the constant term in $ax^2 + bx + c = 0$?

31. $3x^2 + 14x - 5 = 0$

32. $3x^2 - 5 = 0$

33. $9x^2 - 7x = 0$

34. $2x^2 - 6x + 3 = 0$

35. $x^2 - 12x + 28 = 0$

36. $5x^2 = 10 + 2x$

EXTRA!

Real numbers of the form

$$a + b\sqrt{n}$$

where a and b are any integers and n is a positive square-free integer are called *quadratic integers.* The word *quadratic* refers to the fact that you are taking the *square* root of n.

For example,

$$3 + 5\sqrt{2}, \quad -1 + 2\sqrt{7}, \quad 4 - 3\sqrt{14}, \quad \sqrt{30}$$

are quadratic integers.

You can do arithmetic with quadratic integers much as you do with regular integers. For example, the quadratic integer $6 - \sqrt{2}$ can be factored into quadratic integers

$$6 - \sqrt{2} = (2 - \sqrt{2})(5 + 2\sqrt{2}).$$

Some quadratic integers are perfect squares, just as some regular integers such as 4 and 49 are perfect squares. How can you tell if a given quadratic integer is a perfect square? Mainly by noting what form the square of a quadratic integer has. For example,

$$\begin{aligned}(a + b\sqrt{2})^2 &= a^2 + 2ab\sqrt{2} + (b\sqrt{2})^2 \\ &= (a^2 + 2b^2) + 2ab\sqrt{2}.\end{aligned}$$

Problem 1. Is $17 + 12\sqrt{2}$ a perfect square of a quadratic integer?

Solution. By the work above,

$$17 + 12\sqrt{2} = (a + b\sqrt{2})^2$$

if you can find integers a and b such that

$$\begin{cases} a^2 + 2b^2 = 17, \\ 2ab = 12. \end{cases}$$

Thus, $ab = 6$ and $a = 1$, $b = 6$ or $a = 2$, $b = 3$ or $a = 3$, $b = 2$, or $a = 6$, $b = 1$. Since $a^2 + 2b^2 = 17$, only $a = 3$, $b = 2$ works. You conclude that

$$(3 + 2\sqrt{2})^2 = 17 + 12\sqrt{2}$$

or that

$$\sqrt{17 + 12\sqrt{2}} = 3 + 2\sqrt{2}.$$

Problem 2. Is $7 - 4\sqrt{3}$ a perfect square of a quadratic integer?

Solution. If it is a perfect square, you must be able to find integers a and b such that

$$7 - 4\sqrt{3} = (a + b\sqrt{3})^2.$$

Thus, as above

$$\begin{cases} a^2 + 3b^2 = 7 \\ 2ab = -4. \end{cases}$$

The possible choices for a and b are $a = 1, b = -2$ or $a = 2, b = -1$. Of these, $a = 2$, $b = -1$ gives $a^2 + 3b^2 = 7$. Therefore,

$$7 - 4\sqrt{3} = (2 - \sqrt{3})^2$$

or

$$\sqrt{7 - 4\sqrt{3}} = 2 - \sqrt{3}.$$

Which of the following quadratic integers are perfect squares? Find the square root of those that are.

1. $3 - 2\sqrt{2}$

2. $-3 + 2\sqrt{2}$

3. $8 + 2\sqrt{7}$

4. $16 - 6\sqrt{7}$

5. $9 + 4\sqrt{6}$

6. $79 - 30\sqrt{6}$

Find:

7. $\sqrt{41 + 24\sqrt{2}}$

8. $\sqrt{31 - 8\sqrt{15}}$

9. $\sqrt{19 + 6\sqrt{10}}$

10. $\sqrt{126 + 10\sqrt{5}}$

The quadratic integers

$$a + b\sqrt{n}, \quad a - b\sqrt{n}$$

are called conjugates of each other. For example, $3 - 5\sqrt{2}$ is the conjugate of $3 + 5\sqrt{2}$.

11. If a quadratic integer is a perfect square, show that its conjugate is also a perfect square.
12. If a quadratic integer is a perfect square, show that the product of the quadratic integer and its conjugate is the square of an ordinary integer.

Preparation for Section 12–8

1. Solve the following equation by factoring:

$$10x^2 + 21x - 10 = 0.$$

2. Solve by completing the square:

$$2x^2 + 8x - 43 = 0.$$

3. Solve by using the quadratic formula:

$$3x^2 + 10x - 15 = 0.$$

12–8 ADDITIONAL WORD PROBLEMS

Now that you have more mathematical tools to use for computation, it is time to examine some additional word problems.

Problem 1. If you can row upstream to a landing 5 miles away and then row back to your starting point all in 3 hours and 20 minutes, and if the river has an average current of 2 miles per hour, at what rate are you able to row in still water?

Solution. If you let r designate the number of miles per hour which you can row in still water, then $r - 2$ is your rate upstream and $r + 2$ is your rate downstream. Since

$$\text{time} = \frac{\text{distance}}{\text{rate}},$$

the time in hours for the upstream journey is $5/(r - 2)$ and the time in hours for the downstream journey is $5/(r + 2)$. It is stated that the total time for the journey is 3 hours and 20 minutes, or $\frac{10}{3}$ hours. The time up plus the time down equals the total time:

$$\frac{5}{r-2} + \frac{5}{r+2} = \frac{10}{3}.$$

You can solve this equation as follows:

$$\frac{5(r+2)+5(r-2)}{(r-2)(r+2)} = \frac{10}{3},$$

$$\frac{10r}{r^2-4} - \frac{10}{3} = 0,$$

$$\frac{30r - 10(r^2-4)}{3(r^2-4)} = 0,$$

$$-10r^2 + 30r + 40 = 0,$$

$$-10(r^2 - 3r - 4) = 0,$$

$$-10(r-4)(r+1) = 0.$$

Thus, the given equation has the solution set

$$\{4, -1\}.$$

Since a rate of -1 miles per hour does not make sense physically, the only possible answer is 4 miles per hour.

Check. If you can row at the rate of 4 miles per hour in still water, then you can row at the rate of 2 miles per hour upstream and 6 miles per hour downstream. Hence, your time upstream is $\frac{5}{2}$ hours, or 2 hours and 30 minutes; and your time downstream is $\frac{5}{6}$ hour, or 50 minutes. The sum of 2 hours and 30 minutes and 50 minutes is 3 hours and 20 minutes, as stated.

Problem 2. For a certain right triangle, it is known that the hypotenuse is 1 unit longer than one of the legs and 32 units longer than the other leg. How long is each side of this right triangle?

Solution. Let H stand for the length of the hypotenuse. Then

$$H - 1 \quad \text{and} \quad H - 32$$

designate the lengths of the two legs of the triangle. By the pythagorean theorem,

$$H^2 = (H-1)^2 + (H-32)^2.$$

This equation is solved by the following steps.

$$H^2 = H^2 - 2H + 1 + H^2 - 64H + 1024$$
$$0 = H^2 - 66H + 1025$$

By the quadratic formula,

$$H = \frac{66 \pm \sqrt{4356 - 4100}}{2}$$
$$= \frac{66 \pm \sqrt{256}}{2}$$
$$= \frac{66 \pm 16}{2}$$
$$= 33 \pm 8.$$

The solution set is

$$\{25, 41\}.$$

Check. Can the hypotenuse be 25 units long and one leg be 32 units shorter? Obviously not, and therefore, 25 is not an allowable answer to this particular problem. If the hypotenuse is 41 units long, then the two legs are 40 units and 9 units long.

$$41^2 \stackrel{?}{=} 40^2 + 9^2$$
$$1681 \stackrel{\checkmark}{=} 1600 + 81$$

Exercises

For each exercise, write an equation and state clearly what the variable denotes. Discard any solutions of each equation which are not reasonable answers to the question asked.

1. (a) The diagonal of a rectangle exceeds the width of the rectangle by 9 inches and the length by 8 inches. Find the dimensions of the rectangle.

(b) The diagonal of a square exceeds the length of its sides by 7 inches. Find the dimensions of the square.

2. (a) One side of a right triangle is 2 inches shorter than the hypotenuse. The other side is 1 foot and 4 inches shorter than the hypotenuse. Find the area of the triangle.

(b) The hypotenuse of a right triangle is 3 inches longer than one side. The other side is 15 inches shorter than the first side. Find the area of the triangle.

3. (a) If the bottom of a 13-foot ladder is 7 feet from a wall, how high up the wall does the ladder reach?

(b) If a 20-foot ladder reaches 10 foot high up a wall, how far is the bottom of the ladder from the wall?

4. (a) The diagonal of a rectangle is 119 inches, and the length is just 7 inches less than twice the width. Find the perimeter of the rectangle.

(b) The diagonal of a rectangle is 65 inches, and the length is 10 inches more than twice the width. Find the perimeter of the rectangle.

5. (a) A boy is cutting the grass on a lawn 40 feet wide by 90 feet long. If he cuts a border around the outside, how wide must the border be so that one-third of the lawn is cut?

(b) A square carpet has a border that is two feet wide. If two-thirds of its area is within the border, what are the dimensions of the carpet?

6. (a) The lengths of the sides of a right triangle, in inches, are given by consecutive even integers. Find the length of each side.

(b) The lengths of the sides of a right triangle, in inches, are given by consecutive integers. Find the length of each side.

---

7. (a) A motorboat travels 40 miles downstream and 40 miles back in 7 hours and 30 minutes. What is the rate of the current if the boat travels 12 miles an hour in still water?

(b) A plane flies between two points that are 700 miles apart, traveling with a wind of 25 miles an hour when going, and against it when returning. The trip out takes 15 minutes less time than the return flight. What is the speed of the plane in still air?

8. (a) A 4-inch square is cut from each corner of a square piece of tin. The remaining piece of tin is then folded to form an open box. If the volume of the box is 900 cubic inches, what is the size of the piece of tin?

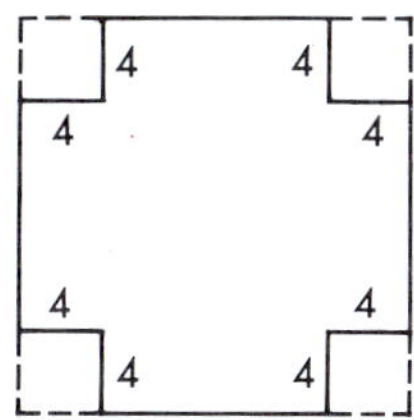

(b) A rectangular paper is 10 inches longer than it is wide. Four-inch squares are cut from each corner, and the ends folded to form an open box whose volume is 44 cubic inches. Find the dimensions of the paper.

---

9. A motorist could have completed a 225-mile trip in half an hour less time if he had driven 5 miles per hour faster. How fast did he drive?

10. A freighter whose speed is 6 knots slower than that of a passenger boat requires 3 days more time to make a trip of 2592 nautical miles. How fast does the freighter travel? (1 knot = 1 nautical mile per hour)

11. The time for the bus trip between two towns is 45 minutes longer than the time for the train trip between the same two towns. If the towns are 165 miles apart and the train's average speed is 10 miles per hour

faster than that of the bus, find one-decimal-place approximations of the average speeds of the bus and of the train.

12. A plane makes a round trip flight in 4 hours 30 minutes, the distance one way being 300 miles. If a wind of 10 miles per hour was with the plane on the trip out and against the plane on the trip back, what was the speed of the plane in still air?

13. A trip of 250 miles requires 8 hours and 45 minutes. If the first 50 miles are driven through city traffic at an average speed of 12 miles per hour less than the speed averaged beyond city traffic, what is the speed in city traffic?

12–9 GRAPHS OF QUADRATIC FUNCTIONS

An equation like

$$y = x^2 - 2x - 3$$

defines y as a function of x, and y is called a *quadratic function* of x. The graph of this function is the graph of set

$$S = \{(x, y) \mid y = x^2 - 2x - 3\}.$$

Seven ordered pairs in set S are given in the following table of values.

x	-2	-1	0	1	2	3	4
y	5	0	-3	-4	-3	0	5

These are plotted in Fig. 12–3.

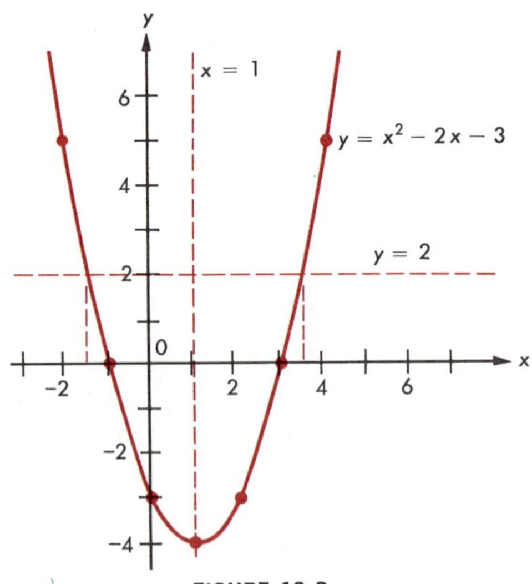

FIGURE 12–3

Obviously, these points do not lie on a straight line. If you plotted many more points on the graph, you would see that each point lies on the curve in the figure. This curve is called a *parabola.* It extends indefinitely upward and to the left beyond the point $(-2, 5)$, and indefinitely upward and to the right beyond the point $(4, 5)$.

A great deal of information about the quadratic function

$$y = x^2 - 2x - 3$$

may be obtained from the graph in Fig. 12–3. Remember, the y-coordinate of each point on the graph is the value of this function at the x-coordinate of that point.

For example, you can find those numbers x for which

$$x^2 - 2x - 3 = 0$$

by finding the points at which the parabola crosses the x-axis. By inspection, these are the points $(-1, 0)$ and $(3, 0)$. Therefore, -1 and 3 are solutions of the quadratic equation

$$x^2 - 2x - 3 = 0.$$

This illustrates a graphical method of solving quadratic equations.

You can also approximate the numbers that give other values of the function. Thus, you can approximate the numbers at which the value of the function is 2 by drawing the line $y = 2$ on the graph. This line cuts the parabola at the points with approximate x-coordinates of -1.5 and 3.5. Hence, -1.5 and 3.5 are approximations of the solutions of the equation $x^2 - 2x - 3 = 2$.

It also appears from the graph that -4 is the least value of the function $y = x^2 - 2x - 3$. This is the value of y when $x = 1$.

To show that -4 is actually the smallest possible value of y, you must find an equation equivalent to $y = x^2 - 2x - 3$ by completing the square.

$$\begin{aligned} y &= x^2 - 2x - 3 \\ y + 3 &= x^2 - 2x \\ (y + 3) + 1 &= x^2 - 2x + 1 \\ y + 4 &= (x - 1)^2 \\ y &= -4 + (x - 1)^2 \end{aligned}$$

Since $(x - 1)^2 \geqq 0$,

$$y \geqq -4 \text{ for every number } x.$$

Thus,

$$y = -4 \quad \text{if } x = 1, \quad y > -4 \quad \text{if } x \neq 1,$$

and -4 is the least value of y.

This least value, namely -4, is called the *minimum value* of the function $y = x^2 - 2x - 3$. If you trace your pencil along the parabola of Fig. 12–3, its path turns from downward to upward as it passes through the point $(1, -4)$. This turning point is commonly called the *vertex* of the parabola.

If you fold the page on which the parabola is drawn along the vertical line passing through the vertex, then the two halves of the parabola will coincide. For this reason, the parabola is said to be *symmetric* with respect to this line, and the line is commonly called the *axis of symmetry* of the parabola. The equation of the axis of symmetry for this parabola is $x = 1$.

The quadratic function

$$y = -x^2 + 4x - 5$$

is different from the one you just looked at because the x^2 term has a negative coefficient. Let us see how this difference affects the graph of the function.

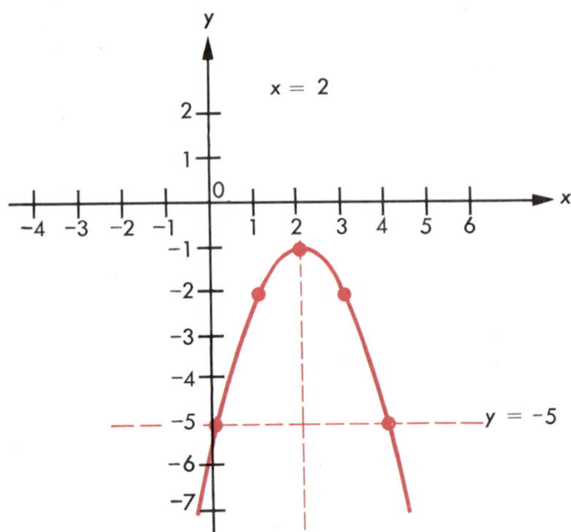

FIGURE 12–4

The graph of this function is sketched in Fig. 12–4 from the following table of values. This graph is also a *parabola*, but it extends indefinitely

x	0	1	2	3	4
y	-5	-2	-1	-2	-5

far downward to the left and to the right. For example, the points $(-20, -485)$ and $(24, -485)$ are on this parabola, even though they do not appear on the incomplete figure.

It is evident from the figure that this parabola does not cross the x-axis. This means that the equation $-x^2 + 4x - 5 = 0$ has no real-number solution.

On the other hand, the line $y = -5$ cuts the parabola in two points, $(0, -5)$ and $(4, -5)$. This means that the quadratic equation $-x^2 + 4x - 5 = -5$ has 0 and 4 as its solutions.

The quadratic function $y = -x^2 + 4x - 5$, unlike the previous one, does not have a least value. However, it appears to have a greatest, or maximum, value of -1. You can prove that such is the case by completing the square.

$$\begin{aligned} y &= -x^2 + 4x - 5 \\ y + 5 &= -x^2 + 4x \\ y + 5 &= -(x^2 - 4x) \\ y + 5 - 4 &= -(x^2 - 4x + 4) \\ y + 1 &= -(x - 2)^2 \\ y &= -1 - (x - 2)^2 \end{aligned}$$

Thus, $y = -1$ when $x = 2$, and $y < -1$ when $x \neq 2$, since $-(x - 2)^2 < 0$ and

$$-1 - (x - 2)^2 < -1 \quad \text{if } x \neq 2.$$

Hence, -1 is the *maximum value* of y.

The turning point on this parabola, $(2, -1)$, is again called the *vertex*, and $x = 2$ is an equation of the *axis of symmetry* of this parabola.

Exercises

In Exercises 1–8, graph each of the quadratic functions of x by making a table of values and plotting at least six points. In each case state the minimum or maximum value of the function y.

1. (a) $y = x^2$ (b) $y = x^2 + 1$

2. (a) $y = -x^2$ (b) $y = -x^2 + 1$

3. (a) $y = x^2 - 1$ (b) $y = -x^2 - 1$

4. (a) $y = -(x - 1)^2$ (b) $y = (x - 1)^2$

5. (a) $y = x^2 - 2x$ (b) $y = x^2 - 3x$

6. (a) $y = 2x - x^2$ (b) $y = 3x - x^2$

7. (a) $y = x^2 - 2x - 4$ (b) $y = x^2 - 2x - 3$

8. (a) $y = -x^2 + 6x - 10$ (b) $y = -x^2 + 4x - 6$

9. (a) Give the coordinates of the vertex and an equation of the axis of symmetry for each parabola in Exercises 1(a), 2(a), 3(a), 4(a), 5(a), 6(a), 7(a), and 8(a).

(b) Give the coordinates of the vertex and an equation of the axis of symmetry for each parabola in Exercises 1(b), 2(b), 3(b), 4(b), 5(b), 6(b), 7(b), and 8(b).

10. (a) (i) Graph the quadratic function

$$y = x^2 - x - 6.$$

(ii) From the graph, find values for x for which $x^2 - x - 6 = 0$.

(iii) From the graph, approximate the values for x at which $x^2 - x - 6 = -1$. (Hint: Draw the line $y = -1$ and tell where it cuts the parabola.)

(iv) Use the quadratic formula to find the exact solutions of the equation given in part (iii). Use the Table of Square Roots to approximate these irrational numbers. Compare with the graphical approximations you found in (iii).

(b) (i) Graph the quadratic function

$$y = x^2 - 2x - 8.$$

(ii) From the graph, find the values for x for which

$$x^2 - 2x - 8 = 0.$$

(iii) From the graph, approximate the values for x at which

$$x^2 - 2x - 8 = 1.$$

(iv) Use the quadratic formula to find the exact solutions of the equation given in part (iii). Use the Table of Square Roots to approximate these irrational numbers. Compare with the graphical approximation you found in (iii).

11. Graph the quadratic function

$$y = 1 - 3x - x^2.$$

(a) From the graph, find the values for x for which $1 - 3x - x^2 = 0$. Solve the equation $1 - 3x - x^2 = 0$, and use the Table of Square Roots to approximate the irrational solutions. Compare your answers with those obtained from the graph.

(b) Where does the line $y = 2$ cut the graph of $y = 1 - 3x - x^2$? When you use the quadratic formula, what solutions do you get for the equation $1 - 3x - x^2 = 2$?

12. (a) If $y = 3 + (x - 2)^2$, what is the minimum value of y?
(b) On the parabola which is the graph of $y = 3 + (x - 2)^2$, where is the vertex? What is an equation of the axis of symmetry?
(c) Plot the point which is the vertex of the parabola with equation

$$y = 3 + (x - 2)^2.$$

Find y when $x = 1$ or 3 and add these two points to the graph. Now find y when $x = 0$ or 4 and plot two more points. What positive value for x will give the same value for y as does $x = -1$? What negative value for x gives the same y value as does $x = 6$?

In Exercises 13–16, find the minimum values of the functions. Graph each of the functions, give the coordinates of the vertex, and write an equation for the axis of symmetry.

13. $y = x^2 + 2x + 3$

14. $y = x^2 + x - 2$

15. $y = x^2 - 7x + 12$

16. $y = x^2 + 3x + 2$

17. (a) Find the greatest value of the function $y = -x^2 + 2x - 4$ and sketch its graph.
(b) Try to solve the equation $-x^2 + 2x - 4 = 0$ by using the quadratic formula. What is your conclusion about the solutions?
(c) How does the graph in (a) reveal the information in (b)?

---

18. (a) Graph each of the following functions on the same set of axes.

$$y = x^2 - 2x, \quad y = x^2 - 2x + 3, \quad y = x^2 - 2x - 1$$

(b) What common feature do these three graphs have? What is the common feature of the functions?
(c) In what way do the graphs differ? How do the functions differ?
(d) Write an equation for a fourth member of this *family of parabolas.* Sketch the graph of your equation to see if it is correct.

19. (a) Graph each of the following functions on the same set of axes.

$$y = 2x^2, \quad y = 2(x - 1)^2, \quad y = 2(x + 1)^2$$

(b) How do your three graphs compare?
(c) Write a fourth function whose graph will belong to this set. Sketch its graph to see whether you are right.

20. (a) Graph each of the following functions on the same set of axes.

$$y = \tfrac{1}{2}(x - 1)^2, \quad y = \tfrac{1}{2}(x - 1)^2 + 2, \quad y = \tfrac{1}{2}(x - 1)^2 - 2$$

(b) Compare the three graphs and their equations. In what respect are they alike? In what respect do they differ?

21. On the same set of axes, graph three members of the family of parabolas expressed by equations of the form $y = x^2 + k$. Replace k by real numbers of your own choice. Describe this family of parabolas, telling in what way they are alike and in what way they differ. Do the same for the families of parabolas expressed by the equations of the form $y = -x^2 + k$.

22. (a) Graph each of the following on the same set of axes.

$$y = x^2 - 3x, \quad y = 3x - x^2$$

(b) How could one of these graphs be obtained from the other? How is the first quadratic in x related to the second quadratic in x?

23. (a) Graph each of the following on the same set of axes.

$$y = \tfrac{1}{2}x^2, \quad y = 2x^2, \quad y = 3x^2$$

(b) On the same set of axes used in (a), sketch the graphs of the following functions.

$$y = -\tfrac{1}{2}x^2, \quad y = -2x^2, \quad y = -3x^2$$

(c) Describe the set or family of parabolas expressed by the equations

$$y = kx^2$$

where $k > 0$.

(d) Describe the family of parabolas expressed by the equations $y = kx^2$ if $k < 0$.

24. In parts (a) and (b), graph both quadratic equations on one set of axes.

(a) $y = x^2, \quad x = y^2$

(b) $y = 1 - x^2, \quad x = 1 - y^2$

(c) In part (a), how could one graph be obtained from the other? How are the equations related? Do your observations about (a) also apply to (b)?

25. For any numbers a, b, and c with $a \neq 0$, the quadratic equation

$$y = ax^2 + bx + c$$

has a parabola as its graph. Which of these numbers determines whether the vertex is the highest point or the lowest point on the parabola?

*12-10 QUADRATIC INEQUALITIES

A parabola separates the plane into two parts just as a line does. Thus, the parabola with equation

$$y = x^2$$

separates the plane into the parts A and B in Fig. 12-5. You might call part A the "inside" and part B the "outside" of the parabola. Thus, every point of the plane is either (1) inside the parabola, or (2) on the parabola, or (3) outside the parabola.

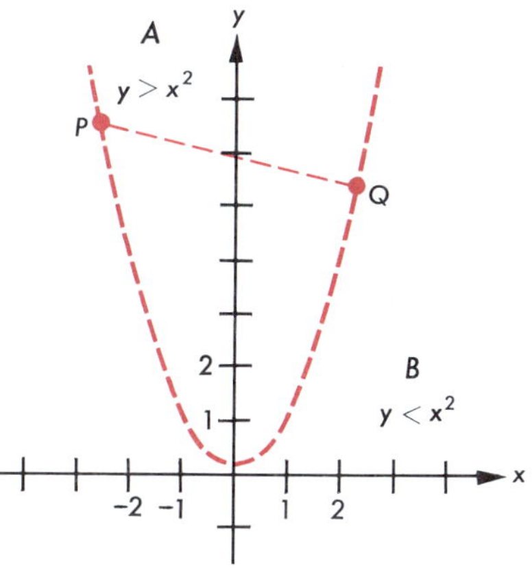

FIGURE 12-5

Part A could be described as consisting of all points lying between P and Q on some chord PQ of the parabola. (If P and Q are points on the parabola, the line segment joining them is called a *chord* of the parabola.) You give a description of part B. You must realize that the parabola extends indefinitely and is not closed like a circle. Thus, the "inside" of it also extends indefinitely.

You might recall that a half-plane was described as the solution set of a linear inequality. Can parts A and B of Fig. 12-5 be described in a similar way? Do you see that each point of part A is vertically above some point on the parabola? Thus, if the point of part A has coordinates (x, y), then it is vertically above the point (x, x^2) on the parabola. Therefore the y-coordinate of the point (x, y) is greater than the y-coordinate of the point (x, x^2):

$$y > x^2.$$

In other words, each point of part A has coordinates (x, y) that are solutions of the inequality $y > x^2$. It is also true that each number-pair solution of the inequality $y > x^2$ is a pair of coordinates of a point in A. Therefore,

$$A = \{(x, y) \mid y > x^2\}.$$

By a similar argument, you can show that

$$B = \{(x, y) \mid y < x^2\}.$$

The parabola itself is in neither A nor B, and therefore, it is illustrated on the graph with a dashed line.

The function

$$y = -x^2 + 4x - 5$$

and its graph were discussed in Section 12–9. For this parabola, the points "above" are "outside," and the points "below" are "inside." (See Fig. 12–4.) Therefore, the graph of

$$\{(x, y) \mid y > -x^2 + 4x - 5\}$$

is the set of points "outside" and the graph of

$$\{(x, y) \mid y < -x^2 + 4x - 5\}$$

is the set of points "inside" the parabola, as shown in Fig. 12–6.

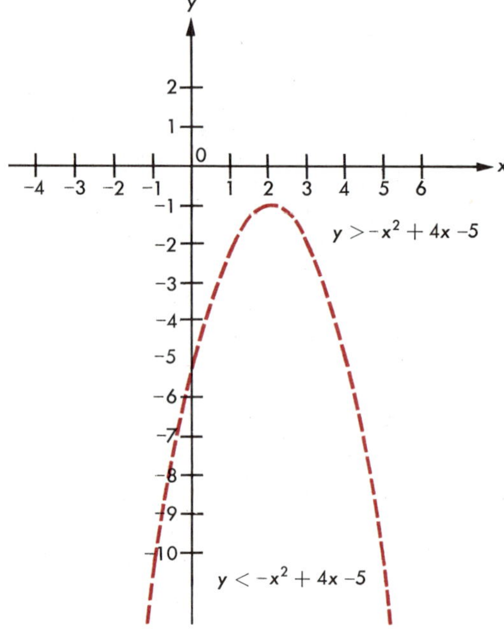

FIGURE 12–6

Exercises

Use inequalities to describe the "inside" and the "outside" of the graph of the following equations.

1. $y = x^2 - 1$

2. $y = -x^2 + 2x - 1$

3. $y = x^2 + 2x + 1$

4. $y = \frac{1}{2}x^2 - x - 4$

Sketch the graph of the following inequalities.

5. $y < x^2 - 3$

6. $y < x^2 - 3x + 1$

7. $y > \frac{1}{2}x^2$

8. $y > -x^2 - 3$

9. $y > x^2 + 6x + 9$

10. $y < -\frac{1}{2}x^2 + 1$

---

11. If $A = \{(x, y) \mid y \leqq x^2 - x - 6\}$,

(a) describe the graph of A and sketch it.

(b) tell which of the following points are in A.

$$(4, 6),\ (0, 9),\ (2, -1),\ (3, -2),\ (-1, -5),\ (1, 4)$$

Describe and sketch the graph of each of the following inequalities.

12. $y > 1 - 4x^2$

13. $y < x^2 - 3x + 4$

14. $y < -3x^2 + 6x - 3$

15. $y < 9 - 2x - \frac{1}{3}x^2$

16. $y \geqq 2x^2 + 3x - 6$

17. $y \leqq \frac{1}{2}x^2 - x - 4$

HISTORICAL NOTE

Problems involving quadratic equations were solved at least 3500 years ago. Such problems and their solutions have been deciphered from ancient Egyptian and Babylonian tablets. Although a general method is not given, it seems likely that the solutions involved completions of squares.

An old Babylonian tablet contains the following solution of the equation

$$x^2 - x = 870.$$

"Take one-half of 1, which is the coefficient of x, and square it. Then add $\frac{1}{4}$ to 870 to obtain $\frac{3481}{4}$. Now take the square root of $\frac{3481}{4}$ to get $\frac{59}{2}$. To this number, add one-half of 1, which is the coefficient of x. The resulting number, 30, is a solution of the equation."†

Needless to say, the solution above is a very loose translation of the original text. If we put words into symbols, we obtain the following steps.

$$x^2 - x + \tfrac{1}{4} = 870 + \tfrac{1}{4}$$
$$(x - \tfrac{1}{2})^2 = \tfrac{3481}{4}$$
$$x - \tfrac{1}{2} = \tfrac{59}{2}$$
$$x = \tfrac{59}{2} + \tfrac{1}{2}, \quad \text{or } 30$$

The other solution of this quadratic equation is a negative number, but such numbers were unknown to the ancient Egyptians and Babylonians.

---

† *Science Awakening*, B. L. Van Der Waerden, P. 69, Oxford University Press, N. Y., (1961).

KEY IDEAS AND KEY WORDS

A **quadratic equation** is a second-degree polynomial equation. The property of factors of zero is used in solving quadratic equations:

$$a \cdot b = 0 \quad \text{if, and only if,} \quad a = 0 \quad \text{or} \quad b = 0.$$

A quadratic equation is called **incomplete** if either its constant term is zero or its first-degree term is missing.

The polynomial

$$x^2 + 2ax + a^2$$

is called a **perfect-square trinomial.**

Completing the square is the process of finding a quadratic equation of the form

$$(x + a)^2 = b$$

which is equivalent to a given quadratic equation.

The number

$$b^2 - 4ac$$

is called the **discriminant** of the quadratic equation

$$ax^2 + bx + c = 0.$$

If $b^2 - 4ac \geqq 0$, then according to the **quadratic formula,** the solutions of the quadratic equation $ax^2 + bx + c = 0$ are the real numbers

$$x = \frac{-b \pm \sqrt{b^2 - 4ac}}{2a}.$$

The graph of a quadratic function of the form

$$y = ax^2 + bx + c, \quad a \neq 0,$$

is called a **parabola.** Every parabola has an **axis of symmetry** and a **vertex.**

CHAPTER REVIEW

Use the method of factoring to solve each of the following quadratic equations.

1. $x^2 - 4x - 5 = 0$

2. $2x^2 + 9x + 10 = 0$

3. $3x^2 - 5x = 0$

4. $25x^2 = 1$

5. $\frac{24}{x} + \frac{x}{3} = x$

6. $x^2 - 2x = 15$

7. $3x^2 + 10 = 17x$

8. $\frac{x}{14} = \frac{21}{x}$

Use the method of completing the square to find the solution set of each quadratic equation.

9. $x^2 - 6x + 4 = 0$

10. $3x^2 - 5x - 12 = 0$

11. $x^2 + 3x + 4 = 0$

12. $2x^2 - 6x + 3 = 0$

Use the quadratic formula to solve each of the following.

13. $6x^2 = 12x - 5$

14. $5x^2 = x + 5$

15. $2x^2 - 6x - 3 = 0$

Simplify each of the following, solve it by a method of your own choosing, and check your solution.

16. $x + \frac{1}{x} = \frac{-5}{2}$

17. $\frac{2x + 1}{2x - 3} + \frac{7x}{9 - 4x^2} = \frac{5}{3} + \frac{x - 4}{2x + 3}$

18. $2x + 4 = \sqrt{1 - x} - 4$

19. $(x + 3)(2x + 7) = 2$

For each of the given quadratic equations, compute the discriminant and describe the solution set without finding it.

20. $15x^2 + 8x - 63 = 0$

21. $6x^2 - x + 5 = 0$

22. $3x^2 - 2x - 6 = 0$

23. $4x^2 - 60x + 225 = 0$

24. Without solving, give the sum and the product of the solutions for each of the following quadratic equations.
(a) $5x^2 - 6x - 9 = 0$
(b) $3x^2 + 7x - 3 = 0$

25. Alice's friends gave her a radio costing \$24 for her sixteenth birthday. If four more friends had contributed, the cost to each would have been \$1 less. How many friends bought the radio?

26. In the auditorium of a school there are 616 seats arranged in rows with 6 more seats per row than the number of rows. How many seats are there in each row?

27. A certain rectangle has a length that is 2 feet greater than its width. If the width is increased by 2 feet, and the length is increased by 3 feet, the area of the larger rectangle is twice the area of the smaller one. Find the dimensions of the smaller rectangle.

28. When a certain number is decreased by 24 times its reciprocal, the remainder is 10. Find the number.

29. If a plane requires 30 minutes less time to travel 600 miles with a tail wind of 30 miles per hour than it requires to cover the same distance against a head wind of 30 miles per hour, what is the speed of the plane in still air?

Graph each of thc following quadratic functions.

30. $y = x^2 - 4$

31. $y = -x^2 + 4$

32. $y = 3 - (x - 2)^2$

33. $y = (x + 1)^2 - 2$

34. $y = 10x - x^2$

35. $y = x^2 - 4x + 3$

36. Give the coordinates of the vertex and an equation of the axis of symmetry of each of the parabolas in Exercises 30–35.

37. Give the maximum or minimum value of each of the functions in Exercises 30–35.

CHAPTER TEST

In Exercises 1–4, find the solution set of each equation.

1. $3x^2 - 14x + 5 = 0$

2. $12x^2 - 9x = 0$

3. $x^2 - 20x + 100 = 0$

4. $3t^2 - 2 = 0$

Use the method of completing the square to solve the following equations in Exercises 5–7.

5. $2x^2 = 12x + 3.$

6. $x^2 + 6x + 4 = 0.$

7. $6x^2 + 5 = 12x$

In Exercises 8–10, use the quadratic formula to find the solution set of the following equations.

8. $8x^2 - 2x - 5 = 0$

9. $5x^2 = 8x - 6$

10. $3x^2 + 10 = 12x$

Without solving, find the discriminant and discuss the solution set of each quadratic equation in Exercises 7 and 8.

11. $2x^2 - 5x - 3 = 0$

12. $3x^2 - 2x + 5 = 0$

13. Find all numbers such that the sum of the number and seven times its reciprocal is eight.

14. A rectangle has a length 4 feet greater than its width. If its length is increased by 3 feet and the width decreased by 2 feet, the area of the new rectangle is 32 square feet less than twice the area of the original rectangle. Find the dimensions of the original rectangle.

Graph each of the following quadratic functions on the same set of axes.

15. $y = \frac{1}{2}x^2$

16. $y = \frac{1}{2}x^2 + 2$

17. $y = -\frac{1}{2}x^2$

18. $y = -\frac{1}{2}x^2 - 2$

19. At what points does the graph of $y = x^2 - 7x + 6$ intersect the x-axis? the y-axis?

20. Find the vertex and the axis of symmetry for the graph of

$$y = x^2 - 7x + 6.$$

21. Does the function $y = x^2 - 7x + 6$ have a maximum value or a minimum value? If so, what is it?

CHAPTER 13

Trigonometry

Objectives . . .

- To use ratio and proportion to express physical relationships.
- To define the basic trigonometric functions.
- To use tables to find values of the trigonometric functions.
- To apply trigonometric ratios to the solution of simple geometric and physical problems.

13–1 PROPORTIONS

An indicated quotient of two numbers or algebraic expressions is called a *ratio,* and an equation stating the equality of two ratios is called a *proportion.* The equation

$$\frac{x}{84} = \frac{50}{32}$$

is an example of a proportion.

Recall from Chapter 6 that two variables, say x and y, *vary directly* with respect to each other if they are related by a linear equation of the form

$$y = kx$$

for some nonzero number k.

It is convenient to use a variable with subscripts to indicate particular values of the variable. Thus, x_1 and x_2 designate two values of the variable x, whereas y_1 and y_2 designate two values of y. Assume that (x_1, y_1) and (x_2, y_2) are two particular solutions of the equation $y = kx$, so that

$$\frac{y_1}{x_1} = k \quad \text{and} \quad \frac{y_2}{x_2} = k.$$

Then

$$\frac{y_1}{x_1} = \frac{y_2}{x_2}.$$

Any two nonzero solutions are said to be *directly proportional.* A proportion can also be written in the form

$$\frac{y_1}{y_2} = \frac{x_1}{x_2}.$$

Problem 1. The weight of a metal rod varies directly with its length. If a 15-foot rod weighs 6 pounds, how much does a 35-foot rod weigh?

Solution. Use the variables W and L for weight and length, respectively. You know that any two particular solutions (W_1, L_1) and (W_2, L_2) of the equation relating W and L are proportional; that is,

$$\frac{W_1}{L_1} = \frac{W_2}{L_2}.$$

If you think of W_1 as 6, L_1 as 15, and L_2 as 35, you must find the number W_2 such that

$$\frac{6}{15} = \frac{W_2}{35}.$$

You find

$$35 \times \tfrac{2}{5} = W_2,$$
$$W_2 = 14.$$

Thus, a rod 35 feet long weighs 14 pounds.

Two variables, say x and y, *vary inversely* with respect to each other if they are related by an equation of the form

$$xy = k$$

for some nonzero number k.

If the variables x and y vary inversely and if (x_1, y_1) and (x_2, y_2) are two particular solutions of the equation $xy = k$, which relates x and y, then both x_1y_1 and x_2y_2 must equal k. Hence, they must be equal to each other:

$$x_1y_1 = x_2y_2.$$

On dividing each side of this equation by x_1y_2, you obtain

$$\frac{x_1y_1}{x_1y_2} = \frac{x_2y_2}{x_1y_2}, \quad \text{or} \quad \frac{y_1}{y_2} = \frac{x_2}{x_1}.$$

According to this equation, x and y are *inversely proportional*, because the order in which the subscripts are written is inverted from one ratio to the other.

Problem 2. The volume of a certain quantity of gas is inversely proportional to its pressure. If its pressure is 15 pounds per square inch (abbreviated psi) when its volume is 3 cubic feet, what will the pressure be when the gas is compressed to a volume of 2 cubic feet?

Solution. Let V designate the volume in cubic feet and P the pressure in psi of the given quantity of gas. Then

$$\frac{V_1}{V_2} = \frac{P_2}{P_1}$$

for any two pairs of values (V_1, P_1) and (V_2, P_2). If you let $V_1 = 3$, $P_1 = 15$, and $V_2 = 2$, you get the equation

$$\frac{3}{2} = \frac{P_2}{15}.$$

To solve this equation, multiply each side by 15:

$$P_2 = \tfrac{45}{2}, \text{ or } 22\tfrac{1}{2}.$$

Thus, the gas exerts a pressure of $22\frac{1}{2}$ psi if its volume is 2 cubic feet.

Exercises

In each proportion, supply the missing number.

1. (a) $\dfrac{2}{6} = \dfrac{10}{?}$ (b) $\dfrac{3}{15} = \dfrac{9}{?}$

2. (a) $\dfrac{7}{8} = \dfrac{?}{16}$ (b) $\dfrac{3}{4} = \dfrac{?}{24}$

3. (a) $\dfrac{?}{4} = \dfrac{18}{24}$ (b) $\dfrac{?}{5} = \dfrac{10}{25}$

4. (a) $\dfrac{14}{?} = \dfrac{98}{35}$ (b) $\dfrac{2}{?} = \dfrac{16}{24}$

5. If x is directly proportional to y and $x = 3$ when $y = 8$,
(a) what is the value of x when $y = 16$?
(b) what is the value of y when $x = 12$?

6. If a and b are directly proportional and $a = 7$ when $b = 13$,
(a) what is the value of b when $a = 91$?
(b) what is the value of a when $b = 91$?

7. If x and y are inversely proportional and $x = 3$ when $y = 8$,
(a) what is the value of x when $y = 16$?
(b) what is the value of y when $x = 6$?

8. If a and b are inversely proportional and $a = 16$ when $b = 9$,
(a) what is the value of b when $a = 18$?
(b) what is the value of a when $b = 4$?

9. (a) The ratio of an object's weight on earth to its weight on Mars is approximately 3 to 1. How much would a man who weighs 210 pounds on earth weigh on Mars?

(b) The ratio of an object's weight on earth to its weight on Neptune is approximately 5 to 7. How much would a man who weighs 180 pounds on earth weigh on Neptune?

10. (a) The number of hours required to grade a set of tests varies inversely with the number of graders if they work at the same rate. If the set can be graded in 3 days by 20 people, how many days would 12 people require?

(b) If 9 men do a job in 4 days, how long do 15 men take?

11. (a) If the cost of transportation is proportional to the distance traveled, and the ticket for a 500-mile trip costs $22.50, how much should the ticket for a 375-mile trip cost?

(b) If it takes 8 hours to make a certain trip going at the rate of 40 miles per hour, how long would it take to make the same trip at the rate of 65 miles per hour?

---

12. If 9 grams of hydrochloric acid neutralize 10 grams of lye, how many grams of lye are neutralized by 270 grams of hydrochloric acid?

13. In a set of rectangles of the same area, the length and width of each rectangle are inversely proportional. If one rectangle is 8 inches wide and 15 inches long, what is the width of a rectangle which is 12 inches long?

14. The time required to make a certain trip is inversely proportional to the speed at which you travel. If the trip takes 5 hours at an average speed of 40 miles per hour, how much time could be saved by averaging 45 miles per hour?

15. The distance at which one should sit from the support of a seesaw is inversely proportional to his weight. If a boy who weighs 80 pounds sits 5 feet from the support, how far from the support should a 100-pound boy sit for them to be in balance?

16. In photographic enlargement or reduction, the dimensions of the new picture are directly proportional to those of the original picture.

(a) What is the height of the enlargement of a photograph which is 3 inches wide and 5 inches high if the width of the enlargement is 5 inches?

(b) A photograph which is 6 inches high and 10 inches wide will be reduced so that a two-column cut of it can be made for a newspaper. If the width of one column is $2\frac{1}{4}$ inches, how high will the cut be?

17. If taxes on property are directly proportional to the assessed value of the property and the tax on a $15,000 house is $480, what is the tax on a $12,000 house?

18. If a belt goes around two pulleys, the number of revolutions per minute of each pulley is inversely proportional to its radius. What is the speed of revolution of an 8-inch wheel if a 5-inch wheel is spinning at 120 revolutions per minute?

13-2 USING PROPORTIONS

Some additional problems involving proportions are given below.

Problem 1. The weight of a solid sphere is directly proportional to the cube of its radius. Suppose that a sphere of radius 2 inches weighs 10 pounds. (a) What is the weight of a sphere of radius 3 inches made out of the same substance? (b) What is the radius of an 80-pound sphere of the same substance?

Solution. If W and R designate the weight in pounds and radius in inches of a sphere, then

$$\frac{W_1}{(R_1)^3} = \frac{W_2}{(R_2)^3},$$

for any two pairs of values (W_1, R_1) and (W_2, R_2). For both parts of the problem, the first weight, W_1, is 10 pounds, and the first radius, R_1, is 2 inches. In part (a), R_2 is 3 inches and W_2 is unknown. Making these replacements, you obtain the equation

$$\frac{10}{2^3} = \frac{W_2}{3^3},$$

which can be solved as follows:

$$\frac{10}{8} = \frac{W_2}{27},$$

$$\tfrac{5}{4} \times 27 = W_2,$$

$$W_2 = \tfrac{135}{4}, \quad \text{or } 33\tfrac{3}{4}.$$

Thus, a sphere of radius 3 inches weighs $33\frac{3}{4}$ pounds.

For part (b), the second weight, W_2, is 80 pounds, and the second radius, R_2, is unknown. Making these replacements in the original proportion, you obtain the equation

$$\frac{10}{2^3} = \frac{80}{(R_2)^3},$$

$$\frac{10}{8} = \frac{80}{(R_2)^3},$$

$$(R_2)^3 = 64,$$

$$R_2 = 4.$$

Therefore, the 80-pound sphere has a radius of 4 inches.

The four parts of a proportion are usually labeled in the following way.

$$\frac{\text{First term}}{\text{Second term}} = \frac{\text{Third term}}{\text{Fourth term}}$$

The first and fourth terms are called the ***extremes,*** and the second and third terms are called the ***means*** of the proportion. The equation above is equivalent to the equation

$$(\text{First term}) \times (\text{Fourth term}) = (\text{Second term}) \times (\text{Third term}).$$

In other words, it is true for every proportion that *the product of the means is equal to the product of the extremes.* If the means are equal, their common value is called a ***mean proportional*** between the extremes.

You may obtain two other proportions equivalent to the one above by interchanging means,

$$\frac{\text{First term}}{\text{Third term}} = \frac{\text{Second term}}{\text{Fourth term}},$$

or extremes,

$$\frac{\text{Fourth term}}{\text{Second term}} = \frac{\text{Third term}}{\text{First term}}.$$

Do you see that, in both cases the *product* of the means and the *product* of the extremes have not changed?

Problem 2. Find the fourth proportional to 6, 8, and 15.

Solution. The fourth proportional is the solution of the equation

$$\frac{6}{8} = \frac{15}{x}.$$

Multiply the means and the extremes to obtain an equivalent equation,

$$6x = 120,$$

which has the solution

$$x = 20.$$

Check.

$$\tfrac{6}{8} \stackrel{?}{=} \tfrac{15}{20}$$

$$\tfrac{3}{4} \stackrel{\checkmark}{=} \tfrac{3}{4}$$

Problem 3. Find a mean proportional between 8 and 18.

Solution. You are asked to solve the equation

$$\frac{8}{x} = \frac{x}{18}.$$

An equivalent equation is

$$8 \cdot 18 = x \cdot x,$$

or

$$x^2 = 144.$$

You know that

$$\{12, -12\}$$

is the solution set of this quadratic equation. Thus, both 12 and -12 are mean proportionals between 8 and 18.

Check.

$$\frac{8}{12} \stackrel{?}{=} \frac{12}{18} \qquad \frac{8}{-12} \stackrel{?}{=} \frac{-12}{18}$$

$$\frac{2}{3} \stackrel{\checkmark}{=} \frac{2}{3} \qquad \frac{-2}{3} \stackrel{\checkmark}{=} \frac{-2}{3}$$

Exercises

In Exercises 1–10, solve each of the following.

1. (a) $\frac{x}{8} = \frac{5}{2}$ (b) $\frac{x}{5} = \frac{4}{6}$

2. (a) $\frac{8}{x} = \frac{5}{2}$ (b) $\frac{5}{x} = \frac{4}{6}$

3. (a) $\frac{5}{8} = \frac{x}{2}$ (b) $\frac{4}{5} = \frac{c}{6}$

4. (a) $\frac{5}{8} = \frac{2}{x}$ (b) $\frac{4}{5} = \frac{6}{c}$

5. (a) $\frac{7}{8} = \frac{d}{9}$ (b) $\frac{8}{9} = \frac{c}{7}$

6. (a) $\frac{2y}{3} = \frac{8}{9}$ (b) $\frac{3x}{2} = \frac{8}{9}$

7. (a) $\frac{x}{7 - x} = \frac{2}{3}$ (b) $\frac{p}{9 - p} = \frac{21}{6}$

8. (a) $\frac{r}{3} = \frac{5}{r - 2}$ (b) $\frac{x}{4} = \frac{6}{x - 2}$

9. (a) $\frac{y}{y-2} = \frac{y-3}{y}$ (b) $\frac{a}{a-4} = \frac{a-1}{a}$

10. (a) $\frac{x-2}{x-4} = \frac{x+3}{x-5}$ (b) $\frac{x-5}{x-3} = \frac{x+4}{x-1}$

Find the fourth proportional to each of the following triples.

11. (a) 3, 4, 6 (b) 4, 9, 7

12. (a) 3, 9, 27 (b) 9, 7, 27

13. (a) 3, $\sqrt{3}$, $\sqrt{6}$ (b) 2, $\sqrt{2}$, $\sqrt{10}$

Find a mean proportional between each of the following pairs of numbers.

14. (a) 4, 9 (b) 2, 8

15. (a) 5, 125 (b) 9, 16

16. (a) $\sqrt{7}$, $\sqrt{63}$ (b) $\sqrt{5}$, $\sqrt{80}$

Solve each of the following equations.

17. $\frac{t}{t+12} = \frac{t+2}{2t+19}$

18. $\frac{3x-4}{x^2-x-3} = \frac{2}{x-3}$

19. $\frac{5y-13}{4y+5} = \frac{4y-5}{5y+13}$

20. $\frac{12}{x+1} = \frac{x+1}{3}$

Find the fourth proportional to each of the following triples.

21. a, a^2, a^3

22. $\sqrt{a}, a, a\sqrt{a}$

Find the mean proportional between each of the following pairs.

23. 3, 5

24. $\frac{4}{15}, \frac{15}{4}$

25. b, b^3

26. If w varies directly with z^2 and $w = 25$ when $z = 3$,
(a) what is the value of w when $z = 4$?
(b) what is the value of z when $w = 4$?

27. If x varies directly with y^3 and $x = 125$ when $y = 4$,
(a) what is the value of x when $y = 2$?
(b) what is the value of y when $x = 27$?

28. If c^2 is directly proportional to d^3 and $c = 8$ when $d = 4$,
(a) what is the value of c when $d = 9$?
(b) what is the value of d when $c = \frac{27}{8}$?

29. If x varies directly with the cube of y,
(a) what is the effect on x if y is doubled? Illustrate your answer.
(b) how does the value of x change if y is trebled? Illustrate your answer.

13-3 SIMILAR TRIANGLES

As you know, a triangle has three points for vertices, three line segments for sides, and three angles for corners. For convenience, the vertices are designated by capital letters, such as A, B, and C. Then the three sides may be designated by $\overline{AB}$, $\overline{AC}$, and $\overline{BC}$, and the three angles by $\angle BAC$, $\angle ABC$, and $\angle ACB$. This labeling is illustrated in Fig. 13-1.

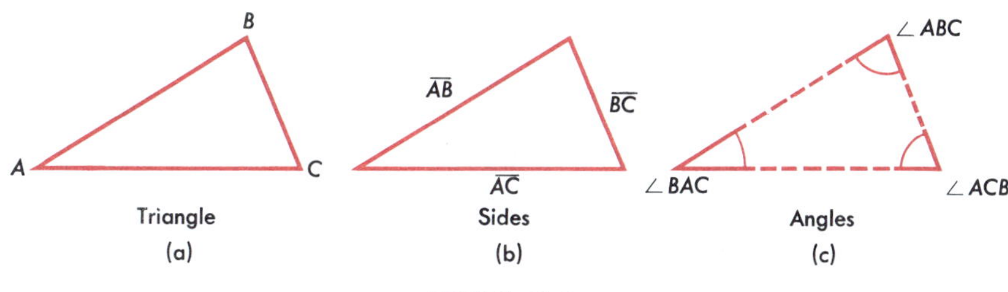

FIGURE 13-1

Each side of a triangle has two *adjacent* angles and one *opposite* angle. Thus, $\overline{AB}$ has adjacent angles $\angle BAC$ and $\angle ABC$, and opposite angle $\angle ACB$. Similarly, each angle of a triangle has two adjacent sides and one opposite side. For example, $\angle ACB$ has adjacent sides $\overline{AC}$ and $\overline{CB}$ and opposite side $\overline{AB}$.

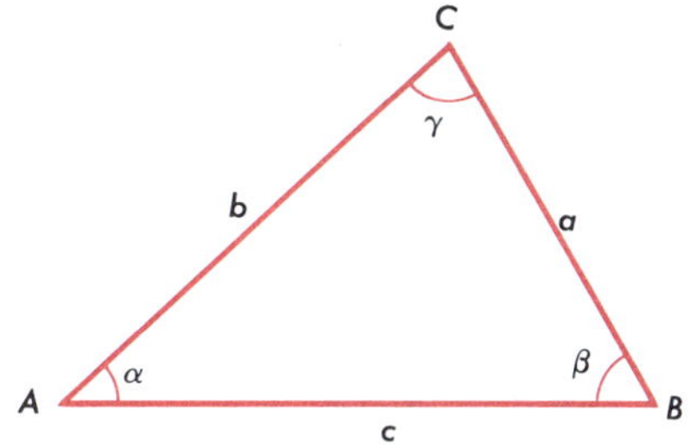

FIGURE 13-2

For convenience, lower-case letters such as a, b, and c will be used to designate the lengths of the sides of a triangle (using some preassigned units of length) and Greek letters such as α, β, and γ (alpha, beta, and gamma) to designate the measures in degrees of the angles. Furthermore, the length of the side opposite vertex A will frequently be labeled by a, the side opposite vertex B by b, and the side opposite vertex C by c. Also, the measure of the angle A is often denoted by α, the measure of the angle at B by β, and the measure of the angle at C by γ, as shown in Fig. 13-2. An interesting fact of geometry is that the sum of the angles of a triangle is two right angles. Therefore,

$$\alpha + \beta + \gamma = 180^\circ.$$

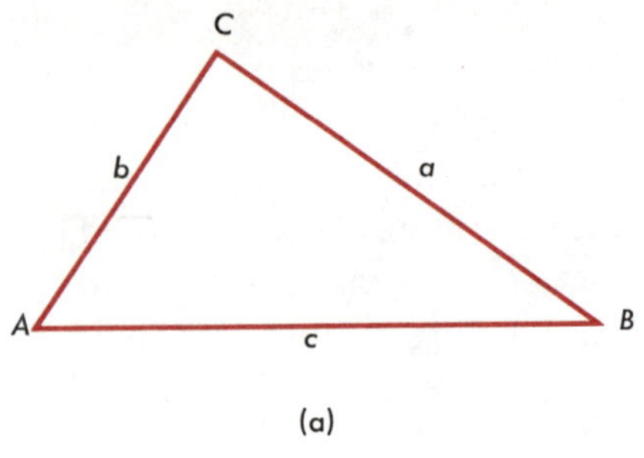

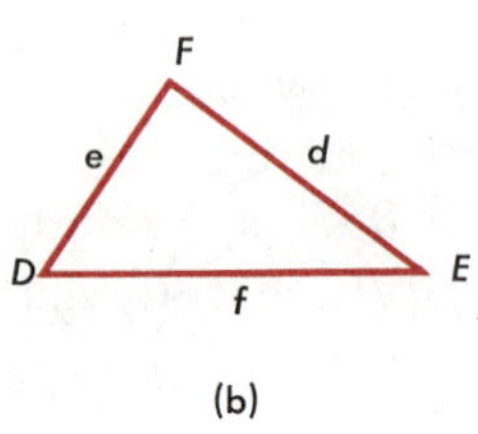

FIGURE 13–3

Although the two triangles shown in Fig. 13–3 are different, they look alike. The reason for this similarity is that the corresponding angles of the two triangles have the same measure. Two triangles with corresponding angles that have the same measure are called *similar triangles.*

Definition of similar triangles

Two triangles are similar if the measures of the corresponding angles are equal.

You will learn in your geometry course that the lengths of corresponding sides of similar triangles are proportional; that is,

$$\frac{a}{d} = \frac{b}{e}, \quad \frac{a}{d} = \frac{c}{f}, \quad \frac{b}{e} = \frac{c}{f}.$$

Conversely, it is also true that if two triangles can be made to correspond so that the ratios of lengths of corresponding sides are equal, then the measures of the corresponding angles are equal and the triangles are similar.

A right triangle has one right angle and two acute angles (an acute angle has a measure of less than 90°). In Fig. 13–4, the angles at A and B are acute angles and the angle at C is a right angle. (In figures, a right angle is usually indicated by placing a small square at its vertex.) According to the pythagorean theorem, the lengths of the sides of a right triangle are related by the equation

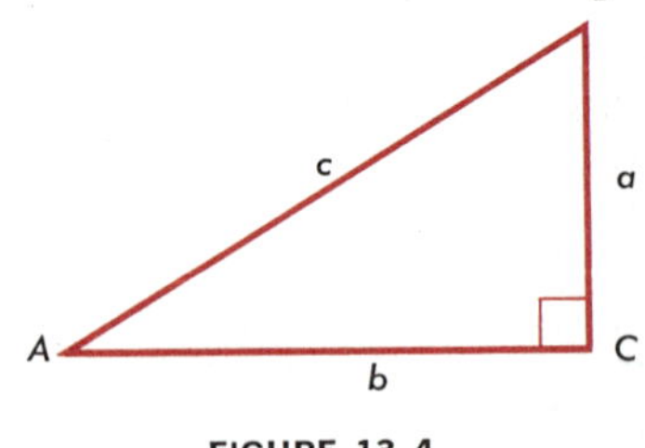

FIGURE 13–4

$$a^2 + b^2 = c^2.$$

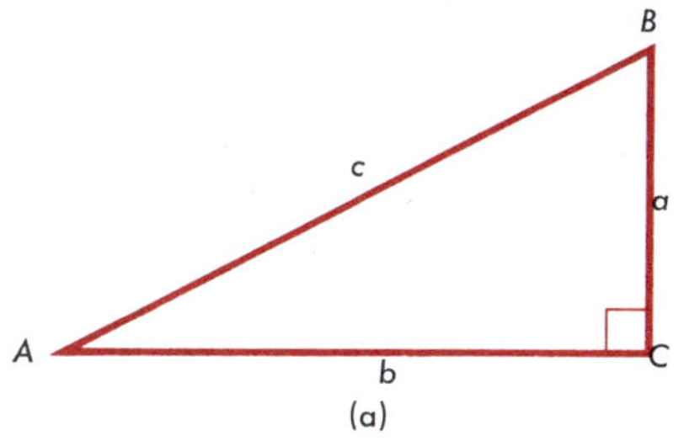

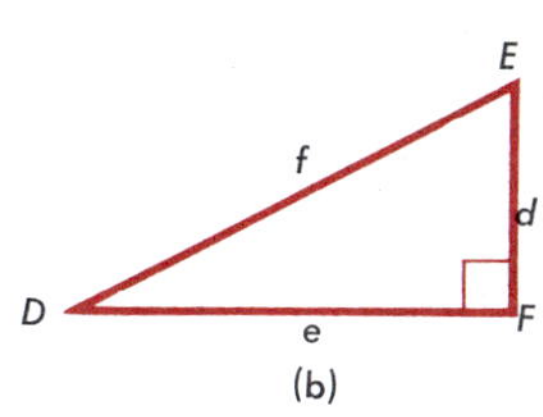

FIGURE 13-5

It should be evident that two right triangles are similar if an acute angle of one equals an acute angle of the other. For example, if in the right triangles shown in Fig. 13-5, the angles A and D have the same measure, then the angles at B and E have the same measure. Hence, you can conclude that the triangles are similar. Because this is so, the corresponding sides have proportional lengths:

$$\frac{a}{d} = \frac{b}{e}, \quad \frac{a}{d} = \frac{c}{f}, \quad \frac{b}{e} = \frac{c}{f}.$$

Problem 1. If two right triangles have an angle in common as indicated in Fig. 13-6 and if the lengths are as shown in the figure, find a, which is the length of one of the legs of the smaller triangle. Also find the length of the hypotenuse of each triangle.

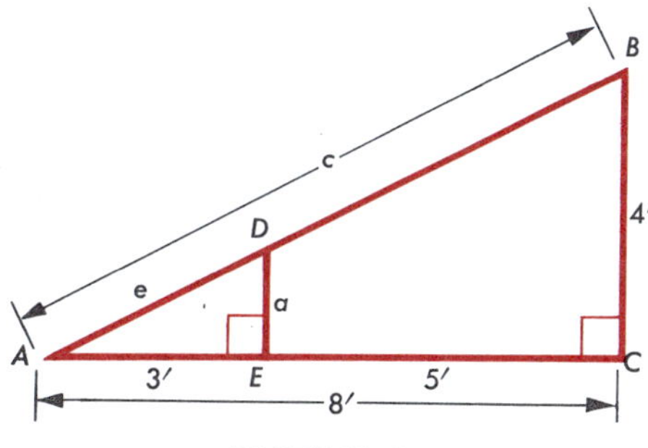

FIGURE 13-6

Solution. The two triangles ABC and ADE in Fig. 13-6 are similar because the angle at vertex A is in each triangle. You know then that corresponding sides must have proportional lengths. In the larger triangle, the side opposite A has length 4 inches; in the smaller triangle, the side opposite A has length a inches. The side opposite B in the larger triangle has length 8 inches; the side opposite the corresponding vertex D in the smaller triangle has length 3 inches. Hence, you can obtain the proportion

$$\frac{4}{a} = \frac{8}{3}.$$

Solve this equation for a as follows:

$$4 \times 3 = a \times 8,$$
$$a = \tfrac{12}{8}, \quad \text{or } \tfrac{3}{2}.$$

Thus, the side opposite A in the smaller triangle has a length of $1\frac{1}{2}$ inches. Using the pythagorean theorem for the smaller triangle, you obtain

$$3^2 + (\tfrac{3}{2})^2 = e^2,$$
$$9 + \tfrac{9}{4} = e^2,$$
$$\tfrac{45}{4} = e^2,$$
$$e = \tfrac{3}{2}\sqrt{5}.$$

Since $\sqrt{5} \doteq 2.24$, $e \doteq 3.36$ inches. Using the pythagorean theorem for the larger triangle, you have

$$8^2 + 4^2 = c^2,$$
$$64 + 16 = c^2,$$
$$80 = c^2,$$
$$c = 4\sqrt{5}.$$

Thus, $c \doteq 8.96$ inches.

Problem 2. Triangle ABC in Fig. 13–7 is a right triangle with right angle at C. Find h, the length of the altitude of the triangle.

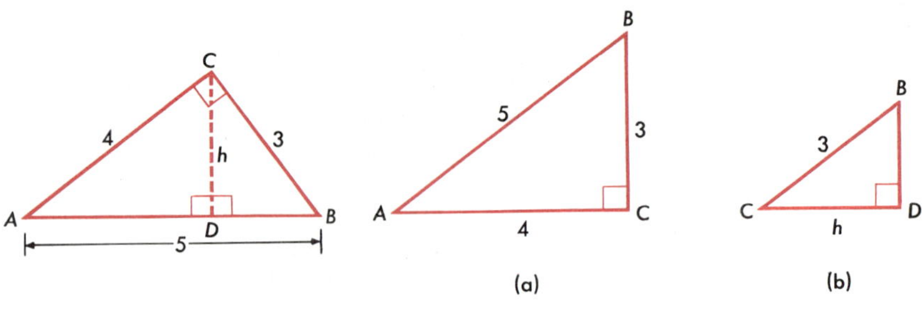

FIGURE 13–7

FIGURE 13–8

Solution. Triangle BCD of the figure is also a right triangle, with right angle at D. Since the right triangles ABC and CBD have a common acute angle at B, the triangles are similar. The triangles are shown in Fig. 13–8, so that corresponding angles and sides can be seen. Thus,

$$\frac{5}{3} = \frac{4}{h},$$
$$5h = 12,$$
$$h = \tfrac{12}{5}, \quad \text{or} \quad 2\tfrac{2}{5}.$$

Exercises

1. (a) If two angles of a triangle measure 40° and 60°, respectively, what is the measure of the largest angle in a similar triangle?
   (b) If two angles of a triangle measure 50° and 30°, respectively, what is the measure of the largest angle in a similar triangle?
2. (a) Two angles of a certain triangle measure 25° and 55°. What is the measure of each angle of a similar triangle?
   (b) Two angles of a certain triangle measure 65° and 15°. What is the measure of each angle of a similar triangle?
3. (a) In right triangles ABC and PQR, the measure of angle BAC equals the measure of angle QPR. Find c, q, and r.

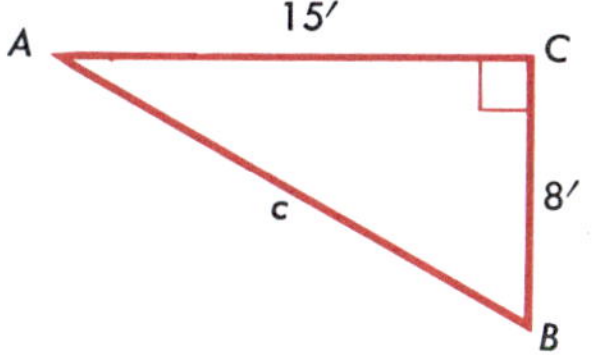

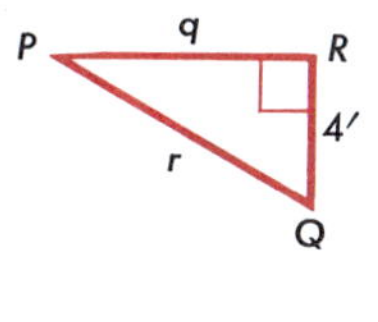

   (b) In right triangles DEF and RST, the measure of angle EDF equals the measure of angle SRT. Find f, s, and t.

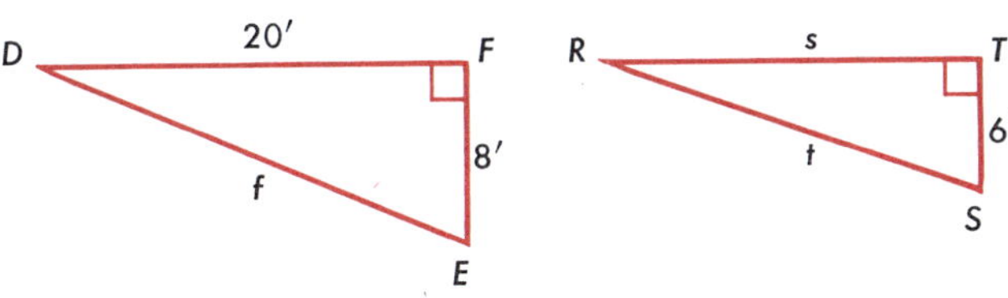

4. (a) In right triangles KMN and ZXY, the measure of angle NMK equals the measure of angle YXZ. Find k, y, and z.

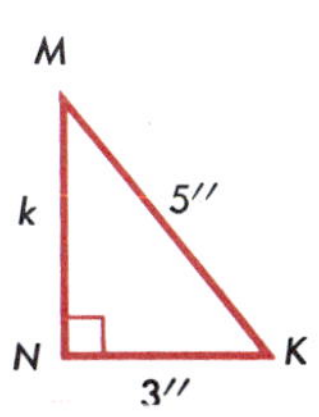

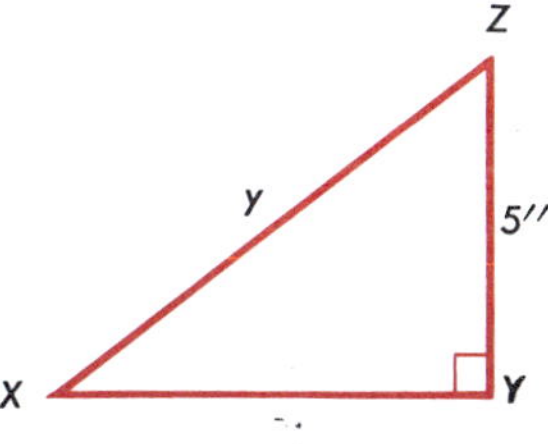

(b) In right triangles MNO and PQN, the measure of angle ONM equals the measure of angle RQP. Find m, r, and p.

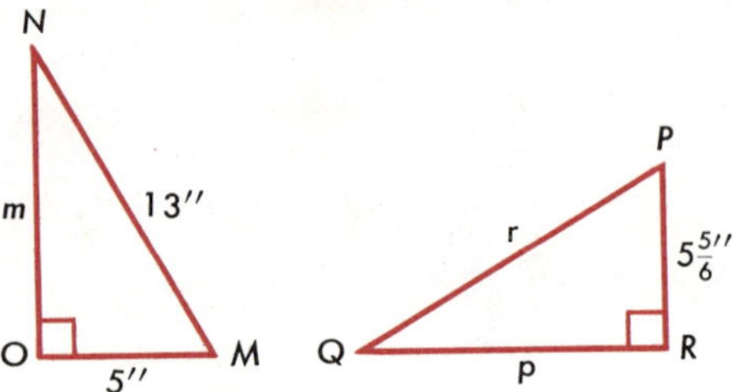

5. (a) In right triangles DEF and UTV, the measure of angle DEF equals the measure of angle UTV. Find d, t, and v.

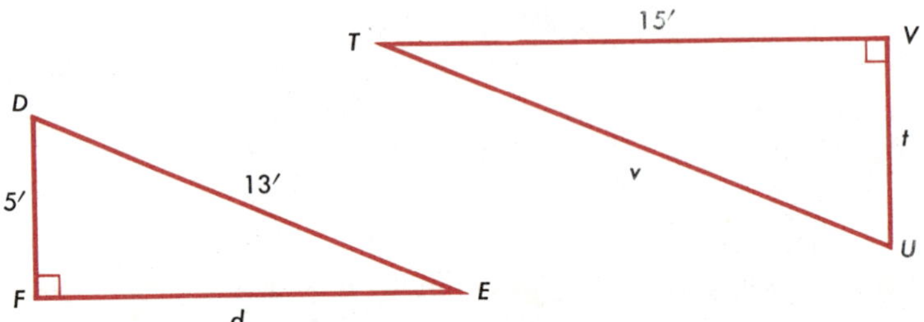

(b) In right triangles ABC and DEF, the measure of angle ABC equals the measure of angle DEF. Find a, e, and f.

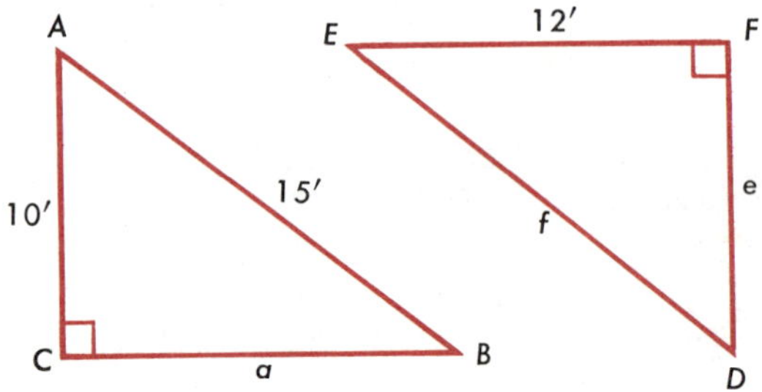

6. Triangles ABC and DEF are similar.

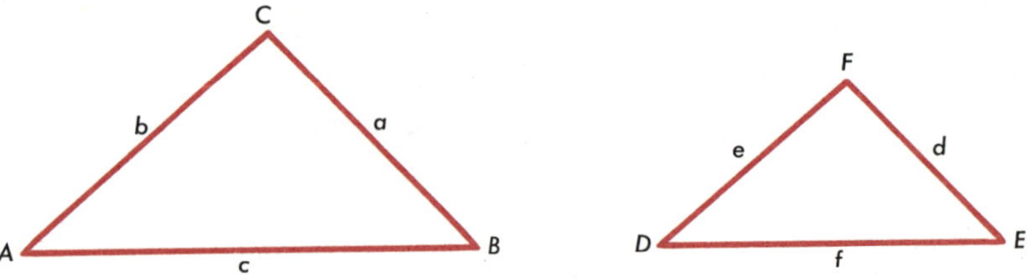

(a) Find a if $b = 3$ inches, $e = 1$ inch, and $d = 4$ inches.
(b) Find d and f, if $a = 16$ inches, $b = 12$ inches, $c = 20$ inches, and $e = 15$ inches.

7. (a) A triangle has sides of 3, 5, and 10 inches. The longest side of a similar triangle is 8 inches. How long are its other sides?
   (b) A triangle has sides of 4, 8, and 18 inches. The shortest side of a similar triangle is 3 inches. How long are its other sides?
8. (a) A man 6 feet tall casts a shadow 7 feet long. What is the height of a pole which at the same time casts a shadow 10 feet long?
   (b) How long would a shadow be if it were cast by a tower 90 feet tall at the same time a pole 15 feet tall casts a shadow of length 3 feet?
9. (a) Find the height of a tree that casts a 30-foot shadow at the same time that a yardstick casts a 9-foot shadow.
   (b) A pole 5 feet high casts a 12-foot shadow, and a tree nearby casts a 30-foot shadow. How high is the tree?

---

10. An acute angle on one right triangle is 55°, and an acute angle of another right triangle is 35°. Are the two triangles similar? List the measures of the angles.

13–4 TRIGONOMETRIC FUNCTIONS

If $\triangle ABC$ and $\triangle DEF$ are similar right triangles, as indicated in Fig. 13–9, you know that

$$\frac{a}{d} = \frac{b}{e}, \quad \frac{a}{d} = \frac{c}{f}, \quad \frac{b}{e} = \frac{c}{f}.$$

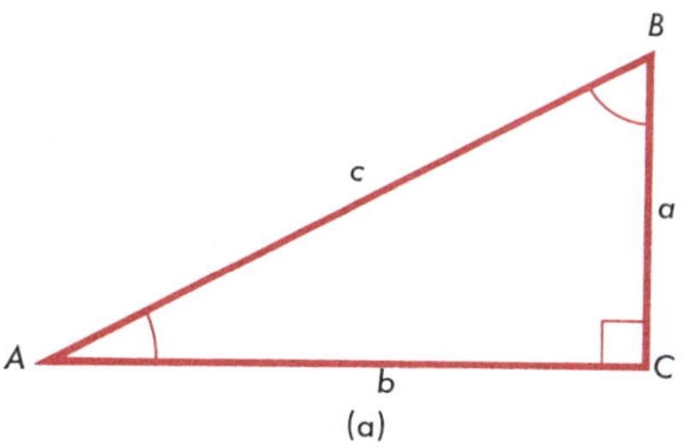

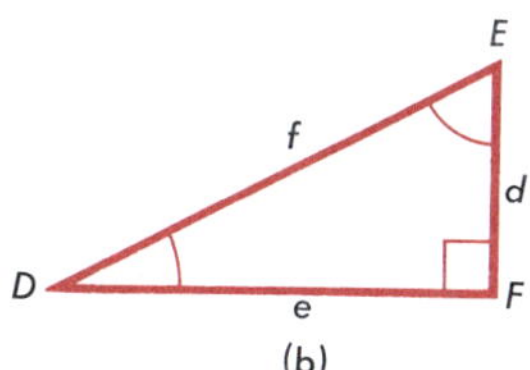

FIGURE 13-9

If you interchange means in each of these proportions, you obtain the equivalent proportions

$$\frac{a}{b} = \frac{d}{e}, \quad \frac{a}{c} = \frac{d}{f}, \quad \frac{b}{c} = \frac{e}{f}.$$

Note that each ratio in these new proportions is a ratio of the lengths of two sides of the same triangle. *If two right triangles are similar, the ratio of the lengths of any two sides of one triangle equals the ratio of the lengths of the two corresponding sides of the other triangle.* Actually, this statement is true for any two similar triangles, whether or not they are right triangles.

If you start with a number α between 0 and 90, $0 < \alpha < 90$, all right triangles having an angle of measure α are similar. Then the ratios of the lengths of the sides are the same for every one of these triangles. In other words, these ratios are functions of α. If $\triangle ABC$ is any right triangle having an angle of measure α (Fig. 13–10), then the functions *sine*, *cosine*, and *tangent* of α are defined in the following way.

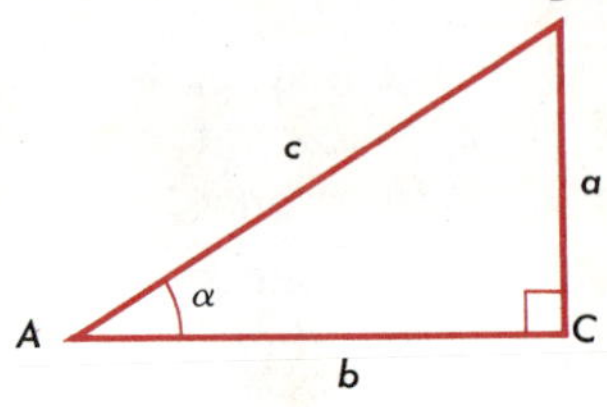

FIGURE 13–10

$$\textit{sine } \alpha = \frac{a}{c}$$

$$\textit{cosine } \alpha = \frac{b}{c}$$

$$\textit{tangent } \alpha = \frac{a}{b}$$

In Fig. 13–10, note that α is the measure of $\angle BAC$, a is the length of the leg opposite $\angle BAC$, b is the length of the leg adjacent to $\angle BAC$, and c is the length of the hypotenuse of $\triangle ABC$. Then you have the following.

$$\textit{sine } \alpha = \frac{\textit{opposite leg}}{\textit{hypotenuse}}$$

$$\textit{cosine } \alpha = \frac{\textit{adjacent leg}}{\textit{hypotenuse}}$$

$$\textit{tangent } \alpha = \frac{\textit{opposite leg}}{\textit{adjacent leg}}$$

The functions sine, cosine, and tangent are called *trigonometric functions.* It is common practice to write "sin α" instead of "sine α," "cos α" instead of "cosine α," and "tan α" instead of "tangent α." From now on, these abbreviations will be used.

There are three other possible ratios of the sides of the $\triangle ABC$ in Fig. 13–10. These ratios lead to the three other trigonometric functions defined below.

$$\textit{secant } \alpha = \frac{c}{b}$$

$$\textit{cosecant } \alpha = \frac{c}{a}$$

$$\textit{cotangent } \alpha = \frac{b}{a}$$

Since these functions of α are simply the reciprocals of $\cos \alpha$, $\sin \alpha$, and $\tan \alpha$, respectively, they are not considered in this introduction to trigonometry.

Problem 1. Find sin 45°, cos 45°, and tan 45°.

Solution. You can find these trigonometric functions of 45° if you can produce a right triangle having a 45° angle. The diagonal of a 3-inch square will cut the square into two such triangles as shown in Fig. 13–11. If c designates the length of the hypotenuse of this triangle, then by the pythagorean theorem,

$$c^2 = 3^2 + 3^2.$$

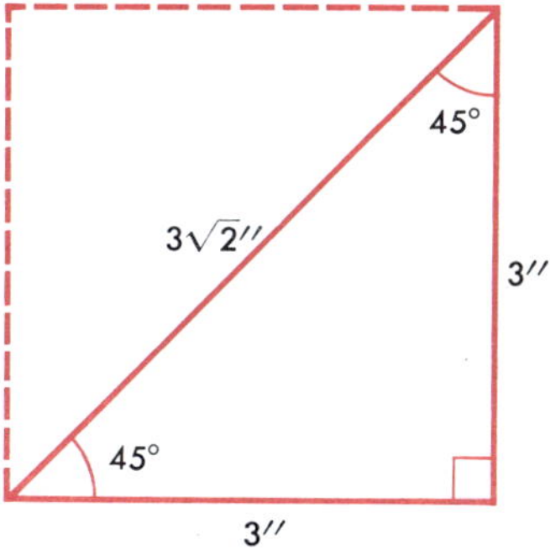

FIGURE 13–11

Hence,

$$c^2 = 18, \quad \text{and} \quad c = \sqrt{18}, \text{ or } 3\sqrt{2}.$$

Now you can find the three trigonometric functions of 45°:

$$\sin 45° = \frac{3}{3\sqrt{2}}, \quad \cos 45° = \frac{3}{3\sqrt{2}}, \quad \tan 45° = \frac{3}{3}.$$

The value of each of these functions may be simplified to yield

$$\sin 45° = \frac{\sqrt{2}}{2}, \quad \cos 45° = \frac{\sqrt{2}}{2}, \quad \tan 45° = 1.$$

Using $\sqrt{2} \doteq 1.414$, you see that

$$\sin 45° \doteq .707, \quad \cos 45° \doteq .707.$$

Problem 2. Find the trigonometric functions of the acute angles of the right triangle with sides 3, 4, 5 in Fig. 13–12.

Solution. By definition,

$$\sin \alpha = \tfrac{3}{5}, \quad \cos \alpha = \tfrac{4}{5}, \quad \tan \alpha = \tfrac{3}{4}.$$

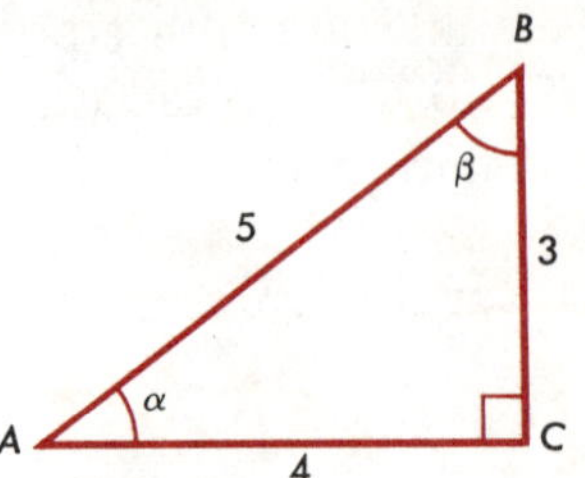

FIGURE 13–12

The leg opposite to the angle of measure β has length 4, while the adjacent leg has length 3. Hence, by definition,

$$\sin \beta = \tfrac{4}{5}, \quad \cos \beta = \tfrac{3}{5}, \quad \tan \beta = \tfrac{4}{3}.$$

Exercises

1. Use the pythagorean theorem to find the missing side in each of the right triangles drawn below.

(a)

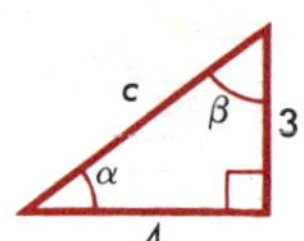

(b)

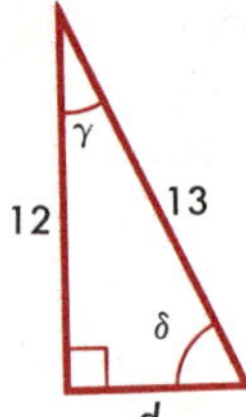

Refer to the right triangles of Exercise 1 and find

2. (a) $\sin \alpha$, $\cos \alpha$, $\tan \alpha$.
(b) $\sin \delta$, $\cos \delta$, $\tan \delta$.

3. (a) $\sin \beta$, $\cos \beta$, $\tan \beta$.
(b) $\sin \gamma$, $\cos \gamma$, $\tan \gamma$.

4. Use the pythagorean theorem to find the missing side in each of the right triangles drawn below.

(a)

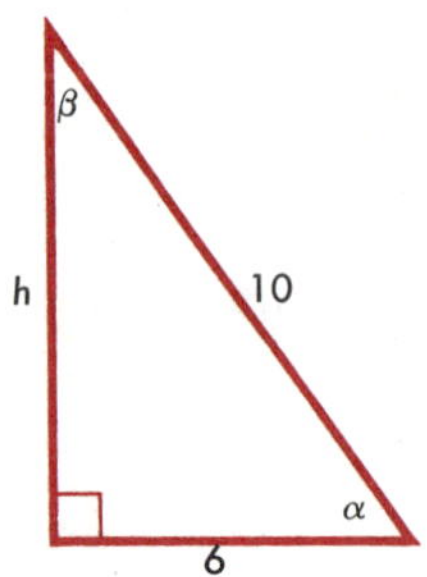

(b)

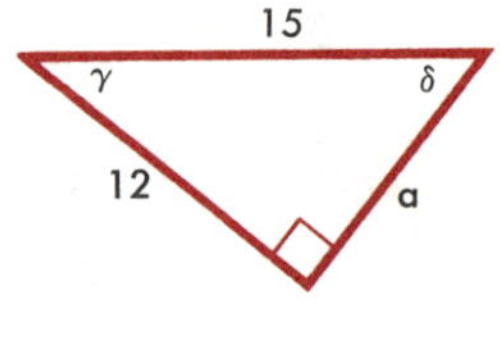

Refer to the right triangles of Exercise 4 and find

5. (a) $\sin \alpha$, $\cos \alpha$, $\tan \alpha$.
(b) $\sin \gamma$, $\cos \gamma$, $\tan \gamma$.

6. (a) $\sin \beta$, $\cos \beta$, $\tan \beta$.
(b) $\sin \delta$, $\cos \delta$, $\tan \delta$.

Refer to the triangle to the right to answer the following questions.

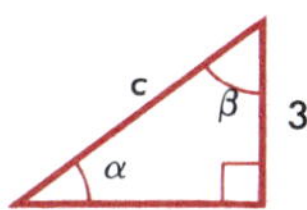

7. What is $\alpha + \beta$?

8. How do $\sin \alpha$ and $\cos \beta$ compare in value?

9. How do $\cos \alpha$ and $\sin \beta$ compare in value?

10. How do $\sin \alpha$ and $\tan \beta$ compare in value?

11. How do $\sin \beta$ and $\tan \beta$ compare in value?

---

12. If $\nu + \mu = 90°$, what relationship do you think exists between $\sin \nu$ and $\cos \mu$? between $\sin \mu$ and $\cos \nu$?

13. Can a leg of a right triangle ever be as long as the hypotenuse of the same triangle?

14. Can the sine of an angle of a right triangle ever be as large as 1?

15. Is there a smallest value for the sine of an angle of a right triangle?

16. What is the range of values for the cosine of an acute angle of a right triangle?

17. How large is ρ if $\tan \rho = 1$? What is the range of values for $\tan \xi$ if $0 < \xi < \rho$? What is the range of values for $\tan \xi$ if $\rho < \xi < 90°$?

13-5 TRIGONOMETRIC TABLES

Using more advanced mathematics, it is possible to approximate the trigonometric functions of angles of 1°, 2°, 3°, and so on, up to 89°. Such a table of three-decimal-place approximations of sine, cosine, and tangent is given in the Appendix. This table may be used to work problems of the type illustrated below.

Problem 1. If an acute angle of a right triangle has a measure of 26°, and if the leg opposite has length 14 feet, approximately how long are the other two sides of the triangle?

Solution. When working a problem like this, you may find it helpful to draw a rough sketch of the triangle and to label the known parts as has been done in Fig. 13-13. Note that b designates the length of the adjacent leg and c designates the length of the hypotenuse. What trigonometric function of 26° involves the known side of 14 feet and b? The tangent relates the two legs as follows:

$$\tan 26° = \frac{14}{b}.$$

By the table, $\tan 26° \doteq .488$. Therefore,

$$.488 \doteq \frac{14}{b},$$

$$.488b \doteq 14,$$

$$b \doteq \frac{14}{.488}, \quad \text{or approximately } 28.8.$$

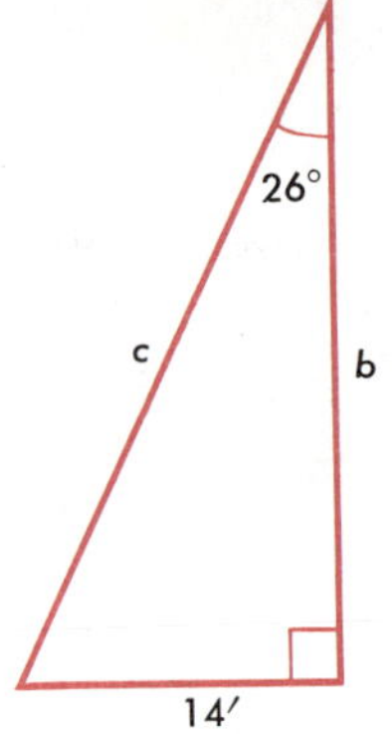

FIGURE 13–13

Thus, b is approximately 28.8 feet.

What trigonometric function of 26° involves the known side of 14 feet and c?

$$\sin 26° = \frac{14}{c}$$

By the table, $\sin 26° \doteq .438$. Therefore,

$$.438 \doteq \frac{14}{c},$$

$$.438c \doteq 14,$$

$$c \doteq \frac{14}{.438}, \quad \text{or approximately } 32.0.$$

Thus, c is approximately 32.0 feet.

Check. Since this is a right triangle, you can use the pythagorean theorem as a check.

$$14^2 + 28.8^2 \stackrel{?}{=} 32^2$$

$$196 + 829 \stackrel{?}{=} 1024$$

$$1025 \doteq 1024$$

Since the trigonometric table gives approximate values (to 3 decimal places), you can only expect your answer to check to 3 decimal plac

Problem 2. Approximate the acute angles of a 3, 4, 5 right triangle, as shown in Fig. 13–14.

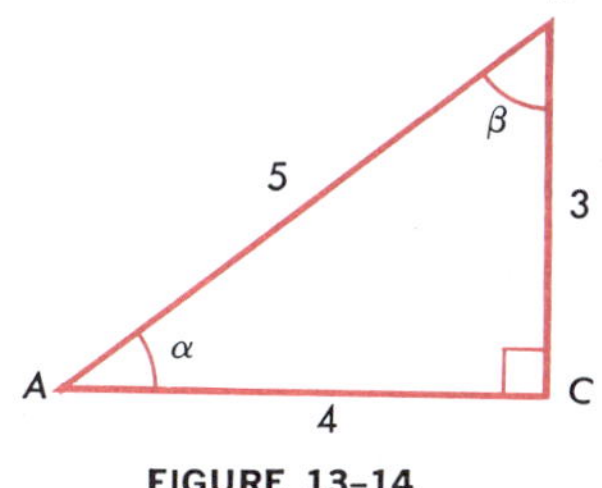

FIGURE 13–14

Solution. You know that

$$\sin \alpha = .6 \quad \text{and} \quad \sin \beta = .8.$$

According to the table,

$$\sin 37° \doteq .602 \quad \text{and} \quad \sin 53° \doteq .799.$$

Therefore,

$$\alpha \doteq 37° \quad \text{and} \quad \beta \doteq 53°.$$

Check. Do the measures of the angles of this triangle add up to 180°?

$$90° + 37° + 53° \stackrel{?}{=} 180°$$
$$180° = 180°$$

Exercises

Use the Table of Values of Trigonometric Functions in the Appendix to find an approximate value of each of the following trigonometric functions.

1. (a) $\sin 27°$ (b) $\sin 54°$

2. (a) $\cos 48°$ (b) $\cos 36°$

3. (a) $\tan 72°$ (b) $\tan 20°$

4. (a) $\sin 82°$ (b) $\sin 60°$

5. (a) $\cos 5°$ (b) $\cos 30°$

6. (a) $\tan 57°$ (b) $\tan 55°$

In each of the following exercises, you are given the approximate value of a trigonometric function of an angle. In each case, find the size of the angle from the tables.

7. (a) $\sin \delta = .174$ (b) $\sin \delta = .342$

8. (a) $\cos \alpha = .545$ (b) $\cos \beta = .899$

9. (a) $\tan \gamma = .404$ (b) $\tan \alpha = 1.600$

10. (a) $\cos \alpha = .276$ (b) $\cos \alpha = .990$

11. (a) $\sin \delta = .990$ (b) $\sin \delta = .966$

12. (a) $\tan \beta = 1.732$ (b) $\tan \alpha = 3.487$

13. (a) $\cos \alpha = .743$ (b) $\cos \gamma = .191$

14. (a) $\tan \beta = .070$ (b) $\tan \alpha = 4.011$

15. (a) $\sin \alpha = .891$ (b) $\sin \alpha = .719$

In Exercises 16–19, find, to the nearest degree, the size of each of the angles described in the exercises below.

16. (a) $\cos \theta = .942$ (b) $\cos \eta = .605$

17. (a) $\sin \omega = .340$ (b) $\sin \phi = .796$

18. (a) $\tan \phi = .582$ (b) $\tan \eta = .330$

19. (a) $\cos \omega = .249$ (b) $\cos \phi = .406$

20. An acute angle of a right triangle has a measure of 50°, and the adjacent leg has length 16 inches. Complete the following steps to approximate the other two sides of the triangle.

(a) $\tan 50^\circ = \dfrac{?}{?}$

Use the Table of Values of Trigonometric Functions to approximate the length of the missing leg.

(b) $\cos 50^\circ = \dfrac{?}{?}$

Use the Table of Values of Trigonometric Functions to approximate the length of the missing hypotenuse.

For each of the right triangles in Exercises 21–26, find the approximate length of the side designated by a letter.

21. (a)

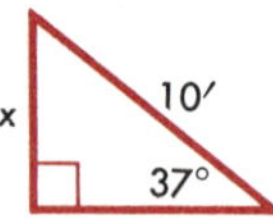

(b)

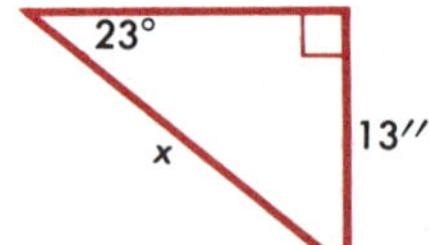

22. (a)

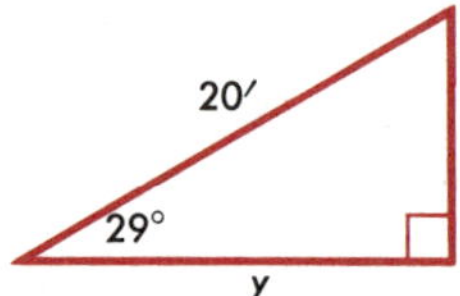

(b)

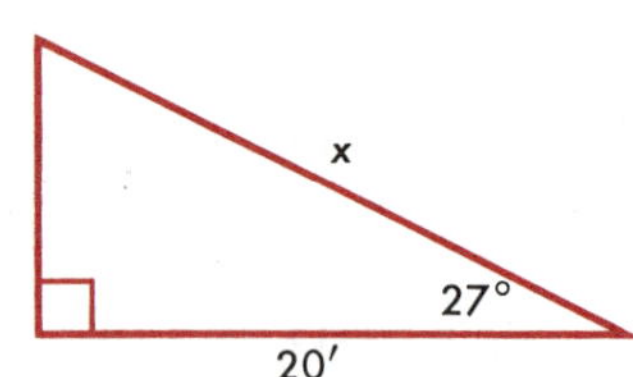

23. (a)

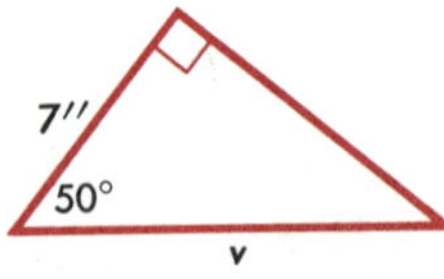

(b)

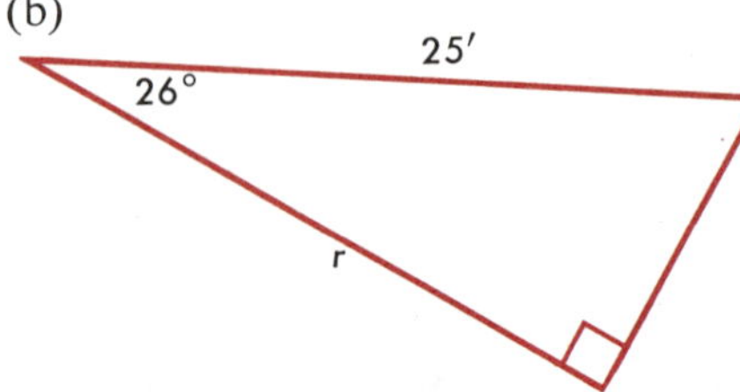

24. (a) (b)

25. (a) (b)

26. (a) (b)

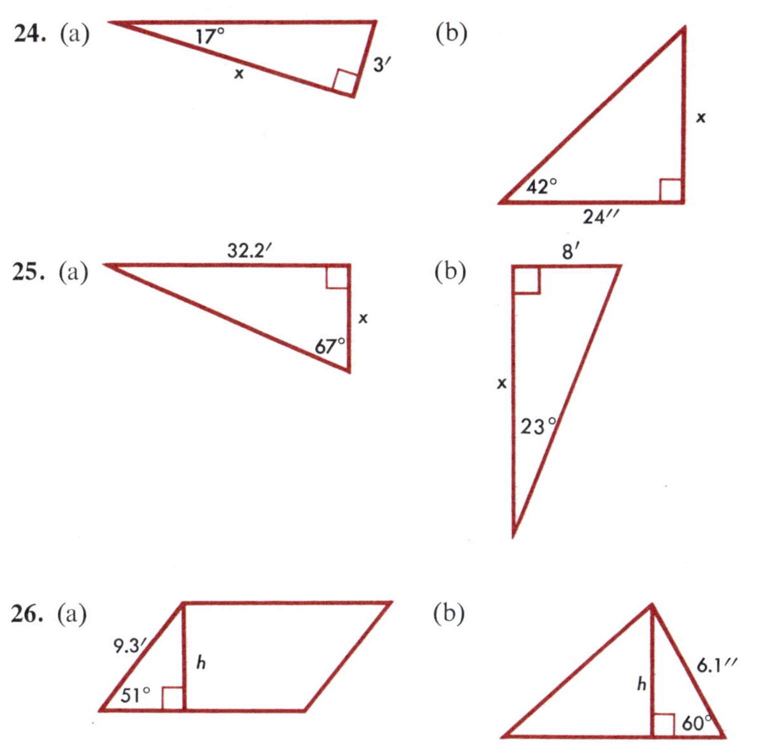

27. Approximate each of the acute angles of a right triangle with sides having length 5, 12, and 13. Check your answers by finding their sum. What should it be?

13-6 APPLICATIONS

Trigonometric tables are helpful in making certain geometric or physical measurements. The first problem is from geometry.

Problem 1. What is the altitude, h, of the triangle drawn in Fig. 13-15?

Solution. The large triangle ABC is cut by the altitude $\overline{BD}$ into two right triangles ABD and BCD, each having a right angle at D. In triangle ABD,

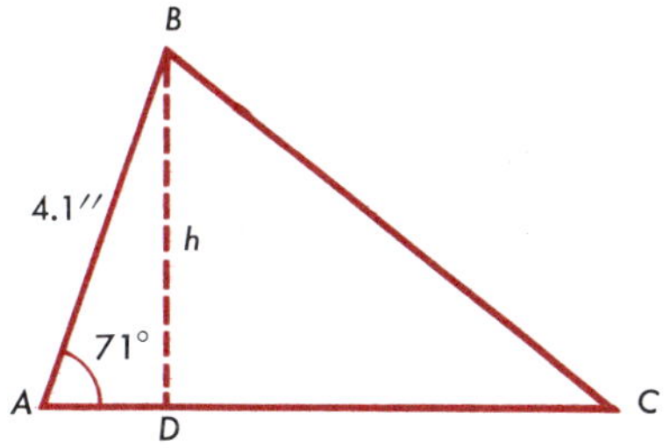

FIGURE 13-15

$$\sin 71^\circ = \frac{h}{4.1}.$$

By the table, $\sin 71^\circ \doteq .946$. Therefore,

$$.946 \doteq \frac{h}{4.1},$$

$$.946 \times 4.1 \doteq h,$$

$$h \doteq 3.9.$$

Thus, the altitude h is approximately 3.9 inches long.

In certain physical measurements, it is necessary to describe the relative positions of two points with respect to the horizontal. If you are at one point looking up at another point, the angle between your line of sight and the horizontal in the same direction is called the *angle of elevation* of the second point from the first. If you have to look down instead of up to see the second point, the angle is called the *angle of depression* of the second point from the first. The two angles are illustrated in Fig. 13–16. The horizontal line drawn from A must intersect the dashed vertical line through B in each case.

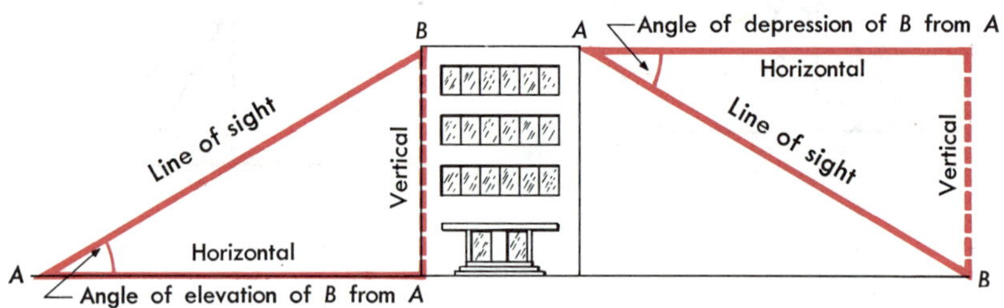

FIGURE 13–16

Problem 2. Find the height of the Federal Building in Fig. 13–17 by using the procedure employed by a surveyor.

Solution. A surveyor finds that the angle of elevation of the top of the Federal Building is 61° when his transit is set up 100 feet from the building in a horizontal line. If x designates the height of the building above his transit, then

$$\tan 61^\circ = \frac{x}{100}.$$

From the Table of Values of Trigonometric Functions, $\tan 61^\circ \doteq 1.804$. Hence,

$$1.804 \doteq \frac{x}{100},$$

$$x \doteq 180.4,$$

and if the height of the transit is 5 feet, the height of the Federal Building is approximately $180.4 + 5$, or 185.4 feet.

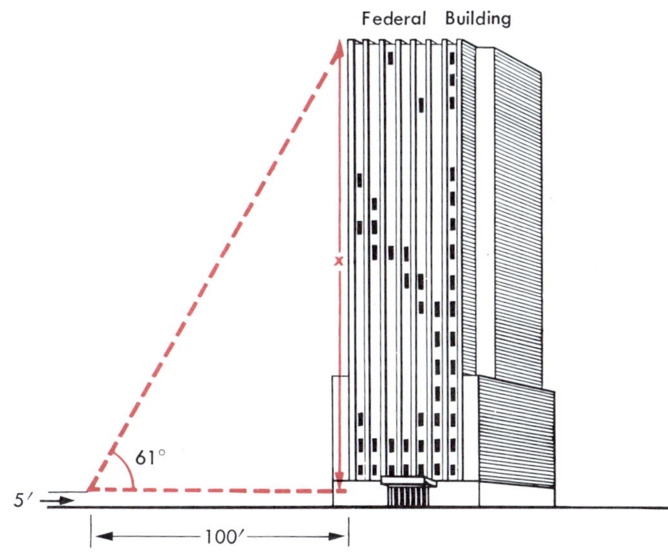

FIGURE 13-17

Problem 3. From an airplane flying at an elevation of 5000 feet over the ocean, the angle of depression of the coast of an island (point B) is 12°. In how many miles will the airplane be over B?

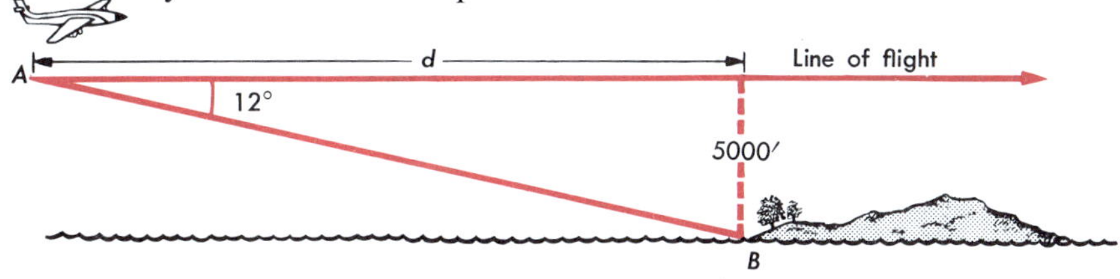

FIGURE 13-18

Solution. You are asked to find the distance d. (See Fig. 13-18.) First,

$$\tan 12^\circ = \frac{5000}{d}.$$

From the table, $\tan 12^\circ \doteq .213$. Hence,

$$.213 \doteq \frac{5000}{d},$$

$$d \doteq \frac{5000}{.213}, \quad \text{or approximately } 23{,}500.$$

Thus, the plane must fly 23,500 feet, or 3.9 nautical miles (a nautical mile is 6080 feet), to reach the island.

Exercises

1. (a) How high is a cliff if the angle of elevation of the top is measured from a point 500 feet from the base and is 41°?
   (b) The angle of elevation of the top of a building is 41° when measured at a distance of 115 feet from the foot of the building. How high is the building?
2. (a) When the angle of elevation of the sun is 28°, find the length of the shadow cast by a flagpole 45 feet high.
   (b) If the angle of elevation of the sun is 40°, how long is the shadow cast by a 60-foot flagpole?
3. (a) From a cliff 90 feet above the surface of a lake, the angle of depression of a sailboat is 4°. How far is the boat from the foot of the cliff?
   (b) From the top of a building 70 feet high, the angle of depression of an automobile on a road is 27°. How far is the automobile from the foot of the building?
4. (a) A guy wire 65 feet long runs from the top of a pole to the ground and makes an angle of 69° with the ground. How tall is the pole?
   (b) How high is a kite if it is held by 500 feet of string, and the string makes an angle of 49° with the ground?
5. (a) How long must a ladder be to reach a window 22 feet above the ground if the ladder makes an angle of 65° with the ground?
   (b) How far from a building must the base of a 30-foot ladder be placed to make an angle of 70° with the ground?
6. (a) Find the length of the diagonal of a rectangle if the diagonal makes an angle of 38° with the adjacent side, whose length is 15 inches.
   (b) Find the altitude of a triangle if a side 18 inches long makes an angle of 52° with the base.
7. (a) A 30-foot ladder leans against a house with its foot 15 feet from the house. Find the angle which the ladder makes with the ground.
   (b) A 10-foot flagpole casts a 20-foot shadow. What is the angle of elevation from the end of the shadow to the top of the flagpole?
8. (a) From the top of a hill 170 feet high, the angle of depression of a boat on the lake is 36°. What is the airline distance from the top of the hill to the boat?
   (b) A lighthouse built at sea level is 120 feet high. From its top the angle of depression of a buoy is 35°. How far is the buoy from the foot of the lighthouse?
9. (a) A broadcasting tower is 255 feet high. How far from the base of the tower is a surveyor who observes the angle of elevation of the top of the tower to be 38°?
   (b) The 12-inch long diagonal of a rectangle makes an angle of 32° with the longer side. Find the length of the longer side.

10. A man whose eyes are 5 feet above the ground observes the angle of elevation of the top of a tower to be 70°. He walks 70 feet farther away and finds the angle of elevation to be 58°. How tall is the tower?

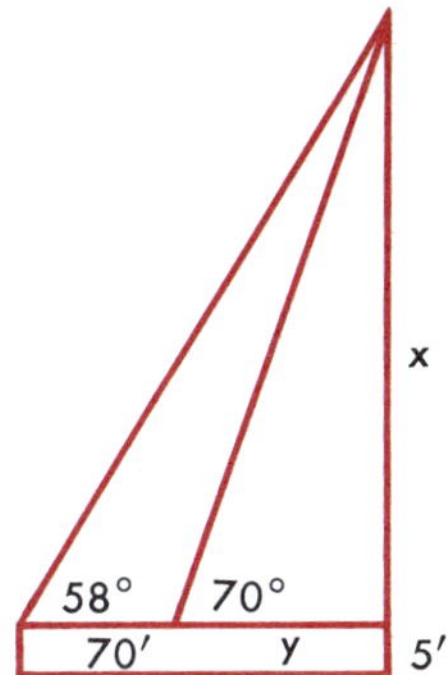

11. Two trees are 100 yards apart. The angles of elevation to their tops from the center of the ground between them are 10° and 15°, respectively. How much higher is the one tree than the other?

12. Find the width of a rectangle whose 10-inch diagonal makes a 15° angle with the longer side.

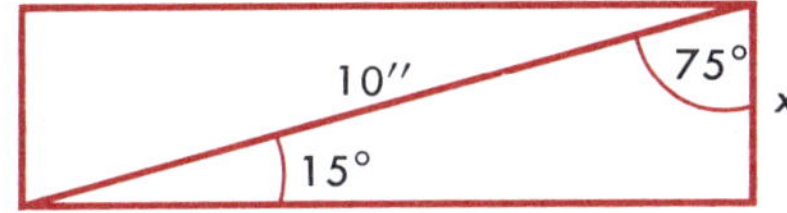

Preparation for Section 13–7

1. What is the slope of the line joining the point (2, 1) to the point (3, 9)?
2. What is the rise from the point (2, 1) to the point $(x, 6)$?

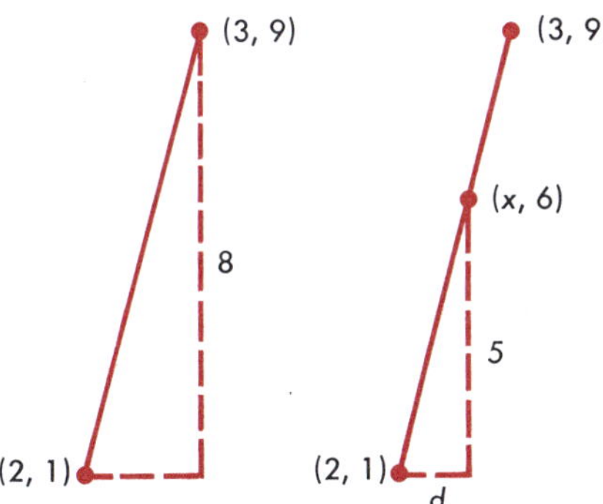

3. If $\frac{5}{d} = \frac{8}{1}$, find d to the nearest tenth.
4. What is a one-decimal-place approximation of $\sqrt{6}$?

13–7 INTERPOLATION

There are problems in navigation and in astronomy, for example, which require measurements of angles to be taken more accurately than to the nearest degree. For these problems, minutes and seconds are frequently used. One minute is defined to be one-sixtieth of a degree, and one second is defined to be one-sixtieth of a minute. The discussion here will be limited to degrees and minutes. The symbol $'$ is used for minutes. For example, $32\frac{3}{4}^\circ$ may be designated by $32^\circ\ 45'$.

The method of linear interpolation (see Chapter 11, Section 11–11), used in conjunction with the Table of Values of Trigonometric Functions, allows you to work with angles measured to the nearest 10 minutes. This method is illustrated in the following problems.

Problem 1. Find $\sin 32^\circ\ 40'$.

Solution. Read from the table that

$$\sin 32^\circ \doteq .530, \quad \sin 33^\circ \doteq .545.$$

Clearly, $\sin 32^\circ\ 40'$ lies between .530 and .545. If you imagine plotting the points (32, .530) and (33, .545) in the plane and joining them with a line segment as indicated in Fig. 13–19, then, by linear interpolation, $\sin 32^\circ\ 40'$ will be the y-coordinate of point A.

By similar triangles,

$$\frac{d}{40} = \frac{.015}{60}.$$

Thus,

$$d = 40 \times \frac{.015}{60}, \quad \text{or } .010.$$

Hence,

$$\sin 32^\circ\ 40' \doteq .530 + .010, \quad \text{or } .540.$$

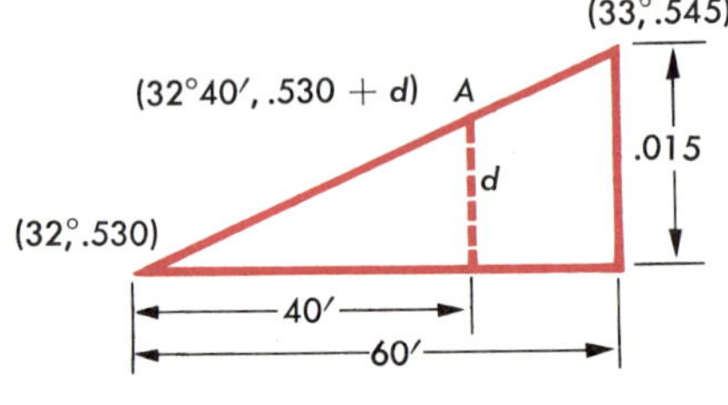

FIGURE 13–19

Problem 2. Find α if $\tan \alpha = 1.824$ (Fig. 13–20).

Solution. Read from the table that

$$\tan 61^\circ \doteq 1.804 \quad \text{and} \quad \tan 62^\circ \doteq 1.881.$$

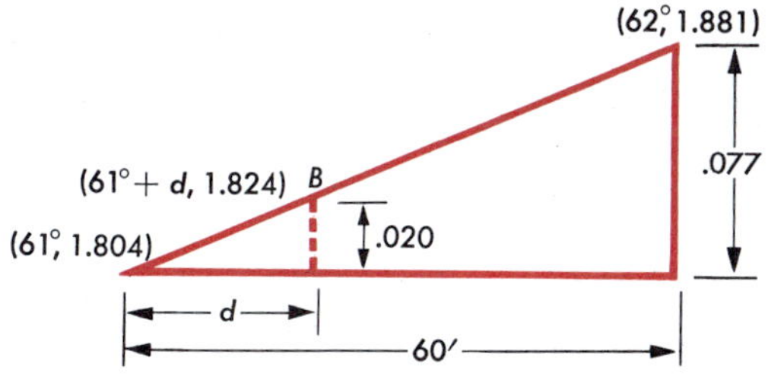

FIGURE 13-20

Clearly, α is between 61° and 62°. If you imagine that α is the x-coordinate of the point B, then by linear interpolation,

$$\frac{d}{.020} = \frac{60}{.077}.$$

Thus, $d = .020 \times \dfrac{60}{.077}$, or 20′ (to the nearest 10′) and $\alpha \doteq 61°\ 20'$.

Exercises

Use linear interpolation to find a three-decimal-place approximation for each of the following trigonometric functions.

1. (a) $\sin 35°\ 20'$ (b) $\sin 25°\ 30'$

2. (a) $\cos 35°\ 20'$ (b) $\cos 25°\ 30'$

3. (a) $\tan 35°\ 20'$ (b) $\tan 25°\ 30'$

4. (a) $\sin 56°\ 30'$ (b) $\sin 42°\ 10'$

5. (a) $\cos 62°\ 40'$ (b) $\cos 72°\ 50'$

6. (a) $\tan 73°\ 10'$ (b) $\tan 40°\ 50'$

Use linear interpolation to find α to the nearest 10′.

7. (a) $\sin \alpha = .410$ (b) $\sin \alpha = .860$

8. (a) $\cos \alpha = .992$ (b) $\cos \alpha = .382$

9. (a) $\tan \alpha = .660$ (b) $\tan \alpha = .845$

10. (a) $\sin \alpha = .720$ (b) $\sin \alpha = .901$

11. (a) $\cos \alpha = .180$ (b) $\cos \alpha = .851$

12. (a) $\tan \alpha = 1.500$ (b) $\tan \alpha = 2.140$

---

13. An airplane flies at 400 miles per hour. If it is flying at an altitude of 15,000 feet, what will its angle of elevation be 4 minutes after passing overhead?

14. Find the angles of an isosceles triangle if its base is 18.6 feet long and the equal sides are 14.4 feet long.

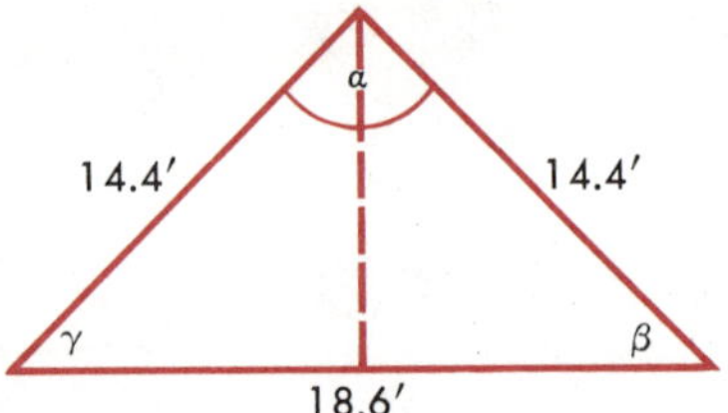

15. An 18-foot ladder leans against a house and makes an angle of 63° 30′ with the ground. How high from the ground is its upper end?

16. What is the angle of depression of the top of a 472-foot building from the top of a 503-foot building if the horizontal distance between the tops is 51 feet?

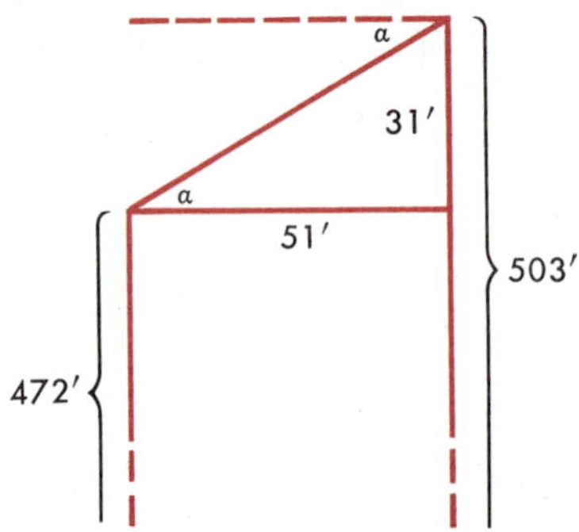

17. What is the angle of elevation of the sun when a pole which is 50 feet high casts a shadow 31.2 feet long?

18. How high above the deck of a sailboat is the masthead if a 32-foot stay makes an angle of 72° 20′ with the deck of the boat?

19. A triangular plot of ground has two boundaries at right angles. One of these boundaries is 75 feet long, and a surveyor measures the adjacent angle as 40° 50′. What is the length of the longest boundary?

20. What is the area of the plot in Exercise 19?

KEY IDEAS AND KEY WORDS

Two number pairs (x_1, y_1) and (x_2, y_2) are said to be **directly proportional** if

$$\frac{x_1}{y_1} = \frac{x_2}{y_2}.$$

Any two solutions of an equation of the form

$$y = kx,$$

where k is a nonzero number, are **directly proportional.**

Two number pairs (x_1, y_1) and (x_2, y_2) are said to be **inversely proportional** if

$$\frac{x_1}{x_2} = \frac{y_2}{y_1}.$$

Any two solutions of an equation of the form

$$xy = k,$$

where k is a nonzero number, are **inversely proportional.**

In any proportion

$$\frac{a}{b} = \frac{c}{d},$$

a and d are called the **extremes,** and b and c are called the **means.** If the means are equal (that is, if $b = c$), their common value is called a **mean proportional** between the extremes. The product of the means is always equal to the product of the extremes.

Each **acute angle** α is an angle of some right triangle ABC. If the parts of a triangle ABC are labeled as in the figure, the six trigonometric functions of α are defined as follows:

$$\sin\alpha = \frac{a}{c}, \quad \cos\alpha = \frac{b}{c},$$

$$\tan\alpha = \frac{a}{b}, \quad \cot\alpha = \frac{b}{a},$$

$$\sec\alpha = \frac{c}{b}, \quad \csc\alpha = \frac{c}{a}.$$

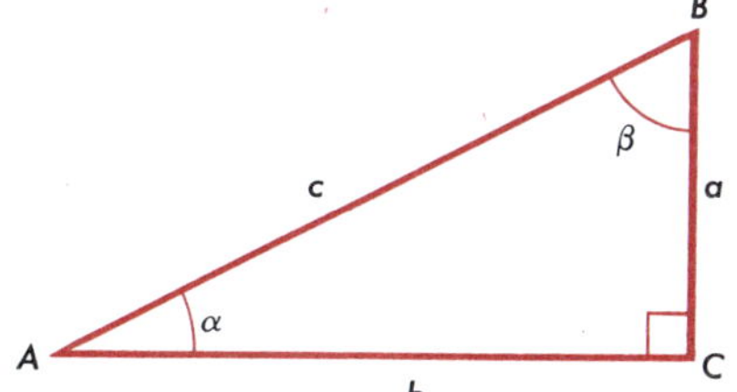

CHAPTER REVIEW

1. Solve each of the following proportions.
   (a) $\frac{x}{5} = \frac{45}{15}$
   (b) $\frac{4}{5} = \frac{5}{c-2}$
   (c) $\frac{2y}{16} = \frac{y+3}{12}$
   (d) $\frac{y-3}{y-1} = \frac{y-2}{y-4}$
2. Find a mean proportional between each of the following pairs.
   (a) 3, 12 (b) 25, 16 (c) $\sqrt{6}, \sqrt{24}$
3. Find the fourth proportional to each of the following triples.
   (a) 5, 7, 35 (b) 6, 10, 21
4. How long is a 60-pound roll of wire which weighs .25 pound per foot?
5. At what rate does \$1500 yield the same annual income as \$2000 at 3%?

6. If y is inversely proportional to x, what change in y doubles x? Illustrate your answer with a numerical example.
7. If y is directly proportional to x, what change in x will double y? Illustrate your answer with a numerical example.
8. If two angles of a triangle measure 65° and 70°, respectively, what is the measure of the smallest angle of a similar triangle?
9. The sides of a triangle measure 8, 9, and 15 inches. If the shortest side of a similar triangle measures 6 inches, what do the other two sides of the triangle measure?
10. Refer to the figure at the right and find
    (a) $\sin \alpha$, $\cos \alpha$, $\tan \alpha$.
    (b) $\sin \beta$, $\cos \beta$, $\tan \beta$.

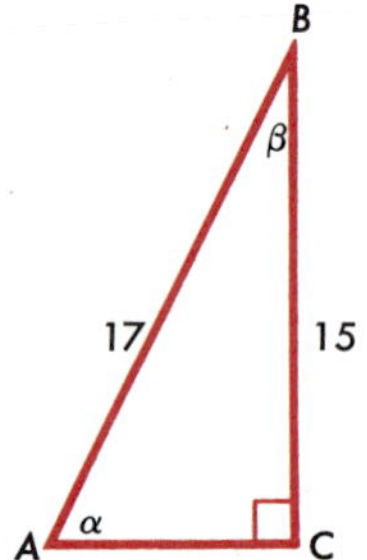

11. Find the trigonometric functions of each of the acute angles of a right triangle with legs measuring 10 inches and 12 inches, respectively.
12. Use the Table of Values of Trigonometric Functions to find approximations for each of the following.
    (a) $\sin 43°$ (b) $\cos 54°$
    (c) $\tan 20°$ (d) $\tan 65°$
    (e) $\sin 75°$ (f) $\cos 14°$
13. Approximate the legs of a right triangle if the hypotenuse measures 20 inches and one acute angle has a measure of 25°.
14. Approximate the hypotenuse and other leg of a right triangle if one acute angle measures 70° and the adjacent leg measures 15 inches.
15. At a point 28 feet from the foot of a tree, the angle of elevation of the top of the tree is 59°. Find the height of the tree to the nearest foot.
16. From a point on the top deck of a ship, the angle of depression of a point at the edge of the dock is 20°. If the top deck is 80 feet above the level of the dock, how far is the ship from the dock?
17. If a road rises 24 feet in a horizontal distance of 300 feet, find to the nearest degree the angle that the road makes with the horizontal.
18. Use linear interpolation to find a three-decimal-place approximation for each of the following trigonometric functions.
    (a) $\sin 54° 54'$ (b) $\tan 31° 23'$
19. Use linear interpolation to find α to the nearest $10'$ if $\tan \alpha = 2.105$.

CHAPTER TEST

1. Solve each of the following proportions.

   (a) $\frac{7}{x} = \frac{x}{28}$ (b) $\frac{\sqrt{5}}{2} = \frac{x}{\sqrt{20}}$ (c) $\frac{x+1}{3} = \frac{5}{x-1}$

2. The sides of a triangle measure 10 inches, 38 inches, and 40 inches. What is the perimeter of a similar triangle in which the longest side measures 90 inches?

3. Refer to the figure at the right to find
   (a) $\sin \alpha$, $\cos \alpha$, $\tan \alpha$.
   (b) $\sin \beta$, $\cos \beta$, $\tan \beta$.

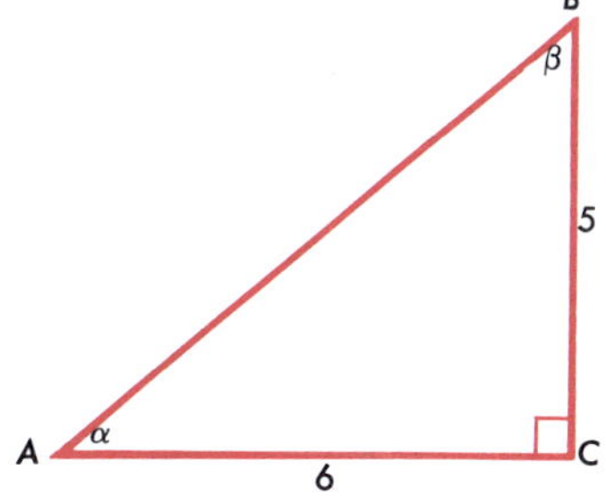

4. Refer to the figure at the right to find
   (a) $\sin \alpha$, $\cos \alpha$, $\tan \alpha$.
   (b) $\sin \beta$, $\cos \beta$, $\tan \beta$.

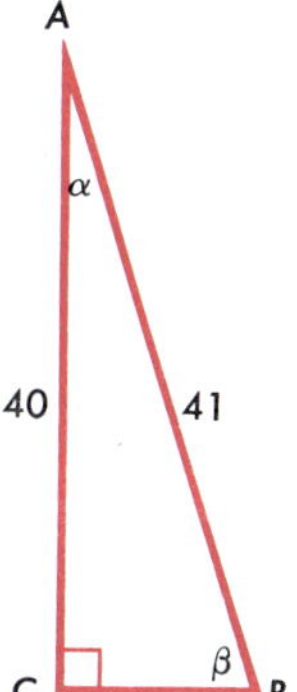

5. Use the Table of Values of Trigonometric Functions to approximate the acute angles of a right triangle with legs measuring 60 inches and 11 inches and an hypotenuse measuring 61 inches.

6. Find the length of h for the figure to the right.

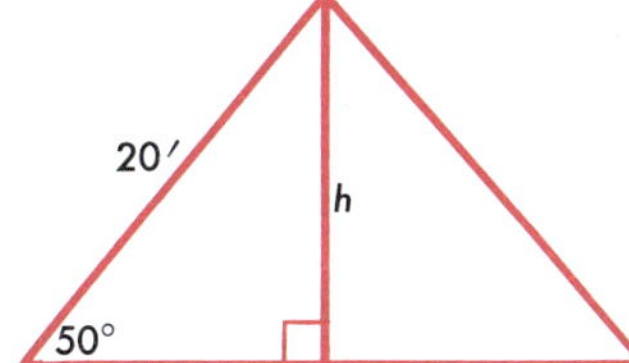

7. The angle of elevation from the tip of its shadow on the ground to the top of a tree is 31°. If the length of the shadow is 86 feet, what is the height of the tree?

CUMULATIVE REVIEW IV

1. Tell which of the following statements are true and which are false. Correct the ones that you marked false.
   (a) The statement $\sqrt{x^2} = x$ is true if $x = -4$.
   (b) The statement $\sqrt{a^2} = |a|$ is true if $a = 5$ or -5.
   (c) The difference between $.\underline{3}$ and 3 is an irrational number.
   (d) The decimal form of $\frac{5}{19}$ is infinite and has a repeating block of at most 18 digits.
   (e) If $x < 0, \sqrt[3]{x}$ is negative.
   (f) A perfect-square rational number between $\frac{1}{2}$ and $\frac{1}{3}$ is $\frac{4}{9}$.
   (g) $\sqrt{9x^4y^6} = 3x^2y^3$ for $y \geqq 0$
   (h) $\sqrt{16 + 9} = 4 + 3$
2. Give an example to substantiate your answer to each of the following questions.
   (a) If $N > 1$, which is greater, N or $\sqrt{N}$?
   (b) If $0 < N < 1$, which is greater, N or $\sqrt{N}$?
   (c) If $N > 1$, which is greater, N or N^2?
   (d) If $0 < N < 1$, which is greater, N or N^2?.
   (e) If $a > b > 0$, is $a^2 > b^2$?
   (f) If $a > b > 0$, is $\sqrt{a} > \sqrt{b}$?

In each of the following exercises, find an equivalent expression having no square root in the denominator.

3. $\dfrac{2}{\sqrt{x}}$ **4.** $\dfrac{3y}{\sqrt{3y}}$ **5.** $\dfrac{5}{\sqrt{a} + 2}$

6. $\dfrac{1}{\sqrt{x} - \sqrt{y}}$ **7.** $\dfrac{\sqrt{x} - \sqrt{y}}{\sqrt{x} + \sqrt{y}}$ **8.** $\dfrac{3\sqrt{x}}{2\sqrt{x} - 5}$

Do the indicated arithmetic and simplify your answers.

9. $3\sqrt{x} \cdot 2\sqrt{x}$ **10.** $\sqrt{2y} \cdot \sqrt{18y}$
11. $\frac{1}{2}\sqrt{12x^3} \cdot \sqrt{2x^5}$ **12.** $\frac{1}{3}\sqrt{18y} \cdot \sqrt{6y^3}$
13. $-\sqrt{5x^2y} \cdot \sqrt{10xy^2}$ **14.** $(-\sqrt{3ab})(-\sqrt{6ab^2})$
15. $(4\sqrt{63a}) \div (6\sqrt{12a})$ **16.** $5\sqrt{40x^3} \div 15\sqrt{4x}$
17. $\sqrt{8y^3} \div \sqrt{2y}$ **18.** $\sqrt{32y^5} \div \sqrt{4y^2}$
19. $\sqrt{2x}(\sqrt{3x} - 5\sqrt{xy})$ **20.** $\sqrt{3a}(2\sqrt{3a} + 9)$
21. $(\sqrt{3x} - \sqrt{xy})^2$ **22.** $(\sqrt{x} + 2\sqrt{y})(\sqrt{x} - 2\sqrt{y})$
23. $5\sqrt{a} - 3\sqrt{b} + 2\sqrt{a} - 7\sqrt{b}$ **24.** $3\sqrt{20xy} - 2\sqrt{45xy} + 7\sqrt{5xy}$

In Exercises 25–44, use only positive rational exponents to express each of the following in simplest form.

25. $\dfrac{-x^{-2}}{2}$ **26.** $\dfrac{r^{-\frac{3}{4}}}{3}$ **27.** $\dfrac{5}{(r-t)^{-3}}$ **28.** $\dfrac{r^{-2}s^{-3}}{t^2}$

29. $\dfrac{7}{-6s^{-2}}$ **30.** $rs^{-3} + r^{-2}s$ **31.** $\dfrac{x^{-\frac{4}{3}}}{x}$ **32.** $\dfrac{(x+y)^{-2}}{3}$

33. $3^{-1}cy^{-2}$ **34.** $(x^4y^2)^{\frac{3}{2}}$ **35.** $\dfrac{x^{-2}y^5}{x^5y^{-2}}$ **36.** $\dfrac{(z^4y^{\frac{3}{4}})^4}{r^{\frac{5}{4}}}$

37. $\left(\dfrac{125y^4}{216y^{-1}}\right)^{\frac{1}{3}}$ **38.** $(36x^{-2})^{-\frac{1}{2}}$ **39.** $(x^{-3})^{-\frac{1}{3}}$ **40.** $\left(\dfrac{9x^{-3}}{4z^2}\right)^{\frac{1}{2}}$

41. $(x^{\frac{3}{4}}y^{\frac{4}{3}})^3$ **42.** $(16r^2t^3)^{\frac{3}{4}}$ **43.** $(125y^9)^{-\frac{1}{3}}$ **44.** $(r^{\frac{2}{3}}s^{\frac{1}{3}})^3$

Solve each of the following equations.

45. $4x + 9 = 1$ **46.** $\dfrac{x}{5} - 5 = -4$

47. $-5(-x - 4) = -5$ **48.** $\dfrac{x+1}{2} = 3$

49. $3(2x - 1) - 2(x + 3) = -17$ **50.** $4(1 + 2x) - 5(1 - x) = -14$

51. $9 - (3y + 4) = 11$ **52.** $x^2 - 2x = 0$

53. $3x^2 - 13x = 10$ **54.** $x^2 + 8x - 20 = 0$

55. $4x^2 + 3x = 0$ **56.** $4x^2 = 3x + 2$

57. $\dfrac{x+1}{3x} + \dfrac{1}{x} = \dfrac{2}{3}$ **58.** $\dfrac{2x+29}{7x} = \dfrac{5}{x}$

59. $\dfrac{4x+5}{8x} - \dfrac{3}{4} = \dfrac{1-x}{2x}$ **60.** $\dfrac{7x+3}{4x} - 2 = \dfrac{4x-5}{5x}$

61. $\dfrac{x}{15} = \dfrac{2x-5}{3x}$ **62.** $\dfrac{3x-5}{6} = \dfrac{x^2-2}{3x}$

Solve each of the following inequalities.

63. $2x - 1 < 4$ **64.** $7x - 1 > 3x + 7$

65. $9x - 5 - 11x < -4$ **66.** $x^2 - 2x - 3 > 0$

67. $x^2 + 9x + 18 \leqq 0$ **68.** $2x^2 - 5x \geqq -2$

In Exercises 69–72, graph the equations and inequalities in a plane.

69. $x > 3$ **70.** $x + 3 - y = 0$ **71.** $x^2 = 9$ **72.** $x^2 + 2x + 1 > 0$

73. Refer to the figure to find the approximate length of c and a, and give the sin, cos, and tan of $\angle BAC$.

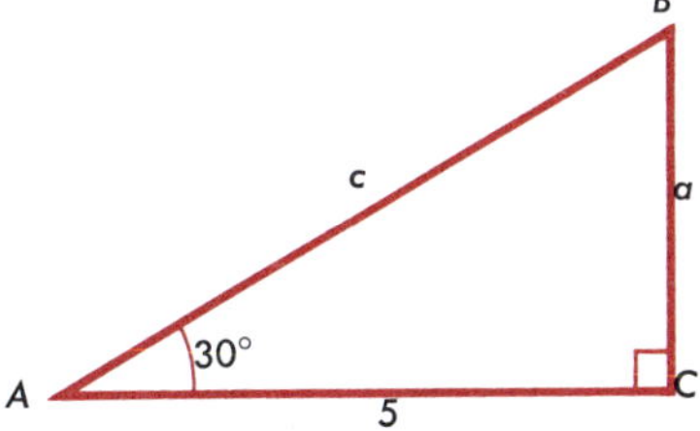

APPENDIX

Table of Squares and Square Roots

N	N^2	$\sqrt{N}$	N	N^2	$\sqrt{N}$
1	1	1	51	2,601	7.141
2	4	1.414	52	2,704	7.211
3	9	1.732	53	2,809	7.280
4	16	2	54	2,916	7.348
5	25	2.236	55	3,025	7.416
6	36	2.449	56	3,136	7.483
7	49	2.646	57	3,249	7.550
8	64	2.828	58	3,364	7.616
9	81	3	59	3,481	7.681
10	100	3.162	60	3,600	7.746
11	121	3.317	61	3,721	7.810
12	144	3.464	62	3,844	7.874
13	169	3.606	63	3,969	7.937
14	196	3.742	64	4,096	8
15	225	3.873	65	4,225	8.062
16	256	4	66	4,356	8.124
17	289	4.123	67	4,489	8.185
18	324	4.243	68	4,624	8.246
19	361	4.359	69	4,761	8.307
20	400	4.472	70	4,900	8.367
21	441	4.583	71	5,041	8.426
22	484	4.690	72	5,184	8.485
23	529	4.796	73	5,329	8.544
24	576	4.899	74	5,476	8.602
25	625	5	75	5,625	8.660
26	676	5.099	76	5,776	8.718
27	729	5.196	77	5,929	8.775
28	784	5.292	78	6,084	8.832
29	841	5.385	79	6,241	8.888
30	900	5.477	80	6,400	8.944
31	961	5.568	81	6,561	9
32	1,024	5.657	82	6,724	9.055
33	1,089	5.745	83	6,889	9.110
34	1,156	5.831	84	7,056	9.165
35	1,225	5.916	85	7,225	9.220
36	1,296	6	86	7,396	9.274
37	1,369	6.083	87	7,569	9.327
38	1,444	6.164	88	7,744	9.381
39	1,521	6.245	89	7,921	9.434
40	1,600	6.325	90	8,100	9.487
41	1,681	6.403	91	8,281	9.539
42	1,764	6.481	92	8,464	9.592
43	1,849	6.557	93	8,649	9.644
44	1,936	6.633	94	8,836	9.695
45	2,025	6.708	95	9,025	9.747
46	2,116	6.782	96	9,216	9.798
47	2,209	6.856	97	9,409	9.849
48	2,304	6.928	98	9,604	9.899
49	2,401	7	99	9,801	9.950
50	2,500	7.071	100	10,000	10

Table of Values of Trigonometric Functions

deg	rad	sin	cos	tan	deg	rad	sin	cos	tan
0	.000	.000	1.000	.000					
1	.017	.017	1.000	.017	46	.803	.719	.695	1.036
2	.035	.035	.999	.035	47	.820	.731	.682	1.072
3	.052	.052	.999	.052	48	.838	.743	.669	1.111
4	.070	.070	.998	.070	49	.855	.755	.656	1.150
5	.087	.087	.996	.087	50	.873	.766	.643	1.192
6	.105	.105	.995	.105	51	.890	.777	.629	1.235
7	.122	.122	.993	.123	52	.908	.788	.616	1.280
8	.140	.139	.990	.141	53	.925	.799	.602	1.327
9	.157	.156	.988	.158	54	.942	.809	.588	1.376
10	.175	.174	.985	.176	55	.960	.819	.574	1.428
11	.192	.191	.982	.194	56	.977	.829	.559	1.483
12	.209	.208	.978	.213	57	.995	.839	.545	1.540
13	.227	.225	.974	.231	58	1.012	.848	.530	1.600
14	.244	.242	.970	.249	59	1.030	.857	.515	1.664
15	.262	.259	.966	.268	60	1.047	.866	.500	1.732
16	.279	.276	.961	.287	61	1.065	.875	.485	1.804
17	.297	.292	.956	.306	62	1.082	.883	.470	1.881
18	.314	.309	.951	.325	63	1.100	.891	.454	1.963
19	.332	.326	.946	.344	64	1.117	.899	.438	2.050
20	.349	.342	.940	.364	65	1.134	.906	.423	2.145
21	.367	.358	.934	.384	66	1.152	.914	.407	2.246
22	.384	.375	.927	.404	67	1.169	.921	.391	2.356
23	.401	.391	.921	.424	68	1.187	.927	.375	2.475
24	.419	.407	.914	.445	69	1.204	.934	.358	2.605
25	.436	.423	.906	.466	70	1.222	.940	.342	2.747
26	.454	.438	.899	.488	71	1.239	.946	.326	2.904
27	.471	.454	.891	.510	72	1.257	.951	.309	3.078
28	.489	.470	.883	.532	73	1.274	.956	.292	3.271
29	.506	.485	.875	.554	74	1.292	.961	.276	3.487
30	.524	.500	.866	.577	75	1.309	.966	.259	3.732
31	.541	.515	.857	.601	76	1.326	.970	.242	4.011
32	.559	.530	.848	.625	77	1.344	.974	.225	4.331
33	.576	.545	.839	.649	78	1.361	.978	.208	4.705
34	.593	.559	.829	.675	79	1.379	.982	.191	5.145
35	.611	.574	.819	.700	80	1.396	.985	.174	5.671
36	.628	.588	.809	.727	81	1.414	.988	.156	6.314
37	.646	.602	.799	.754	82	1.431	.990	.139	7.115
38	.663	.616	.788	.781	83	1.449	.993	.122	8.144
39	.681	.629	.777	.810	84	1.466	.995	.105	9.514
40	.698	.643	.766	.839	85	1.484	.996	.087	11.430
41	.716	.656	.755	.869	86	1.501	.998	.070	14.301
42	.733	.669	.743	.900	87	1.518	.999	.052	19.081
43	.751	.682	.731	.933	88	1.536	.999	.035	28.636
44	.768	.695	.719	.966	89	1.553	1.000	.017	57.290
45	.785	.707	.707	1.000	90	1.571	1.000	.000	—

LIST OF ABBREVIATIONS AND SYMBOLS

ABBREVIATIONS

(A>), (A<) Addition property of inequalities
(A-A) Associative axiom of addition
(A-M) Associative axiom of multiplication
(Add-A) Additive axiom
(C-A) Commutative axiom of addition
(C-M) Commutative axiom of multiplication
(Can-A) Cancellation law of addition
(Can-M) Cancellation law of multiplication
(Clos-A) Closure of P under addition
(Clos-M) Closure of P under multiplication
(D) Distributive axiom
(Def>) Definition of greater than
(Def<) Definition of less than
(Def-Cu. Rt.) Definition of cube root
(Def-Div) Definition of division
(Def-Neg. Exp.) Definition of a negative exponent
(Def-*n*th Rt.) Definition of nth root
(Def-Rat. Exp.) Definition of rational exponents
(Def-Sq. Rt.) Definition of square root
(Def-Sub) Definition of subtraction
(F-0) Factors of zero
(Fr-D) Division of fractions
(Fr-M) Multiplication of fractions
(Id-A) Additive identity axiom
(Id-M) Multiplicative identity axiom
(Inv-A) Additive inverse axiom
(Inv-M) Multiplicative inverse axiom
(LE-1) First law of exponents
(LE-2) Second law of exponents
(LE-3) Third law of exponents
(LE-4) Fourth law of exponents
(LE-5) Fifth law of exponents
(LR-1) First law of radicals
(LR-2) Second law of radicals
(M>), (M<) Positive multiplication property of inequalities
(−M>), (−M<) Negative multiplication property of inequalities
(Mult-A) Multiplicative axiom
(Neg-M) Negative multiplication
(Neg-Sum) Negative of a sum
(Q) Quotient equals zero
(R-A) Rearrangement property of addition
(R-M) Rearrangement property of multiplication
(R-P) Rational power law
(Recip-Prod) Reciprocal of a product
(Ref-A) Reflexive axiom
(T>) Transitive law for greater than
(T<) Transitive law for less than
(Sym-A) Symmetric axiom
(Trans-A) Transitive axiom
(Tri) Trichotomy axiom
(U-Pos. Rad.) Uniqueness of positive radicals
(U-Q) Uniqueness of quotients
(Zero-M) Multiplication by zero

SYMBOLS

$=$ Equal to
$\neq$ Not equal to
$[x]$ Greatest integer function
$>$ Greater than
$<$ Less than
$\{\ \}$ Set
$\subset$ Subset
$\cap$ Intersection
$\cup$ Union
$\emptyset$ Empty set
$\geqq$ Greater than or equal to
$\leqq$ Less than or equal to
$\ngtr$ Not greater than
$\nless$ Not less than
$|x|$ Absolute value

GLOSSARY

Absolute value. $|a| = a$ if $a \geqq 0$ and $|a| = -a$ if $a < 0$.
Additive identity. The additive identity element is 0.
Additive inverse. Each real number x has a unique opposite, $-x$.
Algorithm. The step-by-step procedure for solving a problem which specifically states how to proceed in all circumstances.
Associative axiom of addition. $(a + b) + c = a + (b + c)$.
Associative axiom of multiplication. $(ab) \cdot c = a \cdot (bc)$.

Binary operation. An operation which combines two numbers to produce a unique number. Multiplication and addition are two binary operations.
Binary system. An enumeration system with base 2; the only two digits used are 0 and 1.

Cancellation law of addition. If $a + c = b + c$, then $a = b$.
Cancellation law of multiplication. If $ac = bc$ and $c \neq 0$, then $a = b$.
Cartesian coordinate system. A system of assigning ordered number pairs to points in a plane.
Cartesian product. The set of all ordered pairs (x, y) where x is an element of A and y is an element of B. It is denoted as $A \times B$.
Closed interval. If P and Q are two points on a number line, the closed interval is the set consisting of P and Q and all points between P and Q.
Commutative axiom of addition. $a + b = b + a$.
Commutative axiom of multiplication. $ab = ba$.
Complex fractions. Sums, differences, products, and quotients of rational algebraic expressions.
Constant term. The term of a polynomial that does not contain a variable.
Coordinate system. An identification of points by numbers.

Decimal notation. Our system of enumeration, based on the ten digits 0, 1, 2, 3, 4, 5, 6, 7, 8, 9.
Degree of a polynomial. The highest power of a variable occurring in a polynomial.
Determinant. If

$$A = \begin{pmatrix} a_1 & b_1 \\ a_2 & b_2 \end{pmatrix}$$

is a 2×2 matrix, then the determinant of A is denoted by $|A|$ and is defined as

$$|A| = a_1b_2 - a_2b_1.$$

Difference of two squares. $x^2 - y^2$.
Direct variation. Two variables x and y vary directly if they are related by an equation of the form $y = kx$ for some nonzero number k.
Directly proportional ordered pairs. (x_1, y_1) and (x_2, y_2) are directly proportional if

$$\frac{x_1}{y_1} = \frac{x_2}{y_2}.$$

Discriminant of a quadratic equation. The discriminant is $b^2 - 4ac$ for the quadratic equation $ax^2 + bx + c = 0$.
Distributive axiom. $a(b + c) = ab + ac$.
Divisor. If $a = b \cdot c$, then b and c are divisors, or factors, of a.
Domain of a function. *See* Function.
Domain of a variable. The set of values which can be assigned to the variable.

Empty set. A set with no elements. It is denoted by $\emptyset$.
Equation. A mathematical sentence involving an equals sign.
Equivalent equations. Equations are equivalent if they have the same solution set.
Equivalent expressions. Algebraic expressions in the same variables are equivalent if they have the same value for every set of values of the variables.
Event. The set of outcomes of an experiment. P(event) is a measure of the likelihood of one of the events occurring.
Exponent. A numeral used to indicate the power of a number.
Extremes. In a proportion

$$\frac{a}{b} = \frac{c}{d},$$

a and d are the extremes.

Factor. *See* Divisor.
Family of lines. A set of lines having a common characteristic.
Flow chart. A picture of an algorithm.
Function. If to each value of one variable, there corresponds a unique value of a second variable, the correspondence between these two variables is called a function. If to each value of a variable x there corresponds a unique value of a variable y, then we say that y is a function of x. The set of values of x is called the *domain of the function*, and the set of values of y is called the *range of the function*.

Graph of a set. The set of graphs of the individual elements in the set.
Greater than. $x > y$ if, and only if, $x - y$ is positive.
Greatest common divisor. The greatest integer which is a divisor of both of a pair of integers. It is denoted g.c.d.

Half-open interval. If P and Q are two points on a number line, the half-open interval is the set consisting of either P or Q and all points between P and Q.
Half-plane. Each of two parts into which a line divides a plane.

Inequality. A mathematical sentence not involving an equal sign.
Intersection of A and B. All elements common to both A and B, denoted by $A \cap B$.

Inverse variation. Two variables x and y vary inversely if they are related by an equation of the form

$$y = \frac{k}{x}$$

for some nonzero number k.

Inversely proportional ordered pairs. (x_1, y_1) and (x_2, y_2) are inversely proportional if

$$\frac{x_1}{y_1} = \frac{y_2}{x_2}.$$

Irrational number. A nonrational number, such as π and $\sqrt{2}$.

Least common denominator. The least common multiple of the denominators of two or more rational algebraic expressions.

Least common multiple. The least integer which is a multiple of both of a pair of integers. It is denoted l.c.m.

Less than. $x < y$ if, and only if, $y - x$ is positive.

Linear form in two variables. An algebraic expression of the form

$$(\text{a number})x + (\text{a number})y + (\text{a number}).$$

Linear programming. A branch of modern mathematics by which particular solutions of systems of linear inequalities are found.

Matrix. A square array of numbers.

$$A = \begin{pmatrix} a_1 & b_1 \\ a_2 & b_2 \end{pmatrix}$$

is a 2×2 matrix.

Mean proportional. If the means of a proportion are equal, their common value is called a mean proportional between the extremes.

Means. In a proportion

$$\frac{a}{b} = \frac{c}{d},$$

b and c are called the means.

Monomial in x and y. A number or a product of a number by positive integral powers of x and y.

Multiple. If $a = b \cdot c$, then a is called a multiple of b and c.

Multiplicative identity. The multiplicative identity element is 1.

Multiplicative inverse. If the product of two numbers is 1, each number is a multiplicative inverse, or reciprocal, of the other.

nth root. The nth root of b is a number x such that $x^n = b$.

Number line. A line having a real-number scale on it.

Number theory. The study of properties of positive integers.

Numeral. A symbol used to denote a number.

Open interval. If P and Q are two points on a line, the open interval is the set of all points on the line between P and Q.

Open ray. A ray without its endpoint.

Open sentence. A statement containing one or more variables and having the property that it becomes either true or false when the variables are given specific values from their domains.

Opposites. All real numbers have additive inverses, called opposites.

Order axioms. The set P of positive numbers is closed under addition and under multiplication. The trichotomy axiom holds.

Ordered pair. Any pair of elements (x, y) having a first element x and a second element y.

Outcome. The result of an experiment.

Parabola. The graph of a quadratic function of the form

$$y = ax^2 + bx + c, \quad a \neq 0.$$

Polynomial in x and y. A polynomial in x and y is either a term or a sum of terms; each term is a number or a product of a number by positive integral powers of x and y.

Prime number. An integer greater than 1 which has 1 and itself as its only positive divisors.

Probability of an outcome. The measure of the likelihood of an outcome occurring. It is denoted P(outcome).

Pythagorean theorem. If a and b are the lengths of the legs and c is the length of the hypotenuse of a right triangle, then $a^2 + b^2 = c^2$.

Quadratic equation. A second-degree polynomial equation.

Quadratic formula. If $ax^2 + bx + c = 0$, then

$$x = \frac{-b \pm \sqrt{b^2 - 4ac}}{2a}.$$

Radical expression. An algebraic expression involving roots of other algebraic expressions.

Range of a function. *See* Function.

Rational exponents. $x^{\frac{m}{n}} = \sqrt[n]{x^m} = (\sqrt[n]{x})^m$.

Rational number. Any number that can be expressed as the quotient of two integers.

Ray. Each of the two parts into which a point P divides a line. The ray includes P.

Real numbers. The set consisting of all rational and irrational numbers.

Rearrangement property of addition. The addends of a sum may be rearranged in any order.

Rearrangement property of multiplication. The multiplicands of a product may be rearranged in any order.

Reciprocal. *See* Multiplicative inverse.

Relation. If S is a set then a subset of $S \times S$ is a relation in S.

Scientific notation. The expression of a number as a product of a number between 1 and 10 and a power of 10.

Set. A collection of objects called elements.

Slope. The ratio of rise to run of a line. It is denoted by m in the general equation of a line,

$$y = mx + k.$$

Solution. Values of variables that make an equation true.

Solution set. The set of all solutions.

Subset. If every element of set A is contained in set B, then A is called a subset of B. It is denoted by $A \subset B$.

Transitive law. If $x > y$ and $y > z$, then $x > z$.

Trichotomy axiom. Every real number is either a positive number, a negative number, or 0.

Trigonometric functions. The sine, cosine, tangent, secant, cosecant, and cotangent.

Union. The union of sets A and B consists of all the elements in either set A or set B. It is denoted by $A \cup B$.

Variable. A symbol used to denote any element of a given set.

INDEX

SELECTED ANSWERS FOR EXERCISES

In the side headings below, the hyphenated numerals (**1–1**, for example) refer to the chapter and section numbers and the final numeral (*3*, for example) refers to the first page of an exercise for which answers are given.

1–1, ***3*** **3. (a)** $(2 \times 10{,}000) + (2 \times 1{,}000) + (2 \times 100) + (2 \times 10) + (2 \times 1)$
5. (a) 24,300,000 **7. (a)** .008 **9. (a)** 5^3 **11. (a)** $(\frac{5}{6})^2$ **13. (a)** $(1.2)^2$
15. (a) 2 **17. (a)** 3 **19. (a)** 1.86×10^5 **21. (a)** 4.567×10^2 **23. (a)** 2^7
25. (a) 10^{11} **27.** 5.87×10^{12} miles **29. (a)** 2.325×10^5 miles
31. 6.3×10^9 seconds

1–2, ***8*** **1. (a)** $\sqrt{3}$ is irrational **3. (a)** For example, $\frac{2}{5}, \frac{1}{7}, \frac{1}{2}, -\{\frac{1}{3}\}, \frac{4}{3}$ **5. (a)** .0909 . . .
7. (a) 1.4 **9. (a)** .9696 . . . **11.** .9130434782608695652173 . . . **13.** No.
Irrational **15.** 3.141593 **17.** Irrational **19.** Irrational **21.** Irrational
23. Irrational **25.** Irrational **27. (c)** Infinitely many **29.** Yes. Yes.

1–3, ***11*** **1. (a)** Set of first four positive multiples of 2 **3. (a)** Set of explorers of the New World **5. (a)** {111, 222, 333, 444, 555, 666, 777, 888, 999}
7. (a) {4, 8, 16, 32} **9. (a)** All even positive integers **11. (a)** $\emptyset$ **(b)** $\emptyset$
13. (a) $A \cap B = B \cap A = 6$ **15. (a)** F **17. (a)** T **19.** Yes; Yes; 8
21. 16; {0}, {2}, {4}, {6}, {0, 2}, {0, 4}, {0, 6}, {2, 4}, {2, 6}, {4, 6}, {0, 2, 4}, {0, 2, 6}, {0, 4, 6}, {2, 4, 6}, {0, 2, 4, 6}, $\emptyset$ **23.** $A \cap B = A$; $A \cup B = B$

1–4, ***16*** **1. (a)** 11, 21, 1, 6 **3. (a)** 25, 49, 9, 16 **5. (a)** $N + 9$ **7. (a)** $2 \cdot x^3$
9. (a) Two more than 3 times a number **11. (a)** T **(b)** T **13. (a)** F **(b)** T
15. (a) F, F, T, F, F **17. (a)** (i) 0 (ii) Any other number **19. (a)** (i) 6 (ii) Any other number **21. (a)** (i) Impossible (ii) Any number
23. (a) $C = \{1, 2, 3, 4\}$ **25. (a)** When n is a multiple of 5 **27. (a)** x is an integer greater than 2 **(b)** x a real number greater than 10, or less than $^-10$
(c) x a real number between $\sqrt{110}$ and $\sqrt{1010}$ or between $^-\sqrt{1010}$ and $^-\sqrt{110}$
29. (a) $d + n + q$ **(d)** $\dfrac{10d + 5n + 25q}{100}$ **31.** {55, 60, 65, 70, 75, 80, . . .}
33. {9, 13, 17, 21, 25} **35. (a)** $\{6x \mid x$ an integer between 1 and $17\}$
(b) $\{4^x \mid x$ an integer less than $5\}$ (Assuming integral powers are being considered) **(c)** $\{2x + 1 \mid x$ an integer between 49 and $500\}$

1–5, ***23*** **1. (a)** 1×12, 2×6, 3×4, {1, 2, 3, 4, 6, 12} **3. (a)** 1×18, $2 + 9$, 3×6, {1, 2, 3, 6, 9, 18} **5. (a)** $1 + 84$, $2 + 42$, 3×28, 4×21, 6×14, 7×12, {1, 2, 3, 4, 6, 7, 12, 14, 21, 28, 42, 84} **7. (a)** 1×101, {1, 101}
9. (a) {1, 2, 4, 8, 16, 32} **11. (a)** {1, 3, 9, 27, 81} **13. (a)** {1, 2, 4, 8, 16, 32, 64, 128, 256, 512, 1024, 2048}. **15.** The sum of the digits is divisible by 9; Yes
17. An even number, the sum of whose digits is divisible by 3, is itself divisible by 6. **19.** {4, 9, 25}; No; Yes **21.** {6, 8, 10, 14, 15, 21, 22, 26, 27}; 2^3, 3^3; Product of two primes **23.** 96; Yes

1–6, ***26*** **1. (a)** (ii) 50, 50 **3. (a)** Fair **5. (a)** Fair **7. (a)** Fair **9. (a)** Approximately 10 times **11. (a)** Fair **13. (a)** Fair **15. (a)** Fair

1–7, ***32*** **1. (a)** $\frac{2}{5}, \frac{3}{5}$ **3. (a)** $\frac{1}{4}, \frac{1}{4}, \frac{1}{2}$ **5. (a)** $\frac{2}{5}$ **(b)** $\frac{3}{5}$ **7. (a)** $\frac{1}{2}, \frac{1}{2}, 1, 0$ **9. (a)** $\frac{1}{2}, \frac{1}{2}$
11. (a) $\frac{1}{36}$ **13. (a)** $\frac{1}{12}$ **15. (a)** $\frac{1}{2}$ **17. (a)** (i) $\frac{1}{52}$ (ii) $\frac{1}{51}$ **19. (a)** $\frac{1}{4}$

1–8, ***37*** **1. (a)** $\frac{2}{5}$ **3. (a)** $\frac{4}{7}$ **5. (a)** $\frac{6}{7}$ **11. (a)** 48 **13. (a)** 65 **15. (a)** 143

EXTRA!, 39 **1.** 1×12, 2×6, 3×4, 4×3, 6×2, 12×1 **3.** 60 **5. (a)** 2 **(c)** 4

Chapter Review, 41 **1.** T **3.** F, rational **5.** F, 10 **7.** F, 1.2×10^5 **9.** F, rational **11.** T
13. (a) 3^3 **(b)** 9^2 or 3^4 **(c)** $(\frac{2}{3})^2$ **(d)** $(.4)^2$ **(e)** 10^4, or 100^2 **(f)** 2^5
15. (a) $d = x - y$ **(b)** All integers greater than or equal to 0. **17.** 8; 10; 16; 208; 20,008 **19. (a)** $x + 12$ **(b)** $2x - 3y$ **(c)** $\frac{1}{2}x - 7$ **(d)** $5x + 10y$
21. (a) {4, 9, 14, 19, 24, 29} **(b)** {12} **(c)** {2, 3, 4, 5, 6} **(d)** {4} **23.** .259 . . .
25. (a) {2, 3, 4, 6, 8, 9, 12, 14, 15, 18, 21, 24, 27} **(b)** {6} **27.** $\frac{1}{18}$ **29.** $\frac{1}{2}$
31. (a) $\frac{15}{2}$ **(b)** $\frac{20}{3}$

2–1, *49* **1. (a)** T **3. (a)** F **5. (a)** Open **7. (a)** 54 **9. (a)** 32 **11. (a)** 1500 **13. (a)** 92 **15. (a)** $5a + 17$, 27 **17. (a)** $35a + 60ab$, 190 **19. (a)** 3 **21. (a)** 3 **23.** Transitive axiom **25.** Additive axiom **27.** Additive axiom **29.** Multiplicative axiom

2–2, *53* **1. (a)** Not commutative **3. (a)** $71 + 18$ **5. (a)** 3, $x + 2$ **7. (a)** =, (C-M) **9. (a)** =, (A-M) **11. (a)** =, (C-A), (C-M) **13. (a)** $\neq$ **15. (a)** $(36 + 14) + (8 + 12) = 70$ **17. (a)** $(69 + 31) + (77 + 23) = 200$ **19. (a)** T, (A-M) **21. (a)** T, (R-A) **23. (a)** T, (R-A) **25. (a)** Open

EXTRA!*, *54 **1. (a)** 19 **(b)** 21 **(c)** 2 **(d)** $\frac{13}{6} = 2\frac{1}{6}$ **(e)** 20 **3.** All numbers when $x = y$ **5.** No

2–3, *57* **1. (a)** $52 \times 3 = (50 + 2) \times 3 = (50 \times 3) + (2 \times 3) = 150 + 6 = 156$ **3. (a)** 3, 17 **5. (a)** 8, 17 **7. (a)** $56 + 84$ **9. (a)** $43(7 + 3)$ or $43 \cdot 10$ **11. (a)** 580 **13. (a)** 18 **15. (a)** T, (D) **17. (a)** T, (D) **19.** T, (D) **21. (a)** 2, 3, 4, 6, 12 **(b)** 3, 5, 15 **(c)** 7, 11, 77 **(d)** 17, 13, 221 **(e)** 13, 37, 481 **(f)** 2, 4, 5, 8, 10, 20, 40 **23.** Even integer, even integer, even integer; Always even **25.** 128

2–4, *61* **1. a.** $^{-}6$ **3. (a)** $^{-}\frac{1}{2}$ **5. (a)** $\frac{1}{3}$ **7. (a)** $\frac{1}{15}$ **9. (a)** $\frac{10}{7}$ **11. (a)** $\frac{3}{17}$ **13. (a)** Additive inverses **15. (a)** Not additive inverses **17. (a)** Multiplicative inverses **19. (a)** Not multiplicative inverses **21. (a)** 0 **23. (a)** $\frac{5}{2}$ **25. (a)** 2 **27.** b/a

2–5, *64* **1. (a)** Closed **3. (a)** Closed **5. (a)** Not closed **7. (a)** Not closed **9. (a)** Not closed

2–6, *67* **1. (a)** (i) $^{-}2$ (ii) $^{-}2$ **3. (a)** (i) 1 (ii) 1 **5. (a)** $^{-}5$ **7. (a)** $5° + 6° = 11°$ **9. (a)** 9000 ft + $^{-}2000$ ft = 7000 ft **11.** $^{-}8$ **13.** 3 **15.** $\frac{2}{3}$ **17.** 28 **19. (a)** $^{-}14$ **(b)** $^{-}40$ **(c)** $^{-}40$ **(d)** $^{-}33$

2–7, *70* **1. (a)** 3 **3. (a)** -32 **5. (a)** 27 **7. (a)** 0 **9. (a)** 7 **11. (a)** -135 **13. (a)** 23 **15. (a)** 1.62 **17.** 22 **19.** 7 **21.** 24 **23.** 7 **25.** .74 **27.** 2.2 **29. (a)** -12 **(b)** 11 **(c)** 27 **(d)** -12

Preparation Exercises,* *72 **1.** Additive inverses **2.** Additive inverses **3.** Not additive inverses **4.** Additive inverses **5.** Multiplicative inverses **6.** Not multiplicative inverses **7.** Multiplicative inverses **8.** Multiplicative inverses **9.** F **10.** T, (R-A) **11.** T, (Id-A) **12.** Open **13.** T, (Inv-M) **14.** T, (Inv-A)

2–8, *74* **1. (a)** (i) T, (Neg-Sum) (ii) T, (Def-Sub) **3. (a)** T, (Recip-Prod) **5. (a)** $\neq$ **7. (a)** = **9. (a)** = **11, (a)** = **13. (a)** = **15. (a)** -2 **17. (a)** $2 + 3$, or 5 **19. (a)** $\dfrac{1}{3 \cdot 8}$, or $\dfrac{1}{24}$ **21. (a)** 63 **23.** (a) **25.** (a) **27.** (Def-Sub), (R-A), (Inv-A), (Id-A) **29.** (Recip-Prod), (R-M), (Inv-M), (Id-M)

Prep. Ex.*, *76 **1.** T, (A-A) **2.** T, (C-A) **3.** T, (Neg-Sum) **4.** T, (Def-Sub) **5.** T, (C-M) **6,** T, (D) **7.** T, (R-A) **8.** T, (D) **9.** T, (Id-M) **10.** T, (Id-A)

2–9, *78* **1. (a)** $10x$, 40 **3. (a)** $9xy$, 180 **5. (a)** $\dfrac{7y}{5}$, 7 **7. (a)** $3y^2$, 75 **9. (a)** $11a + ab$ **11. (a)** $30a^3b^3$ **13. (a)** $9xy$ **15. (a)** $6a - 7$ **17. (a)** $5y - 8$ **19. (a)** 1 **21.** Yes **23.** Yes **25.** No **27.** Yes **29.** (A-M) **31.** (D) **33.** (Id-M) **35.** (Id-A) **37.** $c(1 + d)$ **39.** $11b(11c + 5b)$ **41.** $2\pi r(r + h)$ **43.** $19^4(1 + 19)$, or $19^4(20)$ **45.** $3^2(1 + 3 + 3^2)$, or $3^2(13)$ **47. (a)** $(m + 1)(2m + n)$ **(b)** $(5 + x)(1 + y)$

2–10, *81* **1. (a)** 4; (Can-A) **3. (a)** 22; (Can-A) **5. (a)** 30; (Can-M) **7. (a)** x; (Id-A) **9. (a)** -2, -2; (Can-A) **11. (a)** 3, (Can-A); 1, (Can-M) **13. (a)** 9, (Can-A); 4, (Can-A); 2, (Can-M) **15. (a)** -5 **17. (a)** 0 **19.** $\frac{15}{2}$ **21.** 0 **23.** 10 **25.** Impossible

2–11, *85* **1. (a)** 15 **3. (a)** -18 **5. (a)** 28 **7. (a)** 8 **9. (a)** 0 **11. (a)** T **(b)** F **13. (a)** F **(b)** F **15. (a)** F **(b)** F **17. (a)** 1, -7 **(b)** 2, -12 **19. (a)** Impossible **(b)** Impossible **21. (a)** -40 **(b)** -16 **23.** F **25.** F **27.** 0, 1, 2; $A = \{x \mid x \text{ is zero or positive}\}$ **29.** 0, 1, 2; $C = \{x \mid x \text{ is zero or positive}\}$ **31.** 13 or -13 **33.** 42 **35.** 19 or -19 **37.** 6

2–12, *88* **1. (a)** (i) -8 (ii) -8 **3. (a)** 0 **5. (a)** -84 **7. (a)** $-.032$ **9. (a)** 310 **11. (a)** -1 **13. (a)** 90 **15. (a)** -5 **17. (a)** -50 **19. (a)** 42 **21. (a)** Negative **23.** 84 **25.** 0 **27. (a)** $2x - 4y$ **(b)** $-4x + 3y$ **(c)** $-72a + 62b$ **(d)** $-21a + 52b$ **(e)** $2x + 15y$ **(f)** $36x + 9y$

Prep. Ex., 89 **1.** $x \cdot x \cdot x$ **2.** $x \cdot x \cdot x \cdot x \cdot x, x^8$ **3.** (R-M) **4.** (Inv-M) **5.** (Id-M) **6.** (Recip-Prod)

2–13, *92* **1. (a)** T **3. (a)** T **5. (a)** 3 **7. (a)** -6 **9. (a)** -72 **11. (a)** -23 **13. (a)** 0.007 **15. (a)** $-\frac{7}{4}$ **17. (a)** 10 **19. (a)** -16 **21. (a)** -10 **(b)** -27 **(c)** 1 **(d)** 4 **(e)** 10 **(f)** 10 **23. (a)** (Def-Div), (D), (D), (Def-Div) **(b)** (Def-Div), (Recip-Prod), (R-M), (Inv-M), (Id-M) **25.** $\frac{38}{a^2} + \frac{7}{b^2}$ **27.** $\frac{x}{5} + \frac{y}{3}$ **29.** $-\frac{2y^4}{5x^4}$ **31.** $\frac{x^2}{4y^3}$

2–14, *98* **1.** (Def-Div), (R-M), (Recip-Prod), (Def-Div) **3.** (Def-Div), (Mult-A), (R-M), (Inv-M), (Id-M), (Def-Div) **5.** (Add-A), (D), (Mult-A), (R-M), (Inv-M), (Id-M), (Def-Div), (Sym-A)

2–15, *103* **1. (a)** -42 **3. (a)** 378 **5. (a)** -837 **7. (a)** $3\frac{2}{7}$ **9. (a)** $7\frac{6}{7}$ **11. (a)** $4\frac{3}{15}$ **13.** $3\frac{9}{12}$ **17. (a)** 16 **19. (a)** 14

Chapter Review, 107 **1.** $-y$ **3.** Add $-a$ **5.** Commutative, addition **7.** Multiply by $\frac{1}{x}$ **9.** Distributive **11.** T, (Zero-M) **13.** T, (Def-Sub) **15.** F, 4 **17.** F, neither **19.** $6\frac{2}{3}$ **21.** 72 **23.** 9 **25.** -4 **27.** -147 **29.** $-\frac{5}{2}$ **31.** 7 **33.** -12 **35.** 0 **37.** -7 **39.** $14x, -42$ **41.** $-2y^2, -32$ **43.** $-\frac{3}{x} - \frac{16}{y}, -3$ **45.** $6x + 3y, -6$ **47. (a)** $x(2 + y)$ **(b)** $x(1 + 5) = 6x$ **(c)** $7x(1 + 6x)$ **(d)** $x(1 + y)$ **(e)** $23^2(24)$ **(f)** $13xy(y + 2x)$ **49.** $7x - 3$ **51.** $-2x + 26y$ **53.** $17x + 11$ **55.** $\frac{3}{2}h - \frac{1}{3}k$ **57.** $2a$ **59.** 3

3–1, *113* **1. (a)** $\{5\}$ **3. (a)** $\{14\}$ **5. (a)** $\{-1\}$ **7. (a)** $\{6\}$ **9. (a)** $\{\frac{9}{2}\}$ **11. (a)** $\{7\}$ **13. (a)** $\{35\}$ **15. (a)** $\{3\}$ **17. (a)** $\{0\}$ **19.** (Mult-A), (A-M), (Neg-M), (Def-Div), (Inv-M), (Id-M) **21.** $\{-4\}$ **23.** $\{5\}$ **25.** $\{-8\}$ **27.** $\{1\}$ **29.** $\{3\}$ **31.** $\{-2\}$

3–2, *116* **1. (a)** $\{6\}$ **3. (a)** $\{-9\}$ **5. (a)** $\{5\}$ **7. (a)** $\{3\}$ **9. (a)** $\{-11\}$ **11. (a)** $\{-9\}$ **13. (a)** $\{3\}$ **15. (a)** $\emptyset$ **17. (a)** $\{\frac{13}{11}\}$ **19. (a)** $\{6\}$ **21. (a)** $\{6\}$ **23. (a)** $\{\frac{1}{2}\}$ **25. (a)** $\emptyset$ **27. (a)** $\{-16\}$ **29. (a)** $\{-\frac{4}{9}\}$ **31.** $\{x \mid x \text{ any real number}\}$

Prep. Ex., 117 **1.** $\frac{2}{9}$ **2.** $\frac{17}{15}$ **3.** $\frac{3}{4}$ **4.** $\frac{7}{6}$ **5.** $\frac{39}{4}$ **6.** $\frac{19}{2}$

3–3, *120* **1. (a)** $\{27\}$ **3. (a)** $\{3\}$ **5. (a)** $\{-14\}$ **7. (a)** $\{-\frac{20}{7}\}$ **9. (a)** $\{77\}$ **11. (a)** $\{6\}$ **13. (a)** $\{3\}$ **15. (a)** $\{-\frac{1}{2}\}$ **17. (a)** $\{-5\}$ **19. (a)** $\{-\frac{1}{4}\}$ **21. (a)** $\{\frac{15}{14}\}$ **23.** $\{31\}$ **25.** $\{-11\}$ **27.** $\{-12\}$ **29.** $\{9\}$ **31.** $\{4\}$

3–4, *122* **1. (a)** $\frac{1}{9}$ **3. (a)** $\frac{5}{9}$ **5. (a)** 1 **7. (a)** $\frac{29}{30}$ **9. (a)** $\frac{127}{495}$ **11. (a)** $\frac{11}{9}$ **13.** $\frac{37}{100}$ **15.** $\frac{104}{495}$ **17.** $\frac{338}{99}$ **19.** $\frac{100}{999}$ **21.** $\frac{7885}{999}$ **23.** $\frac{3331}{333}$ **25.** $\frac{4115}{33333}$ **27.** $\frac{30371}{3333}$ **29. (a)** 9 **(b)** 9 **(c)** 9, or 3 **(d)** 9

3–5, *126* **1. (a)** $r = .04$ **3. (a)** $S = 2.3$ **5. (a)** $a = 4$ **7. (a)** $y = 144$ **9. (a)** 15 in. **11. (a)** 150 mi **13. (a)** $r = \frac{L - 3s}{2\pi}$ **(b)** $y = \frac{12 - 3x}{2}$ **(c)** $C = ID^2$ **(d)** $l = \frac{Rd^2}{k}$ **15.** $d = 16t^2$; d = distance, t = falling time **(a)** 64 ft **(b)** 400 ft **(c)** 5 sec **(d)** 4 sec **17.** $A = \frac{h}{2}(B + b), \frac{36}{5}$ ft **19.** $A = p(1 + rt)$, \$2386.25 **21. (a)** $C = .35 + .10(4m - 1) + .25t$, \$3.55 **(b)** $2.65 = .35 + .10(4m - 1)$, 6 mi

Prep. Ex., 128 **1.** $\{5\}$ **2.** $\{5\}$ **3.** $\{13\}$ **4.** $\{3\}$ **5.** $\{1\}$ **6.** $\{\frac{7}{3}\}$ **7.** $\{\frac{400}{3}\}$ **8.** $\{100\}$

3–6, *130* **3. (a)** $x - 3$ **5. (a)** $2(x - 5)$ **7. (a)** $3.98x + 4.95(5 - x)$; x, the number of \$3.98 records **9. (a)** Harry's age is 14, Dick's age is 32

11. **(a)** Andrea's age is 9, Pat's age is 18 **13.** **(a)** \$1200 at $4\frac{1}{2}\%$, \$1300 at 5%
15. 11 pennies, 15 nickels **17.** 8 lb at 85¢, 12 lb at \$1.10
19. 6 liters of 50% acid and 4 liters of 75% acid.

Preparation Exercises, 132 **1.** **(a)** $5x$ **(b)** $40x$ **(c)** $\frac{500}{x}$ **(d)** $\frac{3x}{30}$, or $\frac{x}{10}$ **(e)** $\frac{x}{7}$ **(f)** $\frac{525}{x}$
2. $(2n - 1) + (2n) + (2n + 1) = 6n$ **3.** $x + 9$ **4.** $x(x + 1)$

3–7, ***135*** **1.** **(a)** $55x - 48x$, or $7x$
3. **(a)** $x + \left(\frac{x}{2} + 4\right) + 3\left(\frac{x}{2} + 4\right)$, or $3x + 16$
5. **(a)** 5 hrs **7.** **(a)** 51, 52, 53 **9.** **(a)** 79, 81, 83 **11.** No **13.** 2.6 m, 1.4 m
15. 11:30 AM, 10 mi **17.** 1 hr 20 min **19.** Car, 40 mph; Train, 60 mph
21. 62.5 mph **23.** \$60, \$80

3–8, ***139*** **2.** (Add-A), (R-A), (Inv-A), (Id-A) **4.** (Id-M), (C-M), (Def-Div), (D), (Def-Div)

Chapter Review, 141 **1.** $\{7\}$ **3.** $\{1\frac{1}{2}\}$ **5.** $\{-\frac{1}{2}\}$ **7.** $\{1\}$ **9.** **(a)** $\{y \mid y \text{ any real number}\}$ **(b)** $\emptyset$ **11.** $\frac{17}{66}$
13. $\frac{271}{99}$ **15.** **(a)** $y = \frac{x - 2h}{3}$ **(b)** $y = \frac{16t^2 + s}{t}$ **(c)** $y = \frac{-c - ab}{b}$ **17.** 6
19. 8 yr **21.** 5 lb \$1.05 coffee, 15 lb \$0.85 coffee **23.** 68 student tickets, 52 adult tickets **25.** Sue is 15, Mary is 13 **27.** **(a)** 1:30 PM **(b)** Speeds did not vary.

4–1, ***147*** **1.** **(a)** Positive, $>$ **3.** **(a)** Positive, $>$ **5.** **(a)** $<$ **7.** **(a)** $<$ **9.** **(a)** $>$
11. **(a)** Any number greater than 5 **13.** **(a)** Any number greater than -9
15. **(a)** Any number less than -20 **17.** **(a)** Any number less than 0
19. **(a)** 11(a), 11(b), 17(b), 18(a)

Prep. Ex., 148 **1.** 5 **2.** -84 **3.** -39 **4.** 200 **5.** 12 **6.** 10 **7.** -16 **8.** -12 **9.** $\{3, -3\}$
10. $\{11, -1\}$

4–2, ***151*** **1.** **(a)** Right; -1000 **3.** **(a)** T **5.** **(a)** F **(b)** T **7.** **(a)** T **(b)** F
9. $A = C, B = D, E = F$ **(a)** $\{-3, -2, -1, 0, 1, 2, 3\}$ **11.** **(a)** $\{-5, -4, -3, -2, -1, 0, 1, 2, 3, 4, 5\}$ **13.** **(a)** $\{-9, -8, -7, -6, -5, -4, -3, -2, -1, 0, 1, 2, 3, 4, 5, 6, 7, 8, 9\}$ **(b)** E.g. $\{-12, -18, -24, -30, -36, \ldots\}$
(c) E.g. $\{-10, -14, -18, -22, -26, \ldots\}$ **(d)** $\{1, \frac{1}{2}, \frac{1}{3}, \frac{1}{4}, \frac{1}{5}, \frac{1}{6}, \frac{1}{7}, \frac{1}{8}, \frac{1}{9}, \frac{1}{10}\}$
15. $x - 6$ **17.** $a < b$

Preparation Exercises, 152 **1.** Additive inverses **2.** Multiplicative inverses **3.** Multiplicative inverses
4. Additive inverses **5.** Neither **6.** Additive inverses **7.** Additive inverses
8. Neither

4–3, ***155*** **1.** **(a)** 0, 2, -1, -10 are solutions. **3.** **(a)** (ii) (A $<$) (iii) (M $<$) (iv) (A $<$)
(v) ($-$M $<$) **5.** **(a)** F, $-10 < -3$ **7.** **(a)** $x < 8$ **9.** **(a)** $x > -4$
11. **(a)** $x > 4$ **13.** $x > 0$ **15.** $x < -3$ **17.** $x < 4$ **19.** $x < -\frac{2}{3}$
21. $x < \frac{3}{4}$

4–4, ***158*** **1.** **(a)** T, $9 = 9$ **3.** **(a)** T, since $x < 0$ **7.** **(a)** $x \leqq -4$ **9.** **(a)** $x \leqq 4$
11. **(a)** $x \leqq -4$ **13.** **(a)** $x \geqq 100$ **15.** **(a)** E.g. $1, \frac{7}{8}, \frac{1}{2}, -\frac{1}{3}, -\frac{1}{2}, -\frac{3}{71}$

4–5, ***163*** **1.** **(a)** $\{x \mid x < 0\}$; Open ray **3.** **(a)** Ray **5.** **(a)** Open ray
7. **(a)** $A = \{x \mid x > 3\}$ **9.** **(a)** 13 elements **11.** **(a)** Infinite **13.** Positive
15. E.g., 3, 4, 10, **17.** No **19.** Yes, Exercise 16 **31.** $\{10, 9, 8, 7, 6, \ldots\}$
33. $\{-16, -12, -8, -4, 0, \ldots\}$ **35.** $\{-3, -5, -7, -9, -11, \ldots\}$

4–6, ***167*** **1.** **(a)** $D = \{x \mid x > -5\} \cap \{x \mid x < 12\}$; Open interval.
3. **(a)** Closed interval **5.** **(a)** Open ray **7.** **(a)** Half-open interval
9. **(a)** (i) $-3, -2, -1$ (iii) Both are half-open intervals (iv) Yes
11. **(a)** Half-open interval **13.** **(a)** Two rays **15.** **(a)** $A = \{x \mid -7 \leqq x \leqq 7\}$, $C = \{x \mid x \geqq 6\} \cup \{x \mid x \leqq -6\}$, $E = \{x \mid -3 < x < 3\}$

4–7, ***170*** **2.** **(a)** $R = \{5, 10, 15, 20, 25, 30, 35, 40, 45, 50, 55, 60, 65, 70\}$
4. **(a)** $A = \{1, 4, 9, 16, 25, 36, 49, 64, 81\}$ **6.** **(a)** $D = \{2, 4, 6, 8\}$
8. **(a)** $S = \{f^2 \mid 4 \leqq f < 12\}$

Chapter Review, 173 **1. (a)** $<$ **(b)** $>$ **(c)** $>$ **(d)** $<$ **(e)** $>$ **(f)** $>$ **(g)** $>$ **(h)** $<$ **3.** Yes **5.** T **7.** F **9.** T **11.** T **13. (a)** A: Any number less than -5 B: Any number between -3 and 0 **(b)** No; Yes, -6 **(c)** Yes, -1; Yes, -2 **15.** $x > 6$ **17.** $x > 5$ **19.** $x > 2$ **21.** $x \leqq \frac{4}{3}$, greatest element $\frac{4}{3}$ **23.** $x > 1$ **25.** $x \leqq 3$, 3 greatest element **27.** $x \leqq \dfrac{-147}{4}$, $\dfrac{-147}{4}$ greatest element.

Cumulative Review I, 175 **1. (a)** 1×10^2 **(b)** 7.8×10 **(c)** 1.1×10^3 **(d)** 8.957×10^3 **3. (a)** 3^7 **(b)** 2^6 **(c)** 5^7 **5. (a)** $\{2, 4, 7\}$ **(b)** $\{2, 4, 7\}$ **(c)** $\{1, 2, 3, 4, 5, 6, 7, 8, 9, 10\}$ **(d)** $\{1, 2, 3, 4, 5, 6, 7, 8, 9, 10\}$ **7.** No **9. (a)** $x - y$ **(b)** $x \cdot y \cdot z$ **(c)** $x + (x + 1)$ **(d)** $x(y + z)$ **(e)** $5x - 8$ **(f)** $x + \frac{1}{2}x$ **11. (a)** $p + n + d$ **(b)** $p + 5n$ **(c)** $.01p + .05n$ **(d)** $.05n + .10d$ **(e)** $p + 5n + 10d$ **13. (a)** $\frac{2}{5}$ **(b)** $\frac{1}{5}$ **(c)** $\frac{2}{5}$ **(d)** $\frac{3}{5}$ **(e)** 0 **15. (a)** False **(b)** Open **(c)** True **(d)** False **(e)** Open **(f)** False **17.** (Add−A) **19.** (A-A) **21.** (D) **23.** (Inv-M) **25.** (T $>$) **27.** $-2x + 5y$ **29.** $2x^4y^4$ **31.** $2.12x + 3.41y$ **33.** $\frac{5}{7}x^6y^6$ **35.** $22x + 21y + 17$ **37.** T **39.** T **41.** T **43.** $=$ **45.** $=$ **47.** $=$ **49.** $=$ **51.** $=$ **53.** $x - 7y$ **55.** $48b$ **57.** $a - 9c$ **59.** $a - b$ **61.** $54b - 45a$ **63.** $66/xy$ **65.** 1 **67. (a)** No **(b)** No **69.** $\{1\}$ **71.** $\{\frac{3}{2}\}$ **73.** $\{2\}$ **75.** $\{14\}$ **77.** $\{4\}$ **79.** $\{13\}$ **81. (a)** $x > 5$ **(b)** $x < 4$ **(c)** $x \geqq 5$, 5 is the smallest element. **(d)** $x \leqq \frac{5}{3}$, $\frac{5}{3}$ is the greatest element. **83.** 8 yr **85.** 16 **87.** 13 nickels, 7 dimes, 14 quarters **89.** 300 **91.** 131 **93.** 16 **95.** Open ray **97.** Two rays

5–1, *183* **1. (a)** $-\frac{5}{4}$; 2, -1, 1 **3. (a)** 0; 7, $-\sqrt{2}$ **5. (a)** 7, 0, -5, -9, -5, 0, 7 **7. (a)** $\frac{3}{2}x^2 - \frac{1}{2}y^2$ **9. (a)** $3r^4 + 7r^2 - 6$ **11. (a)** $-6u^3 + 3u^2 - 4u + 10$ **13. (a)** $-x^3 + 3x^2 - x - 22$ **15. (a)** Value is the constant term **17.** Additive inverse **(a)** $-3x - 5$ **(b)** $-7x + 2$ **(c)** $-6 + 2x + 3x^2$ **19.** (R-A), (Def-Sub); (D) **21.** $a^2 + 3a + 4$ **23.** $y^2 - \frac{11}{8}y + 5$ **25.** $3x^2 - 13x + 16$ **27.** $-3y + 3$ **29.** $9y^2 - 6y + 15$ **31.** $8x - 5$ **33.** $s - 20$ **35. (a)** Positive, Negative **(b)** Positive, Value > -5

5–2, *188* **1. (a)** x^8 **3. (a)** r^5 **5. (a)** -2^{170} **7. (a)** t^{45} **9. (a)** x^6 **11. (a)** y^{15} **13. (a)** $-12x^{32}$ **15. (a)** 9 **(b)** -9 **(c)** 9 **(d)** -9 **17. (a)** 16 **(b)** -16 **(c)** 16 **(d)** -16 **19.** $-6x^2 + 15$ **21.** $-66y^3 + 77y^2$ **23.** $x^6 - 10x^4 + 3x^2$ **25.** $-a + 3a^2 + 9a^3$ **27.** $11x^2 - 6x$ **29.** $-3x^3 - 9x^2$ **31.** $2y^2 - 10y$ **33.** $8x^{12}$ **35.** 3^{12} **37.** $-320a^6$ **39.** $50y^5$ **41.** $3a^8$ **43.** $2x^{11}$ **45.** $\frac{2}{3}y^{21}$ **47.** $4x^{18}$ **49.** $-7y^4 - 54y^3 + 39y^2 - 43y$ **51.** $-x^4 + 1$ **53.** $36y^4 - 84y^2 + 49$ **55.** No, $2^{15} \cdot 3^8$

5–3, *192* **1. (a)** $2x^2 + x$; 1, 1, 2 **3. (a)** $2x$, -5; $6x^2 - 13x - 5$; 1, 1, 2 **5. (a)** $8x^2 - 29x + 15$; 1, 1, 2 **7. (a)** $21x^2 + 40x - 21$; 1, 1, 2 **9. (a)** $6y^3 + 11y^2 - 2y + 20$; 2, 1, 3 **11. (a)** $9x^2 - 60x + 100$; 1, 1, 2 **13. (a)** $3a^3 - 14a^2 + 36a - 35$; 1, 2, 3 **15. (a)** $2x^6 + 4x^5 - x^4 + x^3 + 2x^2 - x + 1$; 3, 3, 6 **17. (a)** 2 **(b)** 2, 1, 0 **(c)** Yes

5–4, *194* **1. (a)** $20x^2 + 13x - 21$ **3. (a)** $64x^2 - 25$ **5. (a)** $8x^3 - 22x^2 + 29x - 21$ **7. (a)** $64y^3 + 1$ **9. (a)** $4x^4 - x^2 + 2x - 1$ **11. (a)** $42y^4 - 25y^3 - 5y^2 + 5y - 2$ **13. (a)** $x^3 - 4x$ **15.** $x^3 - 3x^2 + 3x - 1$ **17. (a)** $x^3 - 1$ **(b)** $x^4 - 1$ **(c)** $x^5 - 1$ **19. (a)** 2772 **(b)** 3399 **(c)** 4466 **(d)** 9664 **21.** $125x^3 - 75x^2 + 15x - 1$

5–5, *197* **1. (a)** $15x^2 + 31x + 14$ **3. (a)** $x^2 + x - 12$ **5. (a)** $10x^2 + 29x + 21$ **7. (a)** $x^2 + 14x + 49$ **9. (a)** $12x^2 - 13x - 14$ **11. (a)** $4x^2 + 20x + 25$ **13. (a)** $2y^2 - 9y - 180$ **15. (a)** $4x^2 - 25$ **17. (a)** $26x^2 - x - 22$ **19. (a)** $-5x^2 - 60x + 65$ **21. (a)** The products are the same: $x^2 - 10x + 25$ **23.** Two **25.** The square of a binomial is the sum of the squares of each term plus or minus twice their product.

5–6, *201* **1. (a)** $8x(x - 9)$ **3. (a)** $a(9a^3 + 1)$ **5. (a)** $15x^2(x^2 + 4)$ **7. (a)** $5x^2(x^2 - 2x + 3)$ **9. (a)** $x^3(x^2 - 7x + 14)$ **11. (a)** $6y(3y^3 + 5y - 7)$ **13. (a)** $z^3(1 + z^3)$ **15. (a)** $y^2(5y - 1)$ **17. (a)** $3x^2(1 + 4x^3 + 5x^{18})$ **19. (a)** $x^2(4x^2 + 3x + 6)$

Prep. Ex., 201 **1.** $x^2 - 25$ **2.** $36 - x^2$ **3.** $16x^2 - 49$ **4.** $\frac{1}{4}x^4 - 81$ **5.** $100x^6 - 9$ **6.** $1 - 81x^8$

5–7, *203* **1. (a)** No **3. (a)** Yes **5. (a)** $(5 + x)(5 - x)$ **7. (a)** $(14x + 1)(14x - 1)$
9. (a) $(\frac{1}{2} + x)(\frac{1}{2} - x)$ **11. (a)** $(1 + 10x)(1 - 10x)$
13. (a) $(m^3 + 5)(m^3 - 5)$ **15. (a)** $(x^4 - 6)(x^4 + 6)$
17. (a) $(\frac{1}{3}b - \frac{1}{10})(\frac{1}{3}b + \frac{1}{10})$ **19. (a)** 13, 1; 12 **21. (a)** $59 \cdot 61$ **23. (a)** $37 \cdot 43$
25. (a) 100, 100; 1^2; 9999 **27. (a)** 391 **29. (a)** 2464 **31. (a)** $3(7y + 3)(7y - 3)$

5–8, *206* **1. (a)** $(x + 2)^2$ **3. (a)** $(y - 5)^2$ **5. (a)** Not a perfect square
7. (a) $(8x - 3)^2$ **9. (a)** $(3 + 4x)^2$ **(b)** $(5 + 2x)^2$ **11. (a)** $(3x - 2)^2$
13. (a) $(x - 8)^2$ **15. (a)** $(6a + 11)^2$ **17. (a)** $(10y - 1)^2$
19. (a) $5x^3(x + 1)^2$ **(b)** $7y^{10}(y + 6)^2$ **(c)** $3(2y + 5)^2$ **(d)** $x^2(x - 9)^2$
21. (b) (i) $2a - 5$ (ii) $2 - b$ (iii) $(c - 5) + 9$; $(c + 4)^2$ (iv) $5(2d + 1) - 3$; $(10d + 2)^2$ (v) $3(4a + 1) + 5(a - 2)$; $12a + 3 - 5a + 10$; $17a - 7$
23. (a) $(x + 5)(x + 3)$ **(b)** $(8 - x)(3 - 5x)$ **25. (a)** $(x + 4)^2$ **(b)** $(3a + 4)^2$
27. (a) $x^3 + 3x^2 + 3x + 1$ **(b)** $x^3 - 3x^2 + 3x - 1$ **(c)** $x^3 + 6x^2 + 12x + 8$

EXTRA!, 208 **1. (a)** 1 **(b)** 6 **(c)** 12 **(d)** 8 **3. (a)** 8 **(b)** 24 **(c)** 24 **(d)** 8 **5.** Yes. Parts (a), (b), (c) and (d) are contained in the equation.

Prep. Ex., 209 **1.** $x^2 + 16x + 63$ **2.** $x^2 - 16x + 63$ **3.** $x^2 + 2x - 63$ **4.** $x^2 - 2x - 63$ **5.** $y^2 - 17y + 60$ **6.** $x^2 - 6x - 40$ **7.** $x^2 + 14x + 48$ **8.** $x^2 + 6x - 55$

5–9, *211* **1. (a)** $(x + 8)(x + 3)$ **3. (a)** $(a + 2)(a - 28)$ **5. (a)** $-, +$
7. (a) $(y + 2)(y + 1)$ **9. (a)** $(a - 7)(a + 2)$ **11. (a)** $(x - 8)(x - 2)$
13. (a) Not factorable **15. (a)** $(b + 7)(b + 2)$ **17. (a)** Not factorable
19. (a) $(x - 9)(x + 8)$ **21. (a)** $(x + 9)(x + 6)$ **23. (a)** $(b - 7)(b - 6)$
25. (a) $(a - 7)(a + 3)$ **27. (a)** Not factorable **29.** $5(x - 8)(x + 2)$
31. $y^2(y - 16)(y - 10)$ **33.** $7(a + 7)(a - 3)$ **35.** $9x(4x - 25)$
37. $7(x + 1)(x - 1)$

Preparation Exercises, 213 **1.** $8x^2 + 14x + 3$ **2.** $8x^2 - 14x + 3$ **3.** $8x^2 + 10x - 3$ **4.** $8x^2 - 10x - 3$ **5.** $18x^2 - 57x + 35$ **6.** $18x^2 + 57x + 35$ **7.** $18x^2 - 27x - 35$ **8.** $18x^2 + 27x - 35$

5–10, *215* **1. (a)** $+, +$ **3. (a)** $+, -$ or $-, +$ **5. (a)** $(3y - 1)(y + 5)$
7. (a) $(2x + 1)(x + 2)$ **9. (a)** $(3x - 7)^2$ **11. (a)** $(2y + 3)(y - 2)$
13. (a) $(7x + 2)(x - 6)$ **15. (a)** $(9x + 13)(9x - 13)$ **17. (a)** $(2y - 1)(6y + 7)$
19. (a) $(2y + 3)(4y - 5)$ **21. (a)** $(11y + 7)(3y - 2)$ **23. (a)** $6(x + 6)(x - 6)$
25. (a) $-2(4x^2 + 28x - 49)$ **27. (a)** $2(3x - 2)(5x - 3)$
29. $(x + 1)(x + 2)(x - 2)$ **31.** $2(x + 3)(2x + 1)$ **33.** $(15 - 4x)(22 - 5x)$
35. $(3y + 5)(y - 3)$

5–11, *211* **1. (a)** $4x^6$ **3. (a)** $5y^5$ **5. (a)** $x^5/2$ **7. (a)** 1 **9. (a)** $\frac{7}{9}$ **11. (a)** $-5a^8$
13. (a) $-\frac{1}{3}x^4$ **15. (a)** $-\frac{18}{7}y$ **17. (a)** $x + 2$ **19. (a)** $-9y^3 + 4y^2 + 2y$
21. (a) $a^4 - \frac{4}{3}a^2 + 2$ **23. (a)** $-4x^2 + 2x + 1$ **25. (a)** $\frac{4}{7}x^3 + \frac{6}{7}x - \frac{2}{7}$
27. (a) $32x^2 - \frac{1}{2}x - \frac{2}{3}$ **29. (a)** $60x - 7$ **31.** (Def-Div), (D), (LE-2)
33. $-\frac{1}{4}x^2$ **35.** $25x^3$ **37.** $-2y^3 - 7$

5–12, *226* **1. (a)** $8x^3 + 12x^2$, $-12x^2 - 18x$, $18x + 27$; $8x^3 + 27 = (2x + 3)(4x^2 - 6x + 9)$
3. (a) $2x^4 + 2x^3 + 8x^2$, $-2x^3 - 8x^2 - 8x$, $-6x^2 - 6x - 24$; $2x^4 - 3x + 7 = (x^2 + x + 4)(2x^2 - 2x - 6) + (11x + 31)$
5. (a) $8x^3 - 27 = (4x^2 + 6x + 9)(2x - 3)$
7. (a) $27a^3 - 8 = (3a - 2)(9a^2 + 6a + 4)$
9. (a) $b^4 - 16 = (b + 2)(b^3 - 2b^2 + 4b - 8)$
11. (a) $x^2 + 36 = (x + 6)(x - 6) + 72$
13. $x^4 - 2x^3 - 1 - 4x^5 = (2x^3 - 2x^2 + 1)(-2x^2 - \frac{3}{2}x - \frac{5}{2}) + (-3x^2 + \frac{3}{2}x + \frac{3}{2})$ **15.** $y^4 + y^2 + 1 = (y^2 + y + 1)(y^2 - y + 1)$
17. $x^5 + 243 = (x + 3)(x^4 - 3x^3 + 9x^2 - 27x + 81)$

Preparation Exercises, 228 **1.** $(x + 7)^2$ **2.** $(x - 11)^2$ **3.** $(x - 6)(x + 5)$ **4.** $(2x - 3)(x + 2)$ **5.** $x(x + 4)(x - 4)$ **6.** $2x(x + 2)(x + 1)$ **7.** $x = \frac{9}{4}$ **8.** $x = -1$ **9.** $x = 0$ **10.** $x = \frac{1}{8}$

5–13, *231* **1. (a)** $\{-2, -5\}$ **3. (a)** $\{\frac{2}{3}, -\frac{1}{2}\}$ **5. (a)** $\{\frac{4}{5}\}$ **7. (a)** $\{2, -1\}$ **9. (a)** $\{0, \frac{2}{3}\}$ **11. (a)** $\{3, 13\}$ **13. (a)** $\{\frac{2}{5}\}$ **15. (a)** $\{-13, 13\}$ **17. (a)** $\{-\frac{9}{2}, \frac{1}{2}\}$ **19.** $\{-2, 18\}$ **21.** $\{4, 9\}$ **23.** $\{7\}$ **25.** $\{0, -4\}$ **27.** $\{\frac{1}{2}, \frac{3}{5}\}$ **29.** $\{-3, 8\}$ **31.** $\{-\frac{1}{3}, 2\}$ **33.** $\{-\frac{1}{2}, 1\}$ **35.** $\{\frac{1}{3}, 1\}$ **37.** $\{4, -5\}$ **39.** $\{\frac{3}{5}, -\frac{3}{11}\}$ **41.** $\{\frac{1}{7}, 1\}$ **43.** $\{-\frac{1}{8}\}$

Chapter Review, 235 **1.** T **3.** T **5.** F **7.** F **9.** (2) The right-hand side of the equation must be zero. (4) If the coefficients of the fifth-power terms were the same, they would cancel out, leaving a polynomial of degree less than 5. (5) $2^0 = 1$ (7) Binomial of degree 2 (8) $\{0, 9\}$ **11.** $3x^2 + 8x + 20$ **13.** $-\frac{9}{4}x^9$ **15.** $3x^2 - 4x + 5$ **17.** $22t + 11$ **19.** $-72x^8$ **21.** $\frac{1}{3}x^{15}$ **23.** $6x^3 + x^2 - 18x - 55$ **25.** $9y^2 - 15y + 25$ **27.** $9x^2 - 49$ **29.** $14x^2 + 23x - 30$ **31.** $20x^2 - x - 1$ **33.** $49x^2 - 70x + 25$ **35.** $66x^2 - 65x - 14$ **37.** Not factorable **39.** $17(x + 1)^2$ **41.** $10(3y - 5)(y + 2)$ **43.** $x(3x^6 + 1)$ **45.** $(x - 15)(x - 15)$ **47.** $(x - 6)(x + 7)$ **49.** $(2y - 1)(9y + 2)$ **51.** $4(3x^2 - 5)^2$ **53.** $4x^2(x + 11)^2$ **55.** $(x + 11)(x + 16)$ **57.** $10, -9$ **59.** $11, -11$ **59.** $-\frac{4}{3}, 9$ **61.** $-\frac{12}{5}, -2$ **63.** $\frac{4}{3}, -2$ **65.** $0, \frac{5}{7}$

6–1, *240* **1. (a)** Not equal, equal, not equal, not equal **3. (a)** $\{(2, 1), (2, 3), (2, 5), (2, 7), (4, 1), (4, 3), (4, 5), (4, 7), (6, 1), (6, 3), (6, 5), (6, 7)\}$ **5. (a)** $A \times B = \{(3, 2), (3, 4), (6, 2), (6, 4), (9, 2), (9, 4)\}$; $B \times A = \{(2, 3), (2, 6), (2, 9), (4, 3), (4, 6), (4, 9)\}$; $A \times B \neq B \times A$ **7. (a)** $\{(3, 3), (3, 9), (3, 10), (9, 3), (9, 9), (9, 10), (10, 3), (10, 9), (10, 10)\}$ **9. (a)** The set of all ordered pairs of even numbers **11. (a)** $\{(1, 1), (4, 2), (25, 5)\}$ **13. (a)** $\{(2, 2), (2, 4), (2, 6), (4, 4), (4, 6), (6, 6)\}$ **15. (a)** $\{(-1, 1), (0, 0), (1, 1)\}$ **19.** $U \times U = \{(-2, -2), (-2, -1), (-2, 0), (-2, 1), (-2, 2), (-1, -2), (-1, -1), (-1, 0), (-1, 1), (-1, 2), (0, -2), (0, -1), (0, 0), (0, 1), (0, 2), (1, -2), (1, -1), (1, 0), (1, 1), (1, 2), (2, -2), (2, -1), (2, 0), (2, 1), (2, 2)\}$ **(a)** $\{(-1, -2), (0, 0), (1, 2)\}$ **(b)** $\{(-2, -1), (0, 0), (2, 1)\}$ The x-elements in (a) are the y-elements in (b) and the y-elements in (a) are the x-elements in (*b*). **21. (a)** $\{(x, y) \mid y = 2x + 1\}$

6–2, *244* **1. (a)** D: $\{-4, 1, 0\}$; R: $\{-3, 0\}$ **3. (a)** D: $\{2, 3\}$; R: $\{2, 3\}$ **5. (a)** D: $\{-4, -2, 0, 2, 4\}$; R: $\{4, 2, 0\}$ **7. (a)** D: $\{-3, -2, -1, 0, 1\}$; R: $\{3\}$ **9. (a)** D: $\{3x \mid x \text{ in } Z\}$; R: $\{2x - 1 \mid x \text{ in } Z\}$ **11. (a)** $D_n = \{n \mid n \geqq 0, n \text{ an integer}\}$; $R_c = \{15n \mid n \geqq 0, \text{ an integer}\}$ **13. (a)** $D = 45t$; $D_t = \{t \mid t > 0\}$; $R_D = \{D \mid D > 0\}$ **15. (a)** Function; $y = x^2$ **17. (a)** Function; $y = \dfrac{1}{x}$ **19. (a)** $x = \sqrt{A}$

6–3, *247* **1. (a)** $f(-1) = 0$, $f(1) = 2$ **3. (a)** $5, -1$ **5. (a)** $-4, -2, 10$ **7. (a)** $4, -2$ **9. (a)** No **11. (a)** -4 **13. (a)** 7 **15. (a)** -1 **17. (a)** $a^2 - 2$ **19. (a)** 7 **21. (a)** -6 **23. (a)** 1 **25. (a)** $\{-4, 0, 4, 8\}$ **27. (a)** $\{5, 4, 3, 1, 0\}$ **29. (a)** Yes **31. (a)** Yes **35. (a)** $f(-4) = \frac{2}{5}$, $f(0) = -2$, $f(5) = \frac{7}{4}$, $f(10) = \frac{4}{3}$, $f(-2) = 0$ **(b)** $\{x \mid x \text{ a real number and } x \neq 1\}$ **(c)** $\frac{5}{2}$ **(d)** No **(e)** All real numbers except 1 **37. (a)** $2, -2$ **(b)** All real numbers **39.** $y = \dfrac{1}{x}$, $y^2 + x = 1$

6–4, *253* **1. (a)** $A(1, 1)$, $B(1, 3)$, $C(-3, 3)$, $D(8, 0)$, $E(0, -5)$, $F(0, 6)$, $H(3, -3)$, $K(-7, 6)$, $L(3, 3)$, $M(-4, 5)$, $N(-1, -1)$, $P(6, -3)$, $Q(-7, 0)$, $R(-2, -2)$, $S(-2, -5)$, $T(2, -2)$ **3. (a)** 4 sq units **5. (a)** Rectangle **7. (a)** (—, —) **9. (a)** I or IV **11. (a)** I or III **13. (a)** I, II, III or IV **15.** Yes **17. (a)** 50 units

6–5, *258* **1. (a)** $i = 12f$; $D_f = \{f \mid f \geqq 0\}$, $R_i = \{i \mid i \geqq 0\}$
3. (a) $d = 7w$; $D_w = \{w \mid w \geqq 0\}$, $R_d = \{d \mid d \geqq 0\}$
5. (a) $k = 1.6m$; $D_m = \{m \mid m \geqq 0\}$, $R_k = \{k \mid k \geqq 0\}$
13. (a) D: $\{x \mid -5 \leqq x \leqq 5, x \text{ a real number}\}$;
R: $\{y \mid -5 \leqq y \leqq 5, y \text{ a real number}\}$
15. (a) Not a function **17. (a)** D: $\{x \mid -2 \leqq x \leqq 2, x \text{ a real number}\}$; R: $\{y \mid 0 \leqq y \leqq 2, y \text{ a real number}\}$ **19. (a)** Not a function

EXTRA!, 263 **1. (a)** 2 **(b)** -10 **(c)** 3 **(d)** -1

6–6, *265* **9. (a)** $p = 25I$, 25 **11.** Yes **13. (a)** 0, 1.5, 3, 4.5, 6, 7.5; $w = 3.75$; $n = 3.3$; $w = \frac{3}{2}n$; The wages are $1\frac{1}{2}$ times the number of hours worked. **15. (a)** $W = \frac{5}{2}E$ or 50 **17. (a)** $x = ky$ **(b)** $\frac{5}{3}$ **(c)** 225 **(d)** 33 **19. (a)** Yes **(b)** No **(c)** No **(d)** No **21. (a)** $y = 3x^2$ **(b)** $y = x^2 + 3$

Chapter Review, 270 **1.** $\{(4, -4), (4, -6), (6, -4), (6, -6), (9, -4), (9, -6), (12, -4), (12, -6)\}$ **3.** D: $\{-10, -5, 0, 5, 10\}$; R: $\{0, 10\}$ **5.** D: all integers; R: all integers greater than or equal to -3 **7. (a)** 7 **(b)** $\frac{7}{2}$ **(c)** 0 **(d)** 4 **9. (a)** 1, 1, -3, -3, -4, $-3\frac{1}{2}$, $-3\frac{1}{2}$, -3.99, -3.99 **(b)** All real numbers greater than or equal to -4 **11. (a)** $y = x^2 - 3$ **(b)** 1, 1, -2, -2, -3, 22, 22, $-2\frac{3}{4}$, $-2\frac{3}{4}$, $-2\frac{99}{100}$, $-2\frac{99}{100}$ **(c)** $\{y \mid y \geqq -3\}$ **13. (b)** Range: $\{-23, -14, -7, -2, 1, 2\}$ **15.** Directly proportional **17. (b)** 60¢ **(c)** 5 lb **(d)** No **(e)** $5 < w \leqq 6$ **19. (b)** $D_w = \{w \mid 0 < w < 16\}$; $R_p = \{p \mid 6 \leqq p \leqq 34$, p a multiple of $z\}$ **21.** (1, 2), (2, 3), (3, 5), (4, 7), (5, 11), (6, 13), (7, 17), (8, 19), (9, 23), (10, 29)

7–1, *277* **1. (a)** $x + \frac{2}{3}y - 4$, $2b + 3p - 5$, and $-\frac{5}{9}$ are linear forms **3. (a)** 6, 0 **5. (a)** -1, 0 **7. (a)** $(-3, -9), (-1, -5), (0, -3), (\frac{11}{2}, 8)$ **9. (a)** $(0, 0), (2, 2), (-3, -3), (-4, -4), (-\frac{1}{2}, -\frac{1}{2})$ **11. (a)** (3, 10), (3, 2), (3, 1) **13. (a)** E.g., $(-2, \frac{10}{3}), (-1, \frac{8}{3}), (0, 2), (1, \frac{4}{3}), (2, \frac{2}{3})$ **15. (a)** Any value of y will yield $x = 8$ **(b)** All $x = 8$ **(c)** The equation is false except for $x = 8$ **(d)** x is always 8 for any value of y. **17.** Yes; Yes **(a)** $7x + 6y - 4$ **(b)** $-4c + 4d - 5$ **19.** $19x + 4y + 22$ **21.** $13x - 29y - 38$ **23.** $3N - \frac{1}{3}$, $M = 18$ **25.** $J = 2P - 3$

7–2, *283* **1. (a)** $(0, 1), (1, 1), (8, 1), (10^6, 1), (-10^6, 1)$ **3. (a)** $\{(x, 0) \mid x \text{ a real number}\}$ **15. (a)** No, No, No; In Quadrant II, $y > 0$ and $x < 0$. Therefore $x - y < 0$ and cannot be 4 **(b)** Yes **(c)** Yes

7–3, *286* **1. (a)** $\frac{x}{3} - 5y = 1$, $3x = 3y$; They are linear equations. **13. (a)** All have the same direction; point where they cross the x and y-axes **15. (a)** 1, 6, 3, 2, 2; Above the graph **(b)** $-2, -2, -1, -7, -16$; Below the graph **(c)** 0, 0, 0, 0, 0; On the graph

7–4, *290* **1. (a)** $y = 11x + 22$ **3. (a)** $y = x + 5$ **5. (a)** $y = \frac{1}{2}(x + 1)$ **7. (a)** $y = -\frac{1}{3}(x + 2)$ **9. (a)** $y = \frac{1}{4}(x - 3)$ **11. (a)** $x = 18y - 9$ **13. (a)** $x = 3y$ **15. (a)** $x = \frac{1}{3}(4y - 4)$ **17. (a)** $x = \frac{1}{6}(8y - 51)$

7–5, *297* **1. (a)** (ii) $(0, 0), (-1, -4), (5, 5), (1, -\frac{9}{2})$; Ordinates > -5 (iii) $(4, -6)$, $(3, -7), (10, -\frac{11}{2})$; Ordinates < -5 **15. (a)** Half-plane to the right of the y-axis **(b)** Half-plane to the right of and including $x = -5$ **(c)** Half-plane below and including $y = 1$ **(d)** Half-plane above $y = -3$ **17. (a)** All of them, i.e., $\{(x, y) \mid x > 0\}$; $\{(x, y) \mid x > 0\}$ **(b)** $\{(x, y) \mid x > 0\}$ **(c)** $\{(x, y) \mid x \geqq -5\}$ **(d)** $A \subset B$ **19. (a)** $\emptyset$ **(b)** All points to the right of $x = 3$ or to the left of $x = -3$ **(c)** Same as (b) **21. (a)** $\emptyset$ **(b)** $\{(x, y) \mid y \geqq \frac{3}{2} \text{ or } y \leqq -\frac{3}{2}\}$ **(c)** Same as (b) **23. (c)** Distributive axiom

7–6, *301* **1. (a)** $(2, 10), (-4, -8)$; $(1, -1), (2, 4), (-3, -8), (-1, -7), (0, 2)$; $(20, 64), (-10, -26)$ **3. (a)** $(3, -1), (0, \frac{1}{2})$ satisfy $x = -2y + 1$; $(1, \frac{1}{2}), (4, -1), (12, -5)$ satisfy $x > -2y + 1$; $(-3, \frac{1}{2}), (2, -1), (14, -7)$ satisfy $x < -2 + 1$; On, left, left, left, right, right, on, right **5. (a)** Right: $\{(x, y) \mid x > \frac{3}{4}y - 2\}$; Left: $\{(x, y) \mid x < \frac{3}{4}y - 2\}$ **7. (a)** All points above $y = 4x - 3$ **15.** $\{(x, y) \mid y < 0\}$ **17.** $\{(x, y) \mid y = 0\}$ **19.** $\{(x, y) \mid y \geqq 0\}$ **21.** $\{(x, y) \mid x < 0, y > 0 \text{ or } x > 0, y < 0\}$ **23.** $\{(x, y) \mid x < 0 \text{ and } y = 0\}$ **25.** $\{(x, y) \mid x > y\}$

Preparation Exercises, 304 **1. (a)** $y = \frac{1}{4}(3x + 12)$ **(b)** $y = \frac{1}{3}(2x + 6)$ **2. (a)** $x = \frac{1}{3}(4y - 12)$ **(b)** $x = -\frac{1}{2}(6 - 3y)$

7–7, *305* **1. (a)** 0 **3. (a)** $-\frac{7}{2}$ **5. (a)** -5 **7. (a)** y-intercept $= 1$; $y = ax + 1$ **9. (a)** x-intercept $= 2$; $x = by + 2$ **11. (a)** All are parallel; $x + y = b$

13. All the graphs pass through the origin. **15.** II and IV **17. (a)** $y = 2x$, $y = 2x - 2, y = 2x + 4$ **(b)** The coefficient of x is 2 **(c)** The graphs are parallel; they have the same slope. **(d)** $y = 2x + b$

7–8, *310* **1. (a)** Up 6 **3. (a)** 6, 4, $s = \frac{3}{2}$ **5. (a)** $-9, 3, s = -3$ **7. (a)** 4 **(b)** -4 **9.** It is the same **11. (a)** 3 **13. (a)** No slope **15. (a)** $y = -2x + 5$; -2; 5 **17. (a)** $y = \frac{1}{3}x + 3$; $\frac{1}{3}$; 3 **19. (a)** Parallel **21. (a)** 0, 4, 0 **(b)** 0, 7, 0 **(c)** All lines in the family are parallel to the x-axis **23. (a)** $y = \frac{3}{5}x + k$ **(b)** $y = mx + 2$ **(c)** $y = \frac{3}{5}x + 2$ **25.** b, e, h **27.** f, j, k

7–9, *314* **1. (a)** (ii) $k = -\frac{8}{3}$ (iii) Yes **3. (a)** $y = 2x + 5$ **5. (a)** $y = 3$ **7. (a)** $x = 2$ **9. (a)** $y = x$ **11. (a)** $y = \frac{11}{5}x - \frac{3}{5}$ **13. (a)** 1, 0 **15. (a)** $y = -\frac{5}{2}$ **17. (a)** $y = x - 1$ **19. (a)** $y = -\frac{9}{7}x + \frac{17}{14}$ **21.** $y > x$ **23.** $y \leqq x + \frac{1}{2}$ **25.** $y < -\frac{1}{9}x - \frac{5}{3}$

7–10, *319* **7. (a)** $\frac{x}{4} - \frac{y}{3} = 1$ **(b)** $-\frac{x}{5} + \frac{y}{2} = 1$

EXTRA!*, *322 **7.** 1 **8.** 0 **9.** 7 **10.** 3 **11.** 5 **12.** 0 **13.** $x = 3$ **14.** $x = 8$ **15.** $x = 11$ **16.** no solution **17.** 1, 3, 7, 9 **18.** 1, 5, 7, 11

Chapter Review*, *323 **1. (a)** F **(b)** F **(c)** F **(d)** F **(e)** F **(f)** T **3. (a)** $y = \frac{1}{4}(3x + 20)$, $x = \frac{1}{3}(4y - 20)$ **(b)** $y = \frac{1}{3}(-2x - 12)$, $x = -\frac{1}{2}(3y + 12)$ **5.** $-\frac{1}{4}$ **7.** $-\frac{1}{2}$ **9.** y-intercept $= -5$ **11.** $y = -3x + 2$ **13.** $y = 3$ **15.** $y = -\frac{8}{5}x + 6$ **25.** I, $A \cap B$; II, $C \cap B$; III, $C \cap D$; IV, $A \cap D$ **27.** All have slope $= 3$; $y = 3x + k$ **29.** All pass through (0, 0); $y = kx$

Cumulative Review II*, *326 **1.** $9x^4y^4$ **3.** $x^2 + 6x + 9$ **5.** $y^3 + 3x + 4y$ **7.** $x - 7$ **9.** $6(a - 3)(a + 3)$ **11.** $(3x + 6)(x + 1) = 3(x + 2)(x + 1)$ **13.** $(x - 2)^2$ **15.** $(x - 7)(x + 4)$ **17.** $\{-2, -3\}$ **19.** $\{-2, -4\}$ **21.** $\{2, -3\}$ **23. (a)** Both **(b)** y is a function of x **(c)** Neither **(d)** x is a function of y **29.** $\frac{3}{-12} = -\frac{1}{4}$ **31.** The graphs have the same slope **35. (a)** $\{(x, y) \mid x > 1\}$, $\{(x, y) \mid x < 1\}$ **(b)** $\{(x, y) \mid x > -1\}$, $\{(x, y) \mid x < -1\}$ **(c)** $\{(x, y) \mid x > -2\}$, $\{(x, y) \mid x < -2\}$ **(d)** $\{(x, y) \mid y > -x\}$, $\{(x, y) \mid y < -x\}$ **(e)** $\left\{(x, y) \mid y > \frac{6 - 2x}{3}\right\}$, $\left\{(x, y) \mid y < \frac{6 - 2x}{3}\right\}$ **(f)** $\{(x, y) \mid y > 5 - x\}$, $\{(x, y) \mid y < 5 - x\}$ **37.** (b), (c), (d) **39. (a)** $\frac{1}{8}$ **(b)** $\frac{2}{8}$ or $\frac{1}{4}$ **(c)** $\frac{5}{8}$ **(d)** 0 **41.** John's age is 4, Mary Jo's age is 12 **43.** $8x + 2 = 50$; $x = 6$ **45.** $x + 16 = 49$; $x = 33$ **47.** $y = 0, x = 5$, $y = -5, x = 0$; (0, 0), (5, 0), (5, -5), (0, -5) **51.** T **53.** F **55.** F, multiplication or addition **57.** T **59.** F, lower **61.** T

8–1, *333* **1. (a)** (ii) infinite (iii) $\{(17, \frac{8}{5})\}$ **3. (a)** $\{(-3, 6)\}$ **5. (a)** $\{(9, 1)\}$ **7. (a)** $\{(2, 0)\}$ **9. (a)** $\{(2, 8)\}$ **11. (a)** $\{(3, 0)\}$ **13. (a)** $\{(-\frac{3}{2}, 2)\}$ **15.** $\{(1, 17.4)\}$ **17.** $\{(.1, .008)\}$ **19.** $a = \frac{10}{3}, b = \frac{2}{3}$ **21.** $\{(-1, 6)\}$ **23.** $\{(-1, 6)\}$ **27.** $\{(-1, 6)\}$

Prep. Ex.*, *336 **1.** 4, -3 **2.** $\frac{1}{2}$; -1 **3.** (8, 3)

8–2, *337* **11.** $(-1, -2)$ **13. (a)** 3 PM, 240 mi **(b)** 6:30 PM **15.** Over \$600 **17.** The slopes are the same.

8–3, *340* **1. (a)** $y + 5$, 2, $x = 7$, $y = 2$, $s = \{(7, 2)\}$; All graphs have a common point of intersection. **3. (a)** $\{(4, 0)\}$ **5. (a)** $\{(2, 3)\}$ **7. (a)** $\{(-4, -3)\}$ **9. (a)** $\{(\frac{5}{2}, -2)\}$ **11. (a)** $\{(-3, -5)\}$ **13. (a)** $\{(-6, 5)\}$ **15. (a)** $\{3, \frac{1}{3})\}$

Prep. Ex.*, *341 **1.** -1 **2.** $-\frac{6}{5}$ **3.** $\frac{13}{7}$ **4.** $-\frac{6}{35}$

8–4, *343* **1. (a)** $x = \frac{2y - 5}{5}$; $\frac{2y - 5}{5}$, $\frac{2y - 5}{5}$; $y = 0$, $x = \frac{2y - 5}{5}$; $\{(-1, 0)\}$ **3. (a)** $\{(1, 2)\}$ **5. (a)** $\{(-2, 0)\}$ **7. (a)** $\{(-1, -3)\}$ **9.** $\{(4, 14)\}$ **11.** $\{(3, 2)\}$. **13.** $\{(\frac{100}{3}, \frac{64}{3})\}$ **15.** $\{(\frac{1}{2}, 2)\}$ **17.** The graph of $A \cup B$ is a pair of straight lines The graph of $A \cap B$ is the point $(-3, -\frac{3}{2})$.

8–5, *346* **1. (a)** $\{(12, 5)\}$ **3. (a)** $\{(6, 0)\}$ **5. (a)** $\{(0, -\frac{7}{3})\}$ **7. (a)** $\{(\frac{5}{3}, 0)\}$
9. (a) $\{(-2, 2)\}$ **11. (a)** $\{(\frac{2}{3}, \frac{3}{4})\}$ **(b)** $\{(\frac{4}{7}, \frac{18}{7})\}$ **(c)** $\{(3, 2)\}$ **(d)** $\{(0, 22)\}$

Prep. Ex., 347 **1.** $\{(1, 1)\}$ **2.** $\{(1, \frac{1}{2})\}$ **3.** $\{(2, -4)\}$

8–6, *351* **1. (a)** $\{(5, 3)\}$ **3. (a)** $\{(3, 0)\}$ **5. (a)** $\{(-5, 9)\}$ **7. (a)** $\{(3, -2)\}$
9. (a) $\{(4, 0)\}$ **11.** $\{(3, 5)\}$ **13.** $\{(5, 4)\}$ **15.** $\{(0, 0)\}$

8–7, *356* **1. (a)** 20, 16 **3. (a)** 57 **5. (a)** \$1900, \$3100 **7. (a)** 198 adult, 162 child tickets
9. 7 **11.** 31 nickels **13.** 60, 95 **15.** 38 **17.** 35 mph, 50 mph

8–8, *360* **1. (a)** (ii) III **9. (a)** Triangle **(b)** Trapezoid **(c)** Parallelogram **(d)** Square

8–9, *346* **1. (a)** -5 **3. (a)** 15 **5. (a)** $\{(-1, -1)\}$ **7. (a)** $\{(\frac{19}{53}, \frac{1}{53})\}$
9. (a) $\{(-\frac{5}{37}, -\frac{3}{37})\}$ **11. (a)** $\{(\frac{45}{13}, \frac{53}{13})\}$ **13. (a)** $\{(\frac{1}{4}, 1)\}$ **15. (a)** $\{(\frac{22}{73}, \frac{10}{73})\}$

8–10, *368* **1. (a)** $x = 1, y = 1, z = -1$ **3. (a)** $x = y = z = 2$ **5. (a)** $r = -1$, $s = 4, t = 5$ **7.** 435 **9.** Wheat, 80¢; corn, 40¢; rye, 60¢
11. $10\left(1 - \frac{1}{2^n}\right)$

8–11, *372* **1.** Vertex (0, 0): Acres of $A = 0$, Acres of $B = 0$, Seed cost = \$0, Labor = 0 man-hours; Vertex (0, 80): Acres of $A = 0$, Acres of $B = 80$, Seed cost = \$240, Labor = 2400 man-hours; Vertex $(27\frac{3}{11}, 72\frac{8}{11})$: Acres of $A = 27\frac{3}{11}$, Acres of $B = 72\frac{8}{11}$, Seed cost = \$354.55, Labor = 2400 man-hours; Vertex (75, 25): Acres of $A = 75$, Acres of $B = 25$, Seed cost = \$450, Labor = 1350 man-hours; Vertex (90, 0): Acres of $A = 90$, Acres of $B = 0$, Seed cost = \$450, Labor = 720 man-hours **2.** $11A + 22B$

Chapter Review, 373 **1. (a)** Slope $= -\frac{1}{2}$, y-intercept $= 2$; Slope $= 0$, y-intercept $= 4$
(b) Slope $= 3$, y-intercept $= 1$; Slope $= -\frac{1}{2}$, y-intercept $= 1$
3. (a) $\{(\frac{1}{2}, 2)\}$ **(b)** $\{(5, -4)\}$ **5. (a)** $\{(\frac{11}{27}, \frac{1}{27})\}$ **(b)** $\{(-\frac{5}{2}, \frac{7}{2})\}$ **7.** 1, -2
9. 14 lb of 80¢ coffee, 10 lb of \$1.04 coffee **11.** 20 mph **13. (a)** 13 quarters, 18 nickels **(b)** Yes; 16 quarters, 15 pennies **15. (a)** Triangle; (1, 0), (−2, 3), (4, 3), e.g., (1, 1), (2, 3), (−1, 2) **(b)** Trapezoid; (−1, 2), (−3, −2), (3, −2), (1, 2), e.g., (−2, −1), (0, 1), (2, −2)

9–1, *378* **1. (a)** 0 **3. (a)** $14a^2 - 6ab + 4b^2$ **5. (a)** $5a^2 + 4ab - 6b^2$
7. (a) $3x^2 - 5xy - y^2$ **9. (a)** $12x + 4xy$ **11.** $-5x + 4y$ **13.** $4x^2 - 5$
15. $3a + 4b + 4$

Prep. Ex., 379 **1.** x^{19} **2.** m^{10} **3.** a^2b^2 **4.** $-x^5y^3$ **5.** (D); (R-M); (LE-1)
6. $2x^3 + 3x^2 + x + 6$

9–2, *380* **1. (a)** $\frac{1}{27}x^3y^3$ **3. (a)** $10{,}000x^4y^8$ **5. (a)** $7x^3y - 7x^2y^2 - 7xy^2$
7. (a) $x^2 + 2xy - 15y^2$ **9. (a)** $x^2 - y^2$ **11. (a)** $x^2y^2 - 2xy - 63$
13. (a) $6x^3 + 7x^2y + 6xy^2 - 4y^3$ **15. (a)** $343x^3 - y^3$
17. (a) $27a^3 - 18a^2b - 12ab^2 + 8b^3$ **19. (a)** $6x^3 - 7x^2y - 35xy^2 + 6y^3$
21. $x^4 - y^4$ **23.** $x^4 - 16x^2y^2 + 32xy^3 - 16y^4$ **25.** $-24xy$
27. $27a^3 - 75ab^2 - 45a^2b + 125b^3$ **29.** $125x^3 - 75x^2y + 15xy^2 - y^3$
31. $64a^3 + 48a^2b + 12ab^2 + b^3$

9–3, *384* **1. (a)** $x^{100}y^{100}$ **3. (a)** $729x^6y^6$ **5. (a)** $-8x^3y^3$ **7. (a)** x^7 **9. (a)** $\frac{16}{625}x^{28}y^{36}$
11. (a) 1 **13. (a)** $4a^2b^6$ **15. (a)** x^6y^6 **17. (a)** $-a^{40}b^{45}$ **19. (a)** $10^6x^6y^6$
21. (a) x^5y^7 **23. (a)** $a^{56}b^{56}$ **25. (a)** $-\frac{1}{3}a^{29}b^{30}$

Prep. Ex., 384 **1.** $x^2 + 8x + 16$ **2.** $9x^2 - 6x + 1$ **3.** $81 - 36x + 4x^2$ **4.** $4x^2 - 9$

9–4, *386* **1. (a)** $64x^2 - 25y^2$ **3. (a)** $12x^4 + 19x^2y + 5y^2$
5. (a) $\frac{1}{27}x^3 - \frac{4}{3}x^2y + 16xy^2 - 64y^3$ **7. (a)** $4x^2 + 20xy + 25y^2$
9. (a) $9x^4 - 6x^2y + y^2$ **11. (a)** $x^3 + 12x^2y + 48xy^2 + 64y^3$
13. (a) $\frac{1}{9}a^2 - \frac{2}{3}ab + b^2$ **15. (a)** $8x^3 - 12x^2y + 6xy^2 - y^3$
17. (a) $27x^6 - 27x^4y + 9x^2y^2 - y^3$
19. (a) $a^8 + 4a^6b + 6a^4b^2 + 4a^2b^3 + b^4$
21. (a) $81x^4 - 108x^3y + 54x^2y^2 - 12xy^3 + y^4$
23. (a) $4x^2 + 12xy + 9y^2 + 4x + 6y + 1$
25. (a) $9x^4 + 4y^2 + 25z^2 + 12x^2y + 30x^2z + 20yz$
27. $9x^2 + 4y^2 + 25z^2 + 16w^2 + 12xy + 30xz + 24xw + 20yz + 16yw + 40zw$

9–5, *391* **1. (a)** $7x^2(1 - 2y^2)$ **3. (a)** $3x(2y - 3y^2 - x)$ **5. (a)** $4y(7x^2 - 4xy + 16y^2)$
7. (a) $2x^3y(2x^2 - xy + 4y^2)$ **9. (a)** $6xy(4xy + y + 5x)$ **11.** $(2x - 3)(y + 4)$

13. $(4x - y)(x - y)$ **15.** $(3y - x)(2x + 1)$ **17.** $(x + y)(x + y - 2)$ **19.** $(x + 5y)(13x^2 - 7xy + 4y^2)$ **21.** $(2x - y)(2x + y + 1)$ **23.** $(y^2 + x)(y + 3)$ **25.** $(6a + 7)(5a + 9b)$ **27.** $(3y + x)(2y + 1)$

9–6, *394* **1. (a)** $(x + 2y)(x - 2y)$ **3. (a)** $(x + 3y)(x - y)$ **5. (a)** $(a - 7b)(a + 5b)$ **7. (a)** $(a + 7b)(a - 2b)$ **9. (a)** $(6x - y)^2$ **11. (a)** $(3a + 5b)^2$ **13. (a)** $(3x - 2y)(x + y)$ **15. (a)** $(5x^3 + y^2)(5x^3 - y^2)$ **17. (a)** $(7a + 8b)(8a - 7b)$ **19. (a)** $(3x^3 + 2y^3)(2x^3 + 3y^3)$ **21.** $7(a + 3b)(a - 5b)$ **23.** $xy(xy + 1)(xy - 1)$ **25.** $(8x^4 + y^4)(8x^4 - y^4)$ **27.** $x^3(y^2 - 8)(y^2 + 2)$ **29.** $(.1x^4 + y^5)(.1x^4 - y^5)$ **31.** $(x^2y^2 + 9)(xy - 3)(xy + 3)$ **33.** Not factorable **35.** $(x - y - 3)(x - y - 3)$ **37.** $(3x + 3y - 1)^2$ **39.** Not factorable **41.** $(3y + 2x)(x - 2y)$ **43. (a)** $-1(3a - 16b)^2$

EXTRA!*, *397 **1.** $(x + y)(x^4 - x^3y + x^2y^2 - xy^3 + y^4)$ **3.** $(x + y)(x^6 - x^5y + x^4y^2 - x^3y^3 + x^2y^4 - xy^5 + y^6)$ **5.** $[(x + y)(x - y)]^2$ **7.** $(x^2 + y^2)(x^4 - x^2y^2 + y^4)$ **9.** $(x - y)(x + y)(x^2 + y^2)(x^4 + y^4)$

9–7, *400* **1.** 2 **3.** 12 **5.** 11 **7.** 1 **9.** 19 **11.** 13; $5 \cdot 13 \cdot 17 \cdot 41$ **13.** 2^2; $2^5 \cdot 3^4$ **15.** 1; $2 \cdot 3 \cdot 11 \cdot 13 \cdot 17 \cdot 19$ **17.** $11 \cdot 13$; $3 \cdot 11^2 \cdot 13^3 \cdot 17$ **19.** 17^3; $17^5 \cdot 29 \cdot 31$ **21.** g.c.d.: Take as a product all the common factors. l.c.m.: Take as a product to their highest powers all the different factors that are found in the numbers. **23.** $2^5 \times 3^5 \times 11$ or 85,536 **25.** The product is equal to the product of the two integers.

Chapter Review*, *401 **3.** $2x^3 - 13x^2y + 7xy^2 + 12y^3$ **5.** $49x^4 - 14x^2y + y^2$ **7.** $\frac{1}{2}x^6y^{10}$ **9.** $\frac{1}{9}x^6 + \frac{4}{3}x^3y^3 + 4y^6$ **11.** $-5a^2b^2 + 7b^3$ **13.** $-\frac{1}{32}a^5b^5$ **15.** $256a^8b^{12}$ **17.** $27x^3 - 135x^2y + 225xy^2 - 125y^3$ **19.** $a^3b^2 - 3a^2b + a^3b$ **21.** $x^2y^2(1 + xy - x^2y^2)$ **23.** $(x - 1)(2x + 1)$ **25.** $(8x + 3y)(8x - 3y)$ **27.** $(x - 16y)(x - y)$ **29.** $(7x - 3y)(5x + y)$ **31.** $(x + 18y)^2$ **33.** $(5x + y)(x - 3y)$ **35.** $(11x + 3y)(x - 2y)$ **37.** $4x^3(5 - 3y^2)(5 - 3y^2)$ **39.** $5(x + 4y)(x - 4y)$

10–1, *408* **1. (a)** $\frac{15}{16}$ **3. (a)** $\frac{x}{4}$ **5. (a)** $-\frac{1}{3}$ **7. (a)** $\frac{x + y}{3}$ **9. (a)** $\frac{5p}{16}$ **11. (a)** $\frac{4(x - y)}{3}$ **13. (a)** $\frac{5x^3(x + y)}{7y(x - y)}$ **15. (a)** $\frac{2x^2 + 3}{4x + 3}$ **17. (a)** $\frac{1 - 2x}{3x^2 - 8}$ **19. (a)** $\frac{1}{7}$ **21. (a)** $-3(x + 3)$ **23. (a)** $\frac{a - b}{a + b}$ **25. (a)** $\frac{3 - x}{6 + x}$ **27. (a)** 1, -3 **29.** $\frac{5 - x}{5 + x}$, $x \neq 2, -5$ **31.** $\frac{3 + 2x}{3 - 2x}$, $x \neq \frac{5}{3}, \frac{3}{2}$ **33. (a)** The denominator is 0 for $x = 5$ **(b)** $x = 0$

10–2, *412* **1. (a)** $\frac{3}{7}$ **3. (a)** $\frac{21}{ab}$ **5. (a)** $\frac{1}{a}$ **7. (a)** $\frac{x}{12}$ **9. (a)** $-\frac{1}{4x^2}$ **11. (a)** $\frac{1}{x^2 - y^2}$ **13. (a)** $\frac{1}{21}$ **15. (a)** -2 **17. (a)** $\frac{5 + x}{2}$ **19. (a)** 1 **21. (a)** 1 **23. (a)** $\frac{1}{2}$ **25. (a)** $\frac{a - 5b}{a - 6b}$ **28.** 1 **29.** $-x^2$ **31.** -1 **33.** $\frac{(x - 2y)(x^2 - 2xy + 4y^2)}{(x - y)^2}$ **35.** $5a^2$

10–3, *415* **1. (a)** $\frac{x + 3}{3x}$ **3. (a)** $\frac{1}{x^2 + xy + y^2}$ **5. (a)** $\frac{7x}{22}$ **7. (a)** $\frac{2}{3}$ **9. (a)** $\frac{6x}{y}$ **11. (a)** $-3x$ **13. (a)** $\frac{25a^2}{6b}$ **15. (a)** $\frac{10}{3}$ **17. (a)** $\frac{1}{a(b + 2)}$ **19. (a)** $\frac{3}{2}(a - 4)$ **21. (a)** $\frac{1}{a}$ **23.** $-\frac{b^2}{a^2}$ **25.** $\frac{(c + d)^3}{(c - d)^2}$ **27.** 1 **29.** $(x + 4y)^2$ **31.** $\frac{(a - 1)(1 - a)}{5(3 - a)}$

Prep. Ex., 417 **1.** $\frac{1}{70}(21+45)=\frac{33}{35}$ **2.** $12x-3+35x; \dfrac{47x-3}{14x^2}$ **3.** $x+2-7; \dfrac{x-5}{x-3}$

10–4, *419* **1. (a)** $\frac{13}{10}$ **3. (a)** $\dfrac{x}{3y}$ **5. (a)** $\dfrac{a(7-5a)}{4b}$ **7. (a)** $\dfrac{7x+3y}{21}$ **9. (a)** x **11. (a)** x
13. (a) $\dfrac{y+x}{y-x}$ **15. (a)** $\dfrac{x}{x-2}$ **17. (a)** $a+3$ **19. (a)** $\dfrac{1}{x-1}$ **21. (a)** $\dfrac{9}{x-y}$
23. $\dfrac{4x+30}{5(x+5)}$ **25.** $\dfrac{32}{3x(x+1)}$

Prep. Ex., 421 **1.** a^2b^2 **2.** $x(x+2)(x+3)$ **3.** $7x+3y$ **4.** $(2x-y)(x+y)$
5. $4(3-a)(3+a)$ **6.** $(x+4)(x+1)(x+7)$

10–5, *424* **1. (a)** $\dfrac{4b-3a}{2a}$ **3. (a)** $\dfrac{22x-1}{12}$ **5. (a)** $\dfrac{4x-5y^2}{x^2y}$ **7. (a)** $\dfrac{7a-6b}{a^2b^2}$
9. (a) $\dfrac{5a-38}{(a+2)(a-4)}$ **11. (a)** $\dfrac{3(3a-5)}{(a-4)(a+3)}$ **13. (a)** $\dfrac{-4x^2+13x+3}{(4x-1)(4x+1)}$
15. (a) $\dfrac{10a+15}{a(a+5)}$ **17. (a)** $\dfrac{48x+61}{5x+7}$ **19. (a)** $\dfrac{x^2+7}{x}$ **21. (a)** $\dfrac{6x-4y}{x-y}$
23. (a) $\dfrac{12a+8b}{2a+b}$ **25.** $\dfrac{-2(a-a)(a+1)}{(a+a)^2}$ **27.** $\dfrac{x+25}{(x+7)(x+4)(x+1)}$
29. $\dfrac{x+6}{(x-2)(x+2)}$ **31.** $\dfrac{21-5x+5y}{3(x-y)(x+y)}$ **33.** $\dfrac{5(5+a)}{4(3+a)(3-a)}$
35. $\dfrac{3x-y-5}{(x+y)(x-y)}$

10–6, *429* **1. (a)** $\frac{4}{3}$ **3. (a)** $\dfrac{b(x+y)}{a}$ **5. (a)** $x(x-3)$ **7. (a)** $\dfrac{a}{b}$ **9. (a)** $2(x+4)$
11. (a) $\dfrac{2b+3a}{5b-2a}$ **13. (a)** $\dfrac{x+y}{x-y}$ **15. (a)** $\dfrac{x}{x^2+x-1}$ **17. (a)** $\dfrac{y}{3y+x}$
19. (a) $\dfrac{x+2}{x-3}$ **21. (a)** $\dfrac{(4x+3)(3x+1)}{x(2x+1)}$

Prep. Ex., 430 **1.** x^{21} **2.** x^3 **3.** $x^{12}y^{12}$ **4.** x^{108} **5.** x^{11} **6.** a^9b^2 **7.** a^5b^5

10–7, *432* **1. (a)** $\dfrac{x^2}{16}$ **3. (a)** $-\dfrac{8x^3}{125y^3}$ **5. (a)** $-\dfrac{32x^5}{y^5}$ **7. (a)** $\dfrac{b^6c^3}{a^{15}}$ **9. (a)** $-\dfrac{27c^6}{125}$
11. (a) $\dfrac{a^{15}}{8b^3}$ **13. (a)** $-\dfrac{a^4b^{10}}{25c^6}$ **15. (a)** $\dfrac{m^{12}n^{16}}{x^8y^6}$ **17. (a)** $-\dfrac{32x^{10}}{1^{15}b^{15}}$
19. (a) $\dfrac{16xw}{5y}$ **21. (a)** $\dfrac{a^3d^4}{b^3c^4}$ **23. (a)** $\dfrac{7x^{18}}{125y^4}$

10–8, *435* **1. (a)** $\dfrac{1}{x^7}$ **3. (a)** $\frac{1}{9}$ **5. (a)** $\dfrac{1}{(x+y)^3}$ **7. (a)** $\frac{5}{6}$ **9. (a)** $\dfrac{1}{9x^4y^2}$ **11. (a)** x^8
13. (a) $\dfrac{1}{x^{10}}$ **15. (a)** 5 **17. (a)** $\frac{1}{3}$ **19. (a)** $\frac{9}{4}$ **21. (a)** $\dfrac{6}{xy^2}$ **23. (a)** x^{11}
25. (a) x^3 **27.** $\dfrac{1}{9a^2b^4}$ **29.** $\dfrac{3b}{a^2}$ **31.** $\dfrac{3}{a^2b^2}$ **33.** $\dfrac{1}{(x+y)^2}$ **35.** x^3y^9 **37.** $\dfrac{x^{12}}{y^8}$
39. $\dfrac{1}{(x-2y)^2}$ **41.** 1

Prep. Ex., 437 **1. (a)** 1.86×10^5 **(b)** 3.3×10 **(c)** 1.93×10^8 **2. (a)** $2, -2$ **(b)** -1

10–9, *438* **1. (a)** 6.4×10^{-7} **3. (a)** 6.78×10^{-6} **5. (a)** 6.42×10^{-2}
7. (a) 4.981×10^{-1} **9. (a)** 1.0×10^{-7} cm **11. (a)** 1.3×10^{11} cm
13. (a) 8.1×10^{-7} **15. (a)** 1.962×10^{-50}

Prep. Ex., 439 **1.** (Def-Div), (Can-M); $y = -4$ **2.** (Def-Div), (Can-M); $y = -4$

10–10, *442* **1.** **(a)** $-\frac{1}{3}$ **3.** **(a)** $-\frac{5}{2}$ **5.** **(a)** $\frac{3}{2}$ **7.** **(a)** $-\frac{2}{3}$ **9.** **(a)** 80 **11.** **(a)** $-\frac{3}{2}$
13. **(a)** $\frac{29}{42}$ **15.** **(a)** -1 **17.** No solution **19.** 2 **21.** $-4, 3$

10–11, *445* **1.** **(a)** $7\frac{1}{5}$ min **3.** **(a)** $2\frac{2}{9}$ hr **5.** **(a)** 25 **7.** **(a)** $5\frac{1}{3}$ hr **9.** **(a)** $\frac{15}{13}$ hr
11. 45 mph, 35 mph **13.** 5 mph

Chapter Review, 448 **1.** $\frac{3y^3}{10x^2}$ **3.** $-\frac{6}{5}$ **5.** -1 **7.** **(a)** $\frac{33x^3}{16y^4}$ **(b)** $\frac{6x^2}{y}$ **(c)** $\frac{(3x-y)(3x+y)(2x)}{(x+3y)^2}$
(d) $-\frac{1}{3}$ **9.** $\frac{-b-8}{b(b-2)}$ **11.** $\frac{8a^2-21a-1}{(a+1)(2a-5)}$ **13.** $\frac{3-x+y}{x-y}$ **15.** $\frac{1}{x+y}$
17. $\frac{x+48}{x-4}$ **19.** $\frac{5x-3}{x}$ **21.** $-2(2y+3)$ **23.** $-\frac{16x^{12}}{81y^8}$ **25.** $\frac{12}{x^8}$ **27.** $-\frac{343}{64}$
29. $\frac{1}{x^{10}}$ **31.** $\frac{y^6}{25x^{12}}$ **33.** 2.67×10^{-4} **35.** 3 **37.** 10 **39.** $\frac{5}{2}$ or -3 **41.** 120 hr

Cumulative Review 451 **1.** $x - 8$ **3.** $3a - 2b + 5a^2b^3$ **5.** $25x^2 - 30xy + 9y^2$ **7.** $2(2-x)(1-x)$
9. Impossible **11.** $(x+7)(x-4)$ **13.** $(x-2y)(x+8y)$
15. Slope: -3; x-intercept: $\frac{5}{3}$; y-intercept: 5 **17.** $y = 3x - 1$ **19.** $y = \frac{1}{2}x + 2$
21. **(a)** on **(b)** left **(c)** left **(d)** on **(e)** right
23. **(a)** $\frac{15x}{8}$ **(b)** $\frac{x-y}{x+y}$ **(c)** $\frac{x^3-2x^2-2x}{(x-2)(x-3)}$ **(d)** $\frac{y^3+y}{(y-1)(y+1)}$ **25.** y^4
27. x^{10} **29.** 6 **31.** $\frac{1}{a^6}$ **33.** c^{12} **35.** $\frac{3}{x}$ **37.** 2 **39.** $\frac{x^3}{8}$ **41.** $\frac{x^8y^{12}}{81}$
43. $x = -\frac{3}{2}, y = -2$ **45.** $x = 6, y = -9$ **49.** **(a)** Both

11–1, *456* **1.** **(a)** $.375\underline{0}$ **3.** **(a)** $1.\underline{076923}$ **5.** **(a)** $0.\underline{18}$ **7.** **(a)** $4\frac{61}{99}$ **9.** **(a)** $\frac{3}{4}$
11. **(a)** $\frac{4288}{333}$ **13.** **(a)** One unit **(b)** .1, .01, .001 **(c)** Yes, yes, yes

11–2, *459* **1.** **(a)** 9 **3.** **(a)** 10 **5.** **(a)** -21 **7.** **(a)** $(910)^3$ **9.** **(a)** $\frac{7}{17}$ **11.** **(a)** 1.2
13. **(a)** $\frac{8}{15}$ **15.** **(a)** $-\frac{4}{5}$ **17.** **(a)** 10^2 or 100
19. No, no, no. The sum must be a perfect square.

11–3, *461* **1.** **(a)** 8 **3.** **(a)** -2 **5.** **(a)** -5 **7.** **(a)** -5 **9.** **(a)** 7 **11.** **(a)** 50 **13.** **(a)** $\frac{3}{4}$
15. **(a)** $-.4$ **17.** **(a)** $-\frac{2}{3}$ **19.** **(a)** .1 **21.** **(a)** .6 **23.** **(a)** $\{-3\}$
25. **(a)** $\{-4, 4\}$ **27.** **(a)** $\{.2\}$

11–4, *464* **1.** **(a)** $2\sqrt{3}$ **3.** **(a)** $5\sqrt{2}$ **5.** **(a)** $3\sqrt{5}$ **7.** **(a)** $6\sqrt{2}$ **9.** **(a)** $21\sqrt{2}$
11. **(a)** $20\sqrt{42}$ **13.** **(a)** $12\sqrt{3}$ **15.** **(a)** $2\sqrt[4]{5}$ **17.** **(a)** $9x^3$ **19.** **(a)** $7x^2\sqrt{2}$
21. **(a)** $5x^4y^5\sqrt{2xy}$ **23.** **(a)** $8x^3y^3\sqrt{2}$ **25.** **(a)** $11x^4y^5$ **27.** **(a)** $4x^5$
29. **(a)** $xy^2\sqrt[3]{2x^2}$ **31.** **(a)** F, $x \geqq 0$ **(b)** T **(c)** F, $y \geqq 0$ **(d)** T **33.** $-5\sqrt[4]{2}$
35. $6\sqrt[7]{9}$ **37.** $-.3xy^2\sqrt[3]{xy}$ **39.** $4 \cdot 3^3 \cdot x^2\sqrt[7]{x}$

11–5, *467* **1.** **(a)** $3\sqrt{2}$ **3.** **(a)** $7\sqrt{2}$ **5.** **(a)** 21 **7.** **(a)** $6\sqrt{3}$ **9.** **(a)** $13\sqrt{6}$
11. **(a)** $11\sqrt{15}$ **13.** **(a)** $97a$ **15.** **(a)** $10a\sqrt{b}$ **17.** **(a)** $30x^2\sqrt{3}$
19. **(a)** $2x^5y^2\sqrt{5x}$ **21.** **(a)** $10x^7\sqrt{3}$ **23.** **(a)** $20x^6y^3\sqrt{2}$ **27.** $3\sqrt[3]{14}$
29. $2\sqrt[4]{15}$ **31.** $18\sqrt[6]{2}$ **33.** $x\sqrt{17}$ **35.** $-5\sqrt[3]{26}$ **37.** $-5a^3\sqrt[3]{6}$ **39.** $4x^2$
41. $6x^2\sqrt[4]{2}$

11–6, *470* **1.** **(a)** $\frac{\sqrt{2}}{5}$ **3.** **(a)** $\frac{\sqrt{2}}{2}$ **5.** **a** $\frac{3\sqrt{15}}{25}$ **7.** **(a)** $\frac{3\sqrt{10}}{10}$ **9.** **(a)** $\frac{3a^2}{4}$
11. **(a)** $\frac{5}{14b}\sqrt{21ab}$ **13.** **(a)** $\frac{1}{2y}\sqrt{30xy}$ **15.** **(a)** $\frac{2x}{25}\sqrt{10xy}$ **17.** **(a)** $\frac{ab^3}{x^2y^5}\sqrt{aby}$
19. **(a)** $\frac{1}{2x^3y^5}\sqrt{6abxy}$ **21.** **(a)** $\frac{xy^2}{2}\sqrt{5x}$ **23.** **(a)** $\frac{x^2y^3}{5a^2b}\sqrt{15ax}$
25. **(a)** $\frac{2x^2y^2}{15ab^4c}\sqrt{21yzabc}$ **27.** $\frac{\sqrt[3]{20}}{4}$ **29.** $-\frac{\sqrt[4]{6}}{3}$ **31.** $\frac{\sqrt[3]{6ab}}{3b}$ **33.** $-\frac{1}{2x}\sqrt[5]{6x}$
35. $\frac{1}{3y^2}\sqrt[5]{72xy^4}$

11–7, *473* **1. (a)** $7\sqrt{6}$; (D) **3. (a)** $9\sqrt{3}$; (D) **5. (a)** $\frac{\sqrt{6}}{4}$; (D) **7. (a)** $-2\sqrt{5}+\sqrt{6}$; (R-A), (D) **9. (a)** $-\frac{4\sqrt{5}}{5}+7\sqrt{13}$; (R-A), (D) **11. (a)** $-7\sqrt{5}$; (LR-1) **13. (a)** $3\sqrt{7}$; (LR-1) **15. (a)** $\sqrt{6}-4\sqrt{5}$; (LR-1) **17. (a)** $2\sqrt{2}$; (LR-2) **19. (a)** $\frac{13\sqrt{10}}{10}$; (LR-2) **21. (a)** $\frac{-4\sqrt{30}}{15}$; (LR-1), (LR-2) **23. (a)** $20+3\sqrt{10}$ **25. (a)** $-7+7\sqrt{2}$ **27. (a)** $104b-10\sqrt{78b}-18$ **29. (a)** $98-12\sqrt{66}$ **31. (a)** $36-3y$ **33.** $\frac{3\sqrt{30}}{2}$ **35.** $21+2\sqrt{42}+8\sqrt{3}+4\sqrt{14}$ **37.** $x\sqrt{6}-y\sqrt{6}-\sqrt{xy}$ **39.** $6x-2y$ **41.** $a\sqrt{a}+3a\sqrt{b}+3b\sqrt{a}+b\sqrt{b}$ **43.** $63x^3+7x-42x^2$ **45.** $2\sqrt[4]{2}$ **47.** $|xy|\sqrt{21}-4|y|\sqrt{xz}-|yz|\sqrt{21}$

11–8, *477* **1. (a)** 3 **3. (a)** $\frac{3\sqrt{6}}{2}$ **5. (a)** $\frac{\sqrt{7}}{14}$ **7. (a)** $5\sqrt{2}-3\sqrt{7}$ **9. (a)** $41\sqrt{2}-\frac{2\sqrt{42}}{3}$ **11. (a)** $-\frac{\sqrt{2}+\sqrt{3}}{7}$ **13. (a)** $\sqrt{5}+2$ **15. (a)** $4(5+2\sqrt{6})$ **17. (a)** $2\sqrt{5}-\sqrt{15}-2\sqrt{3}+3$ **19. (a)** $\frac{14\sqrt{5}-43}{11}$ **21. (a)** $\frac{8+7\sqrt{2}}{17}$ **23. (a)** $\frac{5-\sqrt{21}}{2}$ **25. (a)** $\frac{\sqrt{3}}{3}$ **(b)** $\frac{\sqrt{5}}{5}$ **(c)** $\frac{\sqrt{17}}{17}$ **(d)** $\frac{\sqrt{23}}{23}$ **27. (a)** $\sqrt{6}$ **(b)** $\sqrt{10}$ **(c)** $\sqrt{21}$ **(d)** $\sqrt{80}=4\sqrt{5}$ **29.** $\frac{\sqrt[3]{4}}{2}$ **31.** $\frac{\sqrt[3]{25}}{5}$

11–9, *482* **1. (a)** 3 **3. (a)** 2 **5. (a)** 27 **7. (a)** 16 **9. (a)** 5 **11. (a)** $\frac{729}{343}$ **13. (a)** $6x$ **15.** $8x^6y^9$ **17. (a)** $\frac{1}{5x^3}$ **19. (a)** $\frac{\sqrt[6]{11}}{11}$ **21. (a)** $2^{\frac{5}{6}}$ **23.** $2^{\frac{7}{6}}$ **25. (a)** $x^{\frac{7}{10}}$ **27. (a)** $7^{\frac{7}{12}}$ **29. (a)** 1 **31. (a)** $\frac{2x}{y^2}$

Preparation Exercises, 483 **1.** T **2.** F, their positive square roots are equal **3.** T **4.** T **5.** T **6.** T **7.** F, $\sqrt{x}=6$ **8. (a)** Yes **(b)** Yes **9.** Yes **10.** ∅

11–10, *485* **1. (a)** 25 **3. (a)** 21 **5. (a)** 58 **7. (a)** -5 **9. (a)** 9 **11. (a)** 60 **13. (a)** 8 **15. (a)** 9 **17. (a)** 45 **19.** $\frac{169}{9}$ **21.** $\frac{49}{16}$ **23.** 5 **25.** 18 **27. (a)** $x=0$ or $y=0$ **(b)** $x=0$ or $y=0$

11–11, *489* **1. (a)** 3.873 **3. (a)** 0.258 **5. (a)** 23.430 **7. (a)** .816 **9. (a)** 2.8 **11. (a)** 1.984 **13. (a)** 27.165 **15. (a)** -20.392 **17. (a)** $-.324$ **(b)** -3.621 **21.** .0001

11–12, *493* **1. (a)** 7.3 **3. (a)** 8.6 **5. (a)** 5.8 **7. (a)** 9.2 **11.** 3.1 **13.** 4.1 **15.** 4.6 **17.** 6.714, or 6.750 **19.** 15.167

11–13, *497* **1. (a)** 10 **3. (a)** 15 **5. (a)** 13.9 **7. (a)** Yes **9. (a)** Yes **11. (a)** 34.6 ft **13.** Yes **15.** Yes

Chapter Review, 501 **3. (a)** $-\frac{4}{9}$ **(b)** $-\frac{29}{30}$ **(c)** $\frac{18107}{33333}$ **5.** 2 **7.** $\frac{4}{5}$ **9.** $\{3,-3\}$ **11.** $2\sqrt[3]{15}$ **13.** $8\sqrt{15}$ **15.** $-3\sqrt[5]{10}$ **17.** $2\sqrt[4]{15}$ **19.** 1 **21.** $\frac{\sqrt[4]{6}}{2}$ **23.** $\frac{\sqrt[3]{50}}{5}$ **25.** $\frac{2\sqrt{15}}{3}$ **27.** $\frac{3\sqrt{5}}{5}$ **29.** 2 **31.** $\frac{\sqrt[3]{9}}{3}$ **33.** 2 **35.** $9\sqrt{6}$ **37.** $\sqrt{2}-8\sqrt{3}$ **39.** $\frac{\sqrt[3]{4}}{2}+\frac{\sqrt[3]{9}}{3}$ **41.** $7ab^3$ **43.** $1.2p^8q^{18}$ **45.** $-5c^4$ **47.** $\frac{2y}{x^2}$ **49.** $\frac{\sqrt{7}}{x^2}$ **51.** $\frac{2}{3xy^2}$ **53.** $\frac{1}{9}$ **55.** $\frac{7}{19}$ **57.** $\frac{7}{2}$ **59.** $\frac{\sqrt{10}}{2}$ **61.** $-30\sqrt{3}+60$ **63.** $131-12\sqrt{119}$

65. $\dfrac{30 - 10\sqrt{2}}{7}$ **67.** $8(\sqrt{7} + \sqrt{6})$ **69.** $\dfrac{3 + \sqrt{35}}{13}$ **71.** $7^{\frac{7}{6}}$ **73.** $9xy^{\frac{3}{2}}$

75. $10xy$ **77.** $\dfrac{5\sqrt{14}}{2} - 2\sqrt{7}$ **79.** $\dfrac{x}{2y}\sqrt{30xy}$ **81.** $\sqrt{ab} - \dfrac{\sqrt{ab}}{ab}$

83. $x = 25$ **85.** 128 **87.** No solution **89. (a)** 3.3 **(b)** 5.7 **(c)** 10.5

12–1, ***506*** **1. (a)** $\{-\frac{1}{2}, -\frac{7}{3}\}$ **3. (a)** $\{0, 2\sqrt{2}\}$ **5. (a)** $\{-4, -2\}$ **7. (a)** $\{2\}$ **9. (a)** $\{-9, 7\}$ **11. (a)** $\{7, -3\}$ **13. (a)** $\{-\frac{1}{2}, \frac{1}{3}\}$ **15. (a)** $\{0, 3\}$ **17. (a)** $\{\frac{5}{3}, 7\}$ **19.** $\{\frac{5}{2}, 4\}$ **21.** $\{0, 15\}$ **23.** $\{12, -9\}$ **25.** $\{-14, 10\}$ **27.** $\{2, 3\}$ **29.** $\{3\}$ **31.** $\{-3\}$, No

Prep. Ex., 508 **1.** $(2x + 1)(2x - 1)$ **2.** $x(3x - 5)$ **3.** $(3x + 5)(3x - 5)$ **4.** $x(10x - 9)$ **5.** $x(x + 3)(x - 3)$

12–2, ***510*** **1. (a)** $\{7, -7\}$ **3. (a)** $\{6, -6\}$ **5. (a)** $\{0, \frac{4}{7}\}$ **7. (a)** $\{0\}$ **9. (a)** $\{\sqrt{5}, -\sqrt{5}\}$ **11. (a)** $\{0, \frac{1}{4}\}$ **13. (a)** $\{-13, 1\}$ **15. (a)** $\{\frac{7}{2}\}$ **17. (a)** $\{5, -1\}$ **19.** $\{\frac{11}{7}, -\frac{13}{7}\}$ **21.** $\{14, -8\}$ **23.** $\{60, -60\}$

Prep. Ex., 511 **1.** $(x - 4)^2$ **2.** $(c + 11)^2$ **3.** $(a + \frac{1}{3})^2$ **4.** $(c + 2\sqrt{5})^2$

12–3, ***515*** **1. (a)** $9, (x + 3)^2$ **3. (a)** $\frac{25}{4}, (z - \frac{5}{2})^2$ **5. (a)** $\{11, 1\}$ **7. (a)** $\{12\}$ **9. (a)** $\{20.39, -.39\}$ **11. (a)** $\{11.08, -1.08\}$ **13. (a)** $\{6.85, .15\}$ **15. (a)** $\{.37, -1.37\}$ **17.** $225, (y^3 + 15)^2$ **19.** $\{1.41\}$ **21.** $\{2.41, .41\}$ **23.** $\{2.24\}$ **25.** $\{1.73, -1.73, 1, -1\}$ **27. (a)** $\{1 + \sqrt{3}, 1 - \sqrt{3}\}$

12–4, ***518*** **1. (a)** $\left\{\dfrac{-1 \pm \sqrt{3}}{2}\right\}$ **3. (a)** $\left\{\dfrac{5 \pm 2\sqrt{10}}{5}\right\}$ **5. a** $\left\{\dfrac{3 \pm \sqrt{7}}{2}\right\}$ **7. (a)** $\left\{\dfrac{-3 \pm 3\sqrt{13}}{2}\right\}$ **9. (a)** $\left\{\dfrac{\sqrt{5} \pm 1}{2}\right\}$ **11.** $\left\{\dfrac{-7 \pm \sqrt{17}}{4}\right\}$ **13.** $\{1 \pm \sqrt{21}\}$ **15.** $\{7\}$ **17.** $\{2, -3\}$ **19.** $\{1, -2\}$ **21.** $\{2\}$ **23.** $\{2\}$ **25.** $(c + \frac{1}{2})^2 = -\frac{3}{4}$ **27.** $(a - \frac{1}{8})^2 = -\frac{15}{64}$ **29.** $\left(c - \dfrac{3\sqrt{3}}{14}\right)^2 = -\frac{1}{196}$

12–5, ***522*** **1. (a)** 13, 14 or $-13, -14$ **3. (a)** 27, 29 or $-29, -27$ **5. (a)** 11, 7 or $-7, -11$ **7. (a)** $\frac{2}{3}$ or $\frac{1}{3}$ **9.** 12 in. $\times$ 8 in. **11.** 60 yd $\times$ 80 yd **13.** 5 ft $\times$ 13 ft **15.** 7 in. **17.** $\dfrac{-17}{-12}$ or $\frac{4}{9}$

12–6, ***527*** **1. (a)** $a = 3, b = -4, c = -1, 28; \left\{\dfrac{2 \pm \sqrt{7}}{3}\right\}$ **3. (a)** $\{-1, -\frac{1}{2}\}$ **5. (a)** $\{7\}$ **7. (a)** $\left\{\dfrac{-1 \pm \sqrt{13}}{6}\right\}$ **9. (a)** $\left\{\dfrac{1 \pm \sqrt{10}}{3}\right\}$ **11. (a)** $\left\{\dfrac{1 \pm \sqrt{241}}{15}\right\}$ **13.** $\{5, -1\}$ **15.** $\{-6, -\frac{1}{2}\}$ **17.** $\{3\}$ **19.** $\{12\}$ **21.** $\left\{\dfrac{-\sqrt{5} \pm 1}{2}\right\}$ **23.** $\left\{\dfrac{-\sqrt{2} \pm \sqrt{14}}{2}\right\}$ **25. (a)** 1, irrational **(b)** 25, irrational **27.** One rational number

12–7, ***530*** **1. (a)** -11, no real solutions **3. (a)** 37, two real solutions **5. (a)** 0, $\{\frac{7}{2}\}$ **7. (a)** 52, $\{5 \pm \sqrt{13}\}$ **9. (a)** 84, $\{-4 \pm \sqrt{21}\}$ **11. (a)** 73, $\left\{\dfrac{-7 \pm \sqrt{73}}{6}\right\}$ **13. (a)** -64, no real solution **15. (a)** $\frac{25}{16}, \{-\frac{3}{4}, \frac{1}{2}\}$ **17.** $\{0\}$ **19.** $\{-1\}$ **21. (a)** $-\dfrac{b}{a}$ **(b)** The sum equals the negative of the coefficient of x divided by the coefficient of x^2. **23.** Yes **25.** $\{2 \pm 2\sqrt{3}\}, 4$ **27.** $\left\{\dfrac{10 \pm \sqrt{35}}{5}\right\}, 4$ **29.** Yes, yes **31.** Yes; $\{-5, \frac{1}{3}\}$ **33.** $\{0, \frac{7}{9}\}$ **35.** $\{6 \pm 2\sqrt{2}\}$

EXTRA!, 533 **1.** Yes, $1 - \sqrt{2}$ **3.** Yes, $1 + \sqrt{7}$ **5.** No **7.** $3 + 4\sqrt{2}$ **9.** $3 + \sqrt{10}$

Prep. Ex., 534 **1.** $\{-\frac{5}{2}, \frac{2}{5}\}$ **2.** $\left\{\frac{-4 \pm \sqrt{102}}{2}\right\}$ **3.** $\left\{\frac{-5 \pm \sqrt{70}}{3}\right\}$

12–8, *536* **1. (a)** 21 in × 20 in. **3. (a)** $2\sqrt{30}$ ft **5. (a)** 5 ft **7. (a)** 4 mph **9.** 45 mph **11.** Bus, 42.2 mph; Train, 52.2 mph **13.** 20 mph

12–9, *541* **1. (a)** Min 0 **3. (a)** Min −1 **5. (a)** Min −1 **7. (a)** Min −5 **11. (a)** $x = -3.3, .3$; $x \doteq -3.303, .303$ **13.** 2, (−1, 2), $x = -1$ **15.** $-\frac{1}{4}$, $(\frac{7}{2}, -\frac{1}{4})$, $x = \frac{7}{2}$ **17. (a)** −3 **(b)** No real solution **(c)** The graph does not cross the x-axis. **19. (b)** Different axes of symmetry and vertices **(c)** $y = 2(x + h)^2$, h is a real number **21.** Same axis of symmetry, shape, but y-intercepts k and vertices differ. **25.** a

12–10, *546* **1.** Inside, $y > x^2 - 1$; Outside, $y < x^2 - 1$ **3.** Inside, $y > x^2 + 2x + 1$; Outside, $y < x^2 + 2x + 1$ **11. (a)** All points outside and including the parabola. **13.** All points outside the parabola. **15.** All points inside the parabola. **17.** All points outside and including the parabola.

Chapter Review, *548* **1.** {5, −1} **3.** $\{0, \frac{5}{3}\}$ **5.** {±6} **7.** $\{\frac{2}{3}, 5\}$ **9.** $\{3 \pm \sqrt{5}\}$ **11.** No real solution **13.** $\left\{\frac{6 \pm \sqrt{6}}{6}\right\}$ **15.** $\left\{\frac{3 \pm \sqrt{15}}{2}\right\}$ **17.** $\left\{-\frac{3}{7}, 3\right\}$ **19.** $\left\{\frac{-13 \pm \sqrt{17}}{4}\right\}$ **21.** −119; No real solution **23.** 0; One rational solution **25.** 8 **27.** $w = 5$, $l = 7$ **29.** 270 mph **37.** Min −4; max 4; max 3; min −2; max 25; min −1

13–1, *555* **1. (a)** 30 **3. (a)** 3 **5. (a)** 6 **7. (a)** $\frac{3}{2}$ **9. (a)** 70 lbs **11. (a)** $16.88 **14.** $\frac{5}{9}$ hr **15.** 4 ft **17.** $384 **18.** 75 rpm

13–2, *559* **1. (a)** 20 **3. (a)** $\frac{5}{4}$ **5. (a)** $\frac{63}{8}$ **7. (a)** $\frac{14}{5}$ **9. (a)** $\frac{6}{5}$ **11. (a)** 8 **13. (a)** $\sqrt{2}$ **15. (a)** ±25 **17.** −8 or 3 **19.** ±4 **21.** a^4 **23.** $\pm\sqrt{15}$ **25.** $\pm b^2$ **27. (a)** $\frac{125}{8}$ **(b)** $\frac{15}{5}$ **29. (a)** x is multiplied by 8 **(b)** x is multiplied by 27

13–3, *565* **1.** 80° **3. (a)** 17 ft, $\frac{15}{2}$ ft, $\frac{17}{2}$ ft **5. (a)** 12 ft, $\frac{25}{4}$ ft, $\frac{65}{4}$ ft **7. (a)** 2.4 in., 4 in. **9. (a)** 10 ft

13–4, *570* **1. (a)** 5 **3. (a)** $\frac{4}{5}, \frac{3}{5}, \frac{4}{3}$ **5. (a)** $\frac{4}{5}, \frac{3}{5}, \frac{4}{3}$ **7.** 90 **9.** $\cos\alpha = \sin\beta$ **11.** $\tan\beta > \sin\beta$ **13.** No **15.** No **17.** 45°; $0 < \tan\xi < 1$; $\tan\xi > 1$

13–5, *573* **1.(a)** .454 **3. (a)** 3.078 **5. (a)** .996 **7. (a)** 10° **9. (a)** 22° **11. (a)** 82° **13. (a)** 42° **15. (a)** 63° **17. (a)** 20° **19. (a)** 76° **21. (a)** 6.0 ft **23. (a)** 10.9 in. **25. (a)** 13.7 ft **27.** 23°, 67°; 90°

13–6, *578* **1. (a)** 434.5 ft **3. (a)** 1285.7 ft **5. (a)** 24.3 ft **7. (a)** 60° **9. (a)** 326.5 ft **11.** 4.6 ft

Prep. Ex., 579 **1.** 8 **2.** 5 **3.** 6 **4.** 2.4

13–7, *581* **1. (a)** .579 **3. (a)** .709 **5. (a)** .459 **7. (a)** 24°10′ **9. (a)** 33°30′ **11. (a)** 79°40′ **13.** 6°10′ **15.** 16.1 ft **17.** 58° **19.** 99.1 ft

Chapter Review, *583* **1. (a)** 15 **(b)** $\frac{19}{2}$ **(c)** 6 **(d)** $\frac{5}{2}$ **3. (a)** 49 **(b)** 35 **5.** 4% **7.** Double x **9.** $11\frac{1}{4}$ in., $6\frac{3}{4}$ in. **11.** $\frac{5\sqrt{61}}{61}, \frac{6\sqrt{61}}{61}, \frac{5}{6}; \frac{6\sqrt{61}}{61}, \frac{5\sqrt{61}}{16}, \frac{6}{5}$ **13.** 8.5 in., 18.1 in. **15.** 47 ft **17.** 5° **19.** 64°30′

Cumulative Review IV, *586* **1. (a)** F, $\sqrt{x^2} = -x$ if $x = -4$ **(b)** T **(c)** F, rational **(d)** T **(e)** T **(f)** T **(g)** T **(h)** F, $\sqrt{16 + 9} = 5$ **3.** $\frac{2\sqrt{x}}{x}$ **5.** $\frac{5(\sqrt{a} - 2)}{a - 4}$ **7.** $\frac{x + y - 2\sqrt{xy}}{x - y}$

9. $6x$ **11.** $x^4\sqrt{6}$ **13.** $-5xy\sqrt{2xy}$ **15.** $\frac{\sqrt{21}}{3}$ **17.** $2y$ **19.** $x\sqrt{6} - 5x\sqrt{2y}$

21. $3x - 2x\sqrt{3y} + xy$ **23.** $7\sqrt{a} - 10\sqrt{b}$ **25.** $-\frac{1}{2x^2}$ **27.** $5(r - t)^3$

29. $-\frac{7s^2}{6}$ **31.** $\frac{1}{x^{\frac{7}{3}}}$ **33.** $\frac{c}{3y^2}$ **35.** $\frac{y^7}{x^7}$ **37.** $\frac{5y^{\frac{5}{3}}}{6}$ **39.** x **41.** $x^{\frac{9}{4}}y^4$ **43.** $\frac{1}{5y^3}$

45. −2 **47.** −5 **49.** −2 **51.** −2 **53.** $-\frac{2}{3}$ or 5 **55.** 0, $-\frac{3}{4}$ **57.** 4 **59.** $-\frac{1}{2}$ **61.** 5 **63.** $x < \frac{5}{2}$ **65.** $x > -\frac{1}{2}$ **67.** $-6 \leqq x \leqq -3$ **73.** $c \doteq 5.8$, $a \doteq 2.9$; .5, .866, .577